CRITICAL VALUES OF STUDENT'S *t* DISTRIBUTION

One-tailed value

W9-BIG-508

DEGREES OF FREEDOM	One-Tailed Value					
	0.25	0.10	0.05	0.025	0.01	0.005
	Two-Tailed Value					
	0.50	0.20	0.10	0.05	0.02	0.01
1	1.000	3.078	6.314	12.706	31.821	63.657
2	0.816	1.886	2.920	4.303	6.965	9.925
3	.765	1.638	2.353	3.182	4.541	5.841
4	.741	1.533	2.132	2.776	3.747	4.604
5	.727	1.476	2.015	2.571	3.365	4.032
6	.718	1.440	1.943	2.447	3.143	3.707
7	.711	1.415	1.895	2.365	2.998	3.499
8	.706	1.397	1.860	2.306	2.896	3.355
9	.703	1.383	1.833	2.262	2.821	3.250
10	.700	1.372	1.812	2.228	2.764	3.169
11	.697	1.363	1.796	2.201	2.718	3.106
12	.695	1.356	1.782	2.179	2.681	3.055
13	.694	1.350	1.771	2.160	2.650	3.012
14	.692	1.345	1.761	2.145	2.626	2.977
15	.691	1.341	1.753	2.131	2.602	2.947
16	.690	1.337	1.746	2.120	2.583	2.921
17	.689	1.333	1.740	2.110	2.567	2.898
18	.688	1.330	1.734	2.101	2.552	2.878
19	.688	1.328	1.729	2.093	2.539	2.861
20	.687	1.325	1.725	2.086	2.528	2.845
21	.686	1.323	1.721	2.080	2.518	2.831
22	.686	1.321	1.717	2.074	2.508	2.819
23	.685	1.319	1.714	2.069	2.500	2.807
24	.685	1.318	1.711	2.064	2.492	2.797
25	.684	1.316	1.708	2.060	2.485	2.787
26	.684	1.315	1.706	2.056	2.479	2.779
27	.684	1.314	1.703	2.052	2.473	2.771
28	.683	1.313	1.701	2.048	2.467	2.763
29	.683	1.311	1.699	2.045	2.462	2.756
30	.683	1.310	1.697	2.042	2.457	2.750
35	.682	1.306	1.690	2.030	2.438	2.724
40	.681	1.303	1.684	2.021	2.423	2.704
45	.680	1.301	1.680	2.014	2.412	2.690
50	.680	1.299	1.676	2.008	2.403	2.678
55	.679	1.297	1.673	2.004	2.396	2.669
60	.679	1.296	1.671	2.000	2.390	2.660
70	.678	1.294	1.667	1.994	2.381	2.648
80	.678	1.293	1.665	1.989	2.374	2.638
90	.678	1.291	1.662	1.986	2.368	2.631
100	.677	1.290	1.661	1.982	2.364	2.625
120	.677	1.289	1.658	1.980	2.358	2.617
∞	.674	1.282	1.645	1.960	2.326	2.576

FIFTH EDITION

STATISTICS

AN INTRODUCTION

FIFTH EDITION

STATISTICS
AN INTRODUCTION

ROBERT D. MASON

DOUGLAS A. LIND

WILLIAM G. MARCHAL

THE UNIVERSITY OF TOLEDO

DUXBURY PRESS

An Imprint of Brooks/Cole Publishing Company
I(T)P™ An International Thomson Publishing Company

Pacific Grove • Albany • Belmont • Bonn • Boston • Cincinnati • Detroit
Johannesburg • London • Madrid • Melbourne • Mexico City • New York
Paris • Singapore • Tokyo • Toronto • Washington

Sponsoring Editor: *Carolyn Crockett*
Marketing Team: *Marcy Perman,*
   Christine Davis, Joy Westberg
Production: *Clarinda Publication Services*
Cover Design: *Terri Wright*
Cover Illustration: *Allees Pietonnieres by*
   Jean-Pierre Stora/The Grand Design,
   Leeds/SuperStock

Design Editor: *Roy R. Neuhaus*
Typesetting: *The Clarinda Company*
Cover Printing: *Phoenix Color Corporation*
Printing and Binding: *The Courier*
   Company, Inc.–Kendallville

For more information, contact Duxbury Press at Brooks/Cole Publishing Company:

BROOKS/COLE PUBLISHING COMPANY
511 Forest Lodge Road
Pacific Grove, CA 93950
USA

International Thomson Publishing Europe
Berkshire House 168-173
High Holborn
London WC1V 7AA
England

Thomas Nelson Australia
102 Dodds Street
South Melbourne, 3205
Victoria, Australia

Nelson Canada
1120 Birchmount Road
Scarborough, Ontario
Canada M1K 5G4

International Thomson Editores
Seneca 53
Col. Polanco
11560 México, D. F., México

International Thomson Publishing GmbH
Königswinterer Strasse 418
53227 Bonn
Germany

International Thomson Publishing Asia
221 Henderson Road
#05-10 Henderson Building
Singapore 0315

International Thomson Publishing Japan
Hirakawacho Kyowa Building, 3F
2-2-1 Hirakawacho
Chiyoda-ku, Tokyo 102
Japan

Printed in the United States of America.

10 9 8 7 6 5 4 3 2 1

Library of Congress Cataloging-in-Publication Data

Mason, Robert Deward, 1919–
   Statistics : an introduction / Robert D. Mason, Douglas A. Lind,
William G. Marchal.—5th ed.
      p.   cm.
   Includes index.
   ISBN 0-534-35379-7 (hardcover)
   1. Statistics.   I. Lind, Douglas A.   II. Marchal, William G.
III. Title.
QA276.12.M36   1998
519.5—dc21   97-34710
                                                                  CIP

To Anita, Jane, and Andrea

Contents

Contents

ANALYSIS OF NOMINAL-LEVEL DATA: THE CHI-SQUARE DISTRIBUTION   495

CHAPTER 16

NONPARAMETRIC METHODS: ANALYSIS OF RANKED DATA   521

CHAPTER 17

QUALITY CHARTING   557

CHAPTER 18

Preface

When we wrote the first edition of this text, locating relevant data was a problem. That has now changed! Today, locating the data is not a problem. The number of items you purchase at the supermarket is recorded automatically. The time you talk on the telephone is automatically recorded. We have medical devices that automatically monitor our heart rate, blood pressure, and record our temperature. CNN flashes "Factoids" that give us interesting information on our environment. We have not even mentioned the volumes of information reported on business (The Dow Jones and NASDAQ averages) or the information reported on sporting events (how many times Cal Ripken has faced Roger Clemens and his batting average against Clemens). How about the World Wide Web?

There are important skills that are necessary to be an intelligent consumer of all this numerical information. We need to be critical consumers of information presented by others. How much will the proposed school tax actually cost you as a homeowner? Second, we need to be able to reduce the large amounts of data available into useful information so we can make sound, effective choices. Test marketing data and demographic data allow business to target small segments of the population that will have a propensity to purchase their product or service.

When we began teaching, only a few students had calculators. Today, all students have a calculator (or two), many a scientific calculator, some a programmable calculator. Many students have a computer at home or in their dorm room. Most have access to a computer in a Computer Lab. Statistical software is widely available. In response to these changes, we have taken out many of the illustrations regarding grouped data and replaced them with reference to statistical software and spreadsheets. In the same vein, we have replaced the emphasis on computation with one on interpretation. In this edition we continue the use of MINITAB software but have added Excel spreadsheet examples because much of our current world uses Microsoft applications.

About ten years ago, perhaps in response to the popularity of Japanese and German automobiles and other electronic equipment, quality became an important competitive priority. In the last several years, quality considerations have also moved into the service sector of the economy. We are now concerned more than ever with the quality of all of our life experiences. For example, hospitals take patient satisfaction surveys. How satisfied were you with the care you received during your last outpatient surgery?

In addition to Excel spreadsheet examples with explanations and a new chapter (18) on Quality Control, the fifth edition includes the following new and improved features:

- The concept and uses of p-values has been integrated.

- More extensive use of tree diagrams, particularly in the development of the binomial distribution.
- A new section on Box Plots has been added.
- Two new examples have been created to improve the description of the Central Limit Theorem.
- The ANOVA chapter has been rewritten to focus more on the underlying idea behind ANOVA.
- The "up to" convention has been used in the development of frequency distributions.
- Additional "Drill Exercises" designed to give students reinforcements of the topics just discussed.
- The narrative Summary has been replaced by a Chapter Outline.

As in previous editions, we want to continue our accessible and motivational style. It is our desire to present statistics in a fashion that is no more difficult than it has to be.

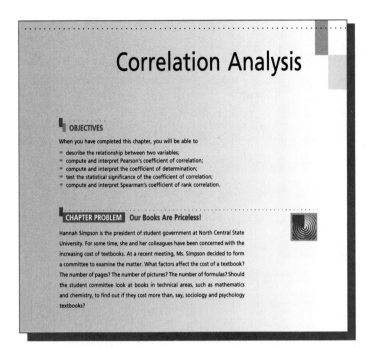

Correlation Analysis

OBJECTIVES

When you have completed this chapter, you will be able to

- describe the relationship between two variables;
- compute and interpret Pearson's coefficient of correlation;
- compute and interpret the coefficient of determination;
- test the statistical significance of the coefficient of correlation;
- compute and interpret Spearman's coefficient of rank correlation.

CHAPTER PROBLEM Our Books Are Priceless!

Hannah Simpson is the president of student government at North Central State University. For some time, she and her colleagues have been concerned with the increasing cost of textbooks. At a recent meeting, Ms. Simpson decided to form a committee to examine the matter. What factors affect the cost of a textbook? The number of pages? The number of pictures? The number of formulas? Should the student committee look at books in technical areas, such as mathematics and chemistry, to find out if they cost more than, say, sociology and psychology textbooks?

Each chapter begins with a set of **Objectives,** designed to provide focus and to motivate the student.

Chapter Problems follow the learning objectives; these are integrative problems that are solved in the chapter when presented with the concept. Designed to garner the student's interest and attention, the problems are identified with a distinctive logo, and this logo is repeated in the portions of the text where the problem is discussed.

At the start of each chapter, we review the important concepts of the previous chapter(s) in a brief **Introduction.** Included in the Introduction is the link between the current and previous chapters.

Definitions of new terms are provided in the margin.

Within each chapter are several **Problem/Solutions.** These Problem/Solutions are designed to illustrate a real-world application of the concept just presented.

INTRODUCTION

In Chapters 10–12, we dealt with hypothesis tests involving means and proportions. The techniques concentrated on only a *single* feature of the sampled item, such as income. This chapter begins a study of the relationship between *two or more* variables. We may want to determine if there is any relationship between the number of years of education completed by federal government employees and their incomes. Or we may want to explore one of these questions: Does the crime rate in inner cities vary with the unemployment rate in those cities? Is there a relationship between the amount of money spent advertising a product, such as a toothpaste, and its sales? Is there any relationship between the number of hours studied and a student's grade on an examination? Note that in each case, there are two separate characteristics—years of education and income, for example.

CORRELATION ANALYSIS

correlation analysis The statistical techniques used to determine the strength of the relationship between two variables.

The study of the relationship between two variables is called **correlation analysis.** The objective of correlation analysis is to study the strength of the association between two variables. Our attention will focus first on the correlation between two interval-scaled variables and then on the relationship between two ordinal-scaled variables.

The Scatter Diagram

One tool that is very useful for visualizing the relationship between two variables is the **scatter diagram.**

Problem Recall from the Chapter Problem that a student committee was formed to look into the factors affecting the cost of textbooks. After some discussion within the committee and consultation with book publishers, the committee decided to first consider the relationship between the cost of the text and its length.

scatter diagram A graphic tool that portrays the relationship between two variables.

Solution The committee selected a sample of six textbooks currently in use at North Central. For each selected text, the committee determined the number of pages and the selling price. The results are reported in Table 13–1.

We call the selling price the dependent variable and the number of pages the independent variable. The dependent variable is thus the variable being estimated, and the independent variable is the variable used as the estimator. It is traditional to put the dependent variable on the vertical axis (Y) and the independent variable on the horizontal axis (X) of the scatter diagram. We plot the paired data for *Introduction to Psychology* (X = 400, Y = 44) by moving to 400 on the X-axis, then moving vertically to a position opposite 44 on the Y-axis, and placing a dot at that intersection (see the scatter diagram in Figure 13–1). This process is continued for all the books in the sample. The completed scatter diagram is shown in Figure 13–1.

TABLE 8–1   Lengths of Service of Six Social Workers

Social Worker	Length of Service (in years)
Dori	4
Gary	3
Sue	3
Bob	2
Kirk	4
Sandra	3

The population mean, μ, is 3.16667:

$$\mu = \frac{4 + 3 + 3 + 2 + 4 + 3}{6}$$
$$= 3.16667$$

Solution   The population consists of 6 employees; 4 are to be randomly selected to remain. We can use the idea of a combination, discussed in connection with the binomial distribution in Chapter 6, to determine the total number of possibilities. Let N be the size of the population—in this case, 6—and n the size of the sample, which is 4. So the number of possible samples (groups of remaining employees) is

$$_NC_n = \frac{N!}{(N-n)!\,n!} = \frac{6!}{(6-4)!\,4!} = 15$$

There are 15 different possible combinations of 4 employees who will remain in the Social Services Department after the cutback. These possible outcomes are listed on the next page, along with the length of service for each retained employee and the mean for each sample (group of employees).

REAL STAT

Is discrimination taking a bite out of your paycheck? Before answering, consider the findings reported in the *Personnel Journal* (December 1995). These findings indicated that good-looking men and women earn about 5% more than average lookers, who in turn earn about 5% more than their plain counterparts. This is true for plain men and even more true for plain women, and it is true in a wide range of occupations from construction work to auto-repair to telemarketing, where it would seem looks should not matter. In another study, lawyers were rated on their appearance using a scale from 1 to 5, and then their earnings were compared. The ugly result: a 12 to 15% pay differential from the highest to the lowest.

REAL STATS are scattered throughout the text, providing interesting uses of statistics and historical insights.

Interspersed in each chapter are a number of **Self-Reviews**. Each Self-Review is closely patterned after the Problem/Solution. The Self-Reviews help students monitor their progress and provide immediate reinforcement. The answers and method of solution are provided at the end of the chapter.

Self-Review 10–3

The admissions officer at a university wants to investigate whether there is any difference between the scores on the mathematics placement test of students who attend classes primarily during the day and those of students who work during the day and attend evening classes.

a. Is this a one-tailed or a two-tailed test? Symbolically, what are the null hypothesis and the alternate hypothesis?
b. The 0.05 level of significance is to be used. State the decision rule:
c. The records of both day and evening students revealed the following:

Student	Mean Score	Standard Deviation	Number in Sample
Day	90	12	40
Evening	94	15	50

Compute z and arrive at a decision.
d. Determine the *p*-value. Interpret.

---

d. Estimate the typical (middle) number of rangers supervised.
e. How many of the respondents supervised 25 up to 30 rangers?
8. An insurance claims adjustor collected the following data on the number of car thefts reported each day for a month. These data were arranged into a stem-and-leaf display.

e. How many days were there in the month?
9. The Quick Change Oil Company advertises that it can change a car's motor oil in 10 minutes. However, business has been very slow lately. A study of the last 25 days revealed the following number of cars serviced each day:

65	98	55	62	79
59	51	90	72	56
70	62	66	80	94
79	63	73	71	85
93	68	86	53	90

Develop a stem-and-leaf display.

Many interesting, real-world **Exercises** are found within the chapter, after the Self-Review and at the end of the chapter. Answers for the even-numbered exercises are given at the back of the book.

Each chapter includes a **Chapter Outline.** This learning aid provides an opportunity for students to review the material, particularly the vocabulary, in an outline format.

Formulas used for the first time are boxed and are numbered for easy reference.

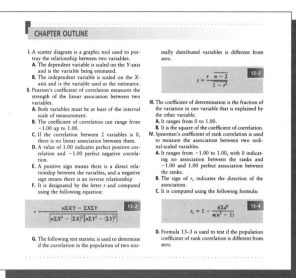

The last two exercises in most chapters are based on two large data sets. The first **Data Exercise** refers to homes sold in Alabama during 1996. Included is such information as the selling price, size, distance from the center of the city, and the number of bedrooms. The second **Data Exercise** includes information on the number of students in each district, average teachers' salary, percent passing the state proficiency examination, percent of families on welfare, and other information on 94 school districts in Northwest Ohio. A complete listing of the data is available in the back of the text and on the data disk.

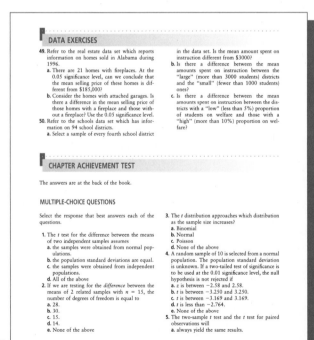

The end-of-chapter **Achievement Test** will help students evaluate their comprehension of the material covered. Answers are given in the back of the book.

CHAPTERS 8, 9, AND 10

. . . .

Unit Review

The last three chapters dealt with the fundamental concepts of statistical inference. They described (1) how we can use sampling to develop a range of values within which a population parameter is likely to occur and (2) the procedures for testing hypotheses about population parameters.

KEY CONCEPTS

1. **Sampling** is important because complete knowledge regarding the population is seldom available. The basic purpose of sampling is to provide an estimate about a population parameter based on sample evidence.

8. A **point estimate** is the value computed from a sample that is used to estimate a population parameter. An **interval estimate** is a range of values within which we have some assurance that the population parameter lies.

9. A **proportion** is a fraction, percentage, or ratio that indicates what part of a sample or population has a certain trait.

10. When the sample constitutes more than 5% of the population, the **finite population correction factor** is used. Its purpose is to reduce the standard error because the sample constitutes a large portion of the population. The term $(N - n)/(N - 1)$ is multiplied by the standard error of the mean or proportion.

A **Unit Review** is included after each group of chapters. The Unit Review includes three parts: **Key Concepts, Key Terms, Key Symbols,** and a **Case Study.**

KEY TERMS

Probability sampling	Sampling distribution of the	Finite population correction
Sample frame	sample proportion	factor
Sample design	Central limit theorem	Null hypothesis
Sample survey	Confidence interval	Alternate hypothesis
Census	Standard error of the sample	Type I error
Control group	mean	Type II error

KEY SYMBOLS

$\sigma_{\bar{x}}$	The standard error of the mean.		π	The proportion of objects or things in the population that possesses a particular trait.
σ_p	The standard error of the sample proportion.		$\bar{p}$	The pooled estimate of the population proportion based on the two samples.
E	Maximum allowable error.		x	The number of successes in a sample of n observation.
H_0	The null hypothesis.			
H_a	The alternate hypothesis.			

CASE STUDY   Excessive Consumer Debt

Banks, financial institutions, and financial planners, as well as psychologists, are concerned with the amount of debt in the United States. In 1996, the amount of credit offered consumers by banks totaled more than $1.02 trillion. The average credit-card holder carries a balance of $3900, and only about one-third of the customers pay their balances at the end of the month. Wendy Lamberg, president of the Genoa Savings Bank, would like to know how customers at her small bank in Tennessee compare with the national statistics given above. Do Genoa Savings customers, on the average, have larger balances? Do fewer customers pay off their balances at the end of the month? A random sample of 50 Genoa customers with VISA cards serviced by the bank revealed the following data: the customers' balances at the end of last month and whether they paid their entire balances the previous month. Write a brief report to Ms. Lamberg summarizing the experience at Genoa. Be sure to report the hypotheses tested, the p-values, and the conclusions.

ANCILLARY MATERIALS

A **Data Disk,** available in the back of each text, provides data for problems in the text and is formatted in ASCII and for MINITAB, Excel, and StataQuest.

A complete ancillary package accompanies the fifth edition. A comprehensive **Study Guide,** written by Walter Lange of The University of Toledo, is organized much like the textbook. Each chapter includes objectives, a brief summary of the chapter, problems and solutions, exercises, and assignments.

An **Instructor's Manual,** prepared by Denise Heban of Penta Vocational High School contains the complete solution to all exercises in each chapter and cases in the textbook as well as those in the study guide. Also included in the Instructor's Manual are syllabi for both a quarter and a semester course. A **Student Solutions Manual** contains a complete solution to the odd-numbered exercises. A **PowerPoint Presentation** developed by Christopher B. Marchal is available for lectures and electronic presentations. The full-color presentations include chapter objectives, an outline of each chapter, additional examples, and definitions of key terms. **Ready Notes** are available separately for student purchase. This booklet contains the printed PowerPoint Presentation and allows the student to annotate the prepared class presentation.

A **Test Bank,** prepared by Professor Gwen H. Terwilliger of the University of Toledo to accompany the fifth edition, includes approximately 500 objective questions. It includes multiple choice and fill-in-the-blank questions for the student to solve.

A TI-83 Manual, developed by Stephen Kokoska of Bloomsburg University, Bloomsburg, Pennsylvania, describes how to use the TI-83 graphing calculator for the techniques described in the text. The examples are keyed to the text.

ACKNOWLEDGMENTS

We wish to express our gratitude to the reviewers of the first edition: John M. Rodgers (California Polytechnic State University, San Luis Obispo), Bayard Baylis (The King's College of New York), William W. Lau (California State University, Fullerton), Kenneth R. Eberhard (Chabot College), and David Mackey (San Diego State University). Their contributions continue to enhance the textbook.

For the Second Edition, we wish to thank Francesca Alexander (California State University, Los Angeles), John R. Anderson (DePauw University), Daniel Cherwein (Cumberland County College), C. Philip Cox (Iowa State University), Geoffrey C. Crosslin (Kalamazoo Valley Community College), Larry Gausen (Bemidji State University), David Macky (San Diego State University), Bruce F. Sloan, Donald Hansen, and Donald Frazee (Bellevue College), Nick Watson (Kankakee Community College), and William W. Wood (Carroll College of Montana).

For the Third Edition, thanks to Dharam Chopra (The Wichita State University), Rudolph J. Freund (Texas A&M University), H. G. Mushenheim

(University of Dayton), James C. Navarra (University of Maryland), and Samuel B. Thompson (University of Maryland) for their reviews of the revised manuscript and their many comments and suggestions.

We thank the following persons for their input to the Fourth Edition: W. P. Abeysinghe (Lock Haven University), Roger Champagne (Hudson Valley Community College), Dan Cherwien (Cumberland Community College), Mahmood Ghamsary (Long Beach City College), Robert Hale (St. Louis University), Mohamad Nayebpour (University of Houston-Clear Lake).

For the Fifth Edition, we wish to thank the following reviewers: Larry D. Blumberg (Washburn University), Roger Champagne (Hudson Valley Community College), Michael Eurgubian (Santa Rosa Junior College), Kay Fulp (North Dakota State College of Science), Mohammed G. Rajah (MiraCosta College), George T. Woodbury Jr. (College of the Sequoias).

At Duxbury Press we wish to thank Carolyn Crockett, Cindy Mazow, Curt Hinrichs, and Alex Kugushev.

We wish to acknowledge the valuable contributions made by friends, students, colleagues, and collaborators, and of course our wives, through the development of five editions.

Finally, are grateful to the Literary Executor of the late Sir Ronald A. Fisher, F.R.S., to Dr. Frank Yates, F.R.S., and to Longman Group Ltd., London, for permission to reprint Table IV from their *Statistical Tables for Biological, Agricultural and Medical Research* (6th ed., 1974).

Robert D. Mason
Douglas A. Lind
William G. Marchal

CHAPTER 1

An Introduction to Statistics

OBJECTIVES

When you have completed this chapter, you will be able to

- explain what is meant by statistics;
- define the terms *descriptive statistics* and *inferential statistics;*
- cite some examples of the application of statistics from the social sciences, business, education, psychology, and other fields;
- explain the differences among nominal, ordinal, interval, and ratio levels of measurement.

INTRODUCTION

Welcome to the world of statistics! Almost daily we see the results of statistical analysis in our lives. For example, to start the day we turn on the shower and let it run a few moments. Then we put our hand in the shower to sample the temperature, and we decide to add more hot water or more cold water or conclude that the temperature is just right and enter the shower. As a second example, suppose we go to the grocery store to buy a frozen pizza. One of the pizza makers has a stand where someone offers us a small wedge of the pizza. After sampling the pizza, we make a decision whether to purchase the pizza or not. In both of these instances, we make a decision and select a course of action based on testing.

On a national level, a candidate for president wants to know what percentage of the voters in California will support him. Here are some strategies to answer that question. He could have his staff call all those who plan to vote in the upcoming election and ask for whom they plan to vote. He could go out on the street in San Diego, stop 10 people that look to be of voting age, and ask for whom they plan to vote. He could select a random cross section of about 2000 voters from the entire state, contact these voters, and make an estimate of the percentage who will vote for him in the upcoming election. Throughout this course, we will show you why the third choice is the best course of action.

What about the word *statistics* itself? What does it mean? We encounter it frequently in our everyday language, but it actually has two meanings. In the more common usage, statistics refers to numerical facts. Examples include the average starting salary of college graduates, the percentage of graduates from your college or university who attend graduate school, the number of cars sold at a dealership during the last month, the number of home runs hit by the Cleveland Indians last season, the Dow Jones Industrial Average, or the number of deaths due to alcoholism last year. In these examples, the word *statistics* refers to numbers. Other examples include the following:

- A recent Harris Poll reveals that 64% of the respondents think that small business owners have good moral and ethical standards. Regarding other professionals, 30% of the respondents think journalists have good moral and ethical standards; business executives, 31%; government officials, 29%; lawyers, 25%; and members of Congress, 19%.
- The National Center for Education Statistics reports that of the 64,465,000 students who use computers, 28,662,000 are in high school and 10,661,000 are college undergraduates.
- The Bureau of the Census projects the population of the United States to be 298,252,000 in 2025.
- According to the U.S. Defense Department's *Selected Manpower Statistics,* there are 112,000 first-time enlisted military personnel in the Army and 111,000 reenlistments.
- The U.S. Department of Labor's *Monthly Labor Review* reports that 249,770 new businesses were started in 1995 and that 57,253 failed. From

the same publication, in March 1996 the average hourly earnings for mining are $11.69; for manufacturing, $12.52; and in retail trade, $7.90.

● According to the U.S. Department of Health and Human Services, the life expectancy of a newborn white female was 40.5 years in 1850; today it is 78.7 years. (See Figure 1-1, which follows.)

Each of the above is an example of a **statistic.** A collection of two or more figures is called **statistics** (plural).

Statistics can appear in graphic form as well as in sentence form. A graph is often used to capture reader attention and portray a large amount of data over an extended period of time. For example, more than 300 data points were used to construct Figure 1–1. However, it takes only a quick glance to discover that the life expectancies of all four groups have risen dramatically since 1900—especially that of the blacks, which has increased by more than 100 percent.

The subject of statistics, as we will explore it in this text, has a much broader meaning than just collecting and publishing numerical information. Statistics encompasses the techniques used to collect, organize, present, analyze, and interpret data. Thus, the first step in investigating a problem is to

> **statistics** Techniques used to collect, organize, present, analyze, and interpret data in order to make better decisions.

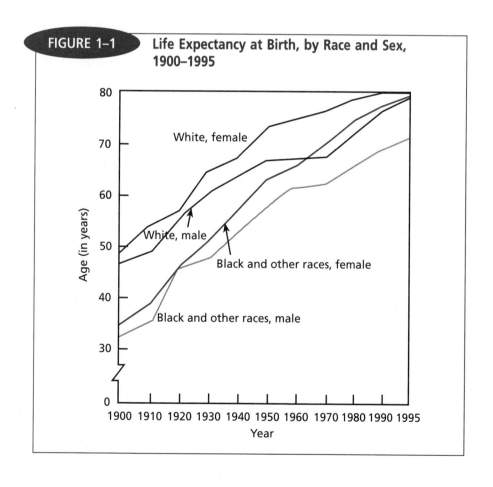

FIGURE 1–1   **Life Expectancy at Birth, by Race and Sex, 1900–1995**

collect pertinent data. It must then be organized in some way and perhaps presented in a chart, such as Figure 1–1. Only then are we able to analyze and interpret the data. Here are some examples of the need for data collection.

1. A life insurance actuary must set premiums for new and unique term insurance policies designed for newborn blacks, whites, males, and females. *Vital Statistics,* published by the U.S. Department of Health and Human Services' National Center for Health Statistics, can provide data on the life expectancies of these four groups. For each group, the actuary needs to know the life expectancy, the commission for the person selling the policy, and so on.

2. In medical research, after defining the problem the next step is to collect data. It may be a problem involving the incidence of cancer or heart attacks in one group compared to another group. For example, a team of Finnish epidemiologists recently explored the relationship between iron in the body and the incidence of heart attacks. They studied 1900 men from eastern Finland for serum levels of ferritin (an iron-storing protein). After 5 years, they found 51 men had suffered heart attacks. An analysis of these data revealed that men with more than 200 micrograms (μg) of ferritin per liter of blood were twice as likely to have heart attacks as men with levels below 200 μg.

We have cited two areas where data are collected and analyzed. One involves vital statistics—life expectancy at birth for four groups since 1900. Another concerns medical research—whether iron in the body is related to the incidence of heart attacks. Are there other disciplines where statistical techniques are applied to analyze data and solve problems?

WHO ELSE USES STATISTICS?

Just as life insurance companies collect, organize, analyze, and interpret data before setting a premium for term, whole life, and other types of policies, so, too, must automobile insurance companies. For example, what kind of information would be needed to set a premium for a Ford Taurus? The automobile's age, the repair record, and the driver's age are some of the data needed. Some examples of how statistics are used in other areas follow:

- Law enforcement agencies are concerned with repeat offenders. Why do some convicted criminals commit a second or third crime, while others do not? If you were doing criminal research, what factors would you consider?
- In medical research, a common way to evaluate the effectiveness of a newly developed medicine is to administer it to one group of people or animals and not to another group, called the "control group." Statistics are collected from both groups, organized, analyzed, and interpreted before making a decision regarding the effectiveness of the new medicine.
- Management in mass production firms must be assured that current production is satisfactory. The quality assurance department has the job of

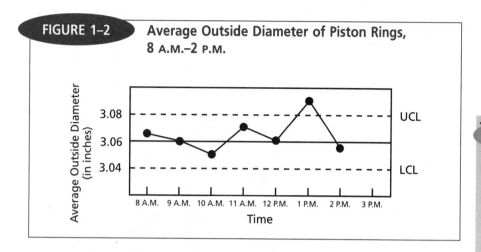

FIGURE 1–2 **Average Outside Diameter of Piston Rings, 8 A.M.–2 P.M.**

selecting, collecting, and evaluating a sample of observations. To make a determination, a quality control inspector might, for example, take a sample of five piston rings every hour, determine the average outside diameter of the rings, and plot it on what is called a "mean chart." Figure 1–2 portrays the average outside diameters for the first seven hours of production.

Note that there is an upper control limit (UCL) and a lower control limit (LCL). If the average outside diameter for the five piston rings is either above the UCL or below the LCL, the process is "out of control." Such is the case at 1 P.M., when the average outside diameter was too large. The quality control engineer reported this to the production manager, and corrective action was taken. By 2 P.M., the process was back "in control."

GRAPHIC PRESENTATION OF DATA

The data (statistics) in the Introduction were in the form of numbers. As an example, we noted that the estimate of the population for the United States in the year 2025 is 298,252,000. In many cases, data are presented in the form of a chart. Charts are easy to read and particularly valuable for visualizing the long-term trend of the data. As an illustration, Figure 1–3 shows the number of home health visits between 1991 and 1995. It is clear that the trend of the number of visits is increasing and that the number of visits has more than doubled from 1991 to 1995.

Figure 1–4 also illustrates the presentation of numerical data in chart form. It shows the number of temporary workers in the work force from 1973 to 1996.

Numerical data are not always available. As an example, no data have been published regarding the opinions of voters in North Dakota on a proposed change in Medicare. Likewise, no data are available about the

Real Stat

Before continuing, we want to introduce a feature we call ***Real Stat.*** Two or three of these are included in most chapters. They do not require any calculations. Instead, read each one to get an appreciation of the wide application of statistics in marketing, nursing, law enforcement, psychology, sports, and other disciplines. This first one is merely an assortment of interesting statistical information, usually from some type of survey, published in *Vitality, USA Today, Business Week,* and other magazines and newspapers.

- Hold your breath. Look over the number carefully. It's not a misprint: $203,010,590. That's what Michael D. Eisner, chairman of Walt Disney Co., made last year.
- Ten percent of U.S. married couples fail in their attempts to conceive a baby.
- Half of American families currently have below $1,000 in net financial assets.
- Thirty percent of accidents occur at home.
- The average American earns $1,235,720 and pays $178,364 in taxes in his or her lifetime.
- Ninety percent of job openings are filled by word of mouth.

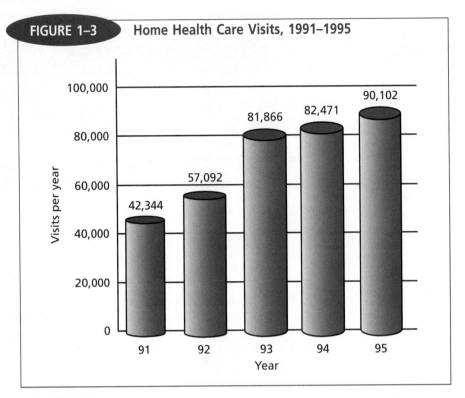

FIGURE 1–3 Home Health Care Visits, 1991–1995

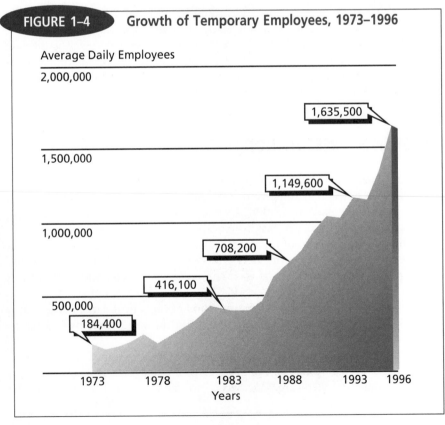

FIGURE 1–4 Growth of Temporary Employees, 1973–1996

acceptance of members of UAW #213 in Detroit of a change in the hourly provisions of a contract proposal presented by management. And we have no data regarding a new lavender face soap with a rose smell developed by Lever Brothers. What is needed in these cases is input from the voters, the auto workers, and the consumers, respectively.

To gather data from individuals or groups it is often necessary to conduct a survey. A compilation of the questions on a survey is referred to as a **questionnaire.**

The questionnaire might be mailed to the respondent, it may take the form of a telephone survey, or personal contact. As an example of survey results, the *Atlanta Journal-Constitution* wanted to find out what football coaches and assistant coaches at NCAA Division IA colleges thought about paying players, as well as other issues. They randomly sampled coaches and found that 71% believed college players should be paid, 61% thought there should be a national championship game, and 69% believed that underclassmen should be able to enter the NFL draft.

Following are two partial survey questionnaires, one by the U.S. Postal Service and the other by Kodalux, a large nationwide firm that processes film. Note that both questionnaires are very simple to complete. All the respondents need to do is check the appropriate boxes.

U.S. Postal Service Questionnaire

1. **Overall Performance** (in the past three months)
   We would like your opinion of the U. S. Postal Service's performance **during the past three months** on some general topics. Use a seven-point scale, where 1 means "Poor," 4 means "Good," and 7 means "Excellent." Please remember that you can mark any box between 1 and 7, or the Don't Know" box.

Please rate the U.S. Postal Service on...	(Poor-Fair)		Good	(Very Good		Excellent)	Don't Know
	1	2	3	4	5	6	7

   a. Its overall performance ☐ ☐ ☐ ☐ ☐ ☐ ☐ ☐
   b. Delivery of the mail in good ☐ ☐ ☐ ☐ ☐ ☐ ☐ ☐
      condition (undamaged)
   c. The length of time it usually ☐ ☐ ☐ ☐ ☐ ☐ ☐ ☐
      takes a letter mailed **in your
      local area** to be delivered in
      your local area
   d. Having conveniently located ☐ ☐ ☐ ☐ ☐ ☐ ☐ ☐
      mail deposit boxes where
      you can mail letters.

Some personal questions will help to evaluate the survey results.

2. Your age:

   ☐ Under 25 years   ☐ 35–44 years   ☐ 55–64 years
   ☐ 25–34 years   ☐ 45–54 years   ☐ 65 or older

3. Highest level of school you completed:

   ☐ Did not finish high school   ☐ Some college/technical school/trade school
   ☐ High school graduate   ☐ College graduate or beyond

Kodalux Questionnaire

Dear Customer: Thank you for your order. In our continuing desire to provide the best possible service to our customers, we would greatly appreciate your answering the following questions, and mailing this postage-free card back to us. Many thanks for your help.

1. What date did you mail your film to us?_____
2. What date did you receive it?_____
3. Was it in good condition when received? Yes____ No____
4. What film type did you use? Slide____ Color Negative____
5. What size film did you use? 35mm____ 110____ Disc____ Other____
6. Do you occasionally send film in multiple batches? Yes____ No____
7. Are your pictures for professional, industrial or commercial use? Yes__No__
8. Are you satisfied with the service you received? Yes____ No____

Additional Comments:_____
_____
_____

PLEASE   City:_____   State:_____
INCLUDE: Zip Code:_____

Kodalux
Processing Services

ROC   Kodalux and design are trademarks of Eastman Kodak Company under license to Qualex, Inc.

TYPES OF STATISTICS

As stated earlier, statistics involves the collection, organization, and presentation of numerical data. Masses of unorganized data stored in a computer are usually of little value. Techniques are available, however, to organize such data into some meaningful form. These aids in organizing, analyzing, and describing (or summarizing) a large collection of numbers are collectively referred to as **descriptive statistics.**

We will discuss one descriptive technique, the frequency distribution, in Chapter 2. In that chapter, we will also examine how data are presented in graphic form. We have already graphically portrayed the life expectancies of males, females, whites, and blacks (Figure 1–1).

An average, also called a measure of central tendency, is another descriptive statistic. There are a number of different averages, but each describes the tendency of a set of data to cluster about its central value. For example:

- According to the U.S. Department of Agriculture, the typical size of a farm in the United States is 281 acres.
- The typical cost of operating an automobile, reports *Motor Vehicle Facts and Figures,* is $.38 per mile. Those cars with high finance charges, high insurance premiums, and so on cost more. An older car with a smaller insurance premium costs less than $.38 per mile.

In Chapter 4, we will use several measures, such as the range, the mean deviation, and the standard deviation, to describe the spread in a set of values.

descriptive statistics
Methods used to describe the data that have been collected.

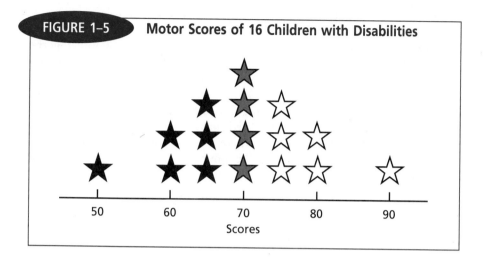

FIGURE 1–5 **Motor Scores of 16 Children with Disabilities**

Figure 1–5, for example, describes the spread in the motor scores of 16 children with disabilities. In describing the spread, the range of the scores is 40, with the scores varying from 50 to 90.

Descriptive statistics is only one facet of the science of statistics. Another is **inferential statistics** or **inductive statistics**. Inferential statistical methods are concerned with determining something about a **population**. Gallup, Harris, and other pollsters do just that when they are hired before an election to estimate how voters (the population) plan to vote on election day.

In the case of political polls, usually about 2000 voters are sampled out of the population of registered voters in the United States. Based on the sample results, an inference is then made about the reaction of *all* voters on election day.

What exactly do we mean by the words *population* and *sample*, as used in our definition of inferential statistics? We normally think of a population as a large group of people: all the residents of Dade County, Florida, for example, or all the employees of the Ford Motor Company assembly plant in Atlanta, Georgia. However, a population can also consist of a group of objects, such as all the Mercury Sables produced in an 8-hour shift at the Ford plant in Atlanta. Some other populations: all the zebras in the Masa Mara game preserve in Kenya, all the *New York Times* newspapers sold last Sunday, and all the inmates at Attica Prison.

It is almost impossible to contact every potential voter in Dade County, Florida, to find out how he or she plans to vote on election day. Nor is it feasible to capture every zebra in the Masa Mara preserve to determine the status of its health or to interview every inmate at Attica regarding his plans after being released. We therefore select a portion, or part, of the population, called a **sample,** and concentrate our efforts on this group.

Examples of inferential statistics include the following:

- A random sample of 1260 accounting graduates of 4-year colleges last year revealed that the mean starting salary was $32,694. We therefore infer that

inferential statistics
Techniques used to make a decision, estimate, prediction, or generalization about a population based on sample evidence.

population A collection, or set, of all individuals, objects, or items of interest.

sample A portion, or part, of the population of interest.

the mean starting salary of all accounting graduates of 4-year colleges is $32,694.

- The Gorski Testing Laboratory selected a random sample of 300 from the lot of 18,000 culture dishes received yesterday. An inspection of the 300 revealed that 18 were defective. We therefore infer that 6% of the total shipment is defective.

- A random sample of 846 prisoners released from the state prison in Smithville between 1990 and 1995 revealed that 27% were sent back to prison for committing another crime. We therefore conclude that 27% of all prisoners released from Smithville will commit another crime and be returned to prison.

The relationship between the sample and the population is portrayed below.

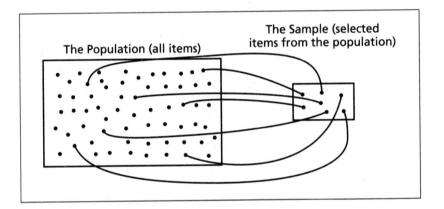

The Population (all items)   The Sample (selected items from the population)

Self-Review 1-1   *This is the first of many Self-Reviews you will encounter throughout the text. They are designed to help you determine immediately whether you comprehend the material. We suggest you complete these Self-Reviews and then check your answers against those provided at the end of the chapter.*

Over a period of a week, the market research department of the Dairy Producers Association randomly selected 300 grocery shoppers and checked their shopping carts for dairy products. Of the 300 carts, 210 contained at least one dairy product.

a. What would the research department report to management regarding the buying habits of all grocery shoppers?
b. Is the 300 a sample or a population?
c. What is the population in this problem?
d. Is this an example of descriptive statistics or inferential statistics?

TYPES OF VARIABLES

There are two basic types of data: (1) those obtained from a qualitative population and (2) those obtained from a quantitative population. When the characteristic or variable being studied is nonnumeric, it is called a **qualitative variable** or an **attribute variable.** Examples of qualitative variables include gender, religious affiliation, state of birth, eye color, type of automobile owned, and movie classifications (G, PG, PG-13, R, and X). When the data being studied are qualitative, we are usually interested in how many or what proportion fall in a certain category. For example, what proportion of those born last month were born in Texas? How many Catholics and how many Protestants are there in Canada? What percentage of the cars sold in the United States last month were Jeeps? Qualitative data are usually summarized in charts and bar graphs, which are discussed in Chapter 2.

When the variable being studied can be reported numerically, the variable is called a **quantitative variable,** and the corresponding population is called a quantitative population. Examples of quantitative variables include the length of an afternoon nap, the balance in your checking account, the number of credit hours you are registered for this semester, the number of children in a family, and the speed of automobiles traveling along I-94 near Milwaukee, Wisconsin.

Quantitative variables are either discrete or continuous. **Discrete variables** can only assume certain values, and there are usually "gaps" between the values. Examples of discrete variables are the number of bedrooms in a house and the number of children in a family. A home can have 2, 3, 4, etc., bedrooms, but it cannot have 3.29 bedrooms. A family can have 1, 2, 3, etc., children, but it cannot have 4.17 children. Thus, there is a "gap" between possible values. Typically, discrete variables are the result of counting. We count the number of bedrooms in a house, and we count the number of children in a family.

Observations of a **continuous variable** can assume any value within a specified range. Examples of continuous variables include the air pressure in a tire, the time to fly from Chicago to Tampa, the number of miles driven between oil changes, and the exact weight of a box of Kellogg's Raisin Bran breakfast cereal. The number of miles driven between oil changes could be reported to the nearest mile, the nearest half mile, and so on. Usually, continuous variables result from measuring something.

The types of variables are summarized in the following diagram.

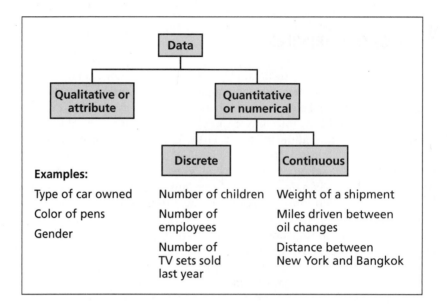

LEVELS OF MEASUREMENT

Data can be classified according to scales or levels of measurement. The level of measurement of the data often dictates the calculations that can be done and the statistical tests that can be performed. To explain, there are six colors of candies in a bag of M&M candies. Suppose we assign the brown a value of 1; yellow, 2; blue, 3; orange, 4; green, 5; and red, 6. From a bag of candies, we add the assigned color values, divide by the number of candies, and report that the mean color is 3.56. Does this mean that the average color is blue or orange? In a high school track meet, there are 8 competitors in the 400-meter run. We report the order of finish and the fact that the mean finish is 4.5. What does the average finish tell us? In both of these instances, we have not properly used the level of measurement and the resulting statements are meaningless. There are actually four levels of measurement: nominal, ordinal, interval, and ratio. The "lowest," or the most primitive measurement, is the nominal scale. The highest, or the scale that gives us the most information about the observation, is the ratio scale.

Nominal Level

With the **nominal level** of measurement, the observations can only be classified or counted. There is no particular order to the categories. The classification of the six colors of M&M candies is an example of the nominal level of measurement. We simply classify the candies by color, and there is no natural order. That is, we could report the brown candies first, the orange first, or any of the colors first. Gender is another example of the nominal level of measurement. Suppose we count the number of students entering a football game with a student ID and report how many are men and how many are women. We could report either the men or the women first. For the nominal

TABLE 1–1	Religions Reported by the Population of the United States 14 Years and Older, 1996	
Religion	**Number**	
Protestant	78,952,000	
Catholic	30,669,000	
Jewish	3,868,000	
Other religion	1,545,000	
No religion	3,195,000	
Religion not reported	1,104,000	
Total	119,333,000	

level of measurement, there is really no measurement involved—only counts. Table 1–1 shows the religions reported by the population of the United States. This is the nominal level of measurement because we merely count the number of people reporting each of the religions.

The arrangement of the religions in Table 1–1 could be changed. That is, we could report Catholic first, Jewish second, and so on. This essentially indicates the major feature of the nominal level of measurement: There is no particular order to the categories. These categories are said to be **mutually exclusive,** meaning, for example, that a person cannot report being a Catholic and a Protestant at the same time.

The categories in Table 1–1 are also said to be **exhaustive,** meaning that every member of the population, or sample, must appear in one of the categories. If a person refused to give her religion, she would be included in the category "Religion not reported." If she said she practices Buddhism, she would be included in the category "Other religion."

In order to process data on religious preference, gender, employment by industry, and so forth, the categories are often coded 1, 2, 3, and so on; with religious preferences, for example, 1 could represent Protestant, 2 could represent Catholic, and so on. This facilitates counting by the computer. However, the fact that we have assigned numbers to the various religions does not give us license to manipulate the numbers arithmetically. For example, $1 + 2$ does not equal 3; that is, Protestant + Catholic does not equal Jewish.

To summarize, nominal data have the following properties.

1. Data categories are mutually exclusive. So an object can belong to only one category.
2. Data categories have no logical order.

In regression and correlation analysis, we are often faced with using nominal-level data. To overcome this problem we resort to dummy or indicator variables. We will discuss variables of this type in Chapter 15.

The next-higher-level of data is the **ordinal level.** Table 1–2 lists the student ratings of the professor in an Introduction to Sociology class in answer to this question: "Overall how did you rate the instructor in this class?" This

mutually exclusive An individual, object, or measurement is included in only one category.

exhaustive Each individual, object, or measurement must appear in a category.

Ordinal Level

TABLE 1–2	Rating of a Sociology Professor	
	Rating	**Frequency**
	Superior	6
	Good	28
	Average	25
	Poor	12
	Inferior	3

illustrates the use of the ordinal scale of measurement. One category is "higher" or "better" than the next one. That is, "Superior" is better than "Good," "Good" is better than "Average," and so on. However, we are not able to determine anything about the *magnitude* of the differences between groups. Is the difference between "Superior" and "Good" the same as the difference between "Poor" and "Inferior"? We cannot tell. If we substitute a 5 for "Superior" and a 4 for "Good," we can conclude that the rating of "Superior" is better than the rating of "Good," but we cannot add a ranking of "Superior" (5) and a ranking of "Good" (4) and obtain meaningful result (9). Further, we cannot conclude that a rating of "Good" (4) is necessarily twice as good as a rating of "Poor" (2). We can conclude only that a rating of "Good" is better than a rating of "Poor," but not how much better.

In summary, ordinal data have the following properties:

1. Data categories are mutually exclusive.
2. Data categories have a logical order.
3. Data categories are scaled according to the amount of the characteristic they possess.

Interval Level

The interval level of measurement is the next-highest-level. It includes all the characteristics of the ordinal scale, but in addition, the difference between values is a constant size. Suppose the high temperatures in Boston on three consecutive days are 28, 31, and 20 degrees Fahrenheit. These temperatures can be easily ranked, but we can also determine the difference between temperatures. This is possible because 1 degree Fahrenheit represents a constant unit of measurement. It is also important to note that 0 is just a point on the scale. It does not represent the absence of the condition. Zero degrees Fahrenheit does not represent the absence of heat, just that it is cold! Equal differences between two temperatures are the same, regardless of their position on the scale. That is, the difference between 10 degrees and 15 degrees is 5 degrees; the difference between 50 and 55 degrees is also 5 degrees.

Interval data have the following properties:

1. Data categories are mutually exclusive.
2. Data categories have a logical order.
3. Data categories are scaled according to the amount of the characteristic they possess.

4. Equal differences in the characteristic are represented by equal differences in the numbers assigned to the categories.
5. The point 0 is just another point on the scale.

The ratio level is the "highest" level of measurement. The ratio level of measurement has all the characteristics of the interval scale, but in addition, the 0 point is meaningful, and the ratio between two numbers is meaningful. Examples of the ratio scale of measurement include money, weight, and height. The major difference between the interval and the ratio scales of measurement is that the ratio scale has a meaningful zero point, so the ratio between two numbers is meaningful. Money is a good illustration. If you have 0 dollars, then you have no money. Weight is another example. If the dial on the scale is at 0, then there is a complete absence of weight. If Jim earns $30,000 per year selling insurance and Rob earns $60,000 per year selling cars, then Rob earns twice as much as Jim.

Ratio data have the following properties:

1. Data categories are mutually exclusive.
2. Data categories have a logical order.
3. Data categories are scaled according to the amount of the characteristic they possess.
4. Equal differences in the characteristic are represented by equal differences in the numbers assigned to the categories.
5. The point 0 reflects the absence of the characteristic.

Table 1–3 illustrates the use of the ratio scale of measurement. It lists the mean salaries reported by pre-MBA students and post-MBA students at some of the top-ranked business schools in the country in 1996. By looking at the table, we can see the amount by which their salaries increased after completing the MBA.

Ratio Level

> **Real Stat**
>
> More and more pressure is being exerted to hire and promote women and minorities to top-level management positions. The federal government appointed a 21-member Glass Ceiling Commission to study ways to remove gender barriers to job advancement. The commission found that, although there are 57 million working women, representing 45% of the total labor force, relatively few are in middle- or top-level positions. The Florida Consumers Federation studied the 750 Publix Supermarket stores and found that less than 3% of all managers and assistant managers are black and less than 2% are women.

TABLE 1–3 **Mean Salaries for Pre-MBA and Post-MBA Students, 1996**

School	Average Pay	
	Pre-MBA	Post-MBA
Vanderbilt	$27,900	$47,230
Cornell	37,820	59,940
MIT	42,630	73,000
Harvard	52,790	84,960
Stanford	51,570	82,860
Duke	39,060	59,870

Self-Review 1-2

What is the level of measurement for each of the following? Give your reasoning.

a. Gallup Polls asked 1000 adults, "Would you say you are financially better off now than you were a year ago?" The responses:

Better off	47%
Worse off	24%
Same	28%
No opinion	1%

b. Every 4 hours the quality control inspector at Cannon Mills selects 100 bedsheets at random and records the intensity of the blue color. There is some latitude, but if the color is too deep, or too light, the sheet is "irregular." The irregular sheets are sold at a discount. The percentages defective for the ten checks Monday through Friday are 1%, 2%, 0%, 1%, 1%, 0%, 0%, 1%, 3%, and 0%.

c. The coaching staff rated each Oklahoma football player who participated in more than five plays in the game against Nebraska.

Rating	Number of Players
Outstanding	4
Excellent	10
Good	8
Fair	5
Poor	7

d. The National Center for Education Statistics cited the following college enrollment figures for 1992 and 2000 (projected):

Sex	1992	2000
Male	6,276,000	6,468,000
Female	7,337,000	7,858,000

Exercises

Answers to the even-numbered Exercises are at the back of the book.

1. Refer to the table at right:
   a. Is this an example of a nominal level of measurement? If so, why?
   b. Are the categories mutually exclusive? Why?
   c. Are the categories exhaustive? Why?

Probable Field of Study of College Freshmen

Probable Field of Study	Percentage
Business	25
Professional	13
Engineering	10
Social sciences	10
Arts and humanities	9
Education	9
Biological sciences	4
Technical	3
Physical sciences	2
Other	15
Total	100

2. Refer to the following table:

Gender and Race of Present Smokers 20 Years Old and Over, 1965–1995 (in percent)

Characteristic	1965	1970	1974	1976	1977	1978	1979	1980	1983	1985	1991	1995
Gender												
Male	50.2	44.3	43.4	42.1	40.9	39.0	38.4	38.5	35.5	33.2	31.5	30.3
Female	31.9	30.8	31.4	31.3	31.4	29.6	29.2	29.0	28.7	28.0	26.2	25.9
Race												
White	40.0	36.5	36.1	35.6	34.9	33.6	33.2	32.9	31.4	29.9	28.3	27.8
Black	43.0	41.4	44.0	41.2	41.8	38.2	36.8	37.2	36.6	36.0	33.5	33.0

   a. Is this an example of a nominal level of measurement? If so, why?
   b. Are the categories mutually exclusive? Why?
   c. Are the categories exhaustive? Why?
   d. Interpret the trend in smoking by gender and by race.
3. Refer to the table at right regarding military ranks:
   a. Why are the data considered an ordinal level of measurement?
   b. Are the categories mutually exclusive? Explain.
   c. Are the categories exhaustive? Explain.

Enlisted Military Rank of Active Duty Personnel, 1996

Rank/Grade	Number (in thousands)
Total	1814.0
Recruit—E-1	128.9
Private—E-2	151.9
Pvt. 1st class—E-3	281.6
Corporal—E-4	456.8
Sergeant—E-5	362.1
Staff Sgt.—E-6	245.7
Sgt. 1st class—E-7	134.0
Master Sgt.—E-8	37.9
Sgt. Major—E-9	15.1

4. Refer to the following table:

Post Secondary Degrees Conferred

Degree Conferred	Number (in thousands)	
	1992	2000*
Total	1,959.7	1,960.3
Associate's	461.2	489.2
Bachelor's	1,062.4	1,036.4
Master's	323.2	327.1
First professional	76.8	71.3
Doctorate	36.1	36.3

*Estimated

a. Why are the data considered an ordinal level of measurement?

b. Are the categories mutually exclusive? Explain.

c. Are the categories exhaustive? Explain.

USES AND ABUSES OF STATISTICS

You have probably heard the old saying that there are three kinds of lies: lies, damn lies, and statistics. This saying is attributed to Benjamin Disraeli and is over a century old. It has also been said that "figures don't lie; liars figure." Both of these statements refer to the abuses of statistics, in which data are presented in ways that are misleading. Many abusers of statistics are simply ignorant or careless, while others intend to mislead the reader by emphasizing data that support their position while leaving out data that may be detrimental to their position. One of our major goals in this text is to make you a more critical consumer of information. When you see charts or data in a newspaper or magazine or reported on TV, always ask yourself these questions: What is the person trying to tell me? Does that person have an agenda? Following are several examples of the abuses of statistical analysis.

The term *average* refers to several different measures of central tendency, which we discuss in Chapter 3. To most people, an average is found by adding the values involved and dividing by the number of values. If a real estate developer tells a client that the average home in the subdivision sold for $150,000, we assume that $150,000 is a representative selling price for all the homes. But suppose there are only 5 homes in the subdivision and they sold for $50,000, $50,000, $60,000, $90,000, and $500,000. We can correctly claim that the average selling price is $150,000, but does $150,000 really seem like a "typical" selling price? Would you also like to know that the same number of homes sold for more than $60,000 as for less than $60,000? Or that $50,000 is the selling price that occurred the most frequently? So what selling price really is the most "typical"? This example illustrates that a reported average can be misleading because it is only one of several numbers that can be used to represent the data. There is really no objective set of criteria that state what average should be reported on each occasion. However, we want to educate you as a consumer of data so you understand that a person or group might report one value that favors their position and exclude other values. We focus on averages, or measures of central tendency, in Chapter 3.

Charts and graphs can also be used to mislead. Suppose the school taxes in a particular school district increased from $100 a year to $200 a year; that is, the taxes doubled. To show this change, the dollar sign below on the right is twice as tall as the one on the left. However, it is also twice as wide, so the area of the dollar sign on the right is four times (not twice) that of the one on the left. The graph is misleading because it makes the increase appear much larger than it really is. We discuss the construction of tables and charts in Chapter 2.

> **$ $** *When we double the dimensions of two-dimensional objects, we increase the area by a factor of four.*

Several years ago a series of TV advertisements reported that "2 out of 3 dentists surveyed indicated they would recommend brand X toothpaste to their patients." The implication is that 67 percent of the dentists would recommend the product to their patients. The trick is that the manufacturer of the toothpaste could take *many* surveys of 3 dentists and report *only* the survey of 3 dentists in which 2 indicated they would recommend brand X. Undoubtedly, a survey of more than 3 dentists is needed, and it must be unbiased and representative of the population of dentists. We discuss sampling methods in Chapter 8.

There can also be misrepresentation of data with regard to the association between variables. In statistical analysis, we often find there is a strong *association* between variables. For example, we find there is a strong association between the number of hours a student studies for an exam and the score he or she receives. Does this mean that studying causes the higher score? No. It means the two variables are related; that is, they tend to act together in a predictable fashion. We study the association between variables in Chapters 13–16.

Sometimes numbers themselves can be deceptive. The mean selling price of homes sold last month in the Tampa, Florida, area is $134,891.58. This sounds like a very precise value and may instill a high degree of confidence in its accuracy. To report that the mean selling price is $135,000 does not convey the same precision and accuracy. However, a statistic that is very precise and carries five or even ten decimal places is not necessarily accurate.

There are many other ways that statistical information can be deceiving. Entire books have been written about the subject. The most famous of these is *How to Lie with Statistics* by Darrell Huff. Understanding these practices will make you a better consumer of statistical information.

A WORD OF ENCOURAGEMENT

If you are a freshman or sophomore studying liberal arts or one of the social sciences, this may be your first college course with a quantitative orientation. Theory and symbols such as Σ, σ, and μ are used extensively here. Also, formulas such as

$$\overline{X} = \frac{\Sigma x}{n}$$

are used throughout. Do not be intimidated by them. They are merely a shorthand that helps to condense the subject matter significantly. As a result, the material is slow reading, and only rarely will you feel you have a complete understanding of it unless you have gone over it more than once.

One of the ways in which you can verify your understanding of the material, as you progress through the book, is to work through each of several **Self-Reviews** that are included in every chapter. Checking your answers against those provided at the end of the chapter allows you to determine immediately whether you understand previously covered subject matter. Further, doing the **Exercises** is a very valuable learning technique. Answers to even-numbered Exercises are at the back of the book. You will also want to complete the **Chapter Achievement Test** found at the end of each chapter. This test will point you to topics you need to review further.

COMPUTER APPLICATIONS

Computer use has accelerated greatly over the last few years, particularly in the field of statistics. Prior to 1940, most of the computations involving statistics were done by hand or on an adding machine. Extensive calculations, like those required in Chapter 15, "Multiple Regression and Correlation Analysis," were very time consuming, and the accuracy of the hundreds of necessary calculations was questionable.

The development of rotary calculators, by such companies as Marchant and Monroe, was the next step in problem solving. These have now been replaced by electronic hand calculators and computers. Many of you have computers at home, and most colleges and universities have PCs available for students. There are many statistical software packages available that will be useful for calculations involving larger data sets. Check with your instructor to see what packages are available to you.

In this text, we use both MINITAB and Excel. MINITAB is a statistical software package and Excel is a spreadsheet product. We chose MINITAB for most of the computer applications in this textbook. We acknowledge the complimentary software provided by MINITAB, INC., for use in preparing this book. It is user friendly, meaning it is easy to operate, and does not require you to learn a programming language. To help you, we give the MINITAB commands at the top of each computer output and highlight them in a second color. Our goal is to show how the computer is used as a tool in statistical analysis. You should view the computer as an aid in performing the calculations. It does not, however, provide any interpretation. That is left to you.

CHAPTER OUTLINE

I. Statistics is the art and the science of collecting, organizing, analyzing, and interpreting data for the purpose of making better decisions.

II. There are two types of statistics.
   A. Descriptive statistics are procedures used to organize and summarize numerical information.
   B. Inferential statistics are procedures used to take a sample from a population and then make estimates about the population based on the sample results.
      1. A population is the total collection of individuals or objects.
      2. A sample is a part or subset of the population.

III. There are two types of variables.
   A. A qualitative variable is nonnumeric.
      1. Usually, we are interested in the number or percentage of the observations in a certain category.
      2. Qualitative data are usually summarized in graphs and bar charts.
   B. There are two types of quantitative variables, and they are usually reported numerically.
      1. Discrete variables can assume only certain values, and there are usually gaps between values.
      2. A continuous variable can assume any value within a specified range.

IV. There are four levels of measurement.
   A. With the nominal level, the data are sorted into categories, with no particular order to the categories.
      1. The categories are mutually exclusive if each individual or object appears in only one category.
      2. The categories are exhaustive if each individual or object appears in at least one of the categories.
   B. The ordinal level of measurement presumes that one category is ranked higher than another.
   C. The interval level of measurement has the ranking characteristic of the ordinal level of measurement plus the characteristic that the distance between values is the same.
   D. The ratio level of measurement has all the characteristics of the interval level, but in addition, there is a meaningful zero point, and the ratio of two values is meaningful.

EXERCISES

5. Explain the difference between qualitative and quantitative data. Give an example of each.

6. Explain the difference between a sample and a population.

7. List the four levels of measurement, and give an example of each that is not included in the text.
8. Define the term *mutually exclusive*.
9. Define the term *exhaustive*.
10. Using actual figures from such publications as the *Statistical Abstract of the United States,* the *World Almanac, Forbes,* or your local newspaper, give examples of the nominal, ordinal, interval, and ratio levels of measurement.
11. A random sample of 300 executives out of 2500 employed by a large firm showed that 270 would move to another location if it meant a substantial promotion. Based on this finding, write a brief note to management regarding all executives in the firm.
12. A random sample of 500 customers was asked to test a new toothpaste. Out of the 500, 400 said it was excellent, 32 thought it was fair, and the remaining customers had no opinion. Based on these sample findings, make an inference about the reaction of all customers to the new toothpaste.
13. Explain the difference between a discrete and a continuous variable. Give an example of each that is not included in the text.

For Exercise 14–17, indicate whether the statement is true or false. If false, give the correct answer.

14. The National Cancer Institute reported the following new cases of cancer last year:

Type	Number (in thousands)
Lung	157
Breast	151
Colon	110
Prostate	106
Bladder	49

The above listing is an ordinal level of measurement.

15. The *Monthly Labor Review* reported the following average hourly earnings by industry: wholesale trade, $11.35; services, $10.52; tobacco products, $17.52; and textile mill products, $8.58. The collection of these data is referred to as statistics.
16. Based on the following chart, we conclude that the sales of educational software declined from 1992 to 1996.

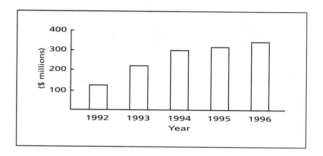

17. For the nominal level of measurement, there is no particular order for the categories.

CHAPTER ACHIEVEMENT TEST

This is the first in a series of Chapter Achievement Tests. These allow you to evaluate your understanding of the material. Answer all the questions. The answers are at the back of the book.

Indicate whether the statement is true or false. If false, give the correct answer.

1. A sample of 1200 senior citizens rated a new cereal, OH Oats, as excellent, good, fair, or poor. The level of measurement is nominal.

2. Inferential and inductive statistics are the same.

3. Ratio-level data are the "lowest level" of measurement, and the data must be mutually exclusive.

4. The National Center for Juvenile Justice reported in their *Juvenile Court Statistics Annual* that last year the juvenile courts disposed of the following delinquency cases (in thousands) by reason of referral: murder, 119; forcible rape, 5; robbery, 33; and motor vehicle theft, 73. This collection of juvenile offenses is called statistics.

5. A sample of 40 former prisoners was studied. Based on their responses, it was said that, had all former prisoners been studied, 20% would be fully adjusted. This is an example of descriptive statistics.

6. There are 14,373 members in the Stone, Clay, and Glass Union. Randomly, 326 were selected for an opinion survey. The 326 can be considered as the population.

7. As reported by the *Sarasota Herald-Tribune*, gold prices soared past $400 an ounce on Tuesday. The $400 is a statistic.

8. In a broad sense, statistics is the collection, organization, presentation, analysis, and interpretation of data for the purpose of making better decisions.

ANSWERS TO SELF-REVIEW PROBLEMS

1-1 a. Of all grocery shoppers, 70% purchase at least one dairy product each time they shop for groceries.
  b. A sample
  c. All grocery shoppers
  d. Inferential statistics, because sample results are being used to infer something about all grocery shoppers

1-2 a. Nominal level, because the categories can be rearranged, as in "worse off," "same," "better off," and "no opinion"
  b. Ratio. The zero point is meaningful.
  c. Ordinal, because "Outstanding" is better than "Excellent", which is in turn better than "Good", and so forth.
  d. Nominal, because gender is simply a label.

CHAPTER 2

Summarizing Data: Frequency Distributions and Graphic Presentation

OBJECTIVES

When you have completed this chapter, you will be able to

- construct a frequency distribution;
- construct a stem-and-leaf chart;
- draw a histogram, a frequency polygon, and a cumulative frequency polygon;
- draw line graphs, bar graphs, and pie graphs;
- Use Excel and MINITAB to create charts and graphs.

CHAPTER PROBLEM   Constructing a Plan for Success

The Reynolds Company wants to further develop the Rafter J Ranch subdivision in Jackson Hole, Wyoming. Charles Reynolds is putting together a prospectus that gives current information on the 80 homes in the subdivision. He checked each of the homes to determine its estimated selling price in today's market. Now he wants to know what a typical home will sell for, what the high and low prices are, where the prices tend to cluster, and so on. In essence, Charles Reynolds wants to describe the price of the homes in the Rafter J Ranch subdivision.

INTRODUCTION

With this chapter, we begin our study of **descriptive statistics.** Recall from Chapter 1 that the purpose of descriptive statistics is to effectively summarize data. Two important elements of descriptive statistics are the arrangement and display of numerical information. Raw numbers alone provide little insight into the underlying pattern of the data from which conclusions can be drawn. For example, we may believe physicians generally have high incomes or young people have lower incomes than old people. However, even if we were given access to Internal Revenue Service files, the income data would be of little value unless arranged, sorted, and condensed. This chapter introduces several techniques used in arranging, sorting, and depicting relevant data—specifically, the **frequency distribution** and various statistical charts.

frequency distribution A grouping of data into categories showing the number of observations in each of the nonoverlapping classes.

FREQUENCY DISTRIBUTIONS

A frequency distribution is a grouping of data into categories showing the number of observations in each of the nonoverlapping classes. Tables 2–1 and 2–2 are examples of frequency distributions.

In its *Motor Vehicle Facts and Figures,* the Motor Vehicle Manufacturers Association annually provides the ages of automobiles currently in use, grouped in the classes "Under 3 years," "3 up to 6 years," and so on. It also gives the number of automobiles in each class—27.1 million in the "Under 3 years" class and so forth. As a second illustration, the U.S. National Center for Health Statistics in its *Vital Statistics of the United States* reports information on the age of mothers at the time of childbirth.

The groupings in Tables 2–1 and 2–2 are referred to as *frequency distributions,* with the categories on the left and the number in each category on the right. These numbers are called the *frequencies.* To interpret Table 2–2, there were 60,700 births last year in which the mother was 15 up to 20 years old.

TABLE 2–1	Ages of Automobiles Currently in Use, 1997
Age (in years)	**Number (in millions)**
Under 3 years	27.1
3 up to 6	30.5
6 up to 9	20.7
9 up to 12	21.1
12 or more	14.0

TABLE 2–2	Age of Mother at the Time of Childbirth	
Age of Mother (in years)	**Number of Mothers (in thousands)**	
10 up to 15	1.4	
15 up to 20	60.7	
20 up to 25	114.6	
25 up to 30	117.4	
30 up to 35	80.2	
35 up to 40	32.5	
40 up to 45	5.9	
45 up to 50	0.3	

We will now construct a frequency distribution using the Rafter J Ranch subdivision data.

Problem   Recall from the Chapter Problem that the Reynolds Company is putting together a prospectus for the further development of the Rafter J Ranch subdivision in Jackson Hole, Wyoming. Charles Reynolds, the developer, had each of the current homes in the Rafter J Ranch area appraised. The appraised value (the estimated selling price) of each of the 80 homes, rounded to the nearest thousand dollars, is shown in Table 2–3.

The numerical information in Table 2–3 is usually called the **raw data.** As such, it has little meaning; that is, it does not reveal much about the value of the homes in the Rafter J Ranch area. We need to construct a frequency distribution to better describe the selling prices.

Constructing Frequency Distributions

TABLE 2–3	Selling Prices of 80 Homes in the Rafter J Ranch Subdivision (in thousands of dollars)								
$ 89	$112	$106	$ 89	$ 96	$ 97	$ 99	$ 85	$113	$ 84
95	102	99	91	102	93	97	98	103	95
105	96	88	90	101	99	101	100	98	100
113	98	95	98	93	86	97	91	108	93
86	85	91	82	94	93	86	98	83	87
105	98	92	98	96	103	93	91	106	94
106	101	98	97	104	83	99	→114	91	97
→ 81	107	104	99	104	93	103	99	97	86
lowest							highest		

Solution   The first step in constructing a frequency distribution is to establish the groupings, called **classes.** In this problem, the first class might contain all the homes with selling prices from $80 thousand up to $85 thousand, the next class might include all the homes selling for $85 thousand up to $90 thousand, and so on. Each class has a lower and an upper limit. The usual practice is to let the lower limit of the smallest class be somewhat smaller than the smallest observation and the upper limit of the largest class be somewhat larger than the largest observation. The classes in a frequency distribution should be **mutually exclusive,** meaning that a particular observation can fit in only one class. For example, a home in the Rafter J Ranch subdivision that sold for $89,000 would fit only in the $85 thousand up to $90 thousand class.

The second step is to tally the selling prices into the appropriate classes. From Table 2–3, the first home sold for $89 thousand, so it is tallied into the $85 up to $90 class. Moving to the right along the first row in Table 2–3, the next home sold for $112 thousand, so it is tallied in the $110 up to $115 class. The other homes are tallied into the various classes. The number of tallies in each class is the **class frequency.** As shown in Table 2–4, the class frequency for the $80 up to $85 class is 5. This means that there are 5 homes with selling prices of $80,000 or more, but less than $85,000. The frequency distribution of all the selling prices of the Rafter J Ranch homes is reported in Table 2–4.

Now that we have organized the data into a frequency distribution, we can summarize the pattern of the selling prices. In Table 2–4, the lowest selling price is about $80,000, and the highest is about $115,000. The highest concentration of selling prices is in the $95,000 up to $100,000 class. More than half the homes have a selling price in the $90,000 to $100,000 range (actually 41 of the 80 homes). If, based on the frequency distribution alone, we wanted to pick one selling price as representative, we would probably select $97,500. Why? It is in the center of the class with the most homes. That is, 26 homes sell for $95,000 up to $100,000, and $97,500 is in the middle of that class.

TABLE 2–4	Frequency Distribution for the Selling Prices of Homes in the Rafter J Ranch Subdivision

Class Selling Price (in thousands of dollars)	Tallies	Class Frequency Number of Homes (f)
$ 80 up to 85	卌	5
85 up to 90	卌 卌	10
90 up to 95	卌 卌 卌	15
95 up to 100	卌 卌 卌 卌 卌 /	26
100 up to 105	卌 卌 ///	13
105 up to 110	卌 //	7
110 up to 115	////	4

In our discussion of frequency distributions, we will frequently refer to the class midpoint and the class interval. The point halfway between the lower and the upper class limits is the **class midpoint.** We also call it the **class mark.** To compute the class midpoint, add the upper and the lower class limits and divide by 2. Referring to Table 2–4, the midpoint of the first class is $82.5, found by ($80.0 + $85.0)/2. The midpoint of $82.5 best represents, or is typical of, the selling prices of the homes in this group. Later in this chapter we use midpoints to draw frequency polygons. They are also used in computing measures of central tendency (in Chapter 3) and measures of dispersion (Chapter 4).

The **class interval** is the difference between the lower limits of consecutive classes. From Table 2–4, the class interval is $5,000, found by $85,000 − $80,000. Assuming the classes in a distribution are the same size, the distance between any two consecutive midpoints may also determine the class interval. For example, the interval using the first two midpoints is found by $87,500 − $82,500 = $5,000.

Here are some suggestions for constructing a frequency distribution.

1. **Overlapping Classes.** Overlapping classes—such as $80–$85, $85–$90, and $90–$95—must be avoided. Otherwise, it is not clear where to tally a selling price of $85 thousand or $90 thousand. Assuming the groupings do not overlap, it is said that the categories are **mutually exclusive,** as they are in Table 2–4. In Chapter 1, we defined mutually exclusive as meaning that each individual or item is included in only one category.
2. **Equal-Sized Classes.** If possible, class intervals should be equal. The use of equal intervals allows computation of averages and measures of dispersion, discussed in the next two chapters. Classes of unequal size are sometimes necessary, however, in order to accommodate all the data, as Table 2–5 illustrates. Had the interval been kept at a constant size of $1000, the frequency distribution would have included at least 1000 classes—obviously too many to make any meaningful analysis. (Incidentally, the Internal Revenue Service used the frequency distribution in Table 2–5 to show the adjusted gross income, before taxes, for over 67 million persons who filed income tax forms during the year.)
3. **Open-Ended Classes.** If possible, open-ended classes should be avoided. Table 2–5 has two such classes—namely, "Under $2000" and "$1,000,000 and over." Strictly speaking, the class "Under $2000" is not open-ended because it has an implied lower limit of zero. However, we usually avoid classes like this because the class midpoint ($1000) is not representative of all the values in the class. If a frequency distribution has an open end, a commonly used measure of central tendency called the arithmetic mean cannot be used. (More about the arithmetic mean in Chapter 3.)
4. **Number of Classes.** No fewer than 5 and no more than 20 classes should be used in the construction of a frequency distribution. Too few, or too many, classes give little insight into the distribution of the data. For example, the following distribution of the selling prices in the Rafter J Ranch,

Suggestions for Constructing Frequency Distributions

> **mutually exclusive categories** Each individual or item, by virtue of being included in one category, is excluded from all others.

TABLE 2–5	Adjusted Gross Income for Individuals Filing Income Tax Returns	
	Adjusted Gross Income Class	**Number of Returns (in thousands)**
	Under $2,000	135
	$2,000–2,999	3,399
	3,000–4,999	8,175
	5,000–9,999	19,740
	10,000–14,999	15,539
	15,000–24,999	14,944
	25,000–49,999	4,451
	50,000–99,999	699
	100,000–499,999	162
	500,000–999,999	3
	$1,000,000 and over	1

subdivision gives very little insight into a typical selling price or the spread of the selling prices:

Selling Price (in thousands of dollars)	Number of Homes
$ 80 up to $100	56
100 up to 120	24

When determining the number of classes, use the smallest integer, k, such that $2^k \geq n$, where n is the total number of observations. Using the 80 selling prices in the Rafter J Ranch data (Table 2–3), first try 6 classes. Thus, $2^6 = 64$, which is less than 80, the total number of observations. With 7 classes, $2^7 = 128$, which is greater than 80, the total number of observations. So the recommended number of classes is 7, which is the number used to in Table 2–4. Based on this 2^k guideline, the following table gives the number of classes recommended for a specific number of observations:

Total Number of Observations	Recommended Number of Classes
9–16	4
17–32	5
33–64	6
65–128	7
129–256	8
257–512	9
513–1,024	10

5. **Class Size.** It is common practice to use intervals that are multiples of 5 or 10, such as 5, 10, 20, 100, or 1,000. If it is decided to have classes of equal size, we approximate the width of the interval by subtracting the lowest from the highest value and dividing by the number of classes. That is,

$$\text{Width of class interval} = \frac{\text{Highest value} - \text{Lowest value}}{\text{Number of classes}} \qquad \boxed{\text{2–1}}$$

Suppose it is decided that the price data in Table 2–3 should be placed into 7 classes. What should the width of the interval be?

$$\text{Width of interval} = \frac{\$114 - \$81}{7} = \$4.71$$

An interval of $4.71 is cumbersome to use. As shown in Table 2–4, a figure of $5 (thousand) is a more convenient interval.

6. **Class Limits.** Frequently, it is easy to make the lower limit of the first class a multiple of the class interval. In this instance, $80 thousand is a multiple of $5 thousand.

- -

Self-Review 2–1

Answers to all Self-Review problems are at the end of the chapter.
  Police files reveal the following ages of persons arrested for purse snatching: 16, 41, 25, 21, 30, 17, 29, 50, 30, and 39.

a. Using 15 years as the lower limit for the first class and an interval of 10 years, organize the age data into a frequency distribution.
b. What are the numbers 16, 41, 25, . . . , called?
c. Based on the data contained only in the frequency distribution, describe the age distribution of the purse snatchers.

- -

Exercises

Answers to the even-numbered Exercises are at the back of the book.

1. In Statistics 101 at Computer University, there were 60 students in the basic statistics course.
   a. How many classes would you recommend to summarize the scores on the first exam?
   b. If the scores ranged from 43 to 97, what would you recommend as a class interval?
   c. Would you round the interval in the above case? If so, to what value?
   d. What value would you recommend as the lower limit of the first class?

2. A sample of 100 AT&T subscribers showed that the amounts they spent on long-distance phone calls last month ranged from $2.15 to $68.73.
   a. How many classes would you recommend to summarize the amounts spent?
   b. What amount would you recommend as a class interval?
   c. What amount would you recommend as the lower limit of the first class?

3. The following data represent the highest amounts spent on a textbook for the current quarter based on a sample of 35 students:

$57	$34	$27	$41	$25	$18
39	33	37	39	38	47
31	42	60	58	31	47
37	16	64	34	41	43
50	46	63	51	41	30
42	37	48	28	34	

Starting with $15 as the beginning number for the first class and using an interval of $10, organize the data into a frequency distribution. Briefly describe the distribution. What is a typical amount spent on books?

4. A survey of 50 amateur photographers included the question "How many rolls of film did you expose during the past month?" The numbers of rolls reported were:

5	3	3	1	4	3	4
3	6	8	4	5	3	4
2	4	7	6	5	9	6
6	6	7	0	11	3	12
4	7	14	0	2	4	4
3	5	15	0	10	4	5
2	3	5	1	8	1	2
12						

Starting with 0 as the beginning number for the first class and using an interval of 3, organize the data into a frequency distribution. Briefly describe

the distribution. What is a typical number of rolls of film?

5. Schneider Nursery employs 30 people. The length of service, in years, for each employee is as follows:

4	1	4	12	10	5	8
14	4	1	8	8	1	4
1	8	8	1	7	3	3
2	7	12	13	11	10	6
2	3					

Construct a frequency distribution. Use 0 as the lower limit of the first class and a class interval of 3 years. What is a typical length of service for employees?

6. The Leona Library (a small rural library) reported the following numbers of patrons using the library in the evening during the past 30 days:

85	81	65	58	47	30	51
92	85	42	55	37	31	82
63	33	44	93	77	57	44
74	63	67	46	73	52	53
47	35					

Construct a frequency distribution. Use a class interval of 10 and a lower limit for the first class of 25. What is a typical number of patrons?

STEM-AND-LEAF CHARTS

Recently, a device known as a **stem-and-leaf chart** has received considerable attention. It is a combination of sorting and graphing, but appears much like a frequency distribution.

The first step in constructing a stem-and-leaf chart is to sort the data from smallest to largest. The stem is the *leading digit* or digits of the number, and the leaf is the *trailing digit*. To illustrate its construction, we will use a class interval of 10 or a multiple of 10. A digit is recorded instead of a tally. For example, the number 52 has a stem value of 5 and a leaf value of 2. The number 867 has a stem value of 86 and a leaf value of 7. An example will clarify the use of stem-and-leaf charts.

Problem   In recent years, more surgical procedures are being done on an outpatient basis. This means the patient enters the hospital on the day of surgery, has the procedure done, and returns home that same day. The director of outpatient surgery at St. Luke's Memorial Hospital gathered the following information on the number of outpatients for the last 20 days:

34	28	32	24	38	16	8	24	46	26
12	18	22	42	36	26	2	26	32	28

Construct a stem-and-leaf chart. Interpret the results.

Solution   A quick glance at the data indicates that the numbers of patients range from 2 to 46 in a single day. The first digit for the number of patients is used as the stem value and the second digit as the leaf value. On the first day, there were 34 patients. Hence, the stem value is 3, and the leaf value is 4. Here is the stem-and-leaf chart after the first three patients are recorded.

Leading Digit	Trailing Digit
0	
1	
2	8
3	42
4	

Organizing all the data:

Leading Digit	Trailing Digit
0	82
1	628
2	84462668
3	42862
4	62

The trailing digits in each row are then rank-ordered to form the stem-and-leaf chart. The first row would appear as

Stem	Leaf
0	28

The leaves for each row when ranked from low to high are

Stem	Leaf
0	28
1	268
2	24466688
3	22468
4	26

Each row in this display represents a stem, and the values in each row are the leaves. The "4" stem has 2 leaves, which are 42 and 46. How do we interpret the stem-and-leaf display? We observe that the largest concentration of patients is between 20 and 29 per day. On 8 of 20 days, there were between 20 and 29 patients in the outpatient surgery area. The values in the "2" stem range from 22 to 28. On 13 of the days, there were between 20 and 39 patients, found by adding the 8 patients in the "2" stem and the 5 patients in the "3" stem. There were 2 days when there were more than 40 patients and 2 days when there were less than 10 patients.

As noted in Chapter 1, a number of computer software packages are available to make setting up and solving statistical problems easier. We use MINITAB to show how the burden of arranging raw data in a stem-and-leaf chart is shifted to the computer. The selling prices of the homes in the Rafter J Ranch subdivision, from Table 2–3, will serve as an example.

To begin, we access the MINITAB system. When MINITAB is ready to run, it issues the command prompt MTB>. The first step is to enter your data. You can do this by typing the command "Set" and then entering "C1". The command Set is a data entry procedure, and the C1 command tells MINITAB to reserve column C1. You can then enter each of the 80 observations from Table 2–3. When all the data are entered, you conclude by typing on the last line the word "end." It is a good idea to name the variables, so type the procedure "Name", followed by the word 'Price' in single quotes. You have now labeled column C1 "Price." Next, using a mouse, click on **Graph,** then **Character Graphs,** and then **Stem-and-Leaf.** When the dialog box appears, select the variable **Price,** click on **Increment** and type the number "5", and then click **OK.** The dialog box appears as follows:

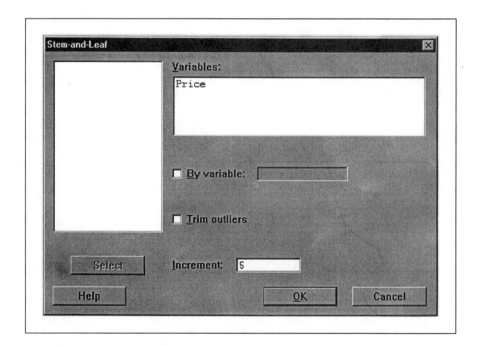

In this text, we will use the shortcut ▶ symbol to indicate the appropriate mouse operation. So we identify the steps to reach the dialog box as

Graph ▶ Character Graphs ▶ Stem-and-Leaf

Then select the variable in the dialog box, indicate the increment is "5", and click **OK.** The output is as follows:

```
MTB > Stem-and-Leaf 'Price';
SUBC>    Increment 5.
```

Character Stem-and-Leaf Display

```
|
Stem-and-leaf of Price     N  = 80
Leaf Unit = 1.0

     5     8 12334
    15     8 5566667899
    30     9 011111233333344
   (26)    9 55566677777788888888999999
    24    10 0011122333444
    11    10 5566678
     4    11 2334
```

The largest concentration of selling prices is between $95,000 and $100,000, and there are 26 homes in that range. The (26) in the left-hand column indicates that there are 26 observations in this stem, which is the largest concentration of data. There were 8 homes with selling prices of $98,000, 4 homes sold for more than $110,000, and 5 sold for less than $85,000.

The MINITAB solution provides some additional information in the column to the left of the stem values. The values 5, 15, 30, and so on are cumulative totals. The number 30, for example, indicates that a total of 30 homes had selling prices of less than $95,000. Similarly, the number 24 indicates that 24 homes had selling prices of $100,000 or more.

The stem-and-leaf display is very flexible. As an example, suppose that, during a two-week period, selected Burger King restaurants sold 6732, 6791, 6823, and 6752 Whoppers. The stem-and-leaf display would be:

Stem	Leaf
67	359
68	2

Notice that the stem includes the thousands and hundreds digits and the leaf contains the tens digit. The units digit is omitted.

Self-Review 2–2   The police records referred to in Self-Review 2–1 on page 31 showed the ages of those arrested for purse snatching to be 16, 41, 25, 21, 30, 17, 29, 50, 30, and 39. Develop a stem-and-leaf display.

Exercises

7. In a survey conducted for the U.S. Forestry Service, each section head was asked to indicate the number of forest rangers under his or her supervision. The data were organized into the following stem-and-leaf chart:

```
Stem-and-leaf of Staff   N = 30
Leaf Unit = 1.0

    1    0   8
    2    1   1
    6    1   7888
    6    2
   12    2   678899
   (4)   3   0014
   14    3   566667
    8    4   2344
    4    4   568
    1    5   1
```

a. How many respondents were there?
b. How many respondents supervised less than 20 people? List the number of people they supervised.
c. What is the largest number of rangers supervised?
d. Estimate the typical (middle) number of rangers supervised.
e. How many of the respondents supervised 25 up to 30 rangers?

8. An insurance claims adjustor collected the following data on the number of car thefts reported each day for a month. These data were arranged into a stem-and-leaf display.

```
Stem-and-leaf of Thefts   N = 30
Leaf Unit = 1.0

    3    5   011
    6    5   233
    9    5   55
   12    5   677
   14    5   89
   15    6
   (2)   6   23
   15    6   55
   13    6   667
   10    6   899
    7    7
    7    7   23
    5    7   55
    3    7   7
    2    7   89
```

a. How many times were there fewer than 55 car thefts?
b. How many observations are there in the first class? List these values.
c. What was the largest number of thefts reported on a single day?
d. Estimate the typical (middle) number of reported thefts.
e. How many days were there in the month?

9. The Quick Change Oil Company advertises that it can change a car's motor oil in 10 minutes. However, business has been very slow lately. A study of the last 25 days revealed the following number of cars serviced each day:

65	98	55	62	79
59	51	90	72	56
70	62	66	80	94
79	63	73	71	85
93	68	86	53	90

Develop a stem-and-leaf display.

10. Robinson TV and Appliance employs 20 salespeople. The number of TVs each salesperson
sold last month is given on the right. Develop a
stem-and-leaf display.

21	34	38	12	6
22	18	36	13	28
19	40	8	14	7
20	18	23	32	14

PORTRAYING FREQUENCY DISTRIBUTIONS GRAPHICALLY

The popularity of magazines such as *Better Homes and Gardens, Time,* and
Ebony and newspapers such as *USA Today* and the *Wall Street Journal* is a
tribute to the Chinese proverb that "a picture is worth a thousand words."

Pictures of a special variety are also used extensively to help hospital ad-
ministrators, business executives, and consumers get a quick grasp of statistical reports. Such pictures are called "graphs" or "charts." Three graphic
forms commonly employed to portray a frequency distribution are the **histogram,** the **frequency polygon,** and the **cumulative frequency polygon,** often
referred to as an **ogive.**

Histograms

A **histogram** is one of the most easily interpreted charts. We will illustrate its
construction by using the distribution of the selling prices of homes in the
Rafter J Ranch subdivision (Table 2–6).

The class frequencies (numbers of homes sold) are plotted on the vertical
axis (Y axis). The variable (selling price) is scaled on the horizontal axis (X
axis).

Five homes sold for $80,000 up to $85,000. The first step in constructing
a histogram is to draw vertical lines from both $80,000 and $85,000 on the
X axis to points opposite 5 on the Y axis and then to connect the tops of
these two lines to form a bar. The area of that bar represents the number of
homes (5) sold in that class.

Figure 2–1 shows how the histogram would appear for the first two classes.
If we insert the lower limits on the X axis and the frequencies (numbers of

TABLE 2–6	Selling Prices of Homes in the Rafter J Ranch Subdivision	
True Limits		**Class Frequencies**
Selling Price		Number of Homes (*f*)
$ 80,000 up to $ 85,000		5
85,000 up to 90,000		10
90,000 up to 95,000		15
95,000 up to 100,000		26
100,000 up to 105,000		13
105,000 up to 110,000		7
110,000 up to 115,000		4

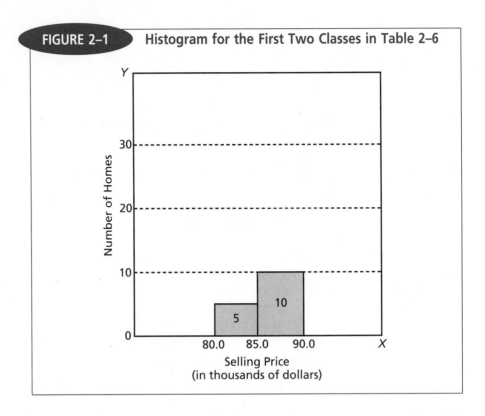

FIGURE 2–1   Histogram for the First Two Classes in Table 2–6

homes) on the Y axis, Figure 2–1 is the result. The completed histogram is shown in Figure 2–2.

To obtain the following histogram from MINITAB, begin by placing the data disk in the A drive, entering MINITAB, and calling up the data file called Rafter. To perform the calculation, click on **Graph** and **Character Graphs,** then **Histogram.** In the dialog box, highlight the data in column C1 and click on **Select.** Type "82.5" in the **First midpoint** box, and then type "5" in the **Interval width** box. Finally, click **OK.**

Self-Review 2–3

The age distribution of part-time prison guards hired by the Nebraska Corrections Bureau in the past two years is as follows:

Age	Number
20 up to 30	2
30 up to 40	13
40 up to 50	20
50 up to 60	12
60 up to 70	3

Construct a histogram for the above prison data.

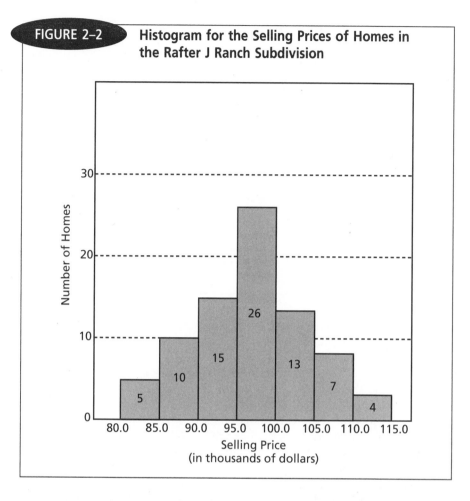

FIGURE 2–2   **Histogram for the Selling Prices of Homes in the Rafter J Ranch Subdivision**

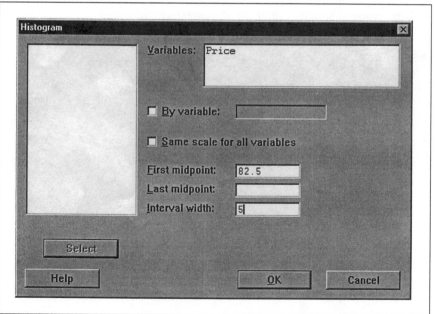

```
MTB > GStd.
MTB > Histogram 'Price';
SUBC>    Start 82.5;
SUBC>    Increment 5.
```

Character Histogram

```
|
Histogram of Price    N = 80

Midpoint   Count
   82.50       5  *****
   87.50      10  **********
   92.50      15  ***************
   97.50      26  **************************
  102.50      13  *************
  107.50       7  *******
  112.50       4  ****
```

Frequency Polygons

To illustrate the construction of a **frequency polygon,** the frequency distribution that lists the selling prices of homes in the Rafter J Ranch subdivision is reintroduced (see Table 2–7).

The class frequencies (numbers of homes in this problem) are plotted on the Y axis, and the class midpoints are on the X axis.

In Table 2–7, there were 5 homes that sold in the $80 up to $85 price range. The midpoint representing that class is $82.5. Therefore, the coordinates of the first plot are $X = 82.5$ and $Y = 5$. The coordinates of the next plot are $X = 87.5$ and $Y = 10$. The dots are connected in order by straight-line segments, as shown in Figure 2–3.

A technical note: The usual practice is to extend the two extremes of the frequency polygon to the X axis. Conventionally, we show this by using dashed, rather than solid, lines for the extension. In Figure 2–3, the left end

TABLE 2–7	Frequency Distribution for the Selling Prices of Homes in the Rafter J Ranch Subdivision	
Class Limits	**Class Midpoint**	
Selling Price (in thousands of dollars)	(in thousands of dollars)	**Number of Homes**
$ 80 up to $ 85	$ 82.5	5
85 up to 90	87.5	10
90 up to 95	92.5	15
95 up to 100	97.5	26
100 up to 105	102.5	13
105 up to 110	107.5	7
110 up to 115	112.5	4

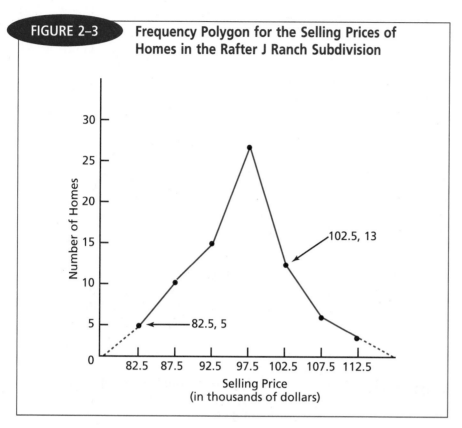

FIGURE 2–3   **Frequency Polygon for the Selling Prices of Homes in the Rafter J Ranch Subdivision**

of the polygon extends to the midpoint of a class below the lowest nonzero class. The upper end of the polygon is treated in a similar way.

Refer to Self-Review 2–3 on page 38. Portray the ages of the guards in the form of a frequency polygon.   **Self-Review 2–4**

Exercises

11. In Exercise 3, you constructed the following frequency distribution:

Amount Spent on Textbooks	Number
$15.00 up to $25.00	2
25.00 up to 35.00	10
35.00 up to 45.00	12
45.00 up to 55.00	6
55.00 up to 65.00	5

a. Portray the distribution in the form of a histogram.

b. Portray the distribution in the form of a frequency polygon.

12. Refer to Exercise 4. Draw a histogram and a frequency polygon using the distribution of the numbers of rolls of film.

13. Refer to Exercise 5. Draw a histogram and a frequency polygon for the lengths of service for the employees.

14. Refer to Exercise 6. Draw a histogram for the numbers of library patrons.

A frequency polygon is ideal for comparing two or more groups of values. As an illustration, recall that the selling prices of the homes in the Rafter J Ranch subdivision ranged from about $80 thousand to about $115 thousand. Suppose the broker at the Reynolds Company wants to compare the Rafter J Ranch selling prices with those in the Indian Paintbrush area. The frequency polygons are in Figure 2–4. It is readily apparent that the selling prices for the Indian Paintbrush homes are significantly higher than those for the Rafter J Ranch homes. The typical price for a Rafter J Ranch home is about $97 thousand and for the Indian Paintbrush area about $135 thousand.

Percentage Frequency Distributions and Percentage Frequency Polygons

The numbers of homes sold in Rafter J Ranch and in Indian Paintbrush were about equal. This made the comparison of the two distributions relatively easy. It is difficult to make comparisons, however, if one group is much larger than the other. As an example, 50 persons were surveyed on controversial subjects, such as abortion and strip mining. Six years later a much larger group of 2000 was asked the same questions. The research report gave the age distributions for both groups (see Table 2–8).

Because the numbers in the 1990 survey are small relative to the numbers in the 1996 survey, the accompanying frequency polygon is difficult to interpret. The 1990 results are compressed at the bottom of Figure 2–5.

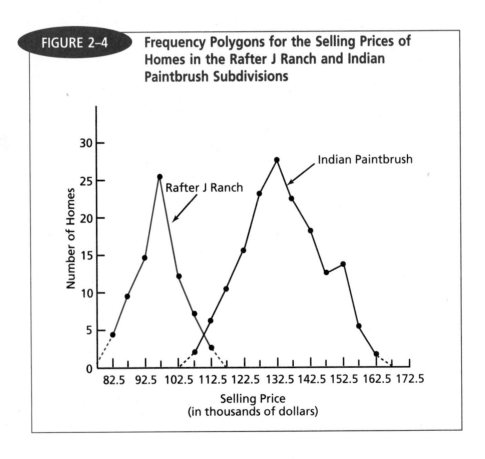

FIGURE 2–4   **Frequency Polygons for the Selling Prices of Homes in the Rafter J Ranch and Indian Paintbrush Subdivisions**

TABLE 2–8	Age Distributions for the 1990 and 1996 Surveys		
Age (in years)	**Number of Persons**		
	1990	1996	
20 up to 30	2	74	
30 up to 40	13	400	
40 up to 50	20	980	
50 up to 60	12	460	
60 up to 70	3	86	
Total	50	2000	

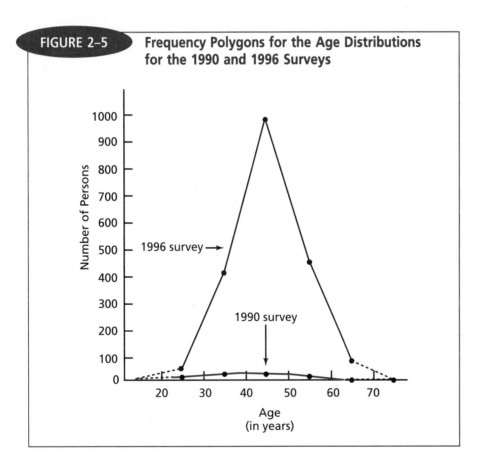

FIGURE 2–5 — **Frequency Polygons for the Age Distributions for the 1990 and 1996 Surveys**

A better approach is to convert the class frequencies to percentages of the total. For example, we can convert the class frequency in the 20 up to 30 class for 1990 to a percentage by dividing 2 by the total number of persons ($2/50 = 0.04 = 4\%$). In terms of a formula, the relative frequency of a class is found by

$$\text{Relative frequency of a class} = \frac{\text{Frequency of a class}}{\text{Total number of frequencies}} \qquad \boxed{\textbf{2-2}}$$

We can now change all the class frequencies to percentages, as shown in Table 2–9.

TABLE 2–9	Percentage Frequency Distributions for the 1990 and 1996 Surveys			
Age (in years)	**Number of Persons**		**Percentage of Total**	
	1990	1996	1990	1996
20 up to 30	2	74	4.0	3.7
30 up to 40	13	400	26.0	20.0
40 up to 50	20	980	40.0	49.0
50 up to 60	12	460	24.0	23.0
60 up to 70	3	86	6.0	4.3
Total	50	2000	100.0	100.0

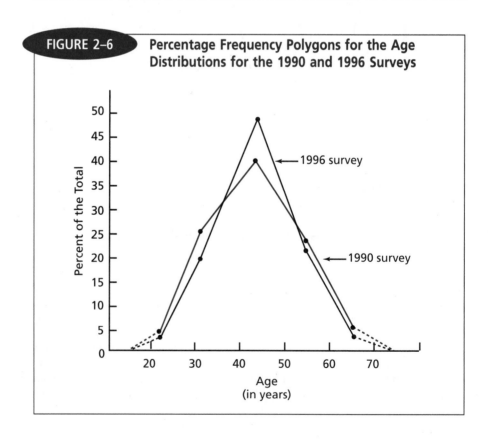

FIGURE 2–6   **Percentage Frequency Polygons for the Age Distributions for the 1990 and 1996 Surveys**

Now notice in Figure 2–6 that the distributions are approximately equal. For example, 24% of those surveyed in 1990 were age 50 up to 60, and in 1996, 23% were in that same category.

Cumulative Frequency Distributions and the Cumulative Frequency Polygons (Ogives)

We may be interested in knowing the percentage of observations that are less than a particular value. Or the question might be; What percentage of the observations are greater than a particular value? For example, what percentage of the Rafter J Ranch selling prices are greater than $100,000? A **cumulative frequency distribution** or a **cumulative frequency polygon,** often referred to as an **ogive,** can be used to answer these questions.

As the name implies, a cumulative frequency distribution requires cumulative class frequencies. There are two cumulative frequency distributions—a "less-than" cumulative frequency distribution and a "more-than" cumulative frequency distribution. To show the construction of both types the selling prices of the Rafter J Ranch homes are repeated in Table 2–10.

To arrive at the Cumulative Frequency column in Table 2–10 we add the frequency in the first class (5) to that in the second class (10) to give a cumulative frequency of 15. Next, 15 + 15 = 30 for the cumulative frequency in the third class. Then 30 + 26 = 56, and so on.

The less-than cumulative frequency distribution in Figure 2–7 is constructed by plotting the cumulative frequencies on the Y axis and in this case the selling prices on the X axis. Note in Figure 2–7 that the cumulative frequencies are on the left vertical axis and the cumulative percentages on the right vertical axis.

Based on this less-than cumulative polygon, we can make statements such as "about half (50%) of the selling prices are less than $97,000" (approximately). We find $97 thousand, as shown, by drawing a horizontal line from the 50% mark on the right vertical axis to the polygon and then dropping the line straight down to the X axis to read the figure $97 thousand.

TABLE 2–10	Less-Than Cumulative Frequency Distribution for the Selling Prices of Homes in the Rafter J Ranch Subdivision		
Selling Price	**Frequency**		**Cumulative Frequency**
$ 80,000 up to $ 85,000	5		5
85,000 up to 90,000	10	add down	15
90,000 up to 95,000	15		30
95,000 up to 100,000	26		56
100,000 up to 105,000	13		69
105,000 up to 110,000	7		76
110,000 up to 115,000	4		80

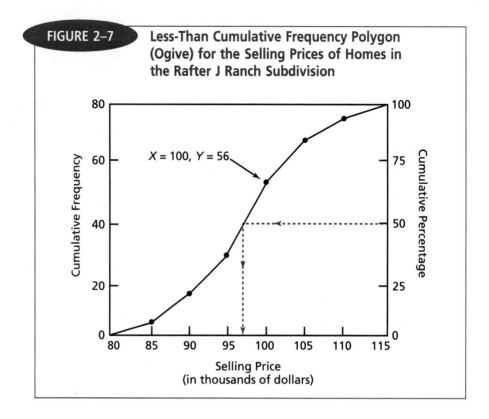

FIGURE 2–7 Less-Than Cumulative Frequency Polygon (Ogive) for the Selling Prices of Homes in the Rafter J Ranch Subdivision

Because no values occur before the smallest class, the corresponding cumulative frequency in that class is zero. No (0) homes sold for less than $80,000 (see Figure 2–7).

We construct the more-than cumulative frequency distribution by adding the frequencies from the *highest* class to the *lowest* class (see Table 2–11). The lower class limits and the corresponding cumulative frequencies are

TABLE 2–11 More-Than Cumulative Frequency Distribution for the Selling Prices of Homes in the Rafter J Ranch Subdivision

Selling Price	Frequency		Cumulative Frequency
$ 80,000 up to $ 85,000	5		80
85,000 up to 90,000	10		75
90,000 up to 95,000	15		65
95,000 up to 100,000	26		50
100,000 up to 105,000	13	add	24
105,000 up to 110,000	7	up	11
110,000 up to 115,000	4		4

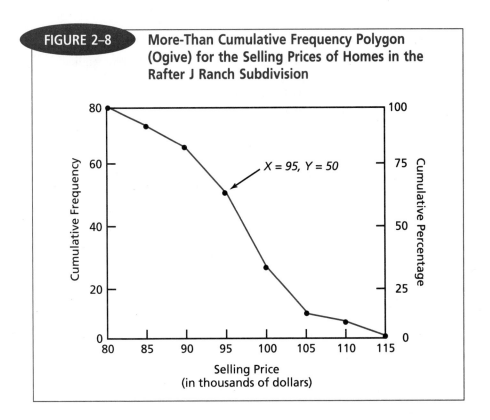

FIGURE 2–8   **More-Than Cumulative Frequency Polygon (Ogive) for the Selling Prices of Homes in the Rafter J Ranch Subdivision**

plotted to construct the more-than cumulative frequency polygon (see Figure 2–8).

Presented below is the age distribution of part-time prison guards from Self-Review 2–3.

Age	Number
20 up to 30	2
30 up to 40	13
40 up to 50	20
50 up to 60	12
60 up to 70	3

a. Construct a less-than cumulative frequency distribution.
b. Draw a less-than cumulative frequency polygon (ogive).
c. Based on the graph, about half the guards hired by the Nebraska Corrections Bureau were less than what age?

Exercises

15. See the following chart:

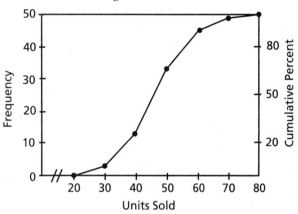

a. What is this chart called?
b. How many observations were studied?
c. What is the class interval?
d. Suppose the horizontal axis refers to the number of units sold per day. On how many days were 13 or less units sold?
e. On 80% of the days, how many units were sold?
f. How many classes are there?
g. Which class has the largest number of observations? How many observations are in that class?

16. See the following chart. Suppose the vertical axis refers to the number of school districts in Vermont and the horizontal axis the size of the teaching staff in the school district.

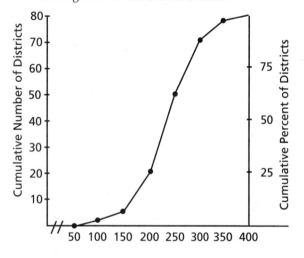

a. How many school districts were studied?
b. What is the class interval?
c. How many classes are there?
d. How many school districts have a teaching staff of 250 or less?
e. Seventy-five percent of the school districts have how many employees or less?
f. Which class has the largest number of observations? How many?
g. Forty of the school districts employ a teaching staff of how many or less?

17. A bottling machine filling bottles with a liquid medicine is tested periodically to determine if it is functioning within certain limits. The frequency distribution below gives the results for 100 observations measured to the nearest hundredth of an ounce.

Class	Frequency
5.00 up to 5.50	1
5.50 up to 6.00	1
6.00 up to 6.50	2
6.50 up to 7.00	5
7.00 up to 7.50	13
7.50 up to 8.00	36
8.00 up to 8.50	31
8.50 up to 9.00	10
9.00 up to 9.50	1

a. Construct a less-than cumulative frequency distribution for these data.
b. Draw a less-than cumulative frequency polygon (ogive).
c. Based on the graph, about one-fourth of the time the machine will yield less than what amount?

18. A psychologist is studying the length of time it takes mice to go through a maze and reach a reward at the end. Time is measured to the nearest tenth of a second, with the results given in the table at the top of the next page.

Time	Frequency
2.0 up to 3.0	3
3.0 up to 4.0	7
4.0 up to 5.0	15
5.0 up to 6.0	29
6.0 up to 7.0	81
7.0 up to 8.0	50
8.0 up to 9.0	10
9.0 up to 10.0	5

a. Construct a more-than cumulative frequency distribution for this data.

b. Draw a more-than cumulative frequency polygon (ogive).

c. Based on the graph, about 90% of the mice take more than what amount of time to complete the maze?

The frequency polygon or ogive and the histogram have strong visual appeal. There are other graphs, however, that are commonly used in government reports, research reports, newspapers, and journals. Several are presented in this section. Others, such as the scatter diagram, are introduced in later chapters.

Other Graphic Techniques

LINE CHARTS.   A **line chart** is particularly useful in portraying data over a period of time. We will use the number of married women in the work force from 1960 to 1995 as an example. Time (years in this case) is *always* plotted on the horizontal axis. The first plot is $X = 1960$, $Y = 12.2$ million; the second plot is $X = 1965$, $Y = 14.9$ million; and so on (see Figure 2–9).

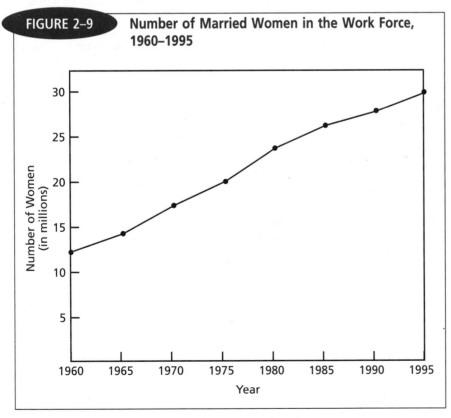

FIGURE 2–9   **Number of Married Women in the Work Force, 1960–1995**

A line chart is relatively easy to interpret. In this case, the number of married women in the labor force has increased steadily from about 12 million in 1960 to over 30 million in 1995.

Self-Review 2–6

The farm population of the United States from 1940 to 1990 is given below:

Year	Farm Population (in millions)
1940	30.5
1950	23.0
1960	15.6
1970	9.7
1980	7.9
1990	7.3

a. Plot the farm population in the form of a line chart.
b. Interpret your chart.

BAR CHARTS.   A **bar chart** is used to portray any one of the four levels of measurement—nominal, ordinal, interval, or ratio. The religions reported by the population of the United States 14 years old and over are used to illustrate the construction of a bar chart. The data in Figures 2–10 and 2–11 are nominal level. The chart may be organized so that the bars are either horizontal, as in Figure 2–10, or vertical, as in Figure 2–11.

Technical notes: As shown in these two bar charts, a small space usually separates each bar. Color is often used—especially in the annual reports of businesses and in magazines such as *Time*. Actual numbers may be placed on the graph, such as 79 million for Protestant, 31 million for Roman Catholic, and so on.

Several time series can be shown on the same chart to depict changes. Figure 2–12 shows the decline in the percentage of players in the National Hockey League that were born in Canada. In the 1969–1970 season, 96.7% of the players were Canadian born, but by 1994–1995, this figure had dropped to 61.2%. At the same time, there had been an increase in the number of players born in Europe and the United States. A recent article in *Sports Illustrated* lamented that hockey is no longer Canada's game.

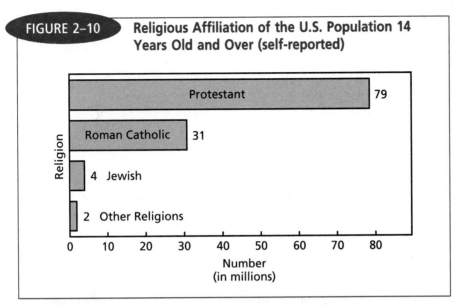

FIGURE 2–10   **Religious Affiliation of the U.S. Population 14 Years Old and Over (self-reported)**

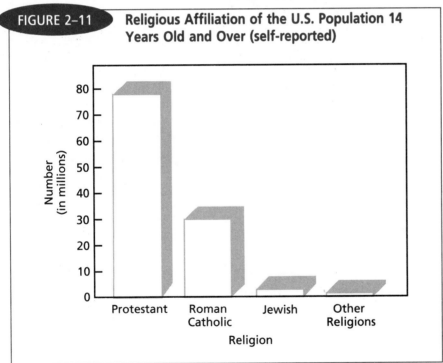

FIGURE 2–11   **Religious Affiliation of the U.S. Population 14 Years Old and Over (self-reported)**

Self-Review 2–7

In Self-Review 2–6, we gave the farm population of the United States in the years 1940–1990 as

Year	Farm Population (in millions)
1940	30.5
1950	23.0
1960	15.6
1970	9.7
1980	7.9
1990	7.3

a. Draw a bar chart placing the bars horizontally.
b. Draw a bar chart placing the bars vertically.

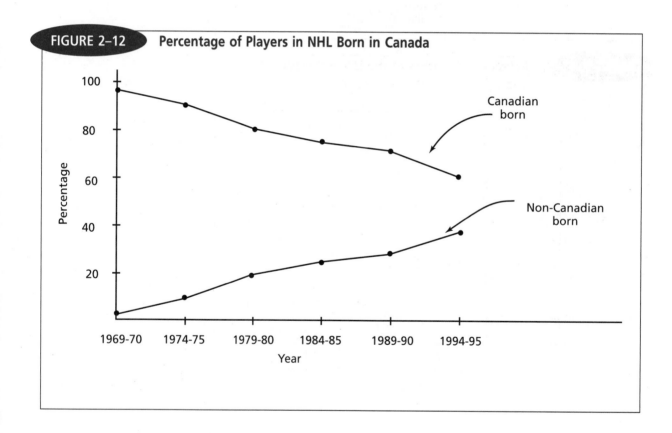

FIGURE 2–12 **Percentage of Players in NHL Born in Canada**

PIE CHARTS.   A very popular graph used to portray parts of the total is called a **pie chart.** As an example, suppose students living on the campus of State University reported their annual expenditures on books, tuition, and so on. The figures were grouped, and typical amounts for each category were determined (see Table 2–12).

To represent tuition and other expenditures, each category must first be converted to a percentage of the total. Tuition, for example, accounts for 40% of the total, and we find it by $3200/$8000 = 0.40 = 40% (see Table 2–12).

To construct a pie chart, we first divide a circle (the pie) into equal "slices," as shown by the notches along the perimeter of the circle in Figure 2–13. Here each slice represents 5% (the total area of the pie is 100%).

TABLE 2–12	Typical Annual Expenditures for Tuition and Other Categories at State University	
Expenditure	Amount	Percentage of Total
Tuition	$3200	40
Room and board	2400	30
Transportation	1200	15
Recreation	800	10
Books	400	5
Total	$8000	100

FIGURE 2–13   **Model for Pie Chart**

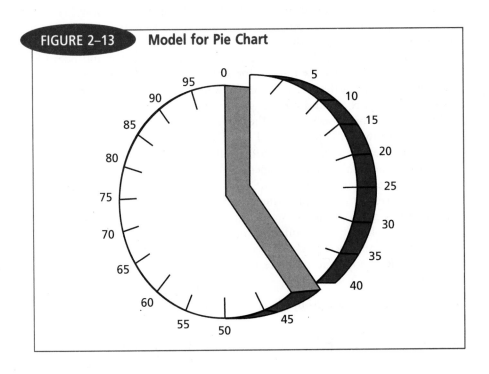

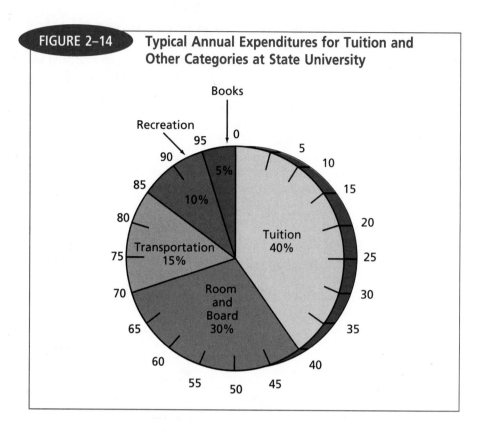

FIGURE 2–14 — **Typical Annual Expenditures for Tuition and Other Categories at State University**

In constructing this pie chart, we plot the 40% spent on tuition first. We draw a line from 0 to the center of the pie and then another line from the center to 40 on the edge of the circle. The enclosed piece of the pie represents expenditures on tuition. Next, we plot expenditures for room and board by adding 30% to the 40% for tuition, for a total of 70%. We draw a line from 70% to the center of the pie. The area between 40% and 70% represents the amount spent on room and board.

We continue this process for all the remaining expenditures. The completed pie chart is shown in Figure 2–14. From Figure 2–14, it is apparent that the largest annual expenditure is for tuition, followed by the expenditure for room and board. The smallest amount spent is for books.

Earlier in the chapter we used MINITAB to develop a stem-and-leaf chart. There are currently many spreadsheet packages available that will also do the computations and output the statistical information. Shown above is a pie chart for the college expense information in Table 2–13. We used Windows95 and Excel, a spreadsheet package available on many home computers, to perform the calculations and generate the chart. The specific Excel steps are as follows:

1. Enter the expenditure groups in a column and the data in the next column.
2. Select

Tools ▶ Data Analysis ▶ Descriptive Statistics.

3. Position the mouse on the upper left cell, hold the left mouse button down, and drag the mouse so as to highlight all the filled cells.

4. From the Tool Bar, click on **Chart Wizard** and point to an empty cell.

5. Click on **Pie** and then click on **Next**.

6. The last step is to select the columns in the Data Series. Use First 1 Column(s) for Pie Slice Labels and First 0 Row(s) for the Chart Title. Click on **Next** and the desired chart will appear.

Self-Review 2–8

The annual expenditures of Cartersville Township during last year were

Expenditure	Amount
Roads	$1,000,000
Schools	600,000
Administration	300,000
Graft	100,000

Portray the expenditures in the form of a pie chart.

Exercises

19. A recent report showed the average value of Iowa farmland from 1986 to 1995. Plot these data in the form of a line chart.

Year	Average Value (per acre)
1986	$1840
1987	1999
1988	1889
1989	1684
1990	1499
1991	1064
1992	841
1993	850
1994	840
1995	795

20. According to the Bureau of the Census, the population of the state of Arizona for the census years 1910–1990 was

Year	Number
1910	204,000
1930	436,000
1950	750,000
1970	1,792,000
1990	2,724,000

Plot the population data in the form of a bar chart.

21. In their *Motor Vehicle Facts and Figures*, the Motor Vehicle Manufacturers Association gives the following data regarding the cost of owning and operating an automobile based on 10,000 miles of travel per year:

Item	Cost per Year
Depreciation	$2094
Insurance	663
Finance charge	626
Gas and oil	520
Maintenance	190
All other	231

Plot the data in the form of a pie chart.

22. The Council of Environmental Quality estimated that the amounts to the right were spent on water pollution by various groups during the most recent 9-year period:

Group	Amount (in billions of dollars)
Industry	$57.0
State and local governments	39.8
Utilities	11.5
Federal government	2.5

Portray the data by drawing a pie chart.

MISUSES OF CHARTS AND GRAPHS

When you purchase a computer for your home or office, it usually includes graphics packages. These packages will produce some very effective charts and graphs. However, you must be careful to not mislead or misrepresent. In this section, we present several examples of charts or graphs that are misleading. Whenever you see a chart or graph, study it carefully, and ask yourself these questions: What is the writer trying to show me? Could the writer have any bias?

One of the easiest ways to mislead the reader is to make the range of the Y axis very small in terms of the units used for that axis. In the chart below,

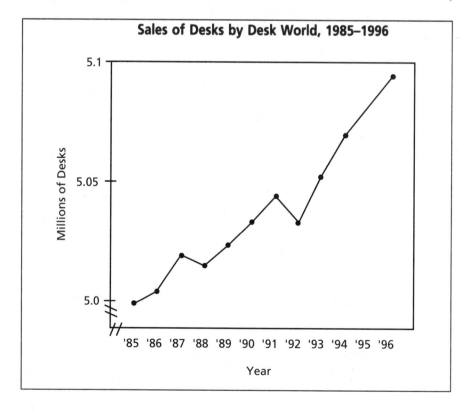

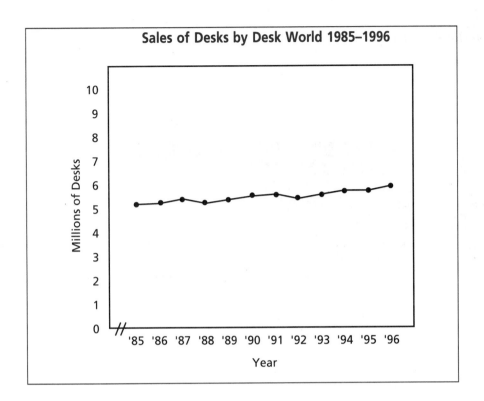

Sales of Desks by Desk World 1985–1996

it looks like there has been a dramatic increase in the number of desks sold by Desk World during the period from 1985 to 1996. However, during the period, sales increased only from 5.0 million to 5.1 million units. This is an increase of 2%, but the chart makes it look like a dramatic increase.

The chart above gives the correct impression of the trend in the sales of desks. That is, sales of desks are almost "flat" from 1985 to 1996. There is almost no change in sales during the 11-year period.

The charts on the next page are examples where the scale is omitted. The first chart was adapted from an advertisement for a new Wilson ULTRA® DISTANCE golf ball. The chart shows that the new ball gets longer distance, but what is the scale on the horizontal axis? Undoubtedly, the horizontal scale is yards, but such a large part of the scale is omitted that it overemphasizes the amount of the difference between golf balls. The ULTRA DISTANCE ball traveled only 6.6 yards farther than the nearest competitor. Is it possible that this difference is due to chance? Also, how were the tests conducted? We will look into these questions in Chapters 9 and 10.

Fibre Tech in Largo, Florida, manufactures and installs fiberglass coatings for swimming pools. The second chart was included in a Fibre Tech brochure. Is the comparison fair? What is the scale for the vertical axis? Is the scale in dollars, percentages, or what?

Again we caution you. When you see a graph or chart, particularly as part of an advertisement, be careful. Look closely at the scales used on the X axis and the Y axis.

Wilson Ultra® Distance Ad

Maybe everybody can't hit a ball like John Daly. But everybody wants to. That's why Wilson® is introducing the new ULTRA® DISTANCE ball. ULTRA DISTANCE is the longest, most accurate ball you'll ever hit.

Wilson has totally redesigned this ball from the inside out, making ULTRA DISTANCE a major advancement in golf technology.

ULTRA® DISTANCE	591.2 Yds.
DUNLOP® DDH IV	584.6 Yds.
MAXFLI MD®	571.2 Yds.
TITLEIST® HVC	569.3 Yds.
TOP-FLITE® Tour 90	565.9 Yds.
TOP-FLITE® MAGNA	564.3 Yds.

Combined yardage with a driver, #5 iron and #9 iron, ULTRA DISTANCE is clearly Measurably Longer.

Fibre Tech Ad

Fibre Tech Reduces Chemical Use, Saving You Time And Money.

- Saves up to 60% on chemical costs alone.
- Reduces water loss, which means less need to replace chemicals and up to 10% warmer water (reducing heating costs, too)
- Fibre Tech pays for itself in reduced maintenance and chemical costs.

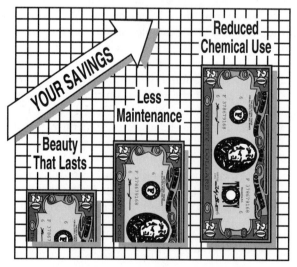

CHAPTER OUTLINE

I. A frequency distribution is a grouping of data into categories showing the number of observations in each of the nonoverlapping classes.
 A. The steps to develop a frequency distribution are as follows:
  1. Decide on the size of the class interval.
   a. If the number of classes has been established, a suggested class interval can be determined by

$$\frac{\text{Highest value} - \text{Lowest value}}{\text{Number of classes}} \qquad \boxed{2\text{--}1}$$

   b. The number of classes can also be estimated from $2^k \geq n$, where n is the total number of frequencies and k is the number of classes.
  2. Tally the raw data into the classes to arrive at the frequency distribution.
  3. Count the number of tallies in each class.
 B. The class frequency is the number of observations in each class.
 C. The class interval or width is the difference between the upper and the lower class limits.
 D. The class midpoint is the point halfway between the lower and the upper limits of a class.
 E. The criteria for constructing a frequency distribution are as follows:
  1. Avoid having fewer than 5 or more than 20 classes.
  2. Avoid open-ended classes.
  3. Keep the class intervals the same size.
  4. Do not have overlapping classes.
II. A percentage frequency distribution shows the percentage of the total observations in each class.

III. A stem-and-leaf chart is an alternative to a frequency distribution.
 A. The leading digit or digits is called the stem and the trailing digit the leaf.
 B. The advantages to a stem-and-leaf chart are as follows:
  1. The identity of the actual observations is not lost.
  2. The digits themselves give a picture of the distribution.
  3. The cumulative frequencies and the center value are reported.
IV. There are two methods to graphically portray a frequency distribution.
 A. A histogram portrays the class frequencies in the form of bars.
 B. A frequency polygon consists of line segments connecting the points formed by the intersection of the class midpoint and the class frequency.
V. A less-than cumulative frequency polygon shows the number of observations below a certain value.
 A. It is also called an ogive.
 B. It is also used to estimate the percentage of observations below a certain value.
VI. Many other types of charts and graphs are used in newspapers and magazines.
 A. Line charts are ideal for showing sales and income trends over a period of time.
 B. Bar charts are similar to line charts and are useful for showing changes in business and economic data over time.
 C. Pie charts are useful for showing what percentages the various components are of a total.

Exercises

23. Citizens' Trust lists the following end-of-month checking account balances for 40 of its customers:

$203	$ 37	$141	$ 43	$ 55
303	252	758	321	123
425	27	72	87	215
358	521	863	284	279
608	302	703	68	149
327	127	125	489	234
498	968	350	57	75
503	712	440	185	404

a. Tally the data into a frequency distribution using $100 class intervals and $0 as the starting point. What is a typical account balance?
b. Draw a less-than and a more-than cumulative frequency polygon.
c. The bank considers a "preferred" customer to be one with an account balance of $500 or more. Estimate the percentage of all customers who are preferred customers.
d. The bank is planning to offer free checking accounts to the 30% of its customers with the largest account balances. Based on the cumulative frequency polygon, what is the minimum balance for the free checking account?

24. The Quickie Change Oil Company has a number of outlets in the metropolitan area. No appointment is required. The numbers of oil changes at the Oak Street outlet in the past 20 days are:

65	98	55	62	79
59	51	90	72	56
70	62	66	80	94
79	63	73	71	85

The data are to be organized into a frequency distribution.
a. How many classes would you recommend?
b. What class interval would you suggest?
c. What lower limit would you recommend for the first class?
d. Organize the numbers of oil changes into a frequency distribution.

e. Comment on the shape of the frequency distribution. What is a typical number of oil changes per day?

25. The annual payments to the 20 employees of a small firm (in thousands of dollars) are given below. Develop a stem-and-leaf display.

$48	$23	$27	$46	$24
28	17	39	43	35
53	45	14	24	21
41	31	23	18	19

26. The numbers of parking tickets issued for the last 25 days in downtown Oxford are shown below. Develop a stem-and-leaf display.

53	42	34	43	35
39	27	17	40	41
32	21	21	37	12
36	19	29	26	14
38	57	45	22	33

27. Following are the percentages of births with low weight for selected states. (A low birth weight means that the child weighs 2800 grams or less at birth.)

	Percentage		Percentage
Maine	5.1	Ohio	6.7
New Hampshire	5.2	Indiana	6.4
Vermont	5.2	Illinois	7.4
Massachusetts	5.8	Michigan	6.9
Rhode Island	6.4	Wisconsin	5.4
Connecticut	6.6	West Virginia	7.0
New York	7.3	North Carolina	7.9
New Jersey	6.5	South Carolina	8.6
Pennsylvania	6.9	Georgia	8.1
Arkansas	7.6	Florida	7.6
Louisiana	8.6	Kentucky	7.1
Oklahoma	6.5	Tennessee	7.9
Texas	6.8	Alabama	8.0
Washington	5.2	Mississippi	8.7
Oregon	5.1	Montana	5.9
California	6.0	Idaho	5.2

Alaska	4.6	Wyoming	6.8
Hawaii	6.9	Colorado	7.7
Delaware	7.4	New Mexico	7.1
Maryland	7.7	Arizona	6.2
Virginia	7.0	Utah	5.4
		Nevada	7.4

a. Organize the percentages in a frequency distribution using 4.5% as the lower limit of the first class and an increment of 1%. What is a typical percentage of low birth weights?
b. Portray the frequency distribution as a histogram.
c. Draw a more-than cumulative frequency polygon.
d. Based on the cumulative frequency polygon, approximately how many states have low birth-weight rates greater than 7.0%?

28. The percentages of the total number of births attributed to unmarried women for each of the 50 states are

19.0	13.9	16.7	19.3	19.8
19.0	29.4	22.9	24.4	23.4
21.0	27.1	19.3	19.6	16.3
15.0	22.5	12.9	17.5	15.5
16.7	27.0	30.5	22.4	20.3
19.5	23.6	27.6	27.2	26.7
20.0	25.3	25.9	34.0	24.0
30.2	18.6	17.7	17.8	11.9
13.9	18.0	27.9	25.6	9.8
16.6	19.8	20.7	26.5	20.8

a. Organize the percentages in a frequency distribution using 9.0% as the lower limit of the first class. You decide on the class interval. What is a typical value?
b. Portray the frequency distribution as a histogram.
c. Draw a less-than cumulative frequency polygon.
d. Based on the cumulative frequency polygon, approximately how many states have less than 20% of their births attributed to unmarried women?

29. The average annual percentage changes in consumer prices for the past five years for selected countries are

United States	5.5	Japan	2.7
Australia	8.3	Luxembourg	6.9
Austria	4.9	New Zealand	12.0
Belgium	7.0	Netherlands	4.2
Canada	7.4	Norway	9.0
Denmark	7.9	Portugal	23.2
Poland	8.6	Spain	12.2
France	9.8	Sweden	9.0
Germany	3.9	Switzerland	4.3
Greece	20.7	Turkey	37.8
Ireland	12.3	United Kingdom	7.2
Italy	13.7		

a. Using 0% as the lower limit of the first class and a class interval of 5%, organize the changes in consumer prices into a frequency distribution.
b. Describe the frequency distribution.
c. Draw a histogram of these data.

30. Kolvey Travel Agency, a nationwide travel agency, offers special rates on certain Caribbean cruises to senior citizens. The president wants additional information on the ages of those people taking cruises. A random sample of 40 customers taking a cruise last year revealed these ages:

77	18	63	84	38	54	50
59	54	56	36	26	50	34
44	41	58	58	53	51	62
43	52	53	63	62	62	65
61	52	60	60	45	66	83
71	63	58	61	71		

a. Organize the data into a frequency distribution using 7 classes and 15 as the lower limit of the first class. What is a typical age?
b. Would you suggest a different interval than indicated in part a? If so, what interval?
c. Describe the distribution. Where do the data tend to cluster?
d. Convert the distribution to a percentage frequency distribution.

31. The following data show the amounts spent on social programs in the United States from 1965 to 1995. Portray the data in a line chart.

Year	Spending (in millions of dollars)
1965	$ 25.0
1970	36.0
1975	73.0
1980	166.0
1985	307.0
1990	459.0
1995	562.0

32. The U.S. Department of Transportation keeps track of the percentages of flights that arrive within 15 minutes of the scheduled time, by airline. Below is the information for May 1996. Construct a stem-and-leaf chart from these data. Summarize your conclusion.

Airline	Percentage on Time
Pan Am	82.7
American West	82.7
Northwest	81.0
USAir	80.1
Southwest	79.7
Alaska	79.7
American	78.1
United	76.4
Delta	76.1
Continental	76.9
British Airways	80.4
Japan Airlines	81.4

33. A breakfast cereal is supposed to include 200 raisins in each box. A sample of 60 boxes produced yesterday showed the following numbers of raisins in each box. Develop a frequency distribution.

200	200	202	204	206
197	199	200	204	195
206	193	196	200	195
202	199	202	200	206
197	202	198	203	201
198	198	200	205	205
206	200	197	203	201
198	202	206	205	207
196	199	199	200	196
205	203	201	200	191
199	200	193	200	198
202	201	193	204	204

What class interval do you suggest? Summarize your findings.

34. Listed below are the numbers of subscribers (in thousands) for the GTE Corporation, by year, for the period 1990 to 1995. Develop an appropriate chart to depict these data. Comment on any trends you note.

Year	Subscribers
1990	594
1991	811
1992	1090
1993	1585
1994	2339
1995	3011

35. Farming has changed from the early 1900s to the 1990s. In the early 20th century, machinery gradually replaced animal power. For example, in 1910, U.S. farms used 24.2 million horses and mules and only about 1000 tractors; by 1960, 4.7 million tractors were in use and only 3.2 million horses and mules. In 1920, there were over 6 million farms in the United States; today there are less than 2 million. Listed below are the numbers of farms (in thousands) for each of the 50 states. Show these data in an appropriate chart or graph, and write a paragraph summarizing your findings.

47	1	8	46	76	26
4	3	39	45	4	21
80	63	100	65	91	29
7	15	7	52	87	39
106	25	55	2	3	8
14	38	59	33	76	71
37	51	1	24	35	86
185	13	7	43	36	20
79	9				

36. Since the early 1980s, the cost of a college education has increased dramatically at both public and private schools. Listed below are the average costs per year for tuition and fees for each academic year from 1980–1981 to 1995–1996. Show these data in an appropriate chart or graph, and write a paragraph summarizing your findings.

Academic year	Public	Private
1980–1981	$ 804	$ 3617
1981–1982	909	4113
1982–1983	1013	4639
1983–1984	1148	5093
1984–1985	1228	5556
1985–1986	1318	6121
1986–1987	1414	6658
1987–1988	1537	7116
1988–1989	1646	7722
1989–1990	1781	8446
1990–1991	1908	9340
1991–1992	2137	10,017
1992–1993	2334	10,449
1993–1994	2527	11,025
1994–1995	2754	11,521
1995–1996	3029	11,982

37. One of the most popular candies in the United States is M&Ms, which are produced by the Mars Company. For many years, the M&M plain candies were produced in six colors: red, green, orange, tan, brown, and yellow. Recently, the tan was replaced with blue, but did you ever wonder how many candies are in a bag and exactly how many of each color? Are there about equal numbers of each color, or are there more of some colors than others? Here are some data from a 1-pound bag of M&Ms plain. It contained a total of 544 candies. There were 135 brown, 156 yellow, 128 red, 22 green, 50 blue, and 53 orange. Develop a chart that summarizes this information, and write a brief report summarizing the results.

38. The following graph is from an article in the *Wall Street Journal* comparing the average selling prices for the Ford Taurus and the Toyota Camry from 1987 to 1995. Write a brief report summarizing the information in the chart. Be sure to include the selling prices of the two cars, the changes in the selling prices, and the direction of the changes over the nine-year period.

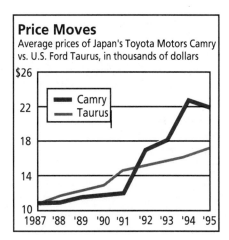

Price Moves
Average prices of Japan's Toyota Motors Camry vs. U.S. Ford Taurus, in thousands of dollars

- Camry
- Taurus

DATA EXERCISES

39. Refer to the real estate data set at the end of the book, which reports the information on homes sold in Alabama in 1996.
  a. Select an appropriate class interval, and organize the selling prices into a frequency distribution.
  b. What is the selling price of a typical home?
  c. Estimate the percentage of homes that sell for less than $200,000.
  d. About 40% of the homes sell for less than what amount?
  e. Organize the selling prices of the homes into a stem-and-leaf chart.
  f. How many homes sell for less than $150,000?

How many homes sell for $190,000 or more?
40. Refer to the school data set which refers to the 94 school districts in northwest Ohio.
   a. Organize the teachers' salaries into a stem-and-leaf chart. Write a brief report indicating the typical salary and the shape of the salary distribution. How many of the school districts have average salaries of more than $30,000?

   b. Organize the information on the percentages of students passing the state proficiency exam into a stem-and-leaf chart. How many of the school districts report that less than 30% of the students pass the exam? How many districts have a passing rate of more than 60%? What is a typical passing rate?

CHAPTER ACHIEVEMENT TEST

Answer all the questions. The answers are at the back of the book.

MULTIPLE-CHOICE QUESTIONS

Select the response that best answers each of the questions.

1. When arranging data into classes, it is suggested that you have
   a. fewer than 5 classes.
   b. between 5 and 20 classes.
   c. more than 20 classes.
   d. between 10 and 40 classes.
2. When constructing a line chart, you plot "time" along
   a. the vertical axis.
   b. the horizontal axis.
   c. either axis.
   d. neither axis.
3. A frequency distribution requires that the data be of at least what scale?
   a. Nominal
   b. Interval
   c. Ordinal
   d. None of these is correct.
4. The class midpoint is the
   a. number of observations in a class.
   b. center of the class.
   c. upper limit of the class.
   d. width of the class.
   The following frequency distribution records the numbers of empty seats on flights from Cleveland to Tampa:

Number of Empty Seats	Frequency
0 up to 5	3
5 up to 10	8
10 up to 15	15
15 up to 20	18
20 up to 25	12
25 up to 30	6

5. The midpoint of the 0 up to 5 class is
   a. 2.
   b. 4.
   c. 2.5.
   d. 0.
6. The lower limit of the 0 up to 5 class is
   a. 0.
   b. −0.5.
   c. 2.0.
   d. 2.5.
7. The size of the class interval is
   a. 5.
   b. 4.
   c. 4.5.
   d. 3.
8. About what percentage of the flights had 19 or fewer empty seats?
   a. 15%
   b. About 29%
   c. About 71%
   d. Cannot be determined from grouped data

COMPUTATION PROBLEMS

9. A hospital administrator has commissioned a study of the length of time (in minutes) a patient spends waiting in the emergency room before receiving treatment. For a typical day, the information is as follows:

12	6	5	23	10
25	19	11	25	18
10	14	12	10	16
5	19	17	17	14
2	21	9	6	21

Construct a frequency distribution using an interval of 5 and a lower limit of 1 minute for the first class.

10. The Bureau of Motor Vehicles of the state of Nevada made a study of the tread depth of tires. Fifty-two passenger cars that had stopped at a highway rest station were tested. The tread depths, measured to the nearest 1/32″, are reported in the table at right.

a. Draw a less-than cumulative frequency polygon.

b. What percentage of the tires have less than 7/32″ tread?

c. Forty percent of the tires have less than how much tread?

Tread Depth (1/32″)	Frequency
0 up to 4	4
4 up to 8	15
8 up to 12	25
12 up to 16	5
16 up to 20	3

11. A family's monthly expenses are as follows: housing, $400; utilities, $140; medical, $25; food, $190; transportation, $150; clothing, $50; savings, $50; miscellaneous, $50. Draw a pie chart showing this information in terms of percentages.

ANSWERS TO SELF-REVIEW PROBLEMS

2-1 a.

Age	Number
15 up to 25	3
25 up to 35	4
35 up to 45	2
45 up to 55	1
	10

b. Raw data

c. Based on the frequency distribution, the ages range from 15 to 55 years. The largest concentration of the data is in the 25 up to 35 class.

2-2

Leading Digit	Trailing Digit
1	67
2	159
3	009
4	1
5	0

2-3

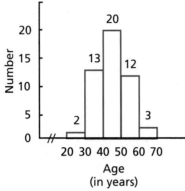

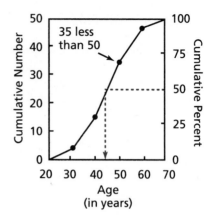

c. About 47 years

2-4

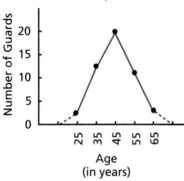

2-6 a.

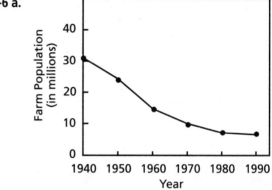

b. The farm population has declined since 1940. In 1990, the farm population was about 25 percent of what it was in 1940. Since 1980, the farm population has stayed about the same.

2-5 a.

Age	Cumulative Frequency
20 up to 30	2
30 up to 40	15
40 up to 50	35
50 up to 60	47
60 up to 70	50

b.

Upper Class Limits	Number Less-Than
20	0
30	2
40	15
50	35
60	47
70	50

2-7 a.

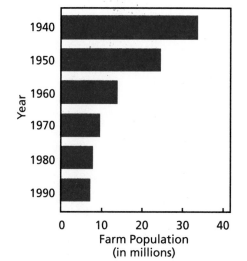

Year

Farm Population
(in millions)

b.

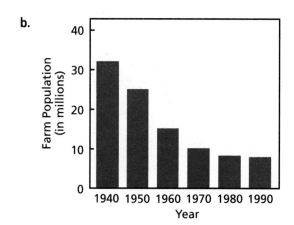

Farm Population
(in millions)

Year

2-8

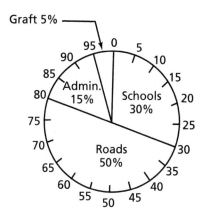

Graft 5%

Descriptive Statistics: Measures of Central Tendency

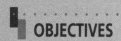

OBJECTIVES

When you have completed this chapter, you will be able to

- compute the mean, median, and mode for both grouped and raw data;
- describe the characteristics of the mean, median, and mode;
- describe the positions of the mean, median, and mode for symmetric and skewed distributions.

CHAPTER PROBLEM   Pricing the Dream

Recall from the Chapter Problem in Chapter 2 that Charles Reynolds, owner of the Reynolds Company, is putting together a prospectus for new homes to be built in the Rafter J Ranch subdivision in Jackson Hole, Wyoming. In Chapter 2, he found the selling price of each of the 80 existing homes and organized them into a frequency distribution. Now Mr. Reynolds wants to determine the selling price of a typical home.

INTRODUCTION

In Chapter 2, the selling prices of homes in the Rafter J Ranch subdivision were organized into a frequency distribution. Next, a histogram or a frequency polygon was drawn to help describe the data. In this chapter, we will describe the selling prices further by computing several **measures of central tendency.** As the name implies, these measures locate the center of a set of observations. The value that either is typical of the set of data or best describes it is often referred to as an **average.**

Here are some examples of averages:

- The average life expectancy in the United States is 74.2 years.
- In 1970, the median age of the population was 27.9 years, and in 1985, it was 31.5 years. By the year 2000, the median age is projected to be 37.2 years.
- The average cost per mile to operate a motor vehicle is 38.2¢, according to *Automobile Facts and Figures.*
- The Bureau of the Census reports that the median household income is $29,100.
- The Institute of International Education reports that there was an average of 6,200,000 students enrolled per year during the last five years in institutions of higher education.
- The average occupancy rate of nursing care facilities is 90.5%.
- A survey by the Hertz Corporation reports that the average annual maintenance and repair cost is $269 for a new car and $565 for a car older than one year.
- In 1995, the team batting average for the Chicago White Sox was .280, and the average player salary was $1,110,770.

As you will soon see, there are several kinds of averages, so you should develop the habit of speaking precisely regarding the various types. What we usually refer to as "averages" should be called "measures of central tendency." Each measure of central tendency has distinct characteristics, advantages, and disadvantages. For the same set of data, all measures of central tendency might have different values. In this chapter, we will examine three commonly used measures of central tendency—the **arithmetic mean,** the **median,** and the **mode.**

THE SAMPLE MEAN ($\overline{X}$)

arithmetic mean The sum of all the values in the sample divided by the number of values.

The most frequently used measure of central tendency is the **arithmetic mean,** or simply the **mean.** In everyday language, the mean is often called the "average"; however, to avoid confusion you should call it the "arithmetic mean."

Because the arithmetic means of many different sets of data will be calculated, it is convenient to use a simple formula in which the components of the arithmetic mean are represented by a few general symbols. First, let X_1,

$X_2, \ldots, X_n$ represent the values of n items or observations in a sample. The arithmetic mean of these sample items, written $\overline{X}$, is then defined as

$$\overline{X} = \frac{X_1 + X_2 + \cdots + X_n}{n}$$ [a shorthand way of saying the sample mean of all values $(\overline{X})$ is equal to the sum of all the values $(X_1, X_2,$ and so on) divided by the number of values]

For simplicity, instead of writing "the sum of" $X_1, X_2, \ldots, X_n$, we use the symbol Σ (which is the Greek capital letter *sigma*, equivalent to our S) to represent a summation, or addition. Hence, the formula for the sample mean of n items is

$$\text{Mean} = \frac{\text{Sum of the values}}{\text{Number of values}}$$

or, in symbolic notation,

$$\overline{X} = \frac{\Sigma X}{n}$$ 3–1

where

$\overline{X}$ (read *X*-bar, or bar *X*) is the designation for the arithmetic mean of a sample.

Σ (the Greek capital letter *sigma*) is the symbol for addition. In this case, it directs us to sum all the X values.

X refers to the individual values in the sample.

n is the total number of items in the sample.

Problem   The estimated ages of 5 rocks uncovered at an excavation site were 5, 4, 9, 2, and 10 million years. What is the arithmetic mean age of this sample of 5 rocks?

Solution   In the above notation, n refers to the number of observations in the sample, which is 5. The values of X are $X_1 = 5$, $X_2 = 4$, $X_3 = 9$, $X_4 = 2$, and $X_5 = 10$. Hence, using formula 3–1, the arithmetic mean of the sample of 5 rocks is 6 million years.

$$\overline{X} = \frac{\Sigma X}{n} = \frac{X_1 + X_2 + X_3 + X_4 + X_5}{n}$$

$$= \frac{5 + 4 + 9 + 2 + 10}{5} = 6 \text{ (million years)}$$

The number 6, which was computed from a sample, is called a **statistic**.

Problems   This is the first in a series of drill problems designed to provide you immediate feedback on your understanding of the topic just discussed. They appear mainly in the first half of the text. We suggest you work them before moving on to the next section. The solutions follow.

REAL STAT

Did you ever meet the "average" American man? Well, his name is Robert (that's a nominal scale of measurement), he is 31 years old (ratio scale), he is 69.5 inches tall (ratio scale), weighs 172 pounds, wears a 9½ shoe, has a 34-inch waist, and wears a size 40 suit. In addition, the average man each year eats 4 pounds of potato chips, watches 2567 hours of TV, eats 26 pounds of bananas, and receives 585 pieces of mail. Also, he sleeps 7.7 hours per night. Is this really an "average" man, or would it be better to call this the "typical" man? Would you expect to find a man with all these characteristics? Oh, one more thing—he is a Dallas Cowboys fan.

statistic A measurable characteristic of a sample.

Determine the arithmetic mean for each of the following sets of data:

a.   3, 5, 7
b.   6, 5, 8, 4
c.   3, 21, 16, 9, 13

Solutions

a.   5
b.   5.75
c.   12.4

| Self-Review 3–1 | Answers to all Self-Review problems are at the end of the chapter.
The numbers of abortions (in thousands) performed last year in six mountain states are listed below. |

State	Number (in thousands)
Montana	4.0
Idaho	3.0
Colorado	23.0
Arizona	16.0
Utah	4.0
Nevada	9.0

a. Give the formula for the arithmetic-mean number of abortions for this sample of six mountain states.
b. Insert the appropriate values in the formula, and compute the arithmetic mean number of abortions per state.

Exercises

Answers to the even-numbered Exercises are at the end of the book.

1. List the three measures of central tendency.
2. Determine the arithmetic mean of the following set of data:

7   3   6   3   2

3. Determine the arithmetic mean of the following set of data:

2   2   2   7   6

4. Determine the arithmetic mean of the following set of data:

8   7   5   7   12
8   13   11   14   11

5. The agility test scores for a sample of 8 children with disabilities are 95, 62, 42, 96, 90, 70, 99, and 70. What is the arithmetic mean score for this sample of children?
6. The following are the high temperatures for a sample of 10 January days in Barrow, Alaska. Determine the arithmetic mean high temperature.

−14   −7   6   24   −12
−1   −11   3   −7   −4

7. The numbers of gallons of gasoline sold to a sample of 8 customers yesterday at Frank's I-75 Marathon were 12.1, 9.7, 5.3, 6.7, 8.4, 9.2, 10.1, and 7.7 gallons. Compute the arithmetic mean number of gallons sold.

8. According to *Motor Vehicle Facts and Figures,* the numbers of motorcycles registered in the past 5 years (in thousands) are 5444, 5262, 4886, 4584, and 4376. What is the arithmetic mean number of motorcycles registered?

THE POPULATION MEAN (μ)

Usually, a statistic is computed for the purpose of estimating an unknown population **parameter**. Hence, the sample mean, $\overline{X}$, is a statistic that may be thought of as an estimate of the population mean from which the sample was drawn. The population mean is a parameter.

> **parameter** A measurable characteristic of a population.

The computation of the population mean is the same as that of a sample, except that the symbols are slightly different. By convention, the population mean is denoted by the Greek letter μ (*mu*). The number of observations in a population is denoted by the capital letter N. Hence, the mean of a population is computed from the following formula:

$$\mu = \frac{X_1 + X_2 + \cdots + X_N}{N} = \frac{\Sigma X}{N} \qquad \boxed{3\text{-}2}$$

Problem For the 4 years of Governor Lamb's administration, the numbers of pieces of legislation enacted by the state legislature were 310, 780, 960, and 842. What is the arithmetic mean number of pieces of legislation per year?

Solution Note that this is a population because it covers all 4 years of the Lamb administration. The population consists of 4 observations, and the population mean is 723 pieces.

$$\mu = \frac{\Sigma X}{N} = \frac{310 + 780 + 960 + 842}{4}$$

$$= \frac{2892}{4} = 723$$

PROPERTIES OF THE ARITHMETIC MEAN

The popularity of the arithmetic mean as a measure of central tendency is not accidental. Not only is it simple, familiar, and easy to calculate, but also it has the following desirable properties:

1. It can be calculated for any set of interval-level or ratio-level data. Therefore, it always exists.
2. A set of data has only one arithmetic mean. Therefore, it is referred to as a "unique" value.

3. It is quite reliable. This property will be discussed in connection with sampling in Chapter 8.

4. All the data items are used in its calculation.

The arithmetic mean has one additional important characteristic—namely, *the sum of the signed deviations from the mean will always be zero.* This concept can be illustrated by a seesaw like those found on most playgrounds. Suppose 3 children weighing 30, 40, and 90 pounds approach the seesaw with the balance point set at 60 pounds. One child stands on the seesaw at 30 (representing 30 pounds), another at 40, and the third at 90. They expect the board to balance so they can go up and down on the seesaw, but it does not. Why? Because 60 pounds is not the arithmetic mean of 30, 40, and 90 pounds (see Figure 3–1).

Suppose the child on the right side of the seesaw keeps moving toward the end of the board to try to make the seesaw balance. Finally, when he reaches 110, it balances, and the children are able to seesaw up and down. Why does it balance? Because the arithmetic mean of 110, 40, and 30 is now at the balance point of 60. Thus, the mean of 60 is the center of gravity of the 3 weights. The sum of the deviations to the left of 60 is equal to the sum of the deviations to the right of 60 (see Figure 3–2).

Note that in Figure 3–2, the distance (deviation) from the balance point to the child standing on the right end of the board ($+50$) is equal in magnitude to the sum of the deviations to the left of the balance point (-20 and -30). The balance point of 60 is the arithmetic mean of the distances $(30 + 40 + 110)/3$. We show this algebraically by letting each observation be represented by X and the mean by $\overline{X}$. In the seesaw problem, $\overline{X} = 60$. The sum of the deviations from the mean is zero. That is, $\Sigma(X - \overline{X}) = -20 + (-30) + 50 = 0$. The mean is the only average for which this is always true.

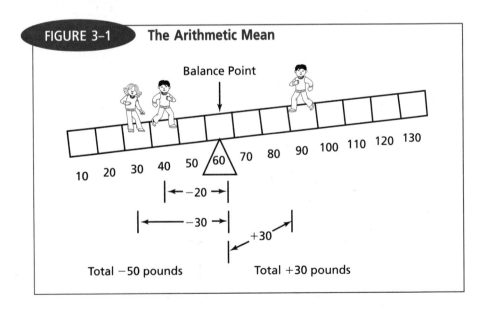

FIGURE 3–1 **The Arithmetic Mean**

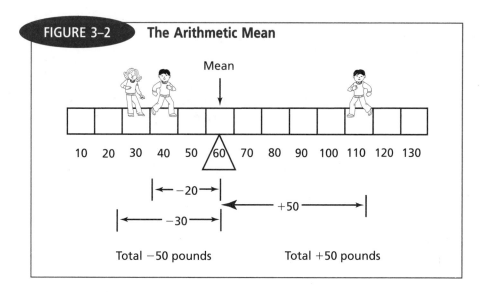

FIGURE 3-2   **The Arithmetic Mean**

- -

A group of football players enrolled in Statistics 166 wants to check whether the sum of the deviations from the arithmetic mean is actually zero. Their weights are 200, 210, 230, and 320 pounds. They stand on the seesaw as shown below.

Self-Review 3-2

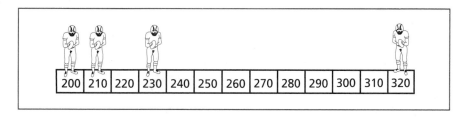

a. Where must the balance point be placed to make the seesaw balance?
b. What is this balance point called?
c. Show that the sum of the deviations from the arithmetic mean equals zero.

- -

Exercises

9. List the characteristics of the arithmetic mean.
10. What is the difference between a sample mean and a population mean?
11. Listed below are the ages of the 6 employees currently on duty at the I-75 Wendy's Restaurant.

| 27 | 23 | 17 | 20 | 17 | 22 |

    a. Determine the arithmetic mean age of the employees.

    b. Show that the sum of the deviations from the arithmetic mean is zero.

12. Listed below are the changes in the Dow Jones Industrial Average for the last 5 days.

| −13 | 66 | 37 | 28 | −28 |

    a. Determine the arithmetic mean change.

    b. Show that the sum of the deviations from the arithmetic mean is zero.

13. A psychologist plans a study of her patients to determine how many sessions were needed before releasing the patient. She initially selected 6 of her former patients and found that 1 needed just 1 session; the others needed 4, 9, 8, 5, and 6 sessions.

   a. Determine the arithmetic mean number of sessions needed.

   b. Show that the sum of the deviations from the arithmetic mean is zero.

14. The following percentages of flights left within 15 minutes of their scheduled departure times from selected cities:

City	Percentage of Departures
Philadelphia	76.0
New York (Kennedy)	78.4
Seattle	80.2
Tampa	83.2
Miami	83.9
Los Angeles	85.9
Memphis	87.4

   a. Find the arithmetic mean percentage of flights departing within 15 minutes of their scheduled departure times.

   b. Show that the sum of the deviations from the mean is equal to zero.

15. An accountant employed by St. Paul's Hospital sampled the unpaid balances of 8 former patients and found these amounts due:

$46	$42	$83	$90
76	84	56	46

   a. Find the arithmetic mean unpaid balance.

   b. Show that the sum of the deviations from the arithmetic mean is equal to zero.

16. The monthly commissions (in thousands of dollars) for a sample of 8 minority executives are $2, $2, $2, $2, $10, $10, $25, and $100.

   a. Find the arithmetic-mean income.

   b. Show that the sum of the deviations from the arithmetic mean is equal to zero.

THE WEIGHTED ARITHMETIC MEAN

Suppose it had been reported that the ages of the senior citizens on the Happy Boys slow-pitch softball team were 60, 70, 80, and 90. It might be concluded that the arithmetic mean age is 75, found by (60 + 70 + 80 + 90)/4. This is true only if the *same* number of senior citizens are aged 60 as are aged 70 and 80 and 90. However, suppose 1 is 60, 1 is 70, 1 is 80, and 9 are 90 years. To find the mean, we **weight** 60 by 1, 70 by 1, 80 by 1, and 90 by 9. The resulting average is aptly called the **weighted arithmetic mean.**

In general, the weighted arithmetic mean of a group of numbers designated $X_1, X_2, X_3, \ldots, X_n$ with corresponding weights $w_1, w_2, w_3, \ldots, w_n$ is computed by

$$\overline{X}_w = \frac{w_1 X_1 + w_2 X_2 + w_3 X_3 + \cdots + w_n X_n}{w_1 + w_2 + w_3 + \cdots + w_n}$$

3–3

or shortened to

$$\overline{X}_w = \frac{\Sigma w \cdot X}{\Sigma w}$$

Problem An accountant wishes to find the typical amount spent per meal for business executives at a firm. He has 3 charge receipts showing costs of $12.50 per person, $10 per person, and $15 per person. These bills are for groups of 2, 4, and 3 people, respectively. What is the arithmetic mean cost per meal?

Solution He might use formula 3–2—namely, (12.50 + 10 + 15)/3—and report $12.50 per person as typical. However, that ignores the fact there were different numbers of people at each meal. In fact, the largest number of people had the smallest check! So $12.50 is probably too high.

A better way to find the arithmetic mean would be

(12.50 + 12.50 + 10 + 10 + 10 + 10 + 15 + 15 + 15)/(2 + 4 + 3)

Using the number of people in each group as "weights," this expression is written simply as

[2(12.50) + 4(10) + 3(15)]/9

This correctly "weighted" mean is $12.22 per person. This captures the fact that the largest group has the smallest cost per person.

He would have the same result if he computed the total bill for each group—that is, 2($12.50) or $25 for the first group, 4($10) or $40 for the second group, and 3($15) or $45 for the third group. Then the total amount spent is $110, found by $25 + 40 + 45. If he divides this total by the number of people (9), the result is also $12.22.

· ·

Calico Pizza sells cola-flavored drinks in 3 sizes: small, medium, and large. The small drink costs $0.50, the medium $0.75, and the large $1.00. Yesterday Calico sold 20 small, 50 medium, and 30 large cola drinks. What is the weighted arithmetic mean selling price of the drinks sold yesterday?

Self-Review 3–3

· ·

Exercises

17. An instructor gives 3 tests in an introductory psychology class. The first test has a weight of 25%, the second 35%, and the third 40%. Steve Sevits scored a 60 on the first test, an 80 on the second, and a 90 on the third. What is Steve's weighted score?

18. The following table shows the percentages of the labor force that are unemployed and the sizes of the labor force in 3 counties in northwest Ohio. Jon Elsas is regional director of economic development. He must make a presentation to several companies that are considering locating facilities in northwest Ohio. What would be an appropriate unemployment rate to show for the entire region?

County	Percentage Unemployed	Size of Work Force
Wood	4.5	15,300
Ottawa	3.0	10,400
Lucus	10.2	150,600

19. A car rental agency purchased 2 tires at $43 each, 5 at $76 each, and 3 at $59 each. What is the weighted arithmetic mean price of a tire? Why can't the 3 prices be summed and the sum divided by 3 to arrive at the mean price per tire?

20. We are studying 8 states with respect to various types of crime reported last year. From the FBI's annual *Uniform Crime Reports,* we discover that, per 100,000 population, 4 states had 28.3 rapes, 3 had 31.7 rapes, and 1 had 17.0 rapes. What is the weighted arithmetic mean number of rapes for the 8 states being studied?

THE MEDIAN

If the data contain an observation that is either very large or very small compared with the other values, that value may make the arithmetic mean unrepresentative. To avoid this possibility, we sometimes describe the "center" of the data with other measures of central value. One measure used to designate the central value of a set of data is the **median.**

median The midpoint of the values after they have been arranged from the smallest to the largest (or the largest to the smallest). There will be as many values above the median as below the median.

Like the mean, the median is unique for any set of data. It is easy to calculate once the data have been arranged according to size. Unlike the mean, however, the median is not affected by extremely large or extremely small observations. Also, while the arithmetic mean requires the interval scale, the median merely requires that the data be at least of ordinal scale. For an *odd* number of values, the median is the middle value after the data have been ordered from low to high.

Problem

a. The test scores of a sample of 5 students are 92, 86, 2, 96, and 90. What is the median test score?

b. The incomes of a sample of 7 local prison wardens are $41,500, $43,900, $47,100, $28,600, $32,500, $44,200, and $44,200. What is the median income?

Solution

a. Arrange the test scores from low to high. Then select the middle value.

```
 2
86
90  ←— the median
92
96
```

The median score is 90.

b. Arrange the incomes of the 7 prison wardens from low to high. Then select the middle value.

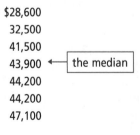

```
$28,600
 32,500
 41,500
 43,900  ←— the median
 44,200
 44,200
 47,100
```

The median income is $43,900. Note that there are the same number of test scores (or incomes) below the median as above it.

For an *even* number of values, as usual we first arrange the values from low to high. It is common practice to determine the median by finding the arithmetic mean of the two center values when there is no unique center value.

Problem

a. A park ranger recorded the lengths of a sample of several rainbow trout caught in the Yellowstone River. The lengths (in inches) were 24½, 15, 10½, 12½, 21, and 26. What was the median length?

b. The weights of 8 collegiate football players are 198, 240, 230, 240, 210, 250, 225, and 188 pounds. What is the median weight?

Solution

a. Arrange the lengths of the rainbow trout from low to high:

10½
12½
15 ← 18
21
24½
26

The median length of the trout is 18 inches, which is halfway between 15 and 21.

b. Arrange the weights of the football players from low to high:

188
198
210
225 ← 227.5
230
240
240
250

The median weight is 227.5 pounds, which is halfway between 225 and 230.

Problems   Determine the median for each of the following sets of data:

a. 2, 9, 4
b. 6, 5, 8, 2
c. 16, 8, 12, 3, 1
d. 1, 2, 2, 1

Solutions

a 4
b. 5.5
c. 8
d. 1.5

Self-Review 3–4

The ages of prisoners in cell block D are 22, 75, 18, 40, and 34.

a. Determine the median age.
b. How many prisoners are above the median? Below it?

In the last 8 games, the Rogers Rams football team scored the following numbers of points: 13, 34, 0, 0, 9, 42, 21, and 7.

c. What is the median number of points scored?
d. In how many games did the Rams score more than the median number of points?

The numbers of telephones in use in the United States for 4 recent years were 149,010,000; 138,290,000; 155,170,000; and 143,970,000.

e. What is the median number of telephones in use?
f. In how many years was the number in use below the median?

Exercises

21. The agility test scores for the children with disabilities in Exercise 5 on page 72 are 95, 62, 42, 96, 90, 70, 99, and 70. What is the median score?

22. The monthly commissions (in thousands of dollars) for several minority executives are $2, $2, $25, $2, $10, $100, $2, and $10. What is the median amount of commission earned?

23. The sales for the last 6 customers at Churchill's Supermarket (rounded to the nearest dollar) were $12, $35, $86, $21, $135, and $72. Compute the median sales per customer.

24. The statistics department at your college offers 7 sections of basic statistics, with individual section enrollments of 43, 41, 52, 13, 21, 39, and 46. What is the median section size?

25. *Modern Healthcare* reported these average patient revenues (in millions of dollars) for various types of hospitals:

Hospital Type	Patient Revenue (in millions of dollars)
Catholic	$46.6
Other church	59.1
Nonprofit	71.7
Public	93.1
For-profit	32.4

What is the median patient revenue? Explain what it indicates.

26. The *Bank Rate Monitor* reported the following savings rates:

Instrument	Savings Rate (in percent)
Money market mutual fund	3.01
Bank money market account	2.96
6-month CD	3.25
1-year CD	3.51
2½-year CD	4.25
5-year CD	5.46

What is the median savings rate? Explain what it indicates.

THE MODE

The **mode** is yet another measure that may be used to describe the central tendency of data.

The mode has two major advantages: It requires no calculation, and it can be determined for all four levels of data—nominal, ordinal, interval, and ratio.

Problem　The lengths of confinement (in days) for a sample of 9 patients in Ward C are 17, 19, 19, 4, 19, 26, 3, 21, and 19. What is the modal length of confinement?

Solution　The mode is 19 days because that value appears most frequently.

There is no mode for these hourly incomes: $4, $9, $7, $16, and $10. There are two modes (76 and 81) for these test scores: 81, 39, 100, 81, 69, 76, 42, and 76. In the last example, the data are considered to be *bimodal*.

mode The value of the observation that appears most often.

CHOOSING AN AVERAGE

The question often arises of what average to employ—the arithmetic mean, the median, or the mode. The mean is probably the most frequently used measure of central tendency. It is the measure most people think of when an "average" is mentioned. The mean is considered the most reliable, or precise, average because the means of several samples taken from a population will not fluctuate as widely as either the median or the mode. (We will return to this discussion later in the book.)

The selection of an average depends in part on the level of data. As noted in Chapter 1 and in the previous section, there are 4 such levels: nominal, ordinal, interval, and ratio. At the very least, an interval level of measurement is required to determine the arithmetic mean. We could, for example, compute the arithmetic mean age of persons between 5 and 59 years old who have died of AIDS since 1982, as given in Table 3–1. (It appears that the mean age is about 32 years. The computations are presented shortly.) Incidentally, the median age and the modal age can also be determined for the AIDS data.

TABLE 3–1	AIDS Deaths Since 1982	
	Age	**Number**
	5 up to 15	182
	15 up to 25	15,044
	25 up to 35	34,790
	35 up to 45	17,001
	45 up to 55	6393

As another example, the arithmetic mean, median, and mode can be computed for the temperatures in Table 3–2.

TABLE 3–2	Interval-Level Data
	Temperature (in °C)
	2
	1
	3
	12
	3
	3

$\overline{X}$ = 4 degrees
Median = 3 degrees
Mode = 3 degrees

The mean of ordinal-level data (see Table 3–3) or nominal-level data would be meaningless. We cannot possibly add warrant officer, lieutenant, captain, major, colonel, and general to determine the mean rank. Nor can we order the political parties in Table 3–4 from low to high to arrive at the mean or median.

TABLE 3–3	Ordinal-Level Data	
	Rank	**Number (in thousands)**
	Warrant Officer—W–1	3.2
	Chief Warrant—W–4	3.1
	2d Lt.—O–1	34.0
	1st Lt.—O–2	42.0
	Captain—O–3	106.2
	Major—O–4	53.6
	Lt. Colonel—O–5	32.4
	Colonel—O–6	14.0
	Brig. General—O–7	0.5
	Major General—O–8	0.4
	Lt. General—O–9	0.1
	General—O–10	(Z)

Z = fewer than 50
Median and modal categories = Captain

TABLE 3–4	Nominal-Level Data	
	Party	**Number**
	Democratic	2561
	Reform	732
	Republican	1602
	Independent	1814

Only modal category can be located.
Modal category = Democratic

ESTIMATING THE ARITHMETIC MEAN ($\overline{X}$) FROM GROUPED DATA

Data on ages, incomes, years of education, and the like are often released by the Bureau of the Census and others in the form of a frequency distribution. In order to estimate the mean of a frequency distribution, it is assumed that *the values are spread evenly throughout each class.* Logically, the mean of all the values in a class is its **midpoint.** The midpoint of a class, therefore, is used to represent the class. These computations are similar to those for finding a weighted mean.

The arithmetic mean of sample data organized in a frequency distribution is estimated by

$$\overline{X} = \frac{\Sigma fX}{n}$$

3–4

where

$\overline{X}$ is the designation for the sample mean.
f is the frequency in each class.
X is the midpoint of each class.
fX is the frequency of each class times its midpoint.
ΣfX directs one to add these products.
n is the total number of observations in the sample.

If the frequency distributions represent a population, the symbols are slightly different, but the computations are the same. The formula for the arithmetic mean of a population is

$$\mu = \frac{\Sigma fX}{N} \qquad \textbf{3-5}$$

where

μ is the population mean.

ΣfX is the sum of the products of the class midpoints times the number of frequencies in each class.

N is the total number of observations in the population.

Problem   Table 3–5 repeats the frequency distribution from Chapter 2 regarding the selling prices of 80 homes in the Rafter J Ranch subdivision.

TABLE 3–5	Selling Prices of 80 Homes in the Rafter J Ranch Subdivision	
	Selling Price	**Number of Homes (f)**
	$ 80,000 up to $ 85,000	5
	85,000 up to 90,000	10
	90,000 up to 95,000	15
	95,000 up to 100,000	26
	100,000 up to 105,000	13
	105,000 up to 110,000	7
	110,000 up to 115,000	4

Estimate the arithmetic mean selling price from the frequency distribution.

Solution   It is assumed that the 5 selling prices in the $80,000 up to $85,000 class averaged $82,500 (which is the midpoint of the class). Thus, the total value of the 5 homes is approximately $412,500, found by 5 · $82,500. The class midpoint of the next larger class is $87,500. This value is used to represent the $85,000 up to $90,000 class. The total value of the 10 homes in that class is 10 · $87,500 = $875,000. This process is continued for all classes. The total value of the 80 homes is estimated to be $7,745,000, and the estimated arithmetic mean is $96,812.50, found by $7,745,000/80. For the details of these calculations, see Table 3–6.

Solving for the estimated sample arithmetic mean using the information in Table 3–6 and formula 3–4,

$$\overline{X} = \frac{\Sigma fX}{n} = \frac{\$7,745,000}{80} = \$96,812.50$$

TABLE 3-6	Calculations Needed for the Arithmetic Mean		
Selling Price	Number of Homes (f)	X	fX
$ 80,000 up to $ 85,000	5	$ 82,500	$ 412,500
85,000 up to 90,000	10	87,500	875,000
90,000 up to 95,000	15	92,500	1,387,500
95,000 up to 100,000	26	97,500	2,535,000
100,000 up to 105,000	13	102,500	1,332,500
105,000 up to 110,000	7	107,500	752,500
110,000 up to 115,000	4	112,500	450,000
			$7,745,000

So we conclude that the typical selling price of a home in the Rafter J Ranch subdivision is $96,812.50. We would probably round this value off and report $96,800 as the typical selling price.

There is usually some discrepancy between the arithmetic mean estimated from grouped data and the mean computed from the raw data. In the case of the selling prices in the Rafter J Ranch subdivision, the mean estimated from the grouped data is $96,812.50. The exact mean of the raw data presented in Table 2–3 is $96,487.50, resulting in a difference of $325. This is a difference of only 0.34%. This loss of accuracy is usually more than offset by the improved ease of calculation of the frequency distribution. But remember that the mean of data grouped into a frequency distribution should be viewed as an *estimate* of the actual mean.

Self-Review 3–5

A sample of television viewers was asked to rate a new soap opera titled "Mother Knows Best." The highest possible score is 100. A score near 0 indicates that the viewer thinks this is a very poor program. A score near 100 indicates that it is an outstanding soap opera. The scores were organized into a frequency distribution:

Rating	Frequency
50 up to 60	2
60 up to 70	6
70 up to 80	10
80 up to 90	8
90 up to 100	4

Determine the arithmetic mean rating for the sample.

Exercises

27. A sample of vehicles traveling on Interstate 75 was clocked by radar, and the following frequency distribution was constructed:

Speed (mph)	Number of Vehicles
40 up to 45	9
45 up to 50	15
50 up to 55	30
55 up to 60	17
60 up to 65	17
65 up to 70	12

a. What is the total number of frequencies?
b. Determine the midpoint for the class 50 up to 55.
c. Find the estimated arithmetic mean speed using the class midpoints.

28. Last month 60 antique cars were sold within the Buffalo metropolitan area. The selling prices were organized into the following frequency distribution:

Selling Price (in thousands of dollars)	Frequency
$40.0 up to $50.0	3
50.0 up to 60.0	6
60.0 up to 70.0	19
70.0 up to 80.0	23
80.0 up to 90.0	9
Total	60

Estimate the arithmetic mean selling price.

29. A sample of light trucks using diesel fuel revealed the following miles per gallon:

Mileage	Number of Trucks
11 up to 14	3
14 up to 17	5
17 up to 20	12
20 up to 23	8
23 up to 26	4

Estimate the arithmetic mean mileage for the sample.

30. A sample of 50 listeners to WRQN was chosen and data on their ages organized into a frequency distribution. Estimate the arithmetic mean age.

Age	Frequency
15 up to 20	10
20 up to 25	12
25 up to 30	14
30 up to 35	9
35 up to 40	5

ESTIMATING THE MEDIAN FROM GROUPED DATA

We noted previously that for some data, the arithmetic mean is not a representative measure of central tendency. Usually, in these cases, the median might be used to represent the typical value. How is the median determined if the data are grouped into a frequency distribution?

We can estimate the median of data organized into a frequency distribution by (1) locating the class in which the median lies and (2) interpolating within that class to arrive at the median.

Problem   The selling prices of homes in the Rafter J Ranch subdivision are reintroduced to illustrate the procedure for estimating the median (see Table 3–7). The cumulative frequencies in the right-hand column will be helpful for locating the class in which the median lies. What is the median selling price?

TABLE 3–7	Selling Prices of 80 Homes in the Rafter J Ranch Subdivision	
Class Limits	**Class Frequency**	**Cumulative Frequency**
Selling Price	Number of Homes (*f*)	CF
$ 80,000 up to $ 85,000	5	5
85,000 up to 90,000	10	15
90,000 up to 95,000	15	30
95,000 up to 100,000	26	56
100,000 up to 105,000	13	69
105,000 up to 110,000	7	76
110,000 up to 115,000	4	80

Solution   Note that there are 80 selling prices. The middle observation is determined by $n/2$; or $80/2 = 40$. [To be consistent with the way the median was located earlier in the chapter, the selling price of the $(n + 1)/2$, or 40.5th, home should be located. However, it is common practice to use the more convenient $n/2$ value. Any error involved is very slight.]

We find the class interval containing the 40th selling price by referring to the Cumulative Frequency column. There are 30 selling prices less than $95,000 and 56 prices less than $100,000. Logically, the 40th selling price is in the $95,000 up to $100,000 class. That particular class contains 26 selling prices.

To compute the median, we again assume the selling prices of the 26 homes in the median class are evenly distributed between the lower and upper limits of that class. Shown in a diagram, that information appears as follows:

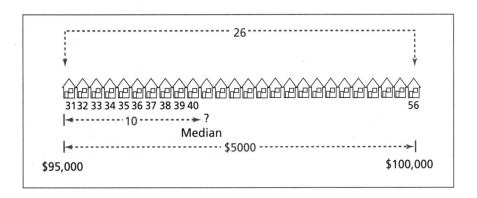

The median can be interpolated from the diagram. There are 26 prices in the class containing the median. And there are 10 selling prices between the 30th price and the 40th price (40 − 30). Therefore, the median price is 10/26 of the amount between $95,000 and $100,000. That amount is $5000. Thus, 10/26 of $5000 is $1923. Adding $1923 to $95,000, we obtain the estimated median of $96,923. Summarizing these calculations:

$$\text{Median} = \$94{,}500 + \frac{10}{26}\,(\$5000) = \$96{,}923$$

Alternatively, we can use the following formula to determine the median:

$$\text{Median} = L + \frac{\dfrac{n}{2} - CF}{f}(i)$$

3–6

where

L is the lower limit of the class containing the median. The median is in the $95,000 up to $100,000 class, and $95,000 is the lower limit of that class.

n is the total number of frequencies. It is 80.

CF is the cumulative number of frequencies in all the classes immediately preceding the class containing the median. The class containing the median is the $95,000 up to $100,000 class, and the cumulative number of selling prices prior to that class is 30.

f is the frequency in the class containing the median. There are 26 in that class.

i is the width of the class in which the median lies. The width is $5000, found by $100,000 − $95,000.

Inserting these values into formula 3–6, we obtain

$$\text{Median} = \$95{,}000 + \frac{\dfrac{80}{2} - 30}{26}\,(\$5000) = \$96{,}923$$

Again, it should be noted that a measure computed using data grouped into a frequency distribution will probably not be identical to the measure we would obtain using the raw data. Here the estimated median using the frequency distribution is $96,923, while the median of the selling prices of Rafter J Ranch homes listed in Chapter 2 is $97,000.

Self-Review 3–6

On arriving in Hawaii, a sample of vacationers is asked their ages by the Hawaii Tourist Bureau. The sample information is organized into the following frequency distribution. (The cumulative frequencies are given to assist in computing the median.)

Age	Number of Vacationers	Cumulative Frequency
20 up to 30	4	4
30 up to 40	9	13
40 up to 50	20	33
50 up to 60	8	41
60 up to 70	5	46
70 up to 80	4	50

Compute the median age.

ESTIMATING THE MODE FROM GROUPED DATA

For data grouped into a frequency distribution, the mode is the *midpoint* of the class containing the largest class frequency. As an example, refer back to Table 3–7. The largest class frequency is 26, and the midpoint of that class is $97,000—the **mode.** This indicates that the largest number of Rafter J Ranch homes sold for $97,000.

Refer to Self-Review 3–6 for the ages of the vacationers visiting Hawaii.

Self-Review 3–7

a. What is the mode of the age distribution of vacationers arriving in Hawaii? This question also could be stated as: What is the *modal* age of the vacationers arriving in Hawaii?
b. Explain what the mode indicates.

Exercises

31. Refer to the frequency distribution of speeds on I-75 in Exercise 27.
  a. Which class interval contains the median value?
  b. What is the lower limit of the class containing the median?
  c. What is the class interval?
  d. Determine the cumulative number of frequencies in all classes smaller than the median class.
  e. Compute the median speed.
  f. What is the mode of the distribution of speeds?
  g. Discuss which measure of central tendency you would use as an average for the data.

32. Refer to the frequency distribution in Exercise 28.
  a. Which class interval contains the median value?
  b. What is the lower limit of the class containing the median?
  c. What is the class interval?
  d. Determine the cumulative number of frequencies in all classes smaller than the median class.
  e. Compute the median price.
  f. Determine the mode of the distribution of prices.

33. A sociologist is studying the impact of TV on the family. Of particular interest is the number of hours of TV school-aged children watch each day. A random sample of 50 homes reveals the following numbers of hours watched on a Wednesday between 4 P.M. and 11 P.M.:

Number of Hours	Frequency
0	9
1	14
2	10
3	9
4	5
5 or more	3
	50

a. Compute the median number of hours of TV viewing.

b. Determine the modal number of hours of TV watched.

34. An advertising campaign is being designed for a local car dealer. One relevant piece of information, the ages of recent purchasers of cars sold by the dealership, is shown in the following frequency distribution:

Age	Frequency
Under 20	7
20 up to 30	13
30 up to 40	26
40 up to 50	15
50 up to 60	6
60 or more	3
Number of cars sold	70

a. Compute the median age.

b. What is the modal age?

c. Explain why the mean cannot be determined.

CHOOSING AN AVERAGE FOR DATA IN A FREQUENCY DISTRIBUTION

If the data being organized into a frequency distribution contain extremely high or extremely low values, the resulting distribution is said to be **skewed.** The frequency distribution in Figure 3–3 is **positively skewed.** It can be identified by the long "tail" on the right.

Note the relative positions of the three averages. The arithmetic mean ($\overline{X}$) is being pulled upward toward the tail by a few very highly paid employees. It is the highest of the three averages. If the skewness is very pronounced, the mean is not an appropriate average to describe the central tendency of the data.

The median is the next higher measure of central tendency for a positively skewed distribution. It, of course, divides the incomes into two parts and may be more typical of the data. The mode is the smallest of the three measures. It occurs at the peak, or apex, of the curve and indicates that the largest number of employees earns $200 a week.

The distribution of the ratings in Figure 3–4 is **negatively skewed.** The arithmetic mean is being pulled down by a few very low ratings, and once again it is not the best average to use. The median is the next higher average, with the mode of 54 the highest of the three averages.

By contrast, Figure 3–5 reveals that there is no skewness in the distribution of the annual incomes of the social workers. This type of distribution is called symmetric. A **symmetric distribution** has the same shape on either side

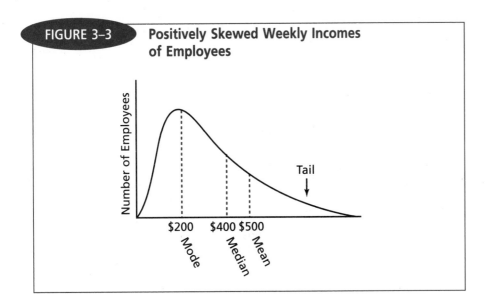

FIGURE 3–3   **Positively Skewed Weekly Incomes of Employees**

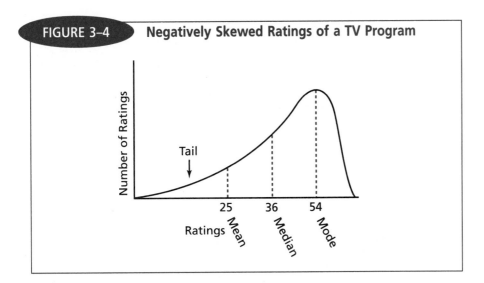

FIGURE 3–4   **Negatively Skewed Ratings of a TV Program**

of the median. That is, if the distribution graph is folded in half at the median value, the 2 halves will be identical. With such a distribution the median and the mean are the same. In Figure 3–5, the value for both is $19,000.

Open-ended frequency distributions pose a problem for the computation of the arithmetic mean. Unless the midpoint of the open-ended class can be approximated, the mean cannot be computed. In the example below, unless the midpoint of the $30,000 and over class can be approximated, the mean of this open-ended distribution cannot be calculated.

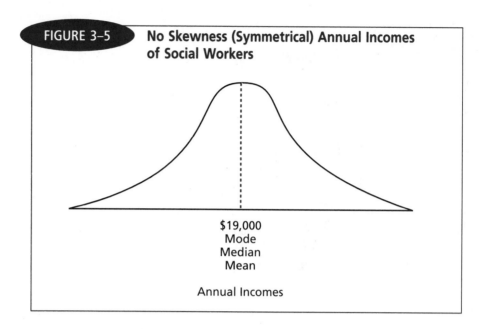

FIGURE 3–5   **No Skewness (Symmetrical) Annual Incomes of Social Workers**

$19,000
Mode
Median
Mean

Annual Incomes

Income	Number
$10,000 up to $20,000	42
20,000 up to 30,000	86
30,000 and over	19

The median or the mode, however, can be used to represent the central value of the income distribution.

CHAPTER OUTLINE

I. A measure of central tendency is a single value used to represent a set of data.

A. The arithmetic mean is the most widely reported measure of central tendency.

  1. It is calculated by adding all the observations and dividing the result by the number of observations.

   a. The formula for the arithmetic mean of a sample is

$$\overline{X} = \frac{\Sigma X}{n} \qquad \text{3–1}$$

   b. The formula for the arithmetic mean of a population is

$$\mu = \frac{\Sigma X}{N} \qquad \text{3–2}$$

c. The weighted arithmetic mean is found by multiplying each observation by its corresponding weight.

$$\overline{X}_w = \frac{w_1X_1 + w_2X_2 + \cdots + w_nX_n}{w_1 + w_2 + \cdots + w_n} \quad \boxed{3\text{-}3}$$

d. For data grouped into a frequency distribution, the formula for the arithmetic mean of a sample is

$$\overline{X} = \frac{\Sigma fX}{n} \quad \boxed{3\text{-}4}$$

2. The major characteristics of the arithmetic mean are as follows:
   a. At least the interval scale of measurement is required.
   b. All the data values are used in the calculation.
   c. A set of data has only one mean; that is, it is unique.
   d. The sum of the deviations from the mean equals 0.
B. The median is the value in the middle of a set of data.

1. The major characteristics of the median are as follows:
   a. At least the ordinal scale of measurement is required.
   b. It is not influenced by extreme values.
   c. Fifty percent of the observations are greater than the median.
   d. It is a unique value for a set of data.
2. The formula for finding the median of grouped data is

$$\text{Median} = L + \frac{\frac{n}{2} - CF}{f}(i) \quad \boxed{3\text{-}6}$$

C. The mode is the value that occurs most often in a set of data.
   1. The mode can be found for all levels of data.
   2. A set of data can have more than one mode.
II. A distribution is symmetric if it is the same shape on either side of the median.
A. In a positively skewed distribution, the mean is larger than the median, and the longer tail of the distribution is to the right.
B. In a negatively skewed distribution, the mean is smaller than the median, and the longer tail of the distribution is to the left.

Exercises

35. Under what conditions would the median be a more representative measure of central tendency than the mean?
36. For the following measures of central tendency, list the required level of measurement:
    a. arithmetic mean.
    b. weighted mean.
    c. median.
    d. mode.
37. Listed below are some measures of central tendency regarding income per household for three counties in the Buffalo, New York, area.

	Erie	Tonawanda	Corry
Arithmetic mean	$36,775	$38,500	$40,500
Median	38,845	38,500	39,700
Mode	43,679	38,500	37,562

a. In which county is the income per household positively skewed?
b. In which county is the income per household symmetric?
c. In which county is the income per household negatively skewed?

38. A small commuter airline recently purchased 5 aircraft. The ages (in years) of the 5 planes are

16	10	11	8	15

a. Determine the mean age of the aircraft.
b. Show that the sum of the deviations from each observation to the mean is 0.
c. Determine the median age of the aircraft.
d. Determine the modal age of the aircraft.

39. The Law Enforcement Assistance Administration reported that on December 31 of each of the past 5 years there were 213.0, 211.0, 196.0, 241.0, and 263.0 prisoners in federal and state institutions (data reported in thousands). Consider this information to be a population.
a. Find the arithmetic mean number of prisoners.
b. Show that the sum of the deviations from the mean is zero.
c. Find the median number of prisoners.

40. The National Science Foundation reported that the median time candidates in the social sciences take to earn a doctorate degree, after receiving a bachelor's degree, is 8.5 years. Explain what the 8.5 years indicates.

41. The surgeon general released these figures for the United States: 54 million smokers smoked 615 billion cigarettes in 1 year. What is the average annual number of cigarettes smoked by persons who smoke? What is the average daily number?

42. A report on the literacy rate for a sample of countries revealed the following percentages:

Country	Literacy Rate (in percent)
Turkey	55
Tunisia	32
Tonga	93
Benin	20
Bangladesh	25
Colombia	47
Iceland	99
Iran	37

a. Find the mean percentage.
b. Find the median percentage.

43. A student is currently enrolled for a 3-credit-hour history course, a 5-credit-hour English course, and a 4-credit-hour statistics course. Grades are reported on a 4-point system where

$A = 4$, $B = 3$, $C = 2$, $D = 1$, and $F = 0$. The student received an A in statistics, a B in English, and a C in history. What is the student's grade point average for the semester? What type of average is this?

44. At Mercy General Hospital, registered nurses (RNs) are paid $18.00 per hour, licensed practical nurses (LPNs) are paid $12.00 per hour, and nurses' aides are paid $9.00 per hour. If the pediatrics wing employs 6 RNs, 4 LPNs, and 4 aides, compute the weighted mean salary for the wing.

45. A study of the mileage per gallon of 35 compact cars of the same make and model revealed the following:

Miles per Gallon	Number of Cars
20 up to 25	2
25 up to 30	7
30 up to 35	15
35 up to 40	8
40 up to 45	3

a. Compute the arithmetic mean mileage.
b. Compute the median mileage.

46. Are the baseball batting averages in the American League higher than those in the National League? Consider these data to be a sample:

Batting Average	American League	National League
.150 up to .200	5	4
.200 up to .250	34	27
.250 up to .300	64	52
.300 up to .350	27	14
.350 up to .400	2	0

a. Estimate the mean batting average for the two leagues.
b. Compare the median average for each league.
c. Determine the mode for each league.
d. Does this show there is much difference in the two leagues?

47. A sample of 7 independently owned service stations in the San Francisco metropolitan area revealed the following cost per pint of transmission fluid:

Andy's Amoco	$0.83
Rossi's Gastown	0.79
Kernie's Mobile	0.92
Mark's Exxon	0.71
Yamamoto's Chevron	0.69
Ali's Sunoco	0.83
Deckers' Union 76	0.74

a. Calculate the mean, median, and modal cost per pint.

b. Discuss which average is most representative of the information or numbers.

48. The American Association of Retired Persons is studying the voting statistics for the most recent election. The following distribution presents a breakdown of voters by age. Compute the median age.

Age	Frequency (in millions)
20 up to 30	22.1
30 up to 40	21.9
40 up to 50	16.4
50 up to 60	15.6
60 up to 70	14.4
70 or more	11.5

49. In a survey of chief executive officers of large companies, the following information was obtained regarding the time they awake each weekday morning:

Time	Percentage of Total
4 up to 5 A.M.	7.0
5 up to 6 A.M.	65.0
6 up to 7 A.M.	24.0
7 up to 8 A.M.	4.0

What is the typical wake-up time?

50. Listed below (in thousands of dollars) are the average player salaries for the 28 major league baseball teams.

$1918	$ 411	$1494	$ 548	$1274
466	1476	1264	1586	865
1206	1071	1344	624	536
1499	635	925	1165	967
1176	808	2000	1534	1223
1138	1370	1030		

a. Determine the mean salary for the 28 teams.

b. Organize the data into a frequency distribution, and compute the mean salary of the grouped data. Compare the mean salary from part a with the mean salary from the grouped data.

c. Determine the median salary of the grouped data.

d. Organize the data into a stem-and-leaf chart. Find the median on the stem-and-leaf chart, and compare it with the median found in part c.

51. The Neal Nail Company manufacturers and packages nails. A package is supposed to contain at least 20 nails. Some packages contain less than 20; others have more than 20. What can a consumer expect? For a sample of 40 packages, the following numbers of nails are found:

Number of Nails	20 or less	21	22	23	24	25 or more
Number of Packages	5	8	11	10	5	1

Determine the median number of nails per package.

52. A study of 109 small and medium-size firms in a large metropolitan area reveals the following about the percentage change in number of persons employed from January 2 last year to January 2 this year.

Percentage Change	Frequency
Less than −20.0	13
−20.0 up to −10.0	21
−10.0 up to 0.0	48
0.0 up to 10.0	21
10.0 up to 20.0	5
20.0 or more	1

Determine the median percentage change in employment.

53. A large taxicab company has a fleet of 45 cabs. Below are the numbers of gallons of gasoline each cab used last month.

123	132	130	119	106	97
121	109	118	128	132	115
130	125	121	127	144	115
107	110	112	118	115	134
132	139	144	104	128	138
114	121	129	128	116	138
129	113	105	142	122	131
126	111	142			

a. Organize the data into a frequency distribution.

b. Determine the mean number of gallons used per taxicab.

c. Determine the median number of gallons used per taxicab.

DATA EXERCISES

54. Refer to the real estate data set, which reports information on homes sold in Alabama during 1996.
   a. Determine the mean and median selling prices. Is one a better measure of central tendency than the other? Defend your choice in a short paragraph.
   b. Describe a typical home. How many bedrooms does it have? How many bathrooms? How far is it from the center of the city?
   c. Compare the mean selling price of a home with an attached garage to that of a home without an attached garage.
   d. Write a brief summary of your findings.

55. Refer to the school data set which refers to the 94 school districts in northwest Ohio.
   a. Determine the mean and the median salaries for teachers in northwest Ohio.
   b. Organize the salary information into a stem-and-leaf chart. Comment on the shape of the distribution of salaries. Is the distribution symmetric or skewed? Also find the median salary from the grouped data in the stem-and-leaf chart. Does it agree with your calculations in part a?
   c. Find the mean and the median percentages on welfare in each of the school districts.
   d. Organize this information into a stem-and-leaf chart. Comment on the distribution.

CHAPTER ACHIEVEMENT TEST

The answers are at the back of the book.

MULTIPLE-CHOICE QUESTIONS

Select the response that best answers each of the questions.

1. The lowest scale of measurement required to calculate the mode is

a. nominal.
b. ordinal.
c. interval.
d. ratio.

2. The lowest scale of measurement required to calculate the mean is
   a. nominal.
   b. ordinal.
   c. interval.
   d. None of the above

3. In a positively skewed distribution, the mean is always
   a. smaller than the median.
   b. equal to the median.
   c. larger than the median.
   d. None of the above

4. Half the observations are always larger than the
   a. mean.
   b. median.
   c. mode.
   d. total.

5. If a frequency distribution has an open-ended class,
   a. a median cannot be computed.
   b. the arithmetic mean and the median will always be exactly equal.
   c. the distribution is always positively skewed.
   d. a mean cannot be computed.

6. Which of the following is *not* a characteristic of the arithmetic mean?
   a. All the observations are used in its calculation.
   b. It is influenced by extreme values.
   c. The sum of the deviations from the mean is zero.
   d. Fifty percent of the observations are always larger than the mean.

7. Which of the following is a true statement about the median?
   a. It is always one of the data values.
   b. It is influenced by extreme values.
   c. The sum of the deviations from it is zero.
   d. Fifty percent of the observations are larger than the median.

8. The median is larger than the arithmetic mean when
   a. the distribution is positively skewed.
   b. the distribution is negatively skewed.
   c. the data are organized into a frequency distribution.
   d. raw data are used.

9. A sample of 10 students is obtained. The students are weighed and ranked according to their weight. The median weight is the
   a. weight of the fifth student.
   b. weight of the sixth student.
   c. The median does not exist.
   d. average weight of the fifth and sixth students.

10. The value that occurs most often in a set of data is called the
    a. mode.
    b. arithmetic mean.
    c. median.
    d. None of the above

COMPUTATION PROBLEMS

11. The weekly incomes for a sample of 4 nurses' aides are $420, $445, $600, and $395.
    a. What is the arithmetic mean weekly income?
    b. What is the median weekly income?

12. The amounts spent on mail advertising in a day by a sample of 5 salespersons are $22.50, $85.90, $18.60, $140.20, and $76.10. What is the median amount spent?

    Problems 13–15 are based on the following frequency distribution of a sample of the weights of king crabs caught in Alaskan waters:

Weights (in pounds)	Number of Crabs
0.0 up to 0.5	3
0.5 up to 1.0	8
1.0 up to 1.5	20
1.5 up to 2.0	10
2.0 up to 2.5	6
2.5 up to 3.0	3

13. What is the arithmetic mean weight (to the nearest hundredth of a pound)?

14. What is the median weight (to the nearest hundredth of a pound)?

15. What is the modal weight (to the nearest hundredth of a pound)?

. . . .

ANSWERS TO SELF-REVIEW PROBLEMS

3-1 a. $\overline{X} = \dfrac{\Sigma X}{n}$

b. $\overline{X} = \dfrac{59}{6}$

   $= 9.8$ abortions (in thousands)

3-2 a. $\overline{X} = \dfrac{960}{4} = 240.0$ pounds

b. The mean

c. $\Sigma(X - \overline{X}) = (200 - 240) + (210 - 240)$
$\qquad\qquad\quad + (230 - 240) + (320 - 240)$
$\qquad\qquad = (-40) + (-30) + (-10) + 80$
$\qquad\qquad = 0$

3-3 $\overline{X}_w = \dfrac{\$.50(20) + \$.75(50) + \$1.00(30)}{100}$

$\qquad = \dfrac{\$77.50}{100} = \$0.775$

3-4 a. Arrange from low to high:

   18
   22
   ㉞ median
   40
   75

b. 2, 2

c. 11 points—halfway between 9 and 13

d. 146,490,000 telephones—halfway between 143,970,000 and 149,010,000

3-5

X	f	fX
55	2	110
65	6	390
75	10	750
85	8	680
95	4	380
		2310

$$\overline{X} = \frac{\Sigma fX}{n}$$

$$= \frac{2310}{30} = 77.0$$

3-6 $\text{Median} = 40 + \dfrac{\dfrac{50}{2} - 13}{20}(10)$

$\qquad\qquad = 40 + \dfrac{12}{20}(10)$

$\qquad\qquad = 46$ years

3-7 a. 45 years

b. The largest number of vacationers was 45 years of age.

CHAPTER 4

Descriptive Statistics: Measures of Dispersion and Skewness

OBJECTIVES

When you have completed this chapter, you will be able to

- describe the spread or dispersion in a set of data;
- calculate and interpret the various measures of dispersion;
- compute the coefficients of variation and skewness.

CHAPTER PROBLEM   Price Disparities

Recall from the Chapter Problem in Chapters 2 and 3 that Charles Reynolds, owner of the Reynolds Company, is putting together a prospectus for new homes to be built in the Rafter J Ranch subdivision in Jackson Hole, Wyoming. In Chapter 2, he found the selling price of each of the 80 existing homes and organized them into a frequency distribution. In Chapter 3, he determined several measures of central tendency to find the estimated selling price of a typical home. Now Mr. Reynolds wants to determine the spread of the estimated selling prices. How can he measure the dispersion in the data?

INTRODUCTION

In Chapter 2, we noted that the usual first step in summarizing raw data is to organize it into a frequency distribution. A histogram or a frequency polygon may be drawn to portray the distribution in graphic form. Chapter 3 described a summary measure called an "average." Three different measures of central tendency or averages were examined: the arithmetic mean, the median, and the mode. An average is a central value around which the data tend to cluster.

A direct comparison of the central value of two or more distributions may be misleading. For example, suppose that a sample of the lengths of imprisonment for armed robbery in Alabama revealed the arithmetic mean length to be 10 years. Suppose further that a similar study of the lengths of imprisonment in neighboring Georgia also revealed a mean length of 10 years. Based on the two sample means, one might conclude that the distributions of the lengths of imprisonment are about the same in both states. Figure 4–1 on the following page, however, reveals this conclusion to be incorrect. Despite the fact that the two means are equal, the lengths of imprisonment in Georgia are spread out, or dispersed, more than those in Alabama.

MEASURES OF DISPERSION FOR RAW DATA

The Range

As Figure 4–1 suggests, the next step in analyzing a set of data is to develop measures that describe the spread, or **dispersion,** of the data. The simplest measure of dispersion is the **range.** The range is the difference between the highest and lowest values in a set of data.

> **Range** = Highest value − Lowest value **4–1**

Problem   Calculate the ranges of the length-of-imprisonment data in Figure 4–1 for both Alabama and Georgia. Compare the two ranges.

Solution   For the Alabama data, the longest stretch in prison is 12 years; the shortest is 8 years. The range is 4 years (12 − 8). The range for the Georgia prisoners is 14 years, or 17 − 3. A comparison of the two ranges indicates the following:

1. There is more spread in the Georgia data because the range of 14 years for Georgia is greater than the range of 4 years for Alabama.
2. The lengths of imprisonment for the Alabama inmates are clustered more closely about the mean of 10 years than those for the Georgia inmates because the range of 4 years is less than the range of 14 years.

Obviously, the range is easy to compute. It does have a serious disadvantage, however. It is based on only two values—the two extreme observations.

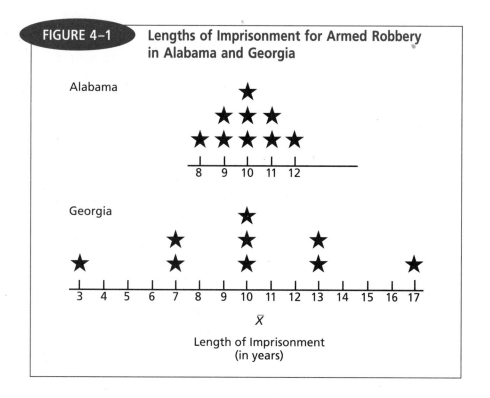

FIGURE 4-1   **Lengths of Imprisonment for Armed Robbery in Alabama and Georgia**

The range therefore does not reflect the variation in the data that lie between the high and low observations. Further, one extremely high (or low) value might give a misleading picture of the dispersion. For example, suppose the ages of a group at a U2 concert were 18, 20, 19, 20, 19, 17, 18, 18, and 78. Reporting the range of 61 years, found by subtracting 17 from 78, would give the impression that there is considerable variation in the ages of the audience. Except for 1 person, however, the distance between the youngest (17 years) and the oldest (20 years) is only 3 years! For that reason, the range can be considered only a rough index of the variation in a set of data.

Problems   Compute the sample range for each of the following sets of values:

a.   2, 9, 1
b.   6, 13, 10, 2
c.   0.1, 5.8, 1.2
d.   7¾, 3³⁄₁₆, 1¼

Solutions

a.   8
b.   11
c.   5.7
d.   6½

| Self-Review 4-1 | Answers to all Self-Review problems are at the end of the chapter. A radar check of several automobiles traveling on the Sunshine Highway recorded these speeds in miles per hour: 56, 40, 61, 80, and 59. Speeds recorded at about the same time on a nearby state highway were 59, 57, 59, 62, 55, and 62. |

a. Compute the range for each set of data.
b. Compare the dispersions for the two sets of speed checks using the ranges.

The Interquartile Range and the Quartile Deviation (QD)

Recall that half of the ordered values are above the median and half are below it. The lower half of the ordered values can be further subdivided into two parts, so that one-fourth of all the ordered values are smaller than a particular value. The point where this subdivision occurs is called the **first quartile** and is designated Q_1. Similarly, the upper half of the ordered distribution can be divided into two equal parts. The **third quartile**, designated Q_3, is the point below which lie three-fourths of the ranked observations. The median, of course, is the same as the second quartile.

The **interquartile range** is the distance between the first quartile and the third quartile.

$$\text{Interquartile range} = Q_3 - Q_1 \qquad \boxed{\textbf{4-2}}$$

Figure 4–2 illustrates this concept; note that the third quartile is 20 days, the first quartile 12 days. The interquartile range is 8 days, which we find by subtracting 12 from 20. Thus, the interquartile range reports the difference between the two values that bound the middle 50% of the observations.

Another measure of dispersion is the **quartile deviation**, abbreviated QD, which is one-half of the interquartile range.

$$QD = \frac{Q_3 - Q_1}{2} \qquad \boxed{\textbf{4-3}}$$

Problems Compute the interquartile range and the quartile deviation for each of the following sets of values:

	Third Quartile	First Quartile
a.	12	8
b.	27	20
c.	103	91
d.	1200	1000

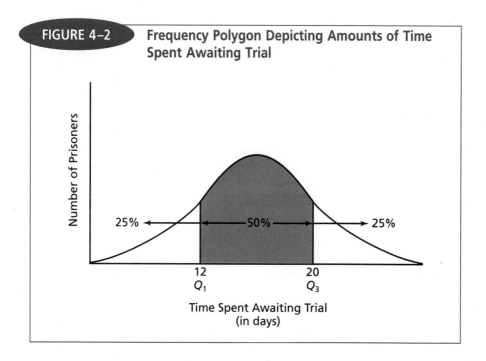

FIGURE 4–2   **Frequency Polygon Depicting Amounts of Time Spent Awaiting Trial**

Solutions

	Interquartile Range	Quartile Deviation
a.	4	2
b.	7	3.5
c.	12	6
d.	200	100

As mentioned earlier, one of the objections to the range as a measure of spread is that its value is based on the two extreme observations in a set of data. An unusually large or small extreme value might give a distorted picture of the spread in the data. The interquartile range and the quartile deviation offset this objection because they are not based on the extreme, or end, values. The first and third quartiles may be more representative of the data. Since the QD is based on the relative positions of Q_1 and Q_3 in the frequency distribution, it may be computed not only for interval-scaled data, but also for ordinal-scaled data. Later in the chapter we will estimate the interquartile range and the quartile deviation for data organized into a frequency distribution.

One of the disadvantages of the range, the interquartile range, and the quartile deviation is that they do not make use of all the observations. However, several measures that incorporate all values do exist; they are based on the de-

The Mean Deviation (*MD*)

viation of each observation from its mean. These measures include the **mean deviation,** the **variance,** and the **standard deviation.**

We compute the **mean deviation,** abbreviated *MD,* by first finding the difference between each observation and the mean. Next, we sum these deviations using their absolute values—that is, disregarding their algebraic signs (+ and −). Finally, we divide the sum obtained by the number of observations to arrive at the mean deviation.

$$MD = \frac{\Sigma|X - \overline{X}|}{n}$$

4–4

where

X is the value of each observation.
$\overline{X}$ is the mean.
n is the number of observations.

(The symbol | | indicates the absolute, or unsigned, value of a number.)

Problem   The test kitchen of Cellu-Lite, a large producer of cake mixes, constantly monitors the weight, moisture content, and flavor of its cakes. The weights (in grams) of a sample of 5 peach upside-down cakes are 498, 500, 503, 498, and 501. What is the mean deviation? How is it interpreted?

Solution   First, we find the arithmetic mean weight of the cakes, which is 500 grams: (498 + 500 + 503 + 498 + 501)/5 = 2500/5 = 500. Next, we find the deviation from the mean for each cake. For example, the first cake selected, weighing 498 grams, deviates 2 from the mean of 500. Calculations for the mean deviation are shown in Table 4–1. Interpreting the result: On

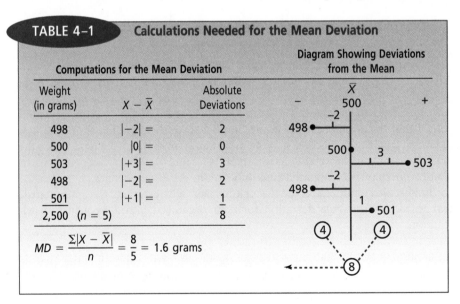

TABLE 4–1   Calculations Needed for the Mean Deviation

Computations for the Mean Deviation			Diagram Showing Deviations from the Mean
Weight (in grams)	$X - \overline{X}$	Absolute Deviations	
498	|−2| =	2	
500	|0| =	0	
503	|+3| =	3	
498	|−2| =	2	
501	|+1| =	1	
2,500 (n = 5)		8	

$$MD = \frac{\Sigma|X - \overline{X}|}{n} = \frac{8}{5} = 1.6 \text{ grams}$$

the average, the weights of the 5 peach upside-down cakes deviate 1.6 grams from the mean of 500 grams.

Problems   Compute the mean deviation of each of the following sets of values:

a.   2, 1, 3
b.   5, 1, 4, 6
c.   10, 9, 7, 14

Solutions

a.   0.67
b.   1.5
c.   2

· ·

The Cellu-Lite test kitchen now wants to compare the variation in the weights of blueberry cakes with that of the peach upside-down cakes studied in this section. The weights of the blueberry cakes are 484, 503, 496, 510, 491, and 516 grams.

Self-Review 4–2

a. Compute the mean and the mean deviation.
b. Interpret the mean deviation.
c. Compare the variation in the weights of the peach upside-down cakes with the variation in the weights of the blueberry cakes.

· ·

Exercises

Answers to the even-numbered Exercises are at the back of the book.

−14	−7	6	24	−12
−1	−11	3	−7	−4

1. Determine the mean deviation for the following set of values:

7	3	6	3	2

2. Determine the mean deviation for the following set of values:

2	2	2	7	6

3. Determine the mean deviation for the following set of values:

8	7	5	7	12
8	13	11	14	11

4. Following are the high temperatures for a sample of ten January days in Barrow, Alaska. Determine the mean deviation of the temperatures.

5. The amounts (in gallons) of fuel oil consumed during the winter heating season for a sample of newer homes are 600, 590, 605, 600, 603, 610, 597, and 595. A second group of older homes used the following amounts: 810, 750, 790, 800, and 850.
   a. What is the range for the newer homes? For the older homes?
   b. What is the mean deviation for the newer homes? For the older homes?
   c. Compare the dispersions in the data for the newer homes and older homes.

6. The numbers of customers entering Romaker's Boutique each day for a 6-day period were 91, 89, 88, 58, 63, and 74.
   a. What is the range?
   b. What is the mean deviation?

7. Many government agencies are authorized to issue tax-exempt bonds (bonds whose interest earnings are tax-exempt to the owner) as a means of generating funds at a rate lower than current interest rates. A sample of 10 such bonds revealed the following interest rates: 9.75, 3.75, 5.50, 7.30, 6.00, 7.30, 7.25, 4.25, 3.30, and 6.60.
   a. What is the range?
   b. What is the mean deviation? Explain what this indicates.

8. A random sample of 15 child custody cases revealed the following yearly amounts paid (in hundreds of dollars) for child support:

$87	$51	$18	$ 52	$58
95	26	70	100	63
33	81	65	32	84

   a. What is the range of the amounts paid?
   b. What is the mean amount paid?
   c. What is the mean deviation of the amount paid?

9. The Greater Boston Bakery Association wanted to know the weekly amount spent on bakery products for households of three or more. A random sample of 20 households revealed the following amounts:

$13.94	$15.81	$ 4.47	$ 6.50	$11.99
8.26	15.75	4.15	1.16	14.09
10.78	15.00	8.12	11.69	2.46
17.88	14.06	7.54	10.65	4.07

   a. What is the range of the amounts spent?
   b. What is the mean amount spent?
   c. What is the mean deviation of the amount spent?

10. The Department of Fisheries stocked two ponds with trout fingerlings. The fish in one pond (pond A) were fed the traditional food. Those in the other pond (pond B) were fed an experimental food. Consider the following to be sample data. After four years, the lengths of the fish were as follows:

Length (in inches)	
Pond A	Pond B
12.5	18.0
14.0	20.0
13.5	12.0
14.5	14.5
15.0	19.0
16.5	13.5
16.0	14.5
12.0	13.0
14.0	20.5
14.0	14.0

   a. Compute the arithmetic mean lengths for both ponds.
   b. Compute the range for both sets of data. Interpret.
   c. Compute the mean deviation for each set. Interpret.

The Variance (σ^2) and the Standard Deviation (σ)

> **variance** The arithmetic mean of the squared deviations from the mean.

The mean deviation has two distinct advantages: It uses all available data in its computation, and it is easy to understand; that is, it is the mean amount by which the observations differ from the mean. As a measure of variation, however, it does have a major flaw: The absolute values used in its computation are difficult to work with.

Two other measures of dispersion—the **variance** and the **standard deviation**—are more versatile than the mean deviation. As a result, they will be used extensively in later chapters. Like the mean deviation, the variance and the standard deviation are based on the deviations from the mean. They differ from the mean deviation, however, in that the sign (+ or −) of the deviation from the mean is *not* ignored in their computation. To compute the variance and the standard deviation, we square the difference between each value and the mean. This squaring eliminates the possibility of negative numbers (since multiplication of two negative numbers yields a positive number). We then total the squared deviations and divide the total by the number of values.

Because of a difference in computation, it is necessary to distinguish between the variance of a *population,* designated by σ^2 (the lowercase Greek *sigma*), and the variance of a *sample,* designated by s^2. Recall that in Chapter 3, we made a distinction between a parameter, which is a measure associated with a population, and a statistic, which is a measure associated with a sample. The formula for the variance of a population is

standard deviation The square root of the variance.

$$\sigma^2 = \frac{\Sigma(X - \mu)^2}{N} \qquad \boxed{\textbf{4-5}}$$

where

μ is the population mean.
X is the value of each observation in the population.
N is the number of observations in the population.

The formula for the variance of a sample is

$$s^2 = \frac{\Sigma(X - \overline{X})^2}{n - 1} \qquad \boxed{\textbf{4-6}}$$

where

X is the value of each observation in the sample.
$\overline{X}$ is the sample mean.
n is the number of observations in the sample.

Note that, when computing the sample variance, we divide the sum of the squared deviations by the quantity $(n - 1)$, rather than n. Although the use of n seems more logical (because then the sample variance would be the average squared distance from the mean), the use of n tends to underestimate the population variance σ^2. The use of the quantity $(n - 1)$ in the denominator provides an appropriate correction for this tendency. Because the primary use of s^2 is to estimate the population variance, σ^2, the quantity $(n - 1)$ is preferred to n when finding the sample variance.

TABLE 4–2	Calculations Needed for the Variance and the Standard Deviation	
Weight (in grams) X	**Finding the Difference** $X - \overline{X}$	**Squaring the Difference** $(X - \overline{X})^2$
498	−2	4
500	0	0
503	+3	9
498	−2	4
501	+1	1
	0	18

sum of the deviations from mean is always zero

Problem Refer back to the weights for a sample of five peach upside-down cakes in Table 4–1. What are the sample variance and the standard deviation?

Solution The data needed to calculate the variance and the standard deviation are included in Table 4–2. The sample mean weight, $\overline{X}$, is 500 grams. To determine the sample variance, we divide the sum of the squared deviations (18) by the number of observations minus one, using formula 4–6:

$$s^2 = \frac{\Sigma(X - \overline{X})^2}{n - 1} = \frac{18}{5 - 1} = 4.5$$

The variance is 4.5. We then determine the standard deviation by taking the square root of the variance. The sample standard deviation is 2.12.

Problems Compute the variance and the standard deviation for each of the following samples:

a. 4, 2, 3
b. 9, 6, 7, 2
c. 2, 6, 1

Solutions

Variance	Standard Deviation
a. 1.0	1.0
b. 8.67	2.94
c. 7.0	2.646

The test kitchen wants to compare the variation in the weights of the blueberry cakes in Self-Review 4–2 with the peach cakes—using the variance and the standard deviation. The weights of a sample of blueberry cakes were 484, 503, 496, 510, 491, and 516 grams.

a. Compute the variance and the standard deviation.
b. Compare the variations in the weights of the peach upside-down cakes and of the blueberry cakes.

Exercises

11. Given the following set of population values:

| 5 | 4 | 5 | 1 | 10 | 5 |

a. Determine the variance.
b. Determine the standard deviation.

12. Given the following set of sample values:

| 11 | | 11 | | 9 | | 8 |
| 6 | | 9 | | 10 | | 16 |

a. Determine the variance.
b. Determine the standard deviation.

13. Refer to the data in Exercise 8, which reports the yearly amounts paid for child support for a sample of 15 cases.
a. Determine the variance.
b. Determine the standard deviation.

14. Refer to the data in Exercise 9, which reports the weekly amounts spent on bakery products for a sample of 20 households.
a. Determine the variance.
b. Determine the standard deviation.

15. The numbers of customers entering Romaker's Boutique each day for a 6-day period, as reported in Exercise 6, were 91, 89, 88, 58, 63, and 74. Consider these to be sample data, not the entire population. Compute the variance and the standard deviation.

16. The interest rates from the tax-exempt bonds in Exercise 7 were 9.75, 3.75, 5.50, 7.30, 6.00, 7.30, 7.25, 4.25, 3.30, and 6.60. Consider these to be sample data, not the entire population. Compute the variance and the standard deviation.

Computing the Variance (σ^2)

Unless the mean is an integer—which is usually not the case—subtracting each observation from the mean and then squaring the differences to arrive at the variance tends to be rather laborious. Rounding errors can also become a problem. A more convenient computational formula for both the variance and the standard deviation is based directly on the raw data. The formulas for the variances of a population and a sample are

$$\sigma^2 = \frac{\Sigma X^2 - \dfrac{(\Sigma X)^2}{N}}{N}$$

4–7

$$s^2 = \frac{\Sigma X^2 - \frac{(\Sigma X)^2}{n}}{n - 1} \qquad \boxed{4\text{-}8}$$

As before, we still derive the standard deviation (σ for a population, s for a sample) by computing the square root of the variance.

Note: ΣX^2 directs us to first square each observation and then sum the squares. This will not yield the same result as $(\Sigma X)^2$, which directs us to sum the numbers first and then square their total.

How do we know this computational formula will give us the same answer as the conceptual formula? Here is the proof.

Recall that

$$\mu = \frac{\Sigma X}{N}$$

$$\sigma^2 = \frac{\Sigma(X - \mu)^2}{N} = \frac{\Sigma\left[X - \frac{(\Sigma X)}{N}\right]^2}{N}$$

$$= \frac{\Sigma\left[X^2 - \frac{2X\Sigma X}{N} + \frac{(\Sigma X)^2}{N^2}\right]}{N}$$

$$= \frac{\Sigma X^2 - \frac{2(\Sigma X)^2}{N} + \frac{(\Sigma X)^2}{N}}{N}$$

Therefore,

$$\sigma^2 = \frac{\Sigma X^2 - \frac{(\Sigma X)^2}{N}}{N}$$

Problem   The numbers of days all the patients currently in St. Luke Hospital's north wing have been confined are 6, 4, 5, 3, and 4 days. Compute the standard deviation of the lengths of confinement using the deviation method and the computational method.

Solution   First, note that the five observations are a population because they include all patients currently in the north wing. The formulas for populations, 4-5 and 4-7, will therefore be used.

Method Using Deviations from the Mean ($\mu = 22/5 = 4.4$)		
X	$X - \mu$	$(X - \mu)^2$
6	1.6	2.56
4	−0.4	0.16
5	0.6	0.36
3	−1.4	1.96
4	−0.4	0.16
22	0.0	5.20

$$\sigma^2 = \frac{\Sigma(X - \mu)^2}{N}$$

$$\sigma^2 = \frac{5.20}{5}$$

$$\sigma^2 = 1.04$$

$$\sigma = \sqrt{1.04} = 1.0198 \text{ days}$$

Method Using Raw Data Only	
X	X^2
6	36
4	16
5	25
3	9
4	16
22	102

$$\sigma^2 = \frac{\Sigma X^2 - \dfrac{(\Sigma X)^2}{N}}{N}$$

$$\sigma^2 = \frac{102 - \dfrac{(22)^2}{5}}{5}$$

$$\sigma^2 = \frac{102 - 96.8}{5} = 1.04$$

$$\sigma = \sqrt{1.04} = 1.0198 \text{ days}$$

Because the variance and the standard deviation use every observation in their computation, they are more reliable measures of spread than the range. Two or more variances (or standard deviations) can be used to compare the variability around their respective means. If the observations are clustered close to the mean, the standard deviation will be small. As the observations become more dispersed from the mean, the variance and the standard deviation become larger.

It should be noted that, if the raw data have a few extreme values, the variance and the standard deviation may not be representative of the dispersion in the data. In such cases, the deviations from the mean will be large, and squaring these deviations will result in an unusually large variance and standard deviation. The standard deviation is especially valuable in sampling theory and statistical inference, topics that will be discussed starting with Chapter 9. The variance will be used extensively in Chapter 12.

Problems   Consider the following sets of data as populations. Determine the variance and the standard deviation.

a.   3, 5, 4
b.   2, 5
c.   4, 2, 1, 3

REAL STAT

Where did the .400 hitters go? Ted Williams batted .406 in 1941, and there has not been a .400 hitter since. The highest batting average from 1942 until 1996 was .394 by Tony Gwynn of the San Diego Padres. However, that was in a strike-shortened 1994 season. The mean batting average for all players has remained almost constant at .260 for almost 100 years, but the standard deviation of the average has declined from .049 to .031. This indicates there is less dispersion in the batting averages today and helps explain why there are fewer .400 hitters in recent times.

Solutions

Variance	Standard Deviation
a. 0.67	0.82
b. 2.25	1.50
c. 1.25	1.12

Self-Review 4–4   Six families live on Merrimac Circle. The numbers of children in each family are 1, 2, 3, 5, 3, and 4.

a. Is this a sample or a population?
b. Apply formula 4–5, which is based on the deviations from the mean, and compute the variance and the standard deviation.
c. As a check, compute the variance and the standard deviation using formula 4–7, which is based on the raw data (the number of children in the family).

Exercises

17. Refer to the data in Exercise 11. Determine the variance and the standard deviation of these population values. Compare these results to those of Exercise 11.

18. Refer to the data in Exercise 12. Determine the variance and the standard deviation of these sample values. Compare these results to those of Exercise 12.

19. Refer to the data in Exercise 8, which reports the yearly amounts paid for child support for a sample of 15 cases. Determine the variance and the standard deviation of these sample values. Compare these results to those of Exercise 13.

20. Refer to the data in Exercise 9, which reports the weekly amounts spent on bakery products for a sample of 20 households. Determine the variance and the standard deviation of these sample values. Compare these results to those of Exercise 14.

21. A study of the time (in minutes) it takes an ambulance to travel from the local fire station to the scene of an accident revealed the following times:

6.1	5.9	4.8	10.2	9.6	6.1

Determine the standard deviation.

22. The amounts of fuel oil (in gallons) consumed for a sample of both newer and older homes is reported below.

Newer homes:	600	590	605	600
	603	610	597	595
Older homes:	810	750	790	800
	850			

a. Determine the variance and the standard deviation for the newer homes.
b. Determine the variance and the standard deviation for the older homes.
c. Compare the dispersions in the consumption of fuel oil for the newer and older homes.

TABLE 4–3	Weekly Amounts Spent on Food by a Group of Newlyweds	
Amount Spent	**Number**	
$40 up to $45	4	
45 up to 50	11	
50 up to 55	20	
55 up to 60	31	
60 up to 65	19	
65 up to 70	11	
70 up to 75	4	

MEASURES OF DISPERSION FOR GROUPED DATA

We estimate the **range** for data in a frequency distribution by finding the difference between the upper and lower class limits. Note that, if there are any open-ended classes in the frequency distribution, the range cannot be computed.

The Range

Problem   Table 4–3 shows the weekly amounts that a sample of young newlyweds spend on food. What is the range?

Solution   The highest value is $75; the lowest is $40. The range is $35 and is found by subtracting $40 from $75.

Self-Review 4–5

A fast-food chain received several complaints about the weight of its new hamburger, "The 125 Gramburger." A check of a sample of hamburgers revealed these weights:

Weight (in grams)	Number
116 up to 119	7
119 up to 122	19
122 up to 125	28
125 up to 128	16
128 up to 131	2

Determine the range for the weights.

The Interquartile Range and the Quartile Deviation

Recall that the **interquartile range** and the **quartile deviation** are based on the distance between the third quartile (Q_3) and the first quartile (Q_1).

$$\text{Interquartile range} = Q_3 - Q_1 \qquad \textbf{4–2}$$

$$\text{Quartile deviation} = \frac{Q_3 - Q_1}{2} \qquad \textbf{4–3}$$

Problem   Table 4–4 reintroduces the familiar frequency distribution of the selling prices of homes in the Rafter J Ranch subdivision. Determine the interquartile range and the quartile deviation of the selling prices.

Solution   The computations for the first and third quartiles are similar to those for the median (which is the second quartile, Q_2), as discussed in Chapter 3. The procedure for locating the first quartile, Q_1, is as follows:

Step 1. Determine how many observations are in the first quartile. Since there are 80 selling prices, one-fourth, or 20, lie in the first quartile.

Step 2. Locate the class in which the first quartile lies. Note that the cumulative column of Table 4–4 contains 15 selling prices in the first two classes and 30 in the first three classes. Clearly, the first quartile is in the third class because 20 is larger than 15, but smaller than 30.

TABLE 4–4	Selling Prices of Homes in the Rafter J Ranch Subdivision		
	Limits	Class Frequency	Cumulative Frequency
	Selling Price	Number of Homes (f)	CF
	$ 80,000 up to $ 85,000	5	5
	85,000 up to 90,000	10	15
first quartile →	90,000 up to 95,000	15	30
	95,000 up to 100,000	26	56
	100,000 up to 105,000	13	69
	105,000 up to 110,000	7	76
	110,000 up to 115,000	4	80
	Total	80	

Step 3. Using the class limits, we have located the first quartile in the $90,000 up to $95,000 class. We must determine how far into that class to go to find the first quartile. We move 5 frequencies into that class, found by $20 - 15$. Since there are 15 selling prices in that class, the first quartile is $5/15$ of the way between $90,000 and $95,000, or $91,666.67. Q_1 is found as follows:

$$Q_1 = \$90,000 + (5/15)(\$5000) = \$91,666.67$$

We can also find the *first quartile*, Q_1, by using the formula

$$Q_1 = L + \frac{\frac{n}{4} - CF}{f}(i) \qquad \text{4-9}$$

where

L is the lower limit of the class containing the first quartile, Q_1. It is in the $90,000 up to $95,000 class.

n is the total number of frequencies. There are 80 selling prices.

CF is the cumulative number of frequencies in all of the classes preceding the class in which the first quartile lies. There are 15 cumulative frequencies in the classes preceding the $90,000 up to $95,000 class.

f is the frequency in the class in which the first quartile falls. There are 15 prices in that class.

i is the width of the class in which the first quartile falls. The width of the $90,000 up to $95,000 class is $5000.

Solving for Q_1 using formula 4–9:

$$Q_1 = L + \frac{\frac{n}{4} - CF}{f}(i)$$

$$= \$90,000 + \frac{\frac{80}{4} - 15}{15}(\$5000)$$

$$= \$91,666.67$$

The formula for the *third quartile*, Q_3, is

$$Q_3 = L + \frac{\frac{3n}{4} - CF}{f}(i) \qquad \text{4-10}$$

The procedure is the same as for the first quartile, except that L, CF, f, and i refer to the values needed for the third quartile.

Three-fourths of 80 is 60. The 60th selling price is in the $100,000 up to $105,000 class. That class contains 13 observations. A total of 56 observations (CF) have accrued prior to the class containing the third quartile. Hence, the 60th observation is 4 observations into the $100,000 up to $105,000 class, found by $60 - 56$. The class interval (i) is $5000. The lower limit of the class containing the third quartile is $100,000. Substituting these values in formula 4–10, we obtain

$$Q_3 = L + \frac{\frac{3n}{4} - CF}{f}(i)$$

$$= \$100,000 + \frac{\frac{3(80)}{4} - 56}{13}(\$5000)$$

$$= \$101,538.46$$

The interquartile range is $9871.79, found by $Q_3 - Q_1 = \$101,538.46 - \$91,666.67$. The quartile deviation is $4935.90, found by $(Q_3 - Q_1)/2$.

Self-Review 4–6

In Self-Review 4–5, a fast-food chain had received complaints about the weight of "The 125-Gramburger." A check of 72 hamburgers had revealed these weights:

Weight (in grams)	Number
116 up to 119	7
119 up to 122	19
122 up to 125	28
125 up to 128	16
128 up to 131	2

a. Determine the first quartile (Q_1).
b. Find the third quartile (Q_3).
c. Determine the interquartile range and the quartile deviation (QD).

Exercises

23. Using the following table on the distribution of per capita state taxes for our 50 states during a recent year, answer the following questions:
   a. Determine the first quartile. Explain what the first quartile indicates.
   b. Determine the third quartile. Explain what the third quartile indicates.
   c. What is the interquartile range?
   d. What is the quartile deviation?

Per Capita Tax	Frequency
$375 up to $450	6
450 up to 525	15
525 up to 600	10
600 up to 675	6
675 up to 750	9
750 up to 825	4
	50

24. A hospital administrator compared the number of emergency admissions on a given Monday with those on the Friday of the same week:

Number of Emergency Admissions	Frequency	
	Monday	Friday
0 up to 5	1	1
5 up to 10	4	4
10 up to 15	15	21
15 up to 20	26	22
20 up to 25	16	13
25 up to 30	7	3
30 up to 35	3	0
	72	64

a. Calculate the quartile deviation for each set of data.

b. Compare the dispersions in the Monday and Friday admissions.

25. The owner of a local movie theater that shows R-rated films tabulated a sample of the ages of customers attending yesterday's showings:

Age	Frequency
15 up to 20	15
20 up to 25	33
25 up to 30	45
30 up to 35	26
35 up to 40	13
40 up to 45	8

a. Compute the first and third quartiles.

b. Compute the quartile deviation.

26. The owner of the Kinzua Inn is studying the occupancy rate at his motel. The inn is near a hunting, fishing, and camping area, so the number of guests is much higher during the summer months. There is a total of 92 nights in the months of June, July, and August. The following frequency distribution shows the number of guests per night. Consider these data to be a sample, not the entire population.

Number of Guests	Frequency
50 up to 60	5
60 up to 70	12
70 up to 80	23
80 up to 90	28
90 up to 100	17
100 up to 110	7

a. Determine the first and third quartiles.

b. Compute the quartile deviation.

The Standard Deviation

If the data are grouped into a frequency distribution, the standard deviation for both a population and a sample may be computed as follows:

$$\sigma = \sqrt{\frac{\Sigma f X^2 - \frac{(\Sigma f X)^2}{N}}{N}}$$

4–11

TABLE 4–5		Calculations Needed for the Standard Deviation			
Selling Price (in thousands of dollars)	**f**	**X**	**fX**	**fX²**
$ 80.0 up to $ 85.0	5	$ 82.50	$ 412.50	$ 34,031.25
85.0 up to 90.0	10	87.50	875.00	76,562.50
90.0 up to 95.0	15	92.50	1387.50	128,343.75
95.0 up to 100.0	26	97.50	2535.00	247,162.50
100.0 up to 105.0	13	102.50	1332.50	136,581.25
105.0 up to 110.0	7	107.50	752.50	80,893.75
110.0 up to 115.0	4	112.50	450.00	50,625.00
	80		$7745.00	$754,200.00

$$s = \sqrt{\dfrac{\Sigma fX^2 - \dfrac{(\Sigma fX)^2}{n}}{n-1}}$$

4–12

where X represents the midpoint of each class and f the number of observations in each class. As usual, n is the sample size and N the population size.

Problem Compute the standard deviation and the variance for the distribution of selling prices of the homes in the Rafter J Ranch subdivision. Assume this information is sample data.

Solution The data and the needed calculations are shown in Table 4–5. Using formula 4–12 to compute the standard deviation of the sample of 80 selling prices (in thousands of dollars), we find

$$s = \sqrt{\dfrac{\Sigma fX^2 - \dfrac{(\Sigma fX)^2}{n}}{n-1}}$$

$$= \sqrt{\dfrac{754,200 - \dfrac{(7745)^2}{80}}{80-1}} = 7.45$$

The sample variance, s^2, is 55.5025, found by $(7.45)^2$.

Self-Review 4–7

The hourly wages, excluding tips, of a sample of restaurant employees are

Hourly Wages	Number
$ 1 up to $ 4	4
4 up to 7	10
7 up to 10	20
10 up to 13	11
13 up to 16	5

a. Determine the sample standard deviation.
b. Compute the sample variance.

Exercises

27. The per capita taxes for the 50 states are repeated from Exercise 23.

Per Capita Tax	Frequency
$375 up to $450	6
450 up to 525	15
525 up to 600	10
600 up to 675	6
675 up to 750	9
750 up to 825	4
	50

a. Is this a sample or a population?
b. Compute the standard deviation.
c. Determine the variance.

28. The numbers of Monday and Friday emergency hospital admissions are repeated below from Exercise 24.

Number of Emergency Admissions	Frequency	
	Monday	Friday
0 up to 5	1	1
5 up to 10	4	4
10 up to 15	15	21
15 up to 20	26	22
20 up to 25	16	13
25 up to 30	7	3
30 up to 35	3	0
	72	64

a. Calculate both the standard deviation and the variance for each set of data.
b. Compare the dispersion in the Monday and Friday emergency admissions.

29. The ages of a sample of customers at a local theater are repeated below from Exercise 25.

Age	Frequency
15 up to 20	15
20 up to 25	33
25 up to 30	45
30 up to 35	26
35 up to 40	13
40 up to 45	8

Compute the standard deviation.

30. The occupancy rates of the Kinzua Inn based on a sample are repeated below from Exercise 26.

Guests	Frequency
50 up to 60	5
60 up to 70	12
70 up to 80	23
80 up to 90	28
90 up to 100	17
100 up to 110	7

Calculate the sample standard deviation.

INTERPRETING AND USING THE STANDARD DEVIATION

The standard deviation is commonly used to measure the spread in two or more sets of observations. For example, the mean amount of tips earned by waiters in the Chicago area is $120 per week, with a standard deviation of $25. The mean amount of tips earned by waiters in the Milwaukee area is also $120, but the standard deviation is $15. Because the standard deviation is smaller for the Milwaukee waiters, we conclude that there is less spread or variation in the amount earned from tips in Milwaukee than in Chicago. We could also say that the mean of $120 is more typical of the amounts earned by the Milwaukee waiters than the Chicago waiters.

A small standard deviation indicates that the values are close to the mean and that the mean is representative of the values. A large standard deviation indicates that the data are spread out. The Russian mathematician P. L. Chebyshev (1821–1894) developed a theorem that allows us to determine the minimum number of observations within a specified number of standard deviations of the mean. To explain, based on **Chebyshev's theorem**, we can report the percentage of the observations within, say, 2 (or any number greater than 1) standard deviations of the mean. This theorem holds for any set of data, regardless of the shape. Thus, we can expect that at least 75% of the observations will be within 2 standard deviations of the mean. How is this determined? We let $k = 2$, so

Chebyshev's theorem For any set of observations (sample or population), the minimum proportion of the values that lie within k standard deviations of the mean is at least

$$1 - \frac{1}{k^2}$$

where k is any constant greater than 1.

$$1 - \frac{1}{k^2} = 1 - \frac{1}{2^2} = 1 - \frac{1}{4} = 0.75$$

Similarly, we would expect at least 89% of the observations to be within 3 standard deviations of the mean.

Problem   In the previous Problem-Solution, we found that the standard deviation of the selling price of homes in the Rafter J Ranch subdivision is 7.45 thousand dollars, or $7450. Earlier in Chapter 3 we found that the estimated mean selling price for data grouped into a frequency distribution is $96,812.50. What percentage of the selling prices will be within 2.5 standard deviations of the mean? What selling prices will be within these limits?

Solution   Chebyshev's theorem will allow us to determine the percentage of the observations within 2.5 standard deviations of the mean.

$$1 - \frac{1}{2.5^2} = 1 - \frac{1}{6.25} = 0.84$$

So we conclude that 84% of the selling prices are within 2.5 standard deviations of the mean. The actual range of selling price is from $78,187.50 to $115,437.50. These limits are computed as follows.

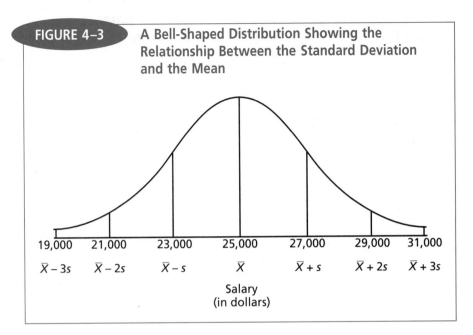

FIGURE 4–3   A Bell-Shaped Distribution Showing the Relationship Between the Standard Deviation and the Mean

Lower limit: $\$96{,}812.50 - 2.5(\$7450) = \$96{,}812.50 - \$18{,}625.00$

$$= \$78{,}187.50$$

Upper limit: $\$96{,}812.50 + 2.5(\$7450) = \$96{,}812.50 + \$18{,}625.00$

$$= \$115{,}437.50$$

Therefore, we expect at least 84% of the selling prices to be between $78,187.50 and $115,437.50.

Chebyshev's theorem is concerned with any set of values; that is, the distribution of values can have any shape. However, for a bell-shaped distribution, such as the one shown in Figure 4–3, we can be more precise in explaining the dispersion about the mean. These relationships involving the standard deviation and the mean are included in the **Empirical Rule,** sometimes called the **Normal Rule.**

Thus, if a study of students graduating in the field of law enforcement showed that the mean starting salary is $25,000, the standard deviation is $2000, and the distribution of salaries is approximately bell-shaped, we would expect that about 95% of the starting salaries would be between $21,000 and $29,000, found by $\$25{,}000 \pm 2(\$2000) = \$25{,}000 \pm \$4{,}000$. Virtually all the salaries would be between $19,000 and $31,000.

> **empirical rule** For a bell-shaped frequency distribution, approximately 68% of the observations will be within plus or minus 1 standard deviation of the mean; about 95% of the observations will be within plus or minus 2 standard deviations of the mean; and practically all will be within plus or minus 3 standard deviations of the mean.

Self-Review 4–8

A recent study of surgical patients that were admitted to the hospital after surgery showed that the mean cost of their hospital stay was $3000, with a standard deviation of $900.

a. At least what percentage of the costs were within 2.25 standard deviations of the mean?

b. What percentage of the costs were between $1650 and $4350?

c. Assume the distribution is bell-shaped. About what percentage of the patient costs were within 3 standard deviations of the mean?

- -

Exercises

31. According to Chebyshev's theorem, at least what percentage of any set of observations will be within 1.8 standard deviations of the mean?

32. The mean of a set of observations is 500, and the standard deviation is 50. According to Chebyshev's theorem, what percentage of the observations will be between 425 and 575?

33. The distribution of the weights of a sample of 1400 cargo containers is bell-shaped, with a mean of 14,000 pounds and a standard deviation of 200 pounds.

    a. About what percentage of the cargo contain-

ers will weigh between 13,600 and 14,400 pounds?

    b. What percentage will weigh more than 14,400 pounds or less than 13,600 pounds?

34. The distribution of amounts spent per month for rent by students attending Computer University is bell-shaped. The mean monthly rental is $450, and the standard deviation $125.

    a. About what percentage of the rentals are between $75 and $825?

    b. What percentage of the rentals are for more than $825 or less than $75?

- -

BOX PLOTS

Previous sections discussed numerical measures of dispersion. For example, we found that the standard deviation of the selling prices of homes in the Rafter J Ranch subdivision was $7.45 (in thousands of dollars). In this section, we turn to a graphical display called a "box plot." It is similar to the stem-and-leaf displays and frequency polygons described in Chapter 3.

A **box plot** is a graphical display that gives us information about the location of various points in a set of data as well as the overall shape of the distribution. It is based on five values in the data set: the minimum value, the maximum value, the median, the first quartile, and the third quartile. An example will help to describe its major features.

Problem   Alexander's Pizza offers free delivery of its pizza within 15 miles of the restaurant. Alex, the owner, wants some information on the time it takes for deliveries. How long does a typical delivery take? Within what range of times will most deliveries fall? For a sample of 20 deliveries, he determined the following information:

$$\text{Minimum value} = 13 \text{ minutes}$$

$$Q_1 = 15 \text{ minutes}$$

$$\text{Median} = 18 \text{ minutes}$$

$$Q_3 = 22 \text{ minutes}$$

$$\text{Maximum value} = 30 \text{ minutes}$$

Develop a box plot for the delivery times. What conclusions can Alex make about the delivery times?

Solution   The first step in drawing a box plot is to create an appropriate scale along the horizontal axis. Next, we draw a box, or rectangle, that starts at Q_1 (15 minutes) and ends at Q_3 (22 minutes). Inside the box, we draw a vertical line that represents the median (18 minutes). Finally, we extend lines from the ends of the box to the minimum value (13 minutes) and the maximum value (30 minutes). See the following diagram:

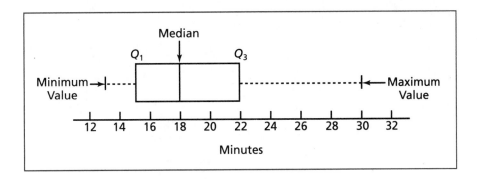

The box plot quickly reveals that most (the middle 50%) of the deliveries take between 15 and 22 minutes. The difference between Q_1 and Q_3 is 7 minutes. This is the interquartile range.

From the box plot, we can also determine that the distribution is positively skewed. How do we know this? The dashed line from 22 minutes (Q_3) to the maximum delivery time of 30 minutes is longer than the dashed line from 15 minutes (Q_1) to the minimum delivery time of 13 minutes. To put it another way, the 25% of the delivery times that are larger than the third quartile are more spread out than the 25 percent of the delivery times that are less than the first quartile. Another indication that the distribution is positively skewed is that the median is not in the center of the box. The distance from the first quartile to the median is shorter than the distance from the median to the third quartile. We know that 25 percent of the delivery times are between 15 and 18 minutes and that another 25 percent are between 18 and 22 minutes. The box plot makes it easy to visualize these relationships.

Box plots placed next to each other are also useful when comparing data sets. Listed below are the ages at which actors and actresses won their Oscars for the last 30 years.

Actor:	32	51	33	61	35	45	55	39	76	37
	42	40	32	60	38	56	48	48	40	43
	62	43	42	44	41	56	39	46	31	47
Actress:	80	26	41	21	61	38	49	33	74	30
	33	41	31	35	41	42	37	26	34	34
	35	26	61	60	34	24	30	37	31	27

We use MINITAB to develop the box plots. The first step is to enter the ages of the actors in one column and the ages of the actresses in another and to name the columns. The mouse commands are

Graphs ▶ Character Graphs ▶ Box Plot

Within the dialog box, we select 'Actor' as the variable and then click on **OK**. We use the same mouse commands and, then select 'Actress' within the dialog box to output the box plot. The output will appear as follows:

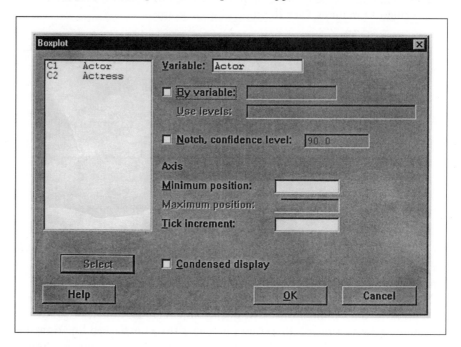

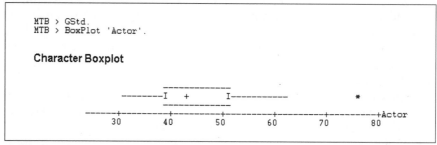

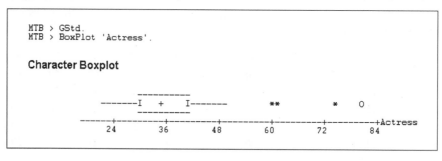

What can we conclude from these box plots? The actresses tend to be younger than the actors. The median age of the actors at the time they won their Oscars is about 43 years. It is about 35 years for the actresses. About 50% of the actors won their awards between ages 39 and 51. For the actresses, about 50% won their awards between ages 30 and 42. Both distributions show a positive skew because for both distributions the tail above Q_3 is larger than the tail below Q_1.

The "*" shows an outlier and the "o" an extreme outlier. An **outlier** is a value that is much larger or smaller than the others. For a box plot, an outlier is an observation that is more than 1.5 times the interquartile range above Q_3 or below Q_1. For the actors, the interquartile range is 12 years, found by 51 − 39. An observation that is more than 18 years above the third quartile or below the first quartile is an outlier. Thus, any value that is more than 69, found by 51, which is Q_3, plus 18, which is 1.5 times the interquartile range, is considered an outlier. For the actors, 1 individual received the award at age 76, which is shown as "*" on the box plot.

An extreme outlier is an observation that is more than 3 times the interquartile range above Q_3 or below Q_1. For the actresses, the interquartile range is also 12 years, so any observation that is more than 36 years above the third quartile is an extreme outlier. An extreme outlier is any observation that is above 78 years old, found by 42, which is Q_3, plus 36, which is 3 times the interquartile range. This point is shown as "o" on the box plot. One actress won her Oscar at age 80, which is an extreme outlier. Outliers are usually from a different population. For example, perhaps the 80-year-old actress won the award because of longevity in the industry, not because of a particular performance.

Self-Review 4–9

The following box plot is given:

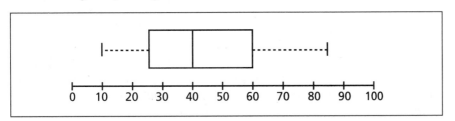

Identify the median, the first and third quartiles, and the largest and smallest values. Is the distribution symmetrical or skewed?

Exercises

35. In a study of the mileage efficiency of automobiles manufactured in 1995, the mean number of miles per gallon of gasoline was 27.5, and the median was 26.8. The smallest value in the study was 12.7 miles per gallon, and the largest was 50.2 miles per gallon. The first and third quartiles were 17.95 and 35.45 miles per gallon, respectively. Develop a box plot and comment on the shape of the distribution of miles per gallon.

36. A sample of 28 private hospitals in the state of Florida revealed the following daily charges for a semiprivate room. For convenience, the data are ordered from smallest to largest. Construct a box plot to represent the data. Comment on the distribution.

$116	$121	$157	$192	$207
209	209	229	232	236
236	239	243	246	260
264	276	281	283	289
296	307	309	312	317
324	341	353		

RELATIVE DISPERSION

A *direct* comparison of two or more measures of dispersion may result in an incorrect conclusion regarding the variation in the data. Also, if one distribution, for example, were in dollars and the other in meters, clearly it would be impossible to compare them. Therefore, it is preferable to use a *relative* measure of dispersion when either

1. the means of the distributions being compared are far apart or
2. the data are in different units.

> coefficient of variation
> The standard deviation divided by the mean.

One such relative measure of dispersion is known as the **coefficient of variation,** abbreviated *CV*.

In order to express the coefficient of variation as a percentage, we multiply $s/\overline{X}$ (for a sample) by 100.

$$CV = \frac{s}{\overline{X}}(100)$$

4–13

Problem   The mean income of a sample of homeowners in Precinct 12 is $40,000, and the standard deviation is $4000. In Precinct 9, the mean income of a sample of homeowners is $24,000, and the standard deviation is $2400. Note that the means are far apart and the standard deviations are considerably different. Compare and interpret the relative dispersions in the two groups of incomes.

Solution   The first impulse is to say there is more dispersion in the incomes in Precinct 12 because the standard deviation associated with that group ($4000) is greater than the $2400 for the Precinct 9 incomes. However, converting the two sets of measurement to relative terms and using the coefficient of variation, we would find that the relative dispersions are the same! Here are the calculations for the coefficients of variation:

Precinct 12

$$CV = \frac{s}{\overline{X}}(100) = \frac{\$4000}{\$40,000}(100) = 10\%$$

Precinct 9

$$CV = \frac{s}{\overline{X}}(100) = \frac{\$2400}{\$24,000}(100) = 10\%$$

Interpreting the results: The variation is ten percent of the mean income in both precincts.

Problem To illustrate the use of the coefficient of variation when two or more distributions are in different units, we will now compare the dispersion in the ages of the homeowners in Precinct 12 with the dispersion in their incomes. The coefficient of variation allows each unlike set of data to be converted to a common denominator (a percentage).

The mean age of the sample of homeowners is 40 years, and the standard deviation is 10 years. Recall that for their incomes $\overline{X} = \$40,000$ and the standard deviation is $4000. Compare the dispersion in their ages and incomes.

Solution

$$\text{Income: } CV = \frac{s}{\overline{X}}(100) = \frac{\$4000}{\$40,000}(100) = 10\%$$

$$\text{Age: } CV = \frac{s}{\overline{X}}(100) = \frac{10}{40}(100) = 25\%$$

There is greater relative dispersion in the ages of the homeowners in Precinct 12 than in their incomes (because 25% is greater than 10%).

Problems Determine the coefficient of variation for each mean and standard deviation.

a. $\overline{X} = 25, s = 5$
b. $\overline{X} = 8, s = 2$
c. $\overline{X} = 50, s = 2$
d. $\overline{X} = 1650, s = 148.5$

Solutions

a. 20%
b. 25%
c. 4%
d. 9%

A sample of Harber High School seniors has recently completed two aptitude tests, one dealing with mechanical aptitude, the other with aptitude for social work. The results are as follows:

Mechanical aptitude: mean 200, standard deviation 30.
Aptitude for social work: mean 500, standard deviation 40.

Compare the relative dispersions in the test results.

Exercises

37. An investor is considering the purchase of 1 of 2 stocks. The yield of Venture Electronics has averaged $105 per share over the past 10 years, with a standard deviation of $15 per share. Aerospace Ltd. has yielded an average of $330 per share during the same period, with a standard deviation of $40. Compare the relative dispersions of the 2 stocks.

38. A recent study of Ohio College of Business faculty revealed that the arithmetic-mean salary for 9 months is $41,000 and the standard deviation of the sample is $4000. The study also showed that the faculty had been employed an average (arithmetic mean) of 15 years, with a standard deviation of 4 years. How does the relative dispersion in the distribution of salaries compare with that of the lengths of service?

39. Radio commercials on the local rock station WNAE average 35 seconds, with a standard deviation of 8 seconds. WLQR, the local easy-listening station, has a mean commercial length of 30 seconds, with a standard deviation of 5 seconds. Compare the relative dispersions of the two stations.

40. A large insurance company offers both homeowner and auto coverage. A study of last year's claims showed that the mean claim settlement for homeowner claims was $1260, with a standard deviation of $425. The mean claim for an auto policy was $875, with a standard deviation of $300. Compare the relative distributions of the two types of claims.

THE COEFFICIENT OF SKEWNESS

An average describes the central tendency of a set of observations, while a measure of dispersion describes the variation in the data. The degree of **skewness** in a distribution can also be measured. Skewness shows the lack of symmetry in a set of data. Recall from Chapter 3 that, if the mean, the median, and the mode are equal, there is no skewness (see Figure 4–4). If the mean is larger than the median and the mode, the distribution is said to have **positive**

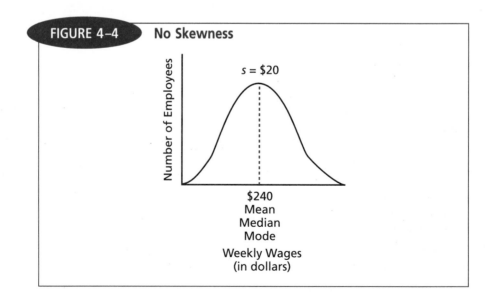

FIGURE 4–4   No Skewness

$s = \$20$

$240
Mean
Median
Mode

Weekly Wages
(in dollars)

Number of Employees

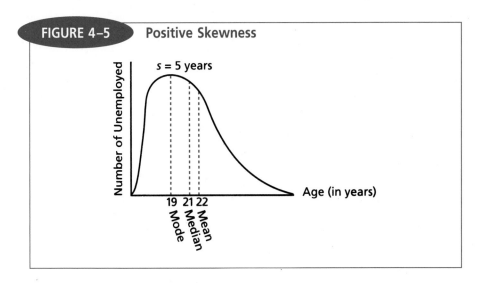

FIGURE 4–5   **Positive Skewness**

skewness (see Figure 4–5). If the mean is the smallest of the three averages, the distribution is **negatively skewed** (see Figure 4–6).

The degree of skewness is measured by the **coefficient of skewness,** abbreviated *sk*. It is found as follows:

$$sk = \frac{3(\text{Mean} - \text{Median})}{\text{Standard deviation}}$$

4–14

Usually, its value ranges from -3 to $+3$.

Problem   What is the coefficient of skewness for the distribution of ages in Figure 4–5?

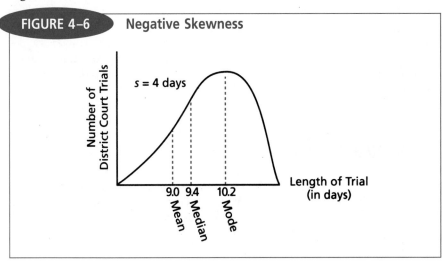

FIGURE 4–6   **Negative Skewness**

Solution   Using formula 4–14:

$$sk = \frac{3(\text{Mean} - \text{Median})}{\text{Standard deviation}} = \frac{3(22 - 21)}{5}$$

$$= +0.6$$

This indicates that there is a slight positive skewness in the age distribution.

Problems   Determine the coefficient of skewness for each of the following sets of data:

a. $\overline{X} = 16$, median = 15, mode = 13, $s = 2$
b. $\overline{X} = \$800$, median = $820, mode = $830, $s = \$50$
c. $\overline{X} = 70$, median = 70, mode = 70, $s = 8$

Solutions

a.   1.5
b.   −1.2
c.   0

Self-Review 4–11

a. Determine the coefficient of skewness for Figure 4–6.
b. Determine the coefficient of skewness for Figure 4–4.
c. Interpret the two coefficients.

Exercises

41. The Flightdeck computed the arithmetic mean dinner check for 2 persons to be $54.00 and the median $50.50. The standard deviation is $3.75. What is the coefficient of skewness? Describe the skewness.

42. The research director of a large oil company conducted a study of consumer buying habits with respect to the amount of gasoline purchased at self-service pumps. The arithmetic mean amount was 11.5 gallons and the median amount was 11.95 gallons. The standard deviation of the sample was 4.5 gallons. Is this distribution neg- atively skewed, positively skewed, or symmetri- cal? What is the coefficient of skewness?

43. At Lander Community College, the mean age of the students is 19.2 years, with a standard de- viation of 1.2 years. The median age is 18.6 years. Compute the coefficient of skewness. De- scribe the skewness.

44. A survey of students at Willard University re- vealed that the mean checking account balance was $376, with a standard deviation of $120. The median balance was $406. Compute the co- efficient of skewness. Describe the skewness.

SOFTWARE EXAMPLE

The work of tallying raw data into a frequency distribution (Chapter 2), com- puting measures of central tendency (Chapter 3), and finding measures of dispersion (this chapter) can be accomplished quickly by a computer. As noted previously, a number of computer packages are used extensively in the social sciences, business, and education. Computer packages are generally easy to use and will save a great deal of time.

The MINITAB system was used to generate the following frequency distribution for the Rafter J Ranch data. (See Table 2–3.) The steps for the mouse operation are

Graphics ▶ Character Graph ▶ Histogram

In the dialog box, select 'Price' as the variable, 82 as the first midpoint, 112 as the last midpoint, and an interval width of 5, and then click **OK.** The dialog box appears as follows:

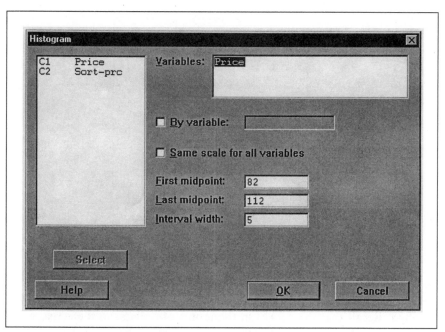

The output follows:

```
MTB > GStd.
MTB > Histogram 'Price';
SUBC>   Start 82 112;
SUBC>   Increment 5.
```

Character Histogram

```
Histogram of Price    N = 80

Midpoint     Count
   82.00        5    *****
   87.00       10    **********
   92.00       15    ***************
   97.00       26    **************************
  102.00       13    *************
  107.00        7    *******
  112.00        4    ****
```

The Excel system was used to compute the following summary measures. From the Excel menu bar, select

Tools ▶ Data Analysis ▶ Descriptive Statistics

Then click **OK.** For the input range, type "a1:a80" (assuming data is in the first column) and select Summary Statistics; then click **OK.** The output is

Column1	
Mean	96.4875
Standard Error	0.844377
Median	97
Mode	98
Standard Deviation	7.552338
Sample Variance	57.03782
Kurtosis	-0.23465
Skewness	0.074977
Range	33
Minimum	81
Maximum	114
Sum	7719
Count	80

From the output, we see that the mean selling price of the 80 homes is 96.4875. Recall that the data is in thousands of dollars, so the actual mean selling price is $96,487.50. The median selling price is $97,000. The standard deviation is $7552. The range of the selling prices is $33,000; the minimum selling price is $81,000, and the maximum is $114,000.

On page 118 and in Table 4–5, we computed the standard deviation of the selling prices of these 80 homes to be 7.45. In the Excel output, we found the standard deviation to be 7.55. With both results reported in thousands of dollars, why the difference? The Excel output works with the raw data, whereas the standard deviation on page 118 is determined from grouped data. The two values are usually quite close, but seldom exactly equal. When we group data, some accuracy is usually lost, and we should consider the values determined from grouped data as estimates of the actual values.

CHAPTER OUTLINE

I. Dispersion indicates the variation, or spread, in a set of data.

II. The range is the difference between the largest and the smallest values in a set of observations.

A. The range is found using the following formula:

$$\text{Range} = \text{Highest value} - \text{Lowest value} \qquad \boxed{\textbf{4-1}}$$

B. It is easy to compute and easy to understand.

C. Only two observations are used in its calculation.

D. It can be distorted by large or small values.

III. The mean deviation is the mean of the absolute differences between each observation and the mean.

A. It is computed by the following formula:

$$MD = \frac{\Sigma |X - \overline{X}|}{n} \qquad \boxed{\textbf{4-4}}$$

B. All values are used in its calculation.

C. It is not influenced by an extreme value.

IV. The variance is the mean of the squared deviations between the mean and each value.

A. The population variance is computed using

$$\sigma^2 = \frac{\Sigma X^2 - \dfrac{(\Sigma X)^2}{N}}{N} \qquad \boxed{\textbf{4-7}}$$

B. The sample variance is computed using

$$s^2 = \frac{\Sigma X^2 - \dfrac{(\Sigma X)^2}{n}}{n - 1} \qquad \boxed{\textbf{4-8}}$$

C. All the values in the sample or the population are used in the calculation.

D. It is not unduly influenced by large or small values.

E. The units are cumbersome to work with because they are the original units squared.

V. The standard deviation is the square root of the variance.

A. It is the most widely used and widely reported measure of dispersion.

B. It is in the same units as the original data.

C. All the values are used in its calculation.

VI. Chebyshev's theorem states that, regardless of the shape of the distribution, at least $1 - (1/k^2)$ of the observations are within k standard deviations of the mean.

VII. The Empirical Rule assumes that the distribution of values is bell-shaped.

A. If the distribution is bell-shaped, then approximately 68% of the observations will be within plus or minus 1 standard deviation of the mean; about 95% of the observations will be within plus or minus 2 standard deviations of the mean; and practically all will be within plus or minus 3 standard deviations of the mean.

B. It is also called the Normal Rule.

VIII. The interquartile range is the difference between the third and the first quartile.

$$\text{Interquartile range} = Q_3 - Q_1 \qquad \boxed{\textbf{4-2}}$$

A. The quartile deviation is the interquartile range divided by 2.

    1. Both the interquartile range and the quartile deviation use the middle 50% of the observations.

    2. They are not influenced by extreme values.

B. The first quartile is the value below which 25% of the observations occur.

$$Q_1 = L + \frac{\dfrac{n}{4} - CF}{f}(i) \qquad \boxed{\textbf{4-9}}$$

C. The third quartile is the value below which 75% of the observations occur.

$$Q_3 = L + \frac{\dfrac{3n}{4} - CF}{f}(i) \qquad \boxed{\textbf{4-10}}$$

IX. The coefficient of variation is a measure of relative dispersion.
  A. It is computed by dividing the standard deviation by the mean.

$$CV = \frac{s}{\overline{X}}(100) \qquad \boxed{4\text{--}13}$$

  B. It reflects the variation in a set of observations relative to the mean.
    1. It is useful when comparing two distributions where the magnitudes of the means are very different.
    2. It is also used to compare distributions in different units.

X. The coefficient of skewness is a measure of the symmetry of a set of observations.
  A. It is computed using the following formula:

$$sk = \frac{3(\text{Mean} - \text{Median})}{\text{Standard deviation}} \qquad \boxed{4\text{--}14}$$

  B. It may range from -3.00 up to 3.00.
  C. A value of 0 indicates no skewness, and the distribution is referred to as symmetric.
  D. If the long tail of the distribution is to the right, it is positively skewed; if the long tail is to the left, it is negatively skewed.

- -

Exercises

45. The ages for a sample of 10 women who have just given birth to their first child are 18, 25, 31, 19, 22, 21, 19, 25, 16, and 27. Compute each of the following descriptive measures:
  a. The range
  b. The average deviation
  c. The variance
  d. The standard deviation
  e. The coefficient of skewness

46. The numbers of yards gained rushing by the Red Hawks football team during the first 7 games of the season are 210, 203, 162, 134, 390, 184, and 211. These data constitute a population. Compute the following descriptive measures:
  a. The range (Interpret it.)
  b. The average deviation (Interpret it.)
  c. The variance
  d. The standard deviation
  e. The coefficient of skewness (Interpret it.)

47. The percentages of minorities enrolled in 12 New Orleans high schools are 32, 62, 58, 26, 61, 34, 46, 27, 56, 61, 64, and 24. Consider the data to be a population and find the following descriptive measures:
  a. The range (Interpret it.)
  b. The average deviation (Interpret it.)
  c. The variance
  d. The standard deviation
  e. The coefficient of skewness (Interpret it.)

48. The Parry-Mutual Insurance Co. is studying the number of group insurance claims submitted each week during the last year. Company records disclosed the following information. Consider these data a population.

Number of Claims per Week	Number of Weeks
0 up to 5	6
5 up to 10	15
10 up to 15	19
15 up to 20	9
20 up to 25	3
	52

  a. Compute the mean.
  b. Compute the median.
  c. Compute the standard deviation.
  d. Compute the quartile deviation.

49. The United States was divided into 25 statistical regions and the percentage of households in which a female is the head of the household was determined for each region. These data were organized into the following frequency distribution:

Percentage of Households	Frequency
5% up to 10%	5
10 up to 15	7
15 up to 20	9
20 up to 25	3
25 up to 30	1

Consider these data a population and determine the following descriptive measures:

a. The range (Interpret it.)
b. The variance
c. The standard deviation
d. The quartile deviation (Interpret it.)
e. The coefficient of skewness (Interpret it.)

50. Two hundred veterans with physical disabilities are enrolled in a physical fitness program. As an initial test of strength, everyone has been asked to lift a weight. The maximum weight each person lifted has been recorded and organized into the following frequency distribution. Consider this data a sample.

Weight (in pounds)	Number of Persons
0 up to 5	10
5 up to 10	37
10 up to 15	50
15 up to 20	68
20 up to 25	30
25 up to 30	5

a. Determine the range. Interpret it.
b. Find the interquartile range and the quartile deviation. Interpret them.
c. Find the standard deviation and the variance.

51. A sample of 50 drivers for the Speedy Package Delivery Service showed they made the following numbers of deliveries in the first hour yesterday:

Number of Packages	Frequency
0 up to 10	10
10 up to 20	12
20 up to 30	14
30 up to 40	9
40 up to 50	5
Total	50

a. Determine the first and third quartiles.
b. Compute the quartile deviation.
c. Compute the standard deviation.

52. A sample of 64 commuters who drive their own cars to work were surveyed to determine the number of minutes they listened to the radio in route. The results are

Number of Minutes	Frequency
10 up to 14	6
14 up to 18	10
18 up to 22	24
22 up to 26	16
26 up to 30	8
Total	64

a. Determine the first and third quartiles.
b. Compute the quartile deviation.
c. Compute the standard deviation.

The following table shows data on major earthquakes by country from 1983 through 1995. The size of the earthquake is measured on the Richter scale.

Country	Size	Deaths
Colombia	5.5	250
Japan	7.7	81
Turkey	7.1	1300
Chile	7.8	146
Mexico	8.1	4200
Ecuador	7.3	4000
India	6.5	1000
China	7.3	1000
Armenia	6.8	55,000
USA	6.9	62
Peru	6.3	114
Romania	6.5	8
Iran	7.7	40,000
Philippines	7.7	1621
Pakistan	6.8	1200
Turkey	6.2	4000
USA	7.5	1
Indonesia	7.5	2000
India	6.4	9748
Indonesia	7.0	215
Colombia	6.8	1000
Algeria	6.0	164
Japan	7.2	5477
Russia	7.6	2000

53. Refer to the data above on earthquakes. Determine the mean and the standard deviation of the sizes of the earthquakes. Because all the earthquakes since 1983 are reported, consider this information to be a population.

54. Refer again to the data on earthquakes. Determine the mean and the standard deviation of the numbers of deaths caused by the earthquakes.

Because all the earthquakes since 1983 are reported, consider this information to be a population.

55. Refer again to the data on earthquakes and your answers to Exercises 53 and 54. Determine the coefficient of variation for the sizes of the earthquakes and the numbers of deaths. Interpret your findings.

56. Refer again to the data on earthquakes. Draw a box plot for the numbers of deaths caused by the earthquakes. Report such values as the median, the interquartile ranges, and the location of the middle 50% of the earthquakes. Are there any outliers? Comment on the box plot.

57. Listed below, in chronological order, are the ages of the U.S. presidents at the time of their inaugurations. From the information, we note that George Washington was 57 years old when he became president and that Bill Clinton was 47.

57	61	57	57	58	57
61	54	68	51	49	64
50	48	65	52	56	46
54	49	51	47	55	55
54	42	51	56	55	51
54	51	60	62	43	55
56	61	52	69	64	47

Determine the range of the ages. Determine the standard deviation of the ages. Interpret your findings.

58. Refer again to the information on the ages of U.S. presidents in the previous problem. Develop a box plot for the ages. Are there any outliers? Determine the first and third quartiles from the box plot. Interpret your findings.

DATA EXERCISES

59. Refer to the real estate data set at the end of the book, which reports information on the homes sold in Alabama during 1996.
a. What is the standard deviation of the selling prices? Using the mean and the median computed earlier, determine the coefficient of variation and the coefficient of skewness.
b. Develop a box plot for the selling prices. In-

terpret the result. Are there any outliers, or do the data seem to be concentrated around the median?

c. Determine the mean, the median, and the standard deviation of the distances from the center of the city. Determine the coefficient of variation and the coefficient of skewness. Compare these results to those of part a. Interpret.

60. Refer to the school data set at the end of the book, which refers to the 94 school districts in northwest Ohio.

a. What is the standard deviation of the salaries for the teachers? Using the mean and the median computed in Chapter 3, determine the coefficient of variation and the coefficient of skewness.

b. Develop a box plot for the teachers' salaries. Interpret the result. Are there any outliers, or do the data seem to be concentrated around the median?

c. Determine the standard deviation of the percentages on welfare in each of the school districts. Using the mean and the median computed in Chapter 3, determine the coefficient of variation and the coefficient of skewness.

d. Develop a box plot for the percentages on welfare in each school district. Interpret the result. Are there any outliers, or do the data seem to be concentrated around the median?

CHAPTER ACHIEVEMENT TEST

The answers are at the back of the book.

MULTIPLE-CHOICE QUESTIONS

Select the response that best answers each of the questions.

1. The major weakness of the range is that
   a. it does not use all the observations in its calculation.
   b. it can be influenced by an extreme value.
   c. Both a and b
   d. None of the above

2. The major weakness of the mean deviation is that
   a. it is based on only two observations.
   b. it is influenced by a large mode.
   c. it employs absolute values, which are often difficult to use.
   d. None of the above

3. The major strength of the standard deviation is that
   a. it uses all the observations in its calculation.
   b. it is not unduly influenced by extreme values.

   c. Both a and b
   d. None of the above

4. The standard deviation is
   a. the square of the variance.
   b. two times the variance.
   c. half the variance.
   d. the square root of the variance.
   e. None of the above

5. If the original data are measured in pounds, the variance is
   a. also measured in pounds.
   b. measured in pounds squared.
   c. measured in "half" pounds.
   d. None of the above

6. The median is equal to the
   a. quartile deviation.
   b. third quartile minus the first quartile.
   c. square of the standard deviation.
   d. second quartile.
   e. None of the above

7. The coefficient of variation is measured in
   a. the same units as the mean and the standard deviation.
   b. percent.
   c. squared units.
   d. None of the above
8. If the mean is larger than the median, the coefficient of skewness will be
   a. zero.
   b. positive.
   c. negative.
   d. None of the above
9. If the "tail" of a frequency distribution is in the positive direction (to the right), the coefficient of skewness is
   a. zero.
   b. positive.
   c. negative.
   d. None of the above
10. The standard deviation of a frequency distribution is $10, the mean is $250, the median is $250, and the mode is also $250. The coefficient of skewness is
   a. zero.
   b. positive.
   c. negative.
   d. measured in dollars.
   e. None of the above

COMPUTATION PROBLEMS

11. A sample of 5 joggers ran the following distances (in miles) yesterday: 2, 6, 3, 4, and 5.
   a. What is the range? Interpret it.
   b. What is the mean deviation? Interpret it.
   c. What is the variance?
   d. What is the standard deviation?
12. The appraised values for a sample of single-family dwellings in the Jefferson tax district are:

Appraised Value (in thousands of dollars)	Number of Dwellings
$20.0 up to $30.0	8
30.0 up to 40.0	15
40.0 up to 50.0	20
50.0 up to 60.0	50
60.0 up to 70.0	18
70.0 up to 80.0	9

   a. Estimate the range.
   b. Estimate the quartile deviation.
   c. Estimate the variance and the standard deviation.
13. The distribution of the appraised values of the single-family dwellings in the Sanford tax district reveals the following: mean, $87,000; median, $84,000; mode, $78,000; standard deviation, $9000.
   a. Compare the relative dispersion in this distribution with that in the Jefferson district in Problem 12.
   b. Compare the skewness of the two distributions.

ANSWERS TO SELF-REVIEW PROBLEMS

4–1 a. Sunshine: 40, found by 80 − 40
   State: 7, found by 62 − 55
   b. There is more dispersion in speeds on Sunshine because 40 is greater than 7. Speeds on the state highway clustered closer to the mean because 7 is less than 40.

4-2 a. $\overline{X} = \dfrac{3000}{6} = 500$

X	$\lvert X - \overline{X}\rvert$
484	$\lvert -16\rvert$
503	$\lvert\ +3\rvert$
496	$\lvert\ -4\rvert$
510	$\lvert +10\rvert$
491	$\lvert\ -9\rvert$
516	$\lvert +16\rvert$
3000	58

$$MD = \frac{58}{6} = 9.7 \text{ grams}$$

b. On the average, blueberry cakes deviate 9.7 grams from the mean of 500 grams.

c. There is greater variation in the weights of blueberry cakes because $9.7 > 1.6$. Peach cakes clustered closer to the mean weight.

4-3 a. $\overline{X} = 3000/6 = 500$ grams

X	$X - \overline{X}$	$(X - \overline{X})^2$
484	-16	256
503	$+3$	9
496	-4	16
510	$+10$	100
491	-9	81
516	$+16$	256
3000	0	718

$$s^2 = \frac{718}{6 - 1} = 143.6$$

The variance is 143.6.

$$s = \sqrt{143.6} = 11.98$$

The standard deviation is 11.98 grams.

b. There is greater variation in blueberry cakes because the standard deviation of $11.98 > 2.12$.

4-4 a. Because there are only six families and all are being studied, this is a population.

b. $\overline{X} = 18/6 = 3$, $\sigma^2 = 10/6 = 1.6667$, $\sigma = \sqrt{1.6667} = 1.29$ children

c.

X	X^2
1	1
2	4
3	9
5	25
3	9
4	16
18	64

$$\sigma^2 = \frac{64 - \dfrac{(18)^2}{6}}{6} = \frac{64 - 54}{6}$$

$$= 1.6667$$

$$\sigma = \sqrt{1.6667} = 1.29 \text{ children}$$

4-5   The range is 15 grams, found by $131 - 116$.

4-6 a. $Q_1 = 119 + \dfrac{\frac{1}{4}(72) - 7}{19}(3) = 120.7$

b. $Q_3 = 125 + \dfrac{\frac{3}{4}(72) - 54}{16}(3) = 125.0$

c. $Q_3 - Q_1 = 125.0 - 120.7 = 4.3$
$QD = 2.15$, found by $4.3/2$

4-7 a.

Wage	f	X	fX	fX^2
\$ 1 up to \$ 4	4	2.5	10.0	25.0
4 up to 7	10	5.5	55.0	302.5
7 up to 10	20	8.5	170.0	1445.0
10 up to 13	11	11.5	126.5	1454.75
13 up to 16	5	14.5	72.5	1051.25
	50		434.0	4278.50

$$s = \sqrt{\frac{4278.50 - \dfrac{(434.0)^2}{50}}{50 - 1}} = 3.23$$

b. The variance is 10.44, found by $(3.23)^2$.

4-8 a. $1 - \dfrac{1}{(2.25)^2} = 0.8025$

b. $1 - \dfrac{1}{(1.5)^2} = 0.5556$

c. Almost all or 99.7%

4-9   Median $= 40$, $Q_1 = 25$, $Q_3 = 60$, largest value $= 85$, smallest value $= 10$
The distribution is positively skewed.

4-10   Mechanical: $\dfrac{30}{200}(100) = 15\%$

Social work: $\dfrac{40}{500}(100) = 8\%$

There is a larger relative dispersion in mechanical aptitude (15%) compared with social work aptitude (8%).

4-11 a. -0.3;

$$sk = \frac{3(9.0 - 9.4)}{4} = -0.3$$

b. 0:

$$sk = \frac{3(240 - 240)}{20} = 0$$

c. Figure 4–6: slight negative skewness
Figure 4–4: no skewness

Unit Review

In these first four chapters, you have learned the basic vocabulary of statistics and the fundamentals of descriptive statistics. Major concepts covered are listed below and are also defined in the Glossary at the back of the book.

KEY CONCEPTS

1. **Statistics** is the collection, organization, presentation, analysis, and interpretation of data for the purpose of making better decisions.
2. Statistics may be divided into two areas, **descriptive** and **inferential**.
3. Data may be classified into four levels of measurement—**nominal, ordinal, interval,** and **ratio**.
4. A **population** is a collection or set of individuals, objects, or measurements. A **sample** is a part of the population. Calculations made from populations are **parameters,** and those made from samples are **statistics.**
5. A **frequency distribution** is a comprehensive summary of a set of observations. It separates the data into classes and shows the number of occurrences in each class.
6. **Histograms, frequency polygons,** and **stem-and-leaf charts** are graphic displays of frequency distributions.
7. A **less-than cumulative frequency distribution, or ogive,** shows the number of observations that are less than or equal to the upper limit of each class. A **more-than cumulative frequency distribution** shows the number of observations that are greater than or equal to the lower limits.
8. An **average** is a single representative value that is typical of the data considered as a whole. Three averages were discussed: **mean, median,** and **mode.**
9. When a data set contains extremely large or extremely small values, the resulting distribution may be **skewed.** A skewed distribution is one that is not **symmetrical.** The skewness can be measured by the **coefficient of skewness.**
10. The **dispersion** of a set of data is the amount of variability or spread in the data.
11. The **standard deviation** is the most useful and widely used measure of dispersion. It is the positive square root of the **variance.** The variance is the mean of the squared differences between the actual observations and the mean of the data set.
12. The **coefficient of variation** is a measure for comparing the relative variability of two sets of data. It is particularly useful when the data sets to be compared are in different units.

KEY TERMS

This list of terms is included in order for you to verify your recall of the material covered in Chapters 1–4. As you read each term, provide its definition in your own words. Then check your answers against the definitions given both in the chapter and in the Glossary at the back of the book.

Descriptive statistics	Stem-and-leaf chart	Third quartile
Inferential statistics	Frequency polygon	Weighted mean
Population	Cumulative frequency	Parameter
Sample	distribution (ogive)	Statistic

Mutually exclusive	Line chart	Range
Exhaustive	Bar chart	Quartile deviation
Nominal scale	Pie chart	Interquartile range
Ordinal scale	Mean	Mean deviation
Interval scale	Median	Variance
Ratio scale	Mode	Standard deviation
Raw data	Bimodal	Coefficient of variation
Classes	Midpoint	Coefficient of skewness
Class frequency	Skewed	Chebyshev's Theorem
Class midpoint	Symmetric distribution	Box plots
Frequency distribution	Dispersion	Empirical Rule
Histogram	First quartile	

. . . .

KEY SYMBOLS

$\overline{X}$	The mean of a sample	Q_1	The value of the first quartile
n	The number of observations in a sample	QD	The quartile deviation
Σ	The symbol that indicates a group of values are to be added	MD	The mean deviation or, as it is sometimes called, the average deviation
μ	The mean of a population	σ^2	The variance of a population
N	The number of observations in a population	σ	The standard deviation of a population
$\overline{X}_w$	The weighted arithmetic mean	s^2	The variance of a sample
Q_3	The value of the third quartile	s	The standard deviation of a sample

. . . .

CASE STUDY • **Senior Medical Expenses**

At a recent meeting of the Southwest Florida Chapter of the American Association of Retired Persons (AARP), discussion centered on the monthly medical expenses for senior citizens. From the comments made, there seemed to be a disparity in typical expenses between persons in the northern part of the region and persons in the southern portion. In addition, there seemed to be more variation for those in the north than those in the south. The president of the chapter would like to investigate the matter thoroughly and present a report to the entire membership. If it is warranted, she would like to discuss the matter with local government officials as well as representatives of the hospitals in the region.

Two committees were formed to gather the data and present the report at the next meeting. The committee investigating the medical expenses in the northern part of the region selected a sample of 40 seniors and found that they spent the following amounts on medical expenses:

$131	$149	$181	$118	$151
115	150	127	149	107
108	106	112	91	123
136	88	127	80	107
112	108	120	111	104
110	135	109	75	88
116	130	133	83	110
137	95	131	126	141

They also presented the following summary statistics:

Variable	N	Mean	Median	Tr Mean	StDev	SE Mean
Northern	40	118.25	115.50	117.86	22.17	3.51

Variable	Min	Max	Q_1	Q_3
Northern	75.00	181.00	107.00	132.50

The committee investigating the medical expenses in the southern part of the region selected a sample of 50 seniors and found that they spent the following amounts on medical expenses:

$ 95	$119	$ 94	$111	$ 95
86	106	104	104	92
95	94	83	99	102
96	95	102	105	117
104	108	94	105	107
97	106	106	97	99
92	90	106	110	100
101	100	90	81	107
104	94	120	112	85
98	91	116	90	95

They presented the following box plot:

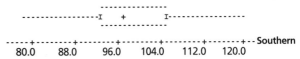

Combine the summary information presented by the two groups. Is there a difference in the typical amounts spent? Is there a difference in the variation in the two groups? What other conclusions can you draw?

CHAPTER 5

An Introduction to Probability

OBJECTIVES

When you have completed this chapter, you will be able to

- define the terms *probability, experiment, outcome,* and *event;*
- describe the classical, empirical, and subjective concepts of probability;
- use the addition, multiplication, and complement rules to calculate probabilities;
- count the number of permutations and combinations.

CHAPTER PROBLEM   Are There Enough Cherry Life Savers?

Johnny Smith has just purchased a roll of Life Savers at The Candy Kounter. As he leaves the store, he peels off the outer wrapping and counts the number of different flavors in his package. Three are cherry, and there are two each of lemon, lime, pineapple, and orange. Two of Johnny's friends, Kerry and Mark, join him outside, and he offers each of them one of his candies. As it turns out, all three prefer cherry, so there are just enough to go around. Did the manufacturer know this was going to happen? Or was it just luck? Can we compute the probability of selecting various candy flavors? What are these probabilities?

INTRODUCTION

Chapters 2, 3, and 4 dealt with descriptive statistics. The emphasis in Chapter 2 was on organizing raw data into a frequency distribution. Recall, for example, the selling prices of homes in the Rafter J Ranch subdivision. When prices were organized into a frequency distribution, the lowest was about $80,000 and the highest $114,000. Furthermore, the largest concentration of selling prices was in the $95,000 up to $100,000 class. In Chapters 3 and 4, measures of central tendency and dispersion were introduced. It was determined, for example, that the average selling price of a Rafter J Ranch home was about $97,000. The main focus in those three chapters was to describe a set of data that had already been observed.

The purpose of this and of subsequent chapters is to examine ways of calculating the probability that some event will occur. For instance, you may be interested in determining the probability that the incumbent governor in your state will be reelected in November or that fewer than 80% of first-time speeding offenders will be fined a second time for this offense. Whatever your purpose, the ability to calculate probability will be a valuable tool in everyday decision making.

Why Study Probability?

As noted in previous chapters, one of the major purposes of the science of statistics is to make it possible to infer, from the results of studies of just one part of a group, general statements about the entire group. As defined in Chapter 1, one portion of the entire group is called a **sample,** while the entire group is called the **population.** The process of reasoning from a set of sample observations to a general conclusion about a population is **statistical inference.** Probability is the "yardstick" used to measure the reasonableness that a particular sample result could also be valid for a certain population. At some point, you may decide that the observed sample result is so improbable that you reject the claim that it applies to a specific population. Hence, in the science of statistics, we "accept" or "reject" statements by calculating how probable or improbable they are.

Probability will help us make decisions with respect to statistical inference. Throughout, we will use case studies like the two that follow to illustrate the potential use of probability in decision making.

CASE 1. Senator Simeon claims that 75% of all tax paying Americans believe additional revision of the income tax structure is necessary. To investigate his contention, 500 tax paying Americans were selected at random and interviewed. Of the 500 sampled, 70%, or 350, said they were in favor of additional income tax reform. Can we challenge the senator's claim as incorrect? Or could his contention still be valid because the difference between his stated percentage (75%) and that of the sample results (70%) is attributable to chance? In a later chapter, we will show how calculating probabilities can help us make this kind of decision.

CASE 2.   Forty percent of the surnames on a very long list of eligible jury members are of Hispanic origin. Each person whose name is on the list has an equal chance of being selected for jury duty. Out of 12 jury members, how many would you expect to have Hispanic surnames? Would you be surprised if no persons with a Hispanic surname were selected for jury duty? You should be! The probability of selecting no Hispanic surnames is $(1 - 0.4)^{12}$, or 0.0022. That is, the chance of selecting no Hispanic surname is only about 2/10 of 1%! In this case, probability is used to show that with a fair method of selection, it would be quite unreasonable to expect no Hispanic-surname jury members to be selected out of 12.

CONCEPTS OF PROBABILITY

All of us have some idea of what a **probability** is—although it may be rather difficult to arrive at a precise definition. The weather forecaster notes that "there is a 30% *chance* of rain." The *likelihood* that humankind will land on Jupiter in the next 4 years is rather remote. Terms such as *chance* and *likelihood* are used interchangeably for the word *probability*.

Three key words are used in the study of probability: *experiment, outcome,* and *event*. While they commonly appear in our language, in statistics these terms have specific meanings.

The statistical definition of **experiment** is more general than the one used in the physical sciences, where we picture researchers manipulating test tubes and microscopes. In statistics, an experiment has two or more possible results, and it is uncertain which will occur.

For example, the tossing of a coin is an experiment. You may observe the coin tossing, but you will be unsure whether it will come up "heads" or "tails." Similarly, asking 500 voters whether or not they intend to vote for a $3.1 million school bond is an experiment. If a coin is tossed, one particular **outcome** might be a "head." Or the outcome might be a "tail." As for the school-bond experiment, one outcome is that 342 people favor its issuance. Another outcome is that only 67 approve of it. Still another outcome is that 203 favor the bond's issue. When one or more of the experiment's outcomes are combined, we call this an **event**.

Consider, for example, the single throw of a die. There are exactly six possible outcomes for this experiment. However, many events may be associated with it—the fact that the number of spots coming face up on the toss is odd, for example, or that the number that comes face up is smaller than 3. With the school bond experiment, the event might be that a majority favor the $3.1 million issue. This event would occur whether the number of people in favor is 251, 252, 253, or any number up to and including 500. Note that an event is not always simply an outcome.

Probabilities may be expressed as fractions (*1/4, 5/9, 7/8*), decimals (*0.250, 0.556, 0.875*), or percentages (*25%, 56%, 88%*). A probability is always between 0 and 1, inclusive. Zero describes the probability of something that cannot happen. If a corn seed is planted, the probability of having an elephant sprout from the seed is 0. At the other extreme, a probability of 1 represents

probability A value between 0 and 1, inclusive, that measures the likelihood that a particular event will occur.

experiment The observation of some activity or the act of taking some measurement.

outcome A particular result of an experiment.

event A collection of one or more possible outcomes of an experiment.

a sure thing. In Portland, Maine, the mean high temperature is sure to be higher in July than in January.

Problems   Indicate whether each of the following statements is true or false:

a.   If a die is rolled, one particular outcome is $\boxed{\cdot}$.
b.   A probability of 1.00 indicates that something cannot happen.
c.   A probability of −100.00 indicates that some event will definitely happen.

Solutions

a.   True
b.   False
c.   False

- -

Self-Review 5–1   Answers to all Self-Review problems are at the end of the chapter.

A politician has recently delivered a major speech. He receives six pieces of mail commenting on it. He is interested in the number of writers who agree with him.

a. What is the experiment?
b. What are the possible outcomes?
c. Describe one possible event that might occur.

- -

Exercises

Answers to the even-numbered Exercises are at the back of the book.

1. Toss a coin. What are the possible outcomes?
2. Toss two coins. What are the possible outcomes?
3. One card is drawn from a well-shuffled, standard 52-card deck. If only the 4 suits (spades, hearts, diamonds, and clubs) are of interest, what are all the possible outcomes? In your deck, hearts and diamonds are red cards, and spades and clubs are black. What is the possibility of selecting a red card called?

4. Three persons are campaigning for mayor. Schwartz is a Democrat, White is a Republican, and Rossi is an Independent. Schwartz and Rossi are men. White is a woman.
a. What is the experiment?
b. Describe possible outcomes with respect to the political party.
c. Describe possible outcomes with respect to the sex of the candidates.

- -

classical concept of probability The number of favorable outcomes divided by the number of possible outcomes.

TYPES OF PROBABILITY

There are three approaches to establishing the probability of events: **classical, empirical,** and **subjective.**

Classical Concept of Probability

The **classical** concept of probability is based on the assumption that several outcomes are *equally likely*.

Problem   Recall the Chapter Problem about the roll of Life Savers purchased by Johnny Smith. If Johnny randomly selects a single candy from the package, what is the probability that it will be cherry?

Solution   Of the 11 candies in the roll, 3 are cherry, so the probability is 3/11, or 0.273, that the life saver Johnny selects will be cherry.

The classical approach is useful when dealing with games of chance, such as dice and card games, and situations in which the outcomes are equally likely to occur. The classical approach is similarly useful when random selection is important—that is, when every outcome has the same chance of occurring. Serious problems develop, however, when the outcomes are not equally likely. If you are an excellent student, for example, there is a good chance you will earn an A in this course. In your case, the outcome A does not have the same likelihood as the outcome F.

Self-Review 5–2

One card is selected at random from a standard deck of 52 cards.

a. What is the probability that the card will be the ace of spades?
b. What is the probability that the card will be an ace?
c. What concept of probability does this problem illustrate?

Empirical Probability Concept

A second approach to probability is based on **empirical** observations. This method uses the frequency of *past* occurrences to develop probabilities for the future.

First, we find the number of times a particular event happened in the past. Then we use this value to determine the likelihood that it will happen again. To illustrate, a mortality table revealed that out of 100,000 men aged 25, 138 die within a year. Based on this experience, a life insurance company would estimate the probability of death for this age as

> **empirical probability concept** The number of times an event occurred in the past divided by the number of observations.

$$P(A) = \frac{138}{100,000} = 0.00138$$

This type of probability is used in actuarial tables to help insurance companies establish the premiums to be charged for various types of life insurance—term, ordinary life, and so on.

In the mortality table example, the probability of an event is denoted by a capital *P*. An abbreviation for the event is then written in parentheses. Commonly, capital letters or numbers are used to denote an event in a concise manner. In this case, $P(A)$ stands for the probability that a 25-year-old male will die during the year.

A political scientist randomly selected 200 eligible voters and determined the number of times they had voted in the last 5 general elections:

Number of Times Voted	Frequency
1	30
2	41
3	40
4	62
5	27
	200

a. What is the probability that a particular voter cast his or her ballot only once in the last 5 elections?
b. What is the probability that a particular voter cast his or her ballot exactly 3 times in the last 5 elections?
c. What concept of probability does this illustrate?

Subjective Probability Concept

subjective probability concept The likelihood of an event assigned on the basis of whatever information is available.

For situations in which there is little or no historical information from which to determine a probability, **subjective** probabilities can be employed. Using this approach, experts given the same information often differ in their estimates of the probability. They differ with respect to the likelihood that another major earthquake will occur in California during this decade, just as they differ in their selection of the team that is most likely to win the National League pennant this coming season. Subjective probability can be thought of as the *probability assigned by an individual or group based on whatever evidence is available.*

a. What probability would you assign to a deep economic recession happening within a year?
b. What concept of probability does this illustrate?

In summary, there are three concepts of probability. The classical viewpoint assumes that the outcomes are equally likely. If the outcomes are not equally likely, the empirical viewpoint is used. If no past experience is available, subjective judgment can be used to assign the probability that an event will occur. Regardless of the viewpoint, the same laws of probability discussed in the following sections will be applied.

Exercises

5. Which concept of probability is used to make each of the following probability statements?

a. The probability of obtaining a 1 when rolling a single die is 1/6.

b. The probability that the high temperature in Tampa, Florida, is lower in January than in September is 0.75.

c. The probability that a nuclear accident will occur at the Davis-Besse plant within the next month is 0.005.

6. Consider the following experiment. Think about flipping 3 coins and focus on the event that exactly 2 "heads" show on the three coins.

a. Before you do this, what is your subjective probability for the event?

b. If the probability of "heads" is 0.5 on each coin, compute the classical probability for this event.

c. Do the experiment 10 times. Count the number of times you get exactly 2 heads. Then find the empirical probability for this event.

PROBABILITY RULES

In applications of probability, often there is a need to combine the probabilities of related events in some meaningful way. In this section, we discuss two of the fundamental methods of combining probabilities—by addition and by multiplication. Generally, events formed using the word *or* are handled by addition; those formed using the word *and* are handled by multiplication.

Addition Rules

THE SPECIAL ADDITION RULE. We apply the **special rule of addition** to calculate the probability of an event consisting of 2 or more mutually exclusive outcomes. Recall from Chapter 1 that *mutually exclusive* means that, when one of the events occurs, none of the other events can occur at the same time. If the result of a single roll of a die is a 4, it cannot be a 6 at the same time. Thus, the outcomes of a 4 and a 6 are mutually exclusive. Similarly, if the respondents to a questionnaire are classified as either male or female, then the events (male respondent, female respondent) are mutually exclusive. The special rule of addition for two events is

$$P(A \text{ or } B) = P(A) + P(B)$$

5–1

Recall that capital letters, such as *A* and *B,* refer to events. This formula can be expanded to any number of events. For example, for three events, it is $P(A$ or B or $C) = P(A) + P(B) + P(C)$.

Problem What is the probability that an even number will result from one roll of a single die?

Solution There are six possible outcomes:

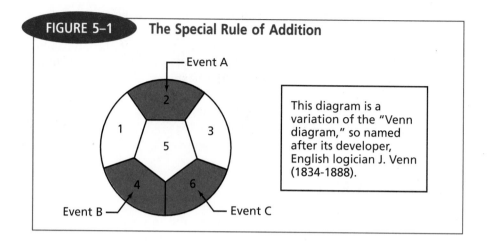

FIGURE 5–1 **The Special Rule of Addition**

Event A

Event B

Event C

This diagram is a variation of the "Venn diagram," so named after its developer, English logician J. Venn (1834-1888).

The event (an even number) is composed of three outcomes; namely:

The outcome of a ⚁ is event A.
The outcome of a ⚃ is event B.
The outcome of a ⚅ is event C.

The probability of each of the outcomes (2, 4, 6) is 1/6. To find the probability of the event "the outcome is an even number," we use formula 5–1 and add the 3 probabilities. That is, $1/6 + 1/6 + 1/6 = 3/6$, or 0.50. In symbols, if A stands for the outcome of a 2, B the outcome of a 4, and C the outcome of a 6, we compute the probability of an even number appearing by

$$P(\text{even}) = P(A) + P(B) + P(C)$$

$$P(A \text{ or } B \text{ or } C) = P(A) + P(B) + P(C)$$

$$= \frac{1}{6} + \frac{1}{6} + \frac{1}{6}$$

$$= 0.50$$

Recall that $P(\ )$ denotes the probability of the event described inside the parentheses.

Addition of the probabilities of the events of a 2, a 4, and a 6 in the previous die-rolling problem is allowed because the events are mutually exclusive. This notion of mutual exclusivity can be shown in diagram form (see Figure 5–1). The 6 regions represent all possible outcomes of an experiment.

Self-Review 5–5

Two hundred randomly selected prisoners in cell block M are surveyed and classified by type of crime committed.

Type of Crime	Number
Murder	48
Armed robbery	42
Rape	101
Kidnapping	7
Other	2

a. What is the probability that a particular prisoner selected in the sample is a convicted murderer?

b. What is the probability that a particular prisoner selected is a convicted kidnapper?

c. What is the probability that a particular prisoner selected is either a convicted murderer or a convicted kidnapper? What rule of probability is employed?

REAL STAT

In the days before airport X-ray machines and metal detectors, there was once a little old lady who carried a bomb with her whenever she traveled. Since she had never heard of *two* bombs on the same plane, she reasoned that hers would be the only one on board!

Exercises

7. $P(X) = 0.5$, $P(Y) = 0.3$, and X and Y are mutually exclusive. Find $P(X \text{ or } Y)$.

8. $P(A) = 0.4$, $P(B) = 0.1$, and A and B are mutually exclusive. Find $P(A \text{ or } B)$.

9. The events A and B are mutually exclusive. Suppose $P(A) = 0.20$ and $P(B) = 0.25$. What is the probability of the event (A or B) occurring?

10. The events X and Y are mutually exclusive. Suppose $P(X) = 0.05$ and $P(Y) = 0.10$. What is the probability of the event (X or Y) occurring?

11. A study was made of 138 children who exhibited evidence of abuse by an adult. The following table shows each abused child's position in the family:

Position	Frequency
Only child	34
Oldest	24
Youngest	50
Other	30

a. What is the probability of randomly selecting a child from the group described above and finding that the child was either the youngest or the oldest child in a family?

b. What is the probability of selecting a child who is not an only child?

12. Shown in the following table are the reported annual deaths of males aged 75 or over from 1 of the 5 leading types of cancer:

Site of Cancer	Number of Deaths
Lung	12,226
Prostate	10,835
Colon or rectum	8426
Stomach	3037
Pancreas	3031
	37,555

For a randomly selected deceased male cancer victim, what is the probability that

a. he died of 1 of the 2 primary causes?

b. he died of cancer of either the stomach or the pancreas?

c. he did not die of lung cancer?

THE GENERAL ADDITION RULE. We apply the **general rule of addition** to calculate the probability of events that are not mutually exclusive. For example, the current Social Security law has both a disability provision and a retirement provision. A welfare worker studying the residents of the Sunshine Retirement Community finds that 20% of the residents are receiving disability payments and 85% are receiving retirement income. If a retiree is randomly chosen for study, what is the probability that the person selected is receiving either disability payments or retirement income (or possibly both)?

The percentages are converted to probabilities:

	Probability
Probability of receiving a disability benefit	0.20
Probability of receiving a retirement benefit	0.85
Total	1.05

Is the probability of 1.05 possible? It is not! A probability of more than 1.0 was ruled out in an earlier section. A probability is always 0 to 1, inclusive. What happened? Some of the people who receive payments have been "double-counted." That is, they receive both disability and retirement incomes.

To overcome this difficulty, the welfare worker must determine the percentage of people who were counted twice and deduct this percent from the total. Suppose the investigator were to find that 15% of the people receive *both* disability and retirement incomes. Subtracting the corresponding probability of 0.15 from the total leaves 0.90. This is the probability of a person receiving at least one of these two types of income. The following table summarizes these calculations:

	Probability
Probability of receiving a disability benefit	0.20
Probability of receiving a retirement benefit	0.85
Probability of receiving *both* a disability benefit and a retirement benefit	−0.15
Probability of receiving one or the other benefit (or possibly both)	0.90

Symbolically, the general rule of addition is written

$$P(A \text{ or } B) = P(A) + P(B) - P(A \text{ and } B) \qquad \boxed{5\text{--}2}$$

where A and B are two events. The word *or* takes into account the possibility that both A and B may occur and may need to be separated. This is sometimes called an "inclusive" *or* because it includes the possibility that both A and B happen and the possibility that either occurs separately.

Apply formula 5–2 to the welfare problem:

Let A stand for the event "the retiree receives a disability benefit."
Let B stand for the event "the retiree receives a retirement benefit."
Let A and B stand for the event "the retiree receives both a disability benefit and a retirement benefit."

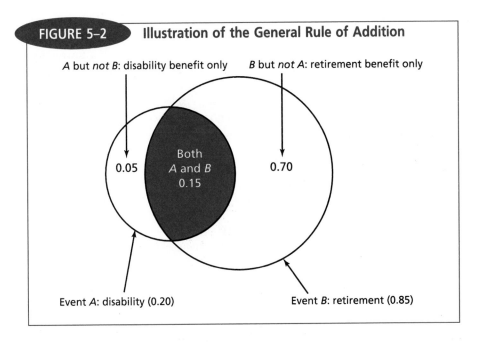

FIGURE 5–2   **Illustration of the General Rule of Addition**

A but *not* B: disability benefit only   B but *not* A: retirement benefit only

0.05   Both A and B 0.15   0.70

Event A: disability (0.20)   Event B: retirement (0.85)

To find the probability of receiving a disability benefit *or* a retirement benefit, use formula 5–2:

$$P(A \text{ or } B) = P(A) + P(B) - P(A \text{ and } B)$$

$$= 0.20 + 0.85 - 0.15$$

$$= 0.90$$

The general rule of addition applied to the welfare problem is illustrated by a Venn diagram (see Figure 5–2). Note the overlapping of events.

Problems

a.   Let $P(X) = 0.50$ and $P(Y) = 0.20$ and $P(X \text{ and } Y) = 0$. What is the $P(X$ or $Y)$? Are the events mutually exclusive?

b.   Let $P(D) = 0.35$ and $P(E) = 0.15$ and $P(D \text{ and } E) = .05$. What is the $P(D$ or $E)$? Are the events mutually exclusive?

Solutions

a.   0.70; yes

b.   0.45; no

Self-Review 5–6

An analysis of the student records at Solid State University revealed that 45% of the students have a grade point average above 3.00. Twenty-five percent of the students are employed. Ten percent of the students are employed *and* have a grade point average above 3.00. What is the probability that a student selected at random will have a grade point average above 3.00 *or* be employed?

Exercises

13. $P(A) = 0.5$, $P(B) = 0.3$, and $P(A$ and $B) = 0.1$. Find $P(A$ or $B)$.

14. $P(X) = 0.4$, $P(Y) = 0.1$, and $P(X$ and $Y) = 0.1$. Find $P(X$ or $Y)$.

15. The probabilities of two events, A and B, are 0.20 and 0.30, respectively. The events are not mutually exclusive. The probability that they both occur is 0.15. What is the probability of either A or B occurring?

16. Let $P(X) = 0.55$ and $P(Y) = 0.35$. Assume that these events are not mutually exclusive and that the probability that they both occur is 0.20. What is the probability of either X or Y occurring?

17. A study of patient records at a public health clinic showed that 15% of the patients had a dental examination, 45% had a general physical examination, and 5% had both. If a patient's record is randomly selected, what is the probability that the patient received either a dental examination or a physical examination?

18. Twenty percent of the members of boards of directors of large corporations are women. Five percent are persons connected with a university. Two percent of large corporations have female board members with university ties. If a corporation is randomly selected, what is the probability that its board of directors will have a member who is either a woman or someone from a university?

19. A student is taking two courses, history and math. The probability that the student will pass the math course is 0.60, and the probability of passing the history course is 0.70. The probability of passing both is 0.50. What is the probability of passing at least one of the two courses?

20. An analysis by the National Weather Service at Toledo, Ohio, showed that the low temperature in January is below freezing on 70% of the days. It snows on 30% of the January days and is below freezing *and* snows on 20% of the days. If tomorrow is January 12, what is the probability that it will snow or that the low temperature will be below freezing?

Multiplication Rules

> **joint probability** Measures the likelihood that two or more events will happen at the same time.

> **conditional probability** The likelihood that an event will occur, given that another event has already occurred.

THE GENERAL MULTIPLICATION RULE. Suppose we want to find the probability of both A and B occurring. For example, we might wish to find the probability of a student having a grade point average above 3.00 and also being employed. Such situations are different from those for which the addition rule can be used. Instead of computing the probability of one of the two outcomes occurring, we wish to find the probability that they *both* happen. This is termed a **joint probability**.

Another probability concept is called **conditional probability**. Recall that probability measures uncertainty. But the degree of uncertainty may change as new data become available. For example, if you find that someone is opposed to gun control, then it increases the likelihood they grew up in the West. Conditional probability is the tool that describes the new probability corresponding to some event B after it is known that some other event A has occurred. Symbolically, it is written $P(B|A)$. The vertical (|) slash does not mean division. Instead, it is read "given," as in "the probability of B given A."

In the Social Security illustration, the probability that a person will receive a retirement benefit, $P(B)$, is 0.85. However, if first we learn that the person is receiving a disability benefit, our estimate of the probability that he or she is also receiving a retirement benefit changes. Only 20% of the total receive disability benefits [$P(A) = 0.20$]. Out of that 20%, 15% also receive a retirement benefit. Hence, the conditional probability is 15/20, or 0.75 [$P(B|A)$].

The knowledge that a person receives a disability benefit reduces the likelihood from 0.85 to 0.75 that he or she receives a retirement benefit. In symbols, this calculation is written as follows:

$$P(B|A) = \frac{P(A \text{ and } B)}{P(A)} \qquad \boxed{5\text{--}3}$$

This same relation gives us a way to compute $P(A \text{ and } B)$ when $P(A)$ and $P(B|A)$ are known. We simply multiply $P(A) \cdot P(B|A)$. This is the **general rule of multiplication**, which states that, if two events, A and B, can both occur in some experiment, then the probability that both A and B occur is $P(A) \cdot P(B|A)$. It is written as

$$P(A \text{ and } B) = P(A) \cdot P(B|A) \qquad \boxed{5\text{--}4}$$

Problem   A study is conducted to determine the possible correlation between the level of education attained and a person's position on abortion. The results of a random sample of 100 individuals are shown in Table 5–1.

A question to be explored: What is the probability of selecting an individual with fewer than four years of high school education who favors abortion? The events occurring at the same time are "fewer than four years of high school" and "favors abortion."

Solution   Applying the general rule of multiplication gives

$$P(B \text{ and } A) = P(A) \cdot P(B|A)$$

where

A stands for the event "a person favors abortion."
B stands for a person with fewer than four years of high school education.

$P(B|A)$ represents the probability of selecting a person who has less than a high school education when it is known (or given) that the person favors abortion. As noted previously, the vertical line is read as "given that." The problem reads, "We find the probability that a person selected has less than

TABLE 5–1	Level of Education and Position on Abortion				
Position on Abortion	Less than 4 Years of High School	High School Graduate	Some College	College Graduate	Total
Favor	5	15	15	25	60
Oppose	10	10	10	10	40
Total	15	25	25	35	100

four years of high school education and favors abortion by multiplying the probability that the person favors abortion by the probability that the person has less than a high school education, given that the individual favors abortion." Referring to Table 5–1 for the probabilities and using formula 5–4, we obtain

$$P(B \text{ and } A) = P(A) \cdot P(B|A)$$
$$= 60/100 \cdot 5/60$$
$$= 0.60 \cdot 0.0833$$
$$= 0.05$$

We could also have computed this probability by determining the probability of selecting an individual with less than a high school education $[P(A) = 15/100 = 0.15]$ and then the probability that the person favors abortion, given that he or she has less than a high school education $[P(B|A) = 5/15 = 0.33]$:

$$P(A \text{ and } B) = P(A) \cdot P(B|A)$$
$$= 0.15 \cdot 0.33$$
$$= 0.05 \text{ (the same as observed and calculated earlier)}$$

If more than two simple events are involved, the general rule of multiplication can be extended. For example, $P(A \text{ and } B \text{ and } C) = P(A) \cdot P(B|A) \cdot P(C|A \text{ and } B)$.

In the previous example about people's attitudes toward abortion, the joint probability could have been read directly from the table. It was pointed out that the general rule of multiplication gives the same answer. Sometimes, however, the information we have is expressed in percentages instead of counts, as in the previous illustration. The general rule of multiplication must be used to solve such problems.

Problem   The police in a small municipality know that 25% of the homeowners leave their doors unlocked. Crime records show that 4% of the homes whose doors are left unlocked are burglarized. What is the probability that a home is both left with its doors unlocked *and* burglarized?

Solution   The general rule of multiplication applies—that is, formula 5–4.

Let *A* represent the event "left with doors unlocked."
Let *B* represent the event "burglarized."

We know that

$$P(A) = 0.25$$
$$P(B|A) = 0.04 \quad \longleftarrow \boxed{\text{4% of the homes } \textit{with unlocked} \text{ } \textit{doors} \text{ are burglarized}}$$

Solving gives

$$P(A \text{ and } B) = P(A) \cdot P(B|A)$$
$$= (0.25)(0.04)$$
$$= 0.01$$

This indicates that 1% of *all* the homes are *both* left with their doors unlocked and burglarized.

Problems

a. For 2 independent events, $P(X) = 0.50$ and $P(Y) = 0.20$. Determine the probability of $P(X \text{ and } Y)$.
b. For 2 dependent events, $P(D) = 0.35$ and $P(E|D) = 0.15$. Determine the probability of $P(D \text{ and } E)$.

Solutions

a. 0.10
b. 0.0525

Self-Review 5–7

A sociologist conducted a study of a sample of 60 students to determine how many of each sex smoke marijuana. The results:

	Male	Female	Total
Smokes marijuana	15	8	23
Does not smoke marijuana	20	17	37
Total	35	25	60

Using the general rule of multiplication, determine the probability of selecting a male who smokes marijuana.

a. Letting *A* be the event "the student smokes marijuana" and *M* the event "a male is selected," supply the appropriate formula.
b. Determine the joint probability.

One technique to show probabilities, joint probabilities, and conditional probabilities is the **tree diagram**. Table 5–1 is repeated here to illustrate the construction of a tree diagram.

Position on Abortion	Less than 4 Years of High School	High School Graduate	Some College	College Graduate	Total
Favor	5	15	15	25	60
Oppose	10	10	10	10	40
Total	15	25	25	35	100

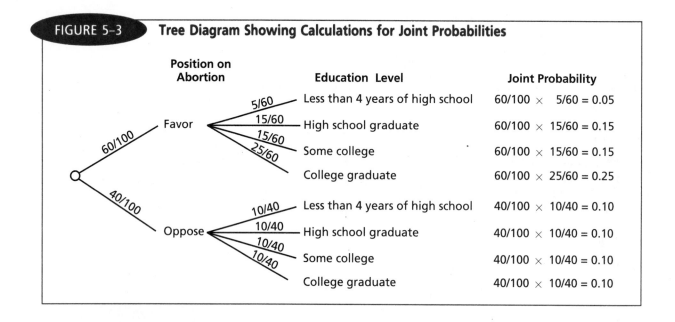

FIGURE 5–3   **Tree Diagram Showing Calculations for Joint Probabilities**

In the tree diagram in Figure 5–3, note that there are two main branches going out from the trunk on the left. The upper branch is labeled "Favor," the lower branch "Oppose." Note that the probability written on the "Favor" branch is 60/100 and on the "Oppose" branch 40/100.

Problem   Using the tree diagram in Figure 5–3, locate the probability that a person favors abortion and has less than a high school education.

Solution

1. Find the upper branch representing the event "favor abortion." As noted before, its probability (60/100) is written on the tree branch.
2. Continuing along the same path, find the branch "Less than 4 years high school." The conditional probability of 5/60 is written on this branch.
3. Multiplying those two probabilities gives 0.05, shown at the end of the path. This is the joint probability of selecting a person who both favors abortion and has less than 4 years of high school education.

Self-Review 5–8   The study from Self-Review 5–7 is repeated below.

	Male	Female	Total
Smokes marijuana	15	8	23
Does not smoke marijuana	20	17	37
Total	35	25	60

a. What is the probability of selecting a female?
b. Assuming the person selected is a female, what is the probability that she smokes marijuana?
c. Draw a tree diagram showing all possible joint probabilities.

. . . .

Exercises

21. $P(A) = 0.5$, and $P(B|A) = 0.4$. Find $P(A \text{ and } B)$.
22. $P(X) = 0.8$, and $P(Y|X) = 0.7$. Find $P(X \text{ and } Y)$.
23. Suppose $P(A) = 0.30$ and $P(B|A) = 0.60$. Determine $P(A \text{ and } B)$.
24. Refer to the following table:

Events	Other Events			Total
	M	N	O	
X	2	1	3	6
Y	1	2	1	4
Total	3	3	4	10

a. Compute $P(Y)$.
b. Compute $P(M|Y)$.
c. Compute $P(M \text{ and } Y)$.

25. A study was undertaken to correlate students' mathematical ability with their interest in statistics. The results:

Math Ability	Interest in Statistics			Total
	Low	Average	High	
Low	40	9	11	60
Average	15	16	19	50
High	6	10	25	41
	61	35	55	151

a. What is the probability of selecting a student with both a low math ability and a low interest in statistics?
b. What is the probability of selecting a student with both an average math ability and a high interest in statistics?
c. Draw a tree diagram showing all possible joint probabilities.

26. Students were classified according to two traits: whether they had won a high school letter for athletics and college grade point average.

High School Athletics	College GPA			
	Low	Average	High	Total
Did win letter	50	30	50	130
Did not win letter	20	30	20	70
	70	60	70	200

a. What is the probability of selecting a student with a low college GPA who won a letter in high school?
b. What is the probability of selecting a student with a high college GPA who won a letter in high school?
c. Draw a tree diagram showing all possible joint probabilities.

27. Thirty-six players enter a tennis tournament. They are classified by gender and age:

	Juniors	Seniors
Males	6	10
Females	8	12

What is the probability that
a. two players are selected and both are senior females?
b. two players are selected and both are junior males?
c. two players are selected and both are females?
d. the first player selected is a senior male and the second is a junior female?

e. the first player selected is male and the second is female?

f. the first four players selected are senior females?

28. A hospital cafeteria menu includes the items shown at right. If a patient must select one item from each group, draw a tree diagram to show the possible luncheon choices.

Entree	Salad	Fruit
Chicken	Lettuce	Apple
Tuna	Gelatin	Banana
Hot dog		Orange

a. How many different outcomes are possible?

b. What is the probability of simultaneously selecting chicken as the entree, lettuce as the salad, and a banana as the fruit?

THE SPECIAL MULTIPLICATION RULE. Two events are "independent" when knowledge one has occurred does not change the probability the other event will occur. This can be written in symbols as $P(A|B) = P(A)$. In other words, knowing event B has happened does not affect the probability event A will happen.

An example illustrating independence will help. Suppose you draw a single card from an ordinary 52-card deck, replace it, and draw another card. The probability you see a "jack" on both draws is $(1/13)(1/13)$ or $1/169$. However, if you draw the second card before you have replaced the first, the probability you see a "jack" on both cards is only $(1/13)(3/51) = 1/221$. The second calculation is different, as the second draw is not **independent** of the first draw. The number of "jacks" remaining in the deck was reduced by the outcome of the first draw.

An advantage of knowing two events are independent is that the general multiplication rule (formula 5–4) can be written more simply as

$$P(A \text{ and } B) = P(A) \cdot P(B) \qquad \boxed{5\text{–}5}$$

Problem There are 6 faces to a die: ⚀ ⚁ ⚂ ⚃ ⚄ ⚅. Suppose we have 2 dice: One is colored blue, the other white. They are rolled on the floor. What is the probability of obtaining a pair of 2s—that is, a 2 on the blue die ⚁ and a 2 on the white die ⚁ ?

Solution The outcome of the white die is not dependent on the outcome of the blue die; the two events are independent. Thus, the special rule of multiplication can be applied.

One of the 6 sides of a die is a 2-spot. The probability of a 2-spot, therefore, is 1/6. The probability of a 2-spot coming face up on the blue die is 1/6, and the probability of a 2-spot coming face up on the white die is also 1/6. The probability of both coming face up is 1/36, found by using formula 5–5:

$$P(A \text{ and } B) = P(A) \cdot P(B)$$

$$= \frac{1}{6} \times \frac{1}{6}$$

$$= 0.0278$$

Figure 5–4 illustrates the concept of independence. Note:

1. There are 36 points (dots).
2. *A* is the event "a 2-spot appears on the blue die."
3. *B* is the event "a 2-spot appears on the white die."
4. The shaded intersection of *A* and *B* is the point at which a 2-spot appears on both the blue and the white dice.

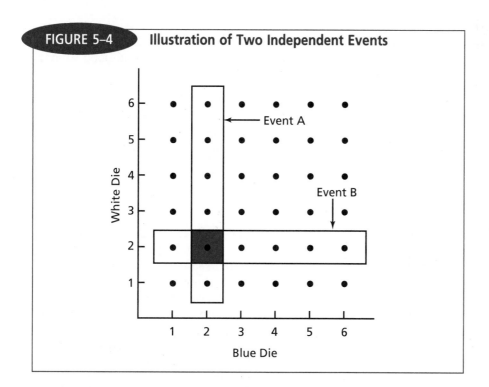

FIGURE 5–4 **Illustration of Two Independent Events**

Note that, while there are 36 points (dots) in Figure 5–4, the shaded intersection includes only 1 point. Hence, 1 out of 36 points meets the stipulations, and the probability is 0.0278, found by 1/36. This agrees with the result we obtained using the special rule of multiplication.

CAUTION. It is easy to confuse *mutually exclusive* with *independent* events. If *A* and *B* are mutually exclusive, $P(A \text{ and } B) = 0$. This means *A* and *B* cannot occur together. If *A* and *B* are independent, $P(A|B) = P(A)$. This means *B* does not affect the probability of *A* occurring. Events *cannot* be both mutually exclusive and independent!

Consider drawing 1 card from a deck. The events of "getting a 5" and "getting a diamond" are independent. If you are dealing with the full deck, the probability of "getting a 5" is 4/52. If you deal only with the diamonds, the probability of "getting a 5" is 1/13, which is the same number. This shows the events are independent. However, since a card could be the 5 of diamonds, these events are independent, but *not* mutually exclusive.

On the other hand, the events "getting a heart" and "getting a club" are mutually exclusive, as a card cannot be both a heart and a club. If the card is a heart, the probability it is a club changes from 1/4 to 0. Also, if the card is not a heart, the probability it is a club changes from 1/4 to 1/3. Thus, the events "getting a heart" and "getting a club" are mutually exclusive, but *not* independent.

Self-Review 5–9

The probability of a newborn being a boy is 1/2; the probability of it being a girl is 1/2. Let *A* be the event "the first child born to Jane and Doug Schmitz is a boy." *B* is the event "the second child is a boy."

a. Are these events independent?
b. Supply the correct formula, and determine the probability that the first 2 children born are boys.
c. What is the probability that the first 3 children are boys?

Exercises

29. $P(A) = 0.6$, $P(B) = 0.4$, and A and B are independent. Find $P(A \text{ and } B)$.
30. $P(A) = 0.6$, $P(B) = 0.4$, and A and B are independent. Find $P(A|B)$.
31. We know that $P(A) = 0.30$ and $P(B) = 0.20$. If events A and B are independent, determine the joint probability of A and B.
32. Refer to the following table:

Events	Other Events M	N	O	Total
X	2	1	3	6
Y	2	1	3	6
Z	4	2	6	12
Total	8	4	12	24

a. Find $P(Y)$.
b. Find $P(M|Y)$.
c. Find $P(M \text{ and } Y)$.
d. Are the events M and Y independent?
e. Are events M and Y mutually exclusive?
33. What is the probability that an archer with 90% accuracy will hit a bull's-eye 4 times in a row?

34. Out of every 100 cars that start in the Texas Grand Prix race, only 60 finish. Two cars are entered by the Penske team in this year's race.
a. What is the probability that both will finish?
b. Are the two events independent?
c. What rule of probability does this problem illustrate?
d. What concept of probability does this problem illustrate?
e. What is the probability that neither of Penske's 2 cars will finish the Grand Prix?
35. The Penn Bank has 2 computers. The probability that the newer one will break down in any particular month is 0.05. The probability that the older one will break down in any particular month is 0.10.
a. Are these events independent?
b. What is the probability that both will break down during the month of October?
c. What is the probability that neither will break down during October?
36. At Mendoza University, 30% of the students live on campus.
a. If 3 students are selected to form a committee, what is the probability that all 3 live on campus?
b. What is the probability that none of the 3 students lives on campus?

We may wish to describe a situation in which an event does *not* happen. For example, we want to estimate the probability that it will not rain next Saturday. It is easier to find the probability that it will rain and then subtract that number from 1. This relationship is termed the **complement rule.** The formula for the complement rule is

$$P(\sim A) = 1 - P(A) \qquad \text{5-6}$$

If A is an event, we can express the complement of A in symbols as $\sim A$. The symbol $\sim$ is called a tilde. It is written in front of the symbol for the first event to denote the second event, which is the "complement" of the first event. This rule is often very effective in reducing the number of calculations needed.

Problem A sample of 50 families revealed the following information about the number of children in each family:

Number of Children	Number of Families
0	20
1	15
2	10
3 or more	5
	50

Compute the probability that a family has at least 1 child.

Solution These events are mutually exclusive. That is, for a particular family, the number of children cannot be both 1 and 2. We can obtain the probability of selecting a family with at least 1 child by using the special rule of addition, formula 5–1. That is, we add the probabilities of having 1, having 2, and having 3 or more:

$$P(\text{at least } 1) = P(1) + P(2) + P(3 \text{ or more})$$

$$= \frac{15}{50} + \frac{10}{50} + \frac{5}{50}$$

$$= 0.60$$

The complement rule, formula 5–6, leads to a more efficient solution:

$$P(\text{at least } 1) = 1 - P(0)$$

$$= 1 - \frac{20}{50}$$

$$= 1 - 0.40$$

$$= 0.60$$

The Complement Rule

REAL STAT

Suppose you have a wallet that contains either a $2 bill or a $20 bill (with equal likelihood), but you don't know which one. You add a $2 bill. Later you reach into your wallet (without looking) and remove a bill. It's a $2 bill. There's 1 bill remaining in the wallet. What are the chances that it is a $2 bill?

Oddly, the chances are 2/3 that the bill remaining in the wallet is a $2 bill. Here's why.

Before you removed the $2 bill, there were two possible combinations of bills in the wallet: (1) $20 (the original bill) and $2 (the added one); or (2) $2 (the original bill) and $2 (the added one).

Now you reach in and remove a bill, which turns out to be a $2 bill. This could happen in 3 different ways, all of them equally likely: It could be the added $2 bill in the first combination, in which case the original $20 bill is left. Or it could be the added $2 bill in the second combination, in which case the original $2 bill is left. Or it could be the original $2 bill in the second combination, in which case the added $2 bill is left. So, in 2 cases out of 3, the bill remaining in your wallet will be a $2 bill.

Problem Recall from the Chapter Problem the roll of Life Savers purchased by Johnny Smith. If Johnny randomly selects two candies from the package, what is the probability that both will be cherry? Neither will be cherry? At least one will be cherry?

Solution The probability that both candies will be cherry is found using the general rule of multiplication. Multiply the probability that the first candy selected will be cherry (3/11) by the conditional probability that the second one chosen will be cherry after it is known that the first selection was cherry (2/10). The result is 6/110, or 0.0545, found by (3/11) (2/10).

The probability that neither will be cherry is also found using the general rule of multiplication, but the events are defined differently. First, find the probability that the first candy selected will not be cherry—that is, 8/11. Next, multiply this probability by the conditional probability that the second candy chosen will not be cherry (7/10). The result is 56/110, or 0.5091.

The complement rule is used to find the probability that at least one candy will be cherry. Use the fact that the probability that *neither* will be cherry is 0.5091, and subtract this value from 1. The probability that at least one will be cherry is .4909, found by 1 − .5091.

Exercises

37. If $P(X) = 0.7$, what is $P(\sim X)$?

38. If $P(A) = 0.4$, what is $P(\sim A)$?

39. If $P(A) = 0.30$, what is $P(\sim A)$?

40. Refer to Exercise 32. Compute $P(\sim Y)$.

41. The postanesthesia care facility at St. Michael's Hospital has 3 beds. The probability that only 1 is occupied is 0.30, the probability that 2 are occupied is 0.25, and the probability that all 3 are occupied is 0.20. The probability that there are no patients in the facility is 0.25. Compute the probability that there is at least 1 patient in the facility.

42. According to the National Weather Service, the probability that it will rain in Pittsburgh, Pennsylvania, on a particular day in September is 0.20. Assume each day is an independent event. What is the probability that it will not rain on the next two Saturdays in September?

 SOME COUNTING PRINCIPLES

When the number of outcomes is small, as in the previous experiments, it is not difficult to list all the possibilities. For example, to list the outcomes for 2 children in a family is quite simple. The possibilities are 2 boys, 1 boy and 1 girl, and 2 girls. Suppose, however, you had to list and count the possible outcomes for a family of 15 children! Obviously, you would need a more efficient method of counting. Three principles of counting that facilitate calculations in such cases will be explored next: (1) the multiplication principle, (2) the permutation formula, and (3) the combination formula.

The multiplication principle refers to the total number of choices possible. An example will serve to explain further:

Problem   The Bullocks art department is designing an advertisement for the new catalog. The artist must show all possible variations on a new outfit consisting of two blouses and three skirts. The blouse options are long sleeve or short sleeve. The skirt options are solid color, plaid, or long evening. How many different interchangeable outfits are there?

Solution   The letter m represents the number of blouse options. The letter n represents the number of skirt options. Using the multiplication principle, we obtain

$$\text{Number of options} = m \times n$$
$$= 2 \times 3$$
$$= 6$$

These six options are shown in Figure 5–5. This counting method can be extended to more than two events.

The Multiplication Principle

multiplication principle If a choice consists of two steps, the first of which can be made in m ways and the second in n ways, there are $m \cdot n$ possible choices.

FIGURE 5–5   **Possible Outfits for Two Blouses and Three Skirts**

Long-sleeved blouse
Solid-color skirt

Short-sleeved blouse
Solid-color skirt

Long-sleeved blouse
Plaid skirt

Short-sleeved blouse
Plaid skirt

Long-sleeved blouse
Long evening skirt

Short-sleeved blouse
Long evening skirt

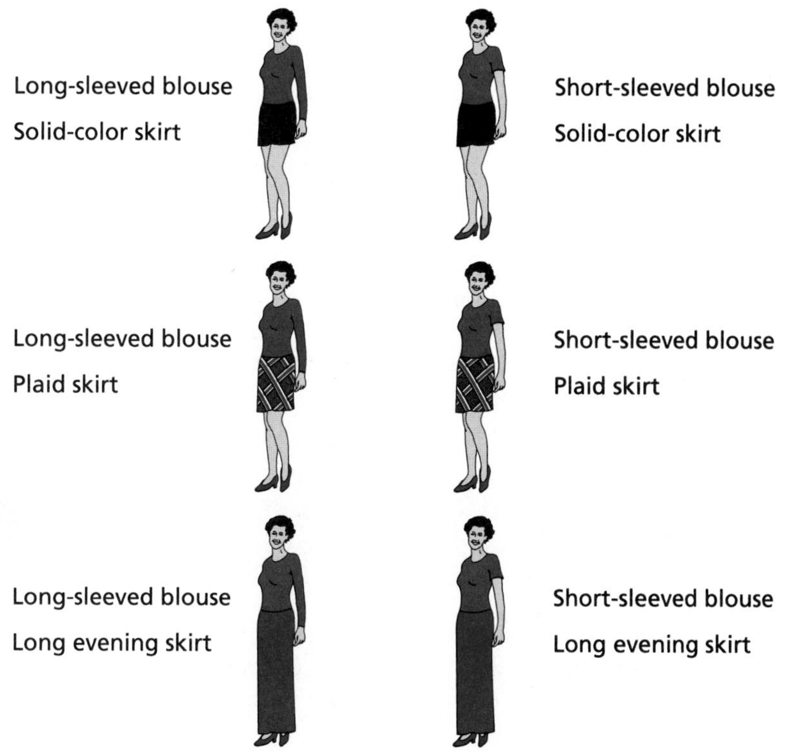

Problem A new-car buyer has a choice of five body styles, two engines, and eight different colors. How many different car choices does the buyer have?

Solution There are $5 \times 2 \times 8 = 80$ different choices among the cars that could be ordered.

Self-Review 5–10

An operating team generally consists of a surgeon, an anesthetist, and a nurse. A team must be scheduled for a delicate operation at Harbor Hospital. There are two surgeons capable of performing the operation. Three anesthetists are available, and five nurses are on call. How many different operating teams could be assembled?

Exercises

43. A coin is flipped and a die is rolled. How many different pairs of outcomes are possible?

44. A card is drawn from an ordinary 52-card deck, and, before it is replaced, a second card is drawn. How many different pairs of outcomes are possible?

45. Data are grouped according to sex (male or female) and income level (low, medium, or high). How many different pairs are there?

46. A restaurant offers five sandwiches, four cold drinks, and six desserts as part of its luncheon special. How many different specials are there?

47. Tex can travel from El Paso to New Orleans by bus, train, or airplane. He can travel from New Orleans to Tampico, Mexico, by airplane or ship. How many choices for the trip from El Paso to Tampico are there? He can continue on to Veracruz by bus, airplane, or ship. How many possible travel choices are there for the trip from El Paso to Veracruz via New Orleans and Tampico?

48. One of Canada's western provinces is considering using only numbers on their automobile license plates. If only 4 digits are to appear on the plates (such as 8313), how many different plates are possible? (*Hint:* The first digit could be any number between 0 and 9.)

Permutations

permutation An ordered arrangement of a group of objects.

In the previous section, one item was chosen from each of several groups (one blouse from the blouses and one skirt from the skirts, for example). In contrast, if more than one item is to be selected from the *same* group and if the **order of the selection is important,** the resulting arrangement is called a **permutation** of the items.

The permutation formula is

$$_nP_r = \frac{n!}{(n - r)!}$$

5–7

where n represents the total number of objects in the group and r represents the number of objects actually selected. For example, if there are 7 persons and 3 are to fill the offices of president, vice president, and secretary, $n = 7$ and $r = 3$.

The notation $n!$ is called "n factorial." It is the product of $n \cdot (n - 1) \cdot (n - 2) \ldots (1)$. So, 4! (four factorial) is 24, found by $(4) \cdot (3) \cdot (2) \cdot (1)$, and $5! = 120$. Zero factorial, written 0!, is treated as a special case and is defined to equal 1.

Problem   Five avid football fans organize the Yellow Jacket Booster Club. The five fans are Smith, Topol, Jackson, Lopez, and McNeil. Three officers are to be selected: a president, a secretary, and a treasurer. One group of officers might consist of President Lopez, Secretary Topol, and Treasurer McNeil. How many distinct slates (permutations) are possible?

Solution

n is the number of avid boosters. (There are five.)
r is the number of offices to be filled. (There are three.)

$$_nP_r = \frac{n!}{(n - r)!}$$

$$_5P_3 = \frac{5!}{(5 - 3)!}$$

$$= \frac{5 \times 4 \times 3 \times \cancel{2} \times \cancel{1}}{\cancel{2} \times \cancel{1}}$$

$$= 60$$

> Since the same numbers are included in the numerator and denominator, they can be cancelled.

Problem   How many three-letter "words" can be made from the letters A, B, C, D, E, and F? No letter may be repeated in a word (such as DBD).

Solution

$$_nP_r = \frac{n!}{(n - r)!}$$

$$_6P_3 = \frac{6!}{(6 - 3)!}$$

$$= \frac{6 \cdot 5 \cdot 4 \cdot \cancel{3} \cdot \cancel{2} \cdot \cancel{1}}{\cancel{3} \cdot \cancel{2} \cdot \cancel{1}}$$

$$= 120$$

· ·

From the four characters *, +, /, and #, make up computer passwords that are three symbols long, without repeating any of the symbols. For example, the three symbols * # + could make up one password. How many distinct passwords of three-unit length are possible from the four characters?

Self-Review 5–11

Exercises

49. If $n = 6$ and $r = 2$, find $_nP_r$.

50. If $n = 8$ and $r = 4$, find $_nP_r$.

51. There are eight different objects. How many permutations of these eight objects can be selected three at a time?

52. Moss has four bowling trophies that he plans to display on a shelf. In how many ways can they be arranged?

53. "The Triple" at the local racetrack consists of correctly picking the order of finish of the first three horses in the eighth race. Ten horses have been entered in today's race. How many "Triple" outcomes are possible?

54. A real estate developer has eight basic house designs. Zoning regulations in his community do not permit look-alike homes on the same street. The developer has five lots on Mallard Road. In how many different ways can the new homes be arranged?

Combinations

> **combination** One particular arrangement of a group of objects or persons selected from a larger group without regard to order.

In the previous section dealing with permutations, the *order* of the outcomes was important. For example, one slate of Yellow Jacket officers might be President Jackson, Secretary Topol, and Treasurer McNeil. Another slate might have President McNeil, Secretary Jackson, and Treasurer Topol. Every time there is a change in the order, the count of possibilities is increased by one.

In another problem, the order may not be important. If, for example, Gonzales, Clay, and Higbee were selected to serve as a social committee, it would be the same committee if the order were Clay, Higbee, and Gonzales. In other words, the committee of Higbee, Clay, and Gonzales is counted just *once*, regardless of the order of the names. Technically, such a group is considered a **combination**.

If order is not important, then we can use the *combination formula* to count the number of combinations:

$$_nC_r = \frac{n!}{(n - r)!r!}$$

5–8

Problem   The Yellow Jacket Booster Club (see page 169) wants to select two persons to serve as a membership committee. How many different committees could be selected? One possible committee might consist of Lopez and McNeil. Of course, that committee would be the same as McNeil and Lopez.

Solution   There are ten possible membership committees:

Lopez, McNeil	McNeil, Smith	Topol, Jackson	Jackson, Smith
Lopez, Smith	McNeil, Topol	Topol, Smith	
Lopez, Topol	McNeil, Jackson		
Lopez, Jackson			

Using the combination formula to determine the total number of possible membership committees of two ($r = 2$) from the five-member booster club ($n = 5$), we find

$$_nC_r = \frac{n!}{(n-r)!r!}$$

$$_5C_2 = \frac{5 \cdot 4 \cdot \cancel{3} \cdot \cancel{2} \cdot \cancel{1}}{(\cancel{3} \cdot \cancel{2} \cdot \cancel{1})(2 \cdot 1)}$$

$$= 10 \text{ (same as listed by name)}$$

Problems

a. A student must register for one of three humanities courses, one of four math courses, and one of three English courses next semester. How many different schedules are possible?

b. There are five different objects. How many permutations of these five objects can be taken two at a time?

c. There are six different objects. How many combinations of the objects can be taken two at a time?

Solutions

a. 36
b. 20
c. 15

A group of seven mountain climbers wishes to form a mountain-climbing team of five. How many different teams could be formed?

Self-Review 5–12

Exercises

55. If $n = 6$ and $r = 2$, find $_nC_r$.

56. If $n = 8$ and $r = 4$, find $_nC_r$.

57. How many different poker hands are possible from a deck of 52 cards? (Note: You are dealt 5 cards.)

58. The director of welfare has just received 20 new welfare cases for investigation. A caseworker will be assigned 8 new cases. How many different groups of 8 cases could caseworker Klein be assigned?

59. Ten employees of Aztec Industries are covered by a union contract. If three of them must be selected to form a bargaining committee, how many distinct committees are possible?

60. The new-car showroom at Berg Motors has room to display four vehicles. The company receives a trailer load of six new models. How many different groups can the manager put out on the floor?

To reemphasize the difference between a permutation and a combination, recall that a **permutation** is an arrangement in which order is important. That is, a, b, c is one permutation; c, a, b is a second permutation; and b, a, c is a third. The permutation formula counts these different orders as different permutations. There are six permutations of the three letters a, b, and c taken three at a time.

Contrasting Permutations and Combinations

If the order of selection is unimportant, the arrangement is called a **combination**. That is, while the three letters can be written a, b, c, or c, a, b, or b, a, c, the combination formula will count these as a single combination. Consequently, there is only one combination of three items taken three at a time.

Self-Review 5–13

A tuna fleet has four flags of different colors. Two flags are arranged on a mast as a signal to the ships in the fleet.

a. If you wish to count the number of different signals that can be constructed using the four flags hoisted two at a time, would you use a permutation or a combination? (Each new order has a different meaning.)
b. A blue flag on top and a yellow one below mean "fish sighted." However, a yellow flag on top and a blue flag below mean "stormy weather ahead." How many different signals can you construct using the four flags two at a time?
c. Suppose it has been decided that any two flags, such as a green flag on top and a red one below, *or* a red flag on top and a green one below, have the same meaning. If you wish to count the number of signals that could be used, would you apply the permutation formula or the combination formula?
d. Refer to part c above. How many different signals can you construct using the four flags two at a time?

CHAPTER OUTLINE

I. Probability is a value between 0 and 1 that indicates the likelihood of an event.
  A. An experiment is what "you do."
  B. An outcome is what "happens."
  C. The event is what you "count."
II. There are three types of probability.
  A. Classical probability is based on equally likely outcomes.
  B. Empirical probability uses relative frequency and is found by dividing the number of times an event has happened by the total number of observations.
  C. Subjective probability is based solely on judgment.
III. There are rules for combining probabilities.
  A. Addition—$P(A \text{ or } B)$
    1. The general rule of addition is used when events are *not* mutually exclusive.

$$P(A \text{ or } B) = P(A) + P(B) - P(A \text{ and } B) \quad \boxed{5\text{-}2}$$

    2. The special rule of addition is used for mutually exclusive events.

$$P(A \text{ or } B) = P(A) + P(B) \quad \boxed{5\text{-}1}$$

  B. Multiplication—$P(A \text{ and } B)$
    1. The general rule of multiplication is used when events are *not* independent.

$$P(A \text{ and } B) = P(A) \cdot P(B|A) \quad \boxed{5\text{-}4}$$

2. The special rule of multiplication is used for independent events.

$$P(A \text{ and } B) = P(A) \cdot P(B) \quad \boxed{5\text{--}5}$$

C. The complement rule finds the probability of an event *not* happening by subtracting the probability of the event happening from 1.

$$P(\sim A) = 1 - P(A). \quad \boxed{5\text{--}6}$$

IV. Three counting formulas are useful.
 A. The multiplication principle: If there are m ways to do one thing and n ways to do a second, there are $m \cdot n$ ways to do both.

B. Use the permutation formula to count when order is important.

$$_nP_r = \frac{n!}{(n-r)!} \quad \boxed{5\text{--}7}$$

C. Use the combination formula when order is not important.

$$_nC_r = \frac{n!}{(n-r)!r!} \quad \boxed{5\text{--}8}$$

- -

Exercises

61. A recent study of 200 high school students revealed the following information on the relationship between age and whether a person understood the message of a television commercial:

	\multicolumn{4}{c}{Age}
	10–12	13–15	16–18	Total
Understood	30	40	50	120
Did not understand	40	30	10	80
Total	70	70	60	200

If an individual is randomly selected,
 a. what is the probability that he or she understood the commercial?
 b. what is the probability that he or she was 10–12 years old and understood the commercial?
 c. given that the individual did not understand the commercial, what is the probability that he or she was 10–12 years old?
 d. what is the probability that an individual either did not understand the commercial or was 10–12 years old?

62. A social club at a large university has 300 members who are registered in 1 of 3 different colleges. Their colleges of registration and their grade point averages, on a 4-point scale, are summarized in the following table:

	\multicolumn{4}{c}{Grade Point Average (GPA)}
College	Greater than 3.0	Between 2.0 and 3.0	Lower than 2.0	Total
Arts and Sciences	20	40	30	90
Business	60	50	10	120
Education	20	60	10	90
Total	100	150	50	300

A student is selected at random from the list of club members.
 a. What is the probability that the student is registered in the College of Business?
 b. What is the probability that the student has a GPA greater than 3.0 *and* is in the College of Education?
 c. What is the probability that the student has a GPA lower than 2.0 *or* is in the College of Education?

63. The small town of Sugar Grove has 2 ambulances. Records show that the first ambulance is in service 80% of the time and the second 70% of the time.

a. What is the probability that both are in service when needed?

b. What is the probability that at least 1 is in service when needed?

64. The security manager of a large building reports that the probability is 0.05 that a fire alarm will not operate when needed. If there are 3 alarms in the building, what is the probability that none of the alarms will operate during a particular fire? What is the probability that at least 1 will operate during a particular fire?

65. An appliance dealer sponsors advertisements on both radio and TV. A study of 200 customers revealed that 80 had seen the advertisement on TV, 120 had heard the radio advertisement, and 40 had both seen the TV ad and heard the radio ad.

a. What is the probability that a customer heard both the radio ad and the TV ad?

b. What is the probability that a customer heard the advertisement *either* on the radio *or* on TV?

66. A sheriff needs new tires for his road patrol cars. The probability that he will purchase Michelin, Goodyear, or Uniroyal tires is, respectively, 0.20, 0.30, and 0.40. What is the probability that he will not purchase any of these brands?

67. According to police records, 70% of reckless drivers are fined, 50% have their driver's licenses revoked, and 40% are both fined and have their licenses revoked. What is the probability that a particular reckless driver will either have her license revoked or be fined?

68. A dealer receives a shipment of four TV sets, one of which is known to be defective. If one of the sets is sold to a customer, what are the possible outcomes?

69. A car dealer has ten new cars in stock. Four are subcompacts, four are compacts, and two are luxury models. A car is sold. What is the experiment? What are the possible events regarding the type of car? If the luxury models are four-door sedans and all the others are two-door, describe the possible outcomes with respect to the number of doors.

70. What concept of probability is used in each of the following events to assign the likelihood of the outcomes?

a. Ford decides to market a new subcompact car. The company has never offered a car in this competitive field before, so no comparable data are available. The probability that Ford will sell 2,000,000 cars next year is 0.70.

b. An outfielder for the Toronto Blue Jays has a season batting average of .285. What is the probability that he will get a base hit the next time he comes to bat?

c. A student is randomly selected from your class. The probability that his or her birthday is in July is 1/12 = 0.083.

71. Mr. Blume and Ms. White are planning a date on Saturday night. The probability that they will see a basketball game, a movie, or a horse race is 0.25, 0.20, and 0.10, respectively. They will not do more than 1 of these activities. Of course, they may decide to do something else entirely. What is the probability that they will do 1 of these 3 things? What is the probability that they will choose 1 of the 2 sporting events?

72. A newly married couple plans to have two children and is thinking about the possible sexes of their offspring.

a. What is the experiment?

b. What are the possible outcomes for the gender of their children?

c. Are the outcomes "female" and "male" mutually exclusive?

d. Are the genders of the two children independent?

73. A positive integer is picked at random—for example, the last digit of the nine digits in your Social Security number. What is the probability that

a. it is larger than four?

b. it is an odd number?

c. it is an odd number larger than four?

74. A review of the records at Lander Community College revealed the following ethnic breakdown of the student population:

Caucasian	1200
Black	640
Hispanic	280

Asian	80
Native American	60
Other	20

a. What is the probability that a randomly selected student is either Caucasian or black?

b. What is the probability that the selected student is not a member of either of the two largest categories?

75. A study by the Florida Tourist Commission revealed that 70% of tourists entering the state visit Disney World, 50% visit Busch Gardens, and 40% visit both. What is the probability that a particular tourist visits at least 1 of the attractions?

76. A certain large city has a morning newspaper, *The Mirror,* and an afternoon paper, *The Observer.* A study shows that 30% of households subscribe to the morning paper and 40% to the afternoon paper. A total of 20% of households subscribe to both.

a. What percentage of households subscribe to at least 1 of the papers?

b. Are these events mutually exclusive?

c. Are these events independent?

77. The owner of a ski resort has been informed by a private weather service that the probability of an abundant snow base is 0.80. If there is an abundant snow base, the probability that the owner will make more than a normal profit is 0.85. What is the probability that there will be an abundant snow base and that the owner will make more than a normal profit?

78. A survey of students at Pemberville Tech revealed the following employment breakdown:

	Full-Time	Part-Time	Unemployed	Total
Male	75	75	50	200
Female	125	75	100	300
	200	150	150	500

a. What is the probability of selecting a female student who works full-time?

b. What is the probability of selecting a male student who works?

c. Given that a male student is selected, what is the probability that he is not employed?

d. Are gender and employment independent?

79. The Department of Labor conducted a study of 29.3 million families with children under 18 years old. Two traits were noted for each family—namely, whether the family was headed by a female and whether the mother worked. These were the results:

Status of Mother	Male Family Head (in millions)	Female Family Head (in millions)
In labor force	10.9	2.4
Not in labor force	14.4	1.6

If a family is selected at random from this population,

a. what is the probability that the mother is in the labor force?

b. what is the probability that the family is headed by a female?

c. what is the probability that the family is headed by a female or that the mother is in the labor force?

d. given that the mother is in the labor force, what is the probability that she heads the family?

80. A recent newspaper article stated that the probability of downing an attacking airplane at each of 4 independent missile stations was 0.25. If an attacking plane had to pass over all 4 stations and the 4 stations were independent, what is the probability that the plane would reach the target?

81. Twenty-eight percent of U.S. households have dogs, 21% have cats, and 10% have both a dog and a cat.

a. What proportion has either a dog or a cat?

b. What is the probability that a household has neither a cat nor a dog?

c. If a household has a cat, what is the probability that it also has a dog?

d. If a household does not have a dog, what is the likelihood that it has a cat?

82. Mr. Benzy just retired at age 65. His wife is 62 years old. If the probability is 0.50 that a man aged 65 will live another 10 years and the probability is 0.70 that a woman aged 62 will live

another 10 years, what is the probability that both Mr. and Mrs. Benzy will live another 10 years? (Assume the husband's life expectancy is independent of the wife's life expectancy.)

83. Titusville Oil Company is currently drilling at 2 new sites. Based on the available geological evidence, the probability of striking oil at Site I (in Oklahoma) is 0.40 and at Site II (in Pennsylvania) is 0.50.
    a. Are these events independent?
    b. What is the probability that both drilling operations will be successful?
    c. What is the probability that neither will be successful?
    d. Use the complement rule to compute the probability that at least 1 operation is successful.

84. A hospital has 2 independent energy sources. Records show there is a 0.98 chance that the primary source will operate during severe weather conditions. The probability that the backup power source will operate under severe weather conditions is 0.95.
    a. What is the probability that both will fail?
    b. What is the probability that the primary source will fail and the backup source will operate properly?
    c. What assumption did you make in order to answer part b?

85. Tests employed in the detection of lung cancer are 90% effective; that is, they fail to detect the disease correctly 10% of the time. If 3 persons, all known to have cancer, are tested, what is the probability that the disease will be detected in all 3 cases? What is the probability that 2 cases will be detected and 1 case will not?

86. A sports newsletter estimates that the probability that Nick Faldo will win the British Open golf tournament next year is 1/9 and the probability that Seve Ballesteros will win is 1/11. If these estimates are correct, what is the probability that
    a. neither of them will win the tournament?
    b. one of them will win the tournament?

87. In the Washington School District, there are 100 voters. Of these voters, 45 are male, and 62 are in favor of annexation by the city of Greenville. Of those in favor of annexation, 22 are male. Find the probability of selecting 1 of these voters and the probability of that voter being in favor of annexation and female.

88. In a recent study, 670 Americans were selected at random. Each person was asked how he got to work yesterday. They were also classified as living in a rural or an urban area. The results are as follows:

Type of Transportation	Type of Worker		
	Urban	Rural	Total
Automobile	400	200	600
Public transportation	50	20	70
Total	450	220	670

If a worker is selected at random, what is the probability that the worker
a. is an urban worker?
b. uses public transportation?
c. is an urban worker or uses public transportation?
d. is an urban worker and uses public transportation?
e. is an urban worker, given that the worker uses public transportation?

89. A recent report on drunk drivers in Ohio indicated the age and number of Ohio first-time and repeat offenders.

Age	First-Time	Repeat	Total
16–20	5311	519	5830
21–25	10,713	4104	14,817
26–30	10,301	5719	16,020
31–35	8246	4344	12,590
36–40	5442	2596	8038
41–45	3474	1719	5193
Total	43,487	19,001	62,488

A driver is selected at random.
a. What is the probability that the driver is a repeat offender?
b. What is the probability that the driver is a repeat offender or a driver under age 21?
c. What is the probability that the driver is a repeat offender and a driver under age 21?

d. What is the probability that the driver is a repeat offender, given that the driver is under age 21?

90. A recent study by the Department of Education showed that 20% of adult males and 17% of adult females cannot read or write. Assume the population is 49% male and 51% female.

a. In the table below, determine the joint probabilities by indicating the fraction of the adult population that falls in each of the categories.

	Illiterate	Not Illiterate	Total
Male			.49
Female			.51
Total			1.00

b. What is the probability that a randomly selected individual cannot read or write?

c. A person who is illiterate is selected at random. What is the probability that the person is male?

DATA EXERCISES

91. Refer to the real estate data set, which reports information on homes sold in Alabama during 1996.

a. Arrange the selling prices into three classes: $0 up to $180,000; $180,000 up to $190,000; and $190,000 or more. Then create a table that shows the grouped selling prices and whether the homes had attached garages.

(1) What percentage of the homes sold for less than $180,000?

(2) What percentage of the homes sold for less than $180,000 and did not have an attached garage?

(3) Given that they did not have attached garages, what percentage of the homes sold for less than $180,000?

(4) What percentage of the homes did not have an attached garage or sold for less than $180,000?

b. Using the grouped selling prices from part a, create a table that shows the grouped selling prices and the numbers of bedrooms.

(1) What percentage of the homes sold for $190,000 or more?

(2) Given that they sold for at least $190,000, what percentage of the homes had 5 bedrooms?

(3) What percentage of the homes sold for at least $190,000 and had 5 bedrooms?

(4) What percentage of the homes sold for at least $190,000 or had 5 bedrooms?

(5) Suppose 2 of the new homeowners are to be contacted by the real estate company. What is the likelihood that both of their homes have 5 bedrooms?

92. Refer to the School Data Set, which has information on 94 school districts in Northwest Ohio.

a. Group the districts based on the percentages of students on welfare: "low" (less than 5%), "moderate" (between 5 and 10%, inclusive), and "high" (more than 10%).

(1) A school district is randomly selected. Then a student within that district is chosen. What is the probability that he or she passed the proficiency exam?

(2) If it is a low welfare district, what is the probability that he or she passed the exam?

(3) What is the probability that the student is from a high welfare district and passed the exam?

(4) Find the probability that the student either is from a moderate welfare district or failed the exam.

b. Now arrange the districts by size: "small" (less than 1000 students), "medium" (between 1000 and 3000 students, inclusive), and "large" (more than 3000 students).

(1) What percentage of the districts are small?

(2) If a district is small, estimate the probability that it is a low welfare district.

(3) What percentage of the districts are both small and low welfare?

(4) What percentage are either small or low welfare?

(5) If 3 districts are randomly chosen, what is the probability that they are all medium in size?

CHAPTER ACHIEVEMENT TEST

The answers are at the back of the book.

MULTIPLE-CHOICE QUESTIONS

Select the response that best answers each of the questions.

1. Which definition of probability includes the condition that the events be equally likely?
   a. Empirical
   b. Subjective
   c. Classical
   d. None of the above
2. Someone's "educated guess" fits which of the following definitions of probability?
   a. Classical
   b. Empirical
   c. Subjective
   d. None of the above
3. The result of an experiment is called
   a. an event.
   b. a subjective probability.
   c. a probability.
   d. an outcome.
4. An experiment has
   a. only one result.
   b. only two results.
   c. two or more results.
   d. None of the above
5. The combination of two or more outcomes results in
   a. an event.
   b. a subjective probability.
   c. an experiment.
   d. a multiplicative rule.

6. For the special rule of addition, the events must be
   a. independent.
   b. mutually exclusive.
   c. equally likely.
   d. not equally likely.
7. If the occurrence of one event does not affect the occurrence of another event, the two events are
   a. mutually exclusive.
   b. independent.
   c. always equal.
   d. conditional.
8. Two events are mutually exclusive if
   a. they overlap on a Venn diagram.
   b. when one event occurs, the other cannot occur.
   c. the probability of one event occurring does not alter the probability that the other will occur.
   d. Both a and b
9. If the order of a set of objects matters, then the set is
   a. a combination.
   b. a permutation.
   c. a sample.
   d. an independent event.
10. Selecting a red ace from a deck of cards could be described as
    a. an outcome.

b. a sample.

c. an event.

d. a complement.

11. The combined events *A* and *B* are shown in a Venn diagram by the area

    a. inside both regions *A* and *B*.

    b. included in region *A* or region *B*.

    c. outside the two regions.

    d. None of the above

12. If two events are mutually exclusive, the probability that both will occur

    a. is 0.

    b. is 1.

    c. is 0.5.

    d. cannot be determined.

13. If the outcome of event *A* is not affected by event *B*, the two events are

    a. mutually exclusive.

    b. exhaustive.

    c. independent.

    d. None of the above

14. We find the number of ways 4 workers can be chosen for a trip from 50 employees by calculating

    a. 4^{50}.

    b. 50!/46!.

    c. 4 × 50.

    d. 50!/4!46!

15. We find the number of distinct ways 6 people can be seated at a table by calculating

    a. 6^6.

    b. 6!.

    c. 6 × 6.

    d. 6!3!3!.

COMPUTATION PROBLEMS

16. Of 500 items produced at Magnum Manufacturing, 200 are produced on the first shift, 175 on the second, and 125 on the third. What is the probability that a randomly selected item is produced on either the first or the third shift?

17. An English class, Introduction to American Poetry, includes 30 students—10 men and 20 women. Five of the men and 2 of the women are out-of-state students.

    a. Summarize this information in a table.

    b. What is the probability of selecting a male student who is from out of state?

    c. What is the probability of selecting a male student, given that the student is from out of state?

    d. What is the probability of selecting a male student or an out-of-state student?

18. A box contains six ham and four cheese sandwiches. Two sandwiches are randomly selected and eaten. What is the probability that both were ham sandwiches?

19. If 35% of the visitors to St. Croix tour Long John Silver's Castle, 25% shop at Cadbury's Department Store, and these events are independent, what proportion of tourists visits *both* attractions?

20. Forty percent of the residents of LaCasa visited the zoo at least once during the last year. What proportion did *not* go to the zoo last year?

21. You have seen three of eight movies showing at the local cinema. You and a friend randomly select a movie to attend tonight. What is the probability that you have not seen the movie?

22. For two events, *A* and *B*, $P(A) = 0.67$, $P(B) = 0.23$, and $P(A \text{ and } B) = 0.12$. Find each of the following values:

    a. $P(\sim A)$

    b. $P(A \text{ or } B)$

23. A contest judge must select a winner, a first runner-up, and a second runner-up from 30 entrants. How many distinct ways can the judge do it?

24. A pollster will select 3 states from a region consisting of 15 states. How many different samples are there?

25. You must match 5 numbers selected at random from the numbers 1 through 30, inclusive, to win the Big Lotto. What is the probability that your ticket will win?

ANSWERS TO SELF-REVIEW PROBLEMS

5-1 **a.** The count of favorable responses
 b. Any integer between 0 and 6, inclusive
 c. Here are 3 possible events:
   1. The majority of comments (4, 5, or 6) are favorable.
   2. Nobody agrees (0).
   3. Everybody liked it (6).

5-2 **a.** 1/52, or 0.0192
 b. 4/52, or 0.0769
 c. The classical concept

5-3 **a.** $P(1) = 30/200 = 0.15$
 b. $P(3) = 40/200 = 0.20$
 c. The empirical concept

5-4 **a.** Our estimate is 0.10; your estimate will no doubt be different.
 b. The subjective concept

5-5 **a.** $48/200 = 0.24$
 b. $7/200 = 0.035$
 c. $55/200 = 0.275$, using the special rule of addition

5-6 "Over 3.00" is event A; "employed" is event B.
$$P(A \text{ or } B) = P(A) + P(B) - P(A \text{ and } B)$$
$$= 0.45 + 0.25 - 0.10$$
$$= 0.60$$

5-7 **a.** $P(A \text{ and } M) = P(A) \cdot P(M|A)$
 b. 0.25, found by $23/60 \times 15/23$

5-8 **a.** 25/60
 b. 8/25

c.

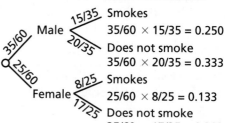

Joint Probabilities

Male — Smokes $35/60 \times 15/35 = 0.250$

Male — Does not smoke $35/60 \times 20/35 = 0.333$

Female — Smokes $25/60 \times 8/25 = 0.133$

Female — Does not smoke $25/60 \times 17/25 = 0.283$

5-9 **a.** Yes
 b. $P(A \text{ and } B) = P(A) \cdot P(B) = 1/2 \times 1/2 = 1/4 = 0.25$
 c. $1/2 \times 1/2 \times 1/2 = 1/8 = 0.1250$

5-10 $2 \times 3 \times 5 = 30$

5-11 $n = 4$
$r = 3$
$$_4P_3 = \frac{4!}{(4-3)!}$$
$$= 4 \times 3 \times 2 = 24$$

5-12 $_7C_5 = \dfrac{7!}{(7-5)!5!} = \dfrac{7 \cdot 6}{2 \cdot 1} = 21$

5-13 **a.** Permutation
 b. $_4P_2 = \dfrac{4!}{(4-2)!} = 12$
 c. Combination
 d. $_4C_2 = \dfrac{4!}{(4-2)!2!} = 6$

CHAPTER 6

Probability Distributions

OBJECTIVES

When you have completed this chapter, you will be able to

- define a random variable;
- compute the mean and variance of a random variable;
- define a probability distribution;
- distinguish between discrete and continuous probability distributions;
- list the characteristics of the binomial probability distribution and compute probabilities associated with it;
- list the characteristics of the Poisson probability distribution and compute probabilities associated with it.

CHAPTER PROBLEM   A Taxing Concern

Lewis Gibson is the Conewango County representative to the Pennsylvania state legislature. He prides himself on his ability to keep in touch with his constituents. Of particular concern at this time is state income tax reform. Mr. Gibson estimates that 80% of voters in his county favor some type of reform, but decides to randomly contact 6 voters to see how many actually support it. What is the likelihood that exactly 5 of the 6 favor income tax reform? What is the likelihood that a majority of the 6 favor reform?

◼ INTRODUCTION

With Chapter 5, we started our investigation of statistical inference. The main objective in statistical inference is to be able to make decisions about an entire population based on a study of just part—that is, a sample—of that population. We illustrated that calculating the probability of an event happening is a very useful technique in this type of decision making. Recall that a probability is the likelihood that a particular outcome will happen.

In the previous chapter, we also calculated the probability of specific events. For example, we determined the probability of selecting no persons with a Hispanic surname for a jury of 12. Now we will study the entire range of events that may result from an experiment. To describe the likelihood of each outcome of this range of events, we use a probability distribution.

◼ WHAT IS A PROBABILITY DISTRIBUTION?

probability distribution
A listing of the outcomes of an experiment that may occur and their corresponding probabilities.

To draw accurate conclusions about a population from which a sample was taken we need to know the probability of the various outcomes of a sample. What is needed is a distribution of the probabilities for specific outcomes. This distribution is called a **probability distribution.** It gives all the outcomes of some random event and their corresponding likelihoods. It is similar to the percentage frequency distribution, described in Chapter 2, which reports the percentage of observations in each class. However, instead of just describing what has occurred, a probability distribution is used to project into the future and describe what will probably occur.

For example, if a coin is tossed twice, the possible outcomes are

First Toss	Second Toss
H	H
H	T
T	H
T	T

where H represents a "head" and T represents a "tail." Table 6–1 describes a situation in which only the *number* of tails that will occur is of interest, and not the order in which they occur. The probability distribution for the number of tails that may occur on two tosses of a coin is shown graphically in Figure 6–1. The number of tails (X) is shown on the horizontal axis and the probability of X on the vertical axis.

As stated, then, a probability distribution lists the values that may occur (0, 1, or 2 on the X axis) and their corresponding probabilities (0.25 or 1/4, 0.50 or 1/2, and 0.25 or 1/4 on the Y axis). As we shall see in later chapters, probability distributions provide the theory on which is built statistical inference—that is, reasoning from a sampled portion of a population to charac-

TABLE 6–1	Probability Distribution for the Number of Tails Resulting from Two Tosses of a Coin	
	Number of Tails X	**Probability** $P(X)$
	0	$1/4 = 0.25$
	1	$1/2 = 0.50$
	2	$1/4 = 0.25$
	Total	$4/4 = 1.00$

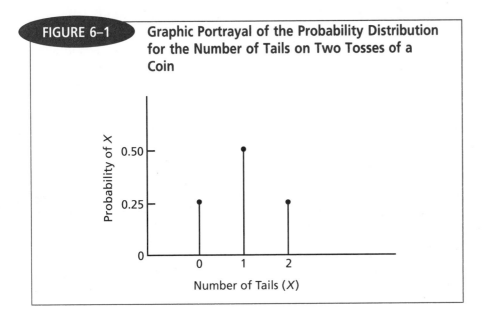

FIGURE 6–1 Graphic Portrayal of the Probability Distribution for the Number of Tails on Two Tosses of a Coin

teristics of the entire population. Two important features of a probability distribution are as follows:

1. The probability of an outcome, X, must always be between 0 and 1, inclusive.
2. The sum of the probabilities of all possible mutually exclusive outcomes is 1.

In the coin-tossing experiment, each outcome has a probability between 0 and 1, inclusive, and the sum of the probabilities of 1/4, 1/2, and 1/4 is 4/4, or 1.00.

Self-Review 6–1

Answers to all Self-Review problems are at the end of the chapter.
 As an experiment, a coin is to be tossed three times. Some of the possible outcomes are

	Toss	
1	2	3
H	H	H
H	H	T
H	T	H

a. Complete the list of possible outcomes.
b. Develop a probability distribution showing the number of tails possible.
c. Portray the probability distribution in a graph.

DISCRETE AND CONTINUOUS RANDOM VARIABLES

The number of tails that may occur when a coin is tossed twice is also an example of a **random variable.** The random variable, in that instance, was designated by the letter X and could assume any 1 of the 3 numbers (0, 1, or 2) assigned to it.

A random variable may take two forms: discrete or continuous. A **discrete random variable** can assume only certain distinct values. The number of male children in a family with 5 children can be only 0, 1, 2, 3, 4, or 5; likewise, the number of highway deaths in Kansas during the Thanksgiving weekend can be counted only in increments of 1. Both are examples of discrete random variables. Note that there can be 0, 1, 2, 3, . . . , but there cannot be 2¼ children or 17.375 deaths. This does not rule out the possibility that a discrete random variable would assume fractional values. However, the fractional values must have distance between them. For example, mortgage interest rates and stock prices are discrete random variables that are commonly expressed in dollar fractions. The price of IBM stock, for example, could be 98, or 98¼, or 98¾, but not 98.52174.

Continuous random variables will be discussed in detail in Chapter 7. One example of them would be the weight of young men in your class. A weight might be 176 pounds, or 176.1 pounds, or 176.13 pounds, or 176.134 pounds, and so on, depending on the accuracy of the scale.

How do you tell the difference between a discrete and a continuous random variable? Discrete random variables generally result from counting, whereas continuous random variables are the result of some type of measurement.

A **discrete probability distribution** is based on a discrete random variable. The difference between a random variable and a probability distribution is that the random variable lists only the possible outcomes, whereas *the probability distribution includes both the list of possible outcomes and the probability of each outcome.*

random variable A quantity that assumes a numerical value as a result of an experiment.

Self-Review 6–2

State which of the following random variables are discrete and which are continuous:

Experiment	Random Variable
a. The number of base hits by a baseball player who bats 5 times during a game.	He can have 0, 1, 2, 3, 4, . . . base hits.
b. The number of patients who enter the emergency room of a hospital for treatment during a day.	There can be 0, 1, 2, 3, . . . patients.
c. The time it takes to fly by commercial airline from Detroit to Sarasota, Florida.	It can take 4 hours 26 minutes, or 4 hours 26 minutes 32⅔ seconds, and so on, depending on the accuracy of the timing device.
d. The number of printing errors in 100 pages of text.	There can be 0, 1, 2, . . . errors.
e. The number of fish caught on a fishing trip.	There can be 0, 1, 2, . . . fish.
f. The length of the first fish caught each day of a fishing trip.	The length can be 12¼", 12.2478", and so on.
g. The length of a power outage.	The time can be 10 minutes, 10.5 minutes, 10.53 minutes, and so on.

Why Study Probability Distributions?

We need to calculate the likelihood of a particular sample outcome so that we can draw conclusions about the entire population. The probability of each possible outcome of some experiment is organized into a probability distribution. The following example gives a hint of the manner in which a probability distribution can be used to make an inference about, or "test" for,

TABLE 6–2	Number of Black Families in a Sample of 10	
Number X	$P(X)$	
0	0.028	
1	0.121	
2	0.233	Very probable
3	0.267	
4	0.200	
5	0.103	
6	0.037	
7	0.009	Almost impossible
8	0.001	
9	0.000	
10	0.000	

racial bias. Consider a community composed of 70% white families and 30% black families. Suppose 10 families have been randomly selected to be interviewed on the advisability of rezoning a school district. The probability distribution for the number of black families among the 10 selected is shown in Table 6–2.

Observe that the most probable events are 2, 3, and 4 black families. In other words, most of the time the selection process will result in 2, 3, or 4 black families being included. The event that 7, 8, 9, or 10 black families out of the 10 will be selected is quite improbable.

THE MEAN (μ) AND THE VARIANCE (σ^2) OF A PROBABILITY DISTRIBUTION

In Chapters 3 and 4, measures of location and spread for a frequency distribution were discussed. The mean is a measure of location and the variance is a measure of spread. In an analogous fashion, a probability distribution can be summarized by its mean, denoted by the Greek letter mu (μ), and its variance, denoted by the Greek letter sigma (σ^2).

The Mean

The mean is a value that is typical of the probability distribution, and it also represents the long-run average value of the random variable. The mean for a probability distribution is also called the **expected value.** The expected value is denoted $E(X)$ and is a "weighted" average. The possible values of the probability distribution are weighted by their corresponding probabilities.

The symbol $P(X)$ is the probability of a particular value X. The mean (μ or expected value) of the probability distribution is computed by

$$\mu = E(X) = \Sigma[X \cdot P(X)] \qquad \boxed{6\text{–}1}$$

In words, the formula directs you to multiply each value, X, by its probability, $P(X)$, and then add these products.

Problem   The Pizza Palace offers 3 sizes of cola—small, medium, and large—to go with its pizza. The colas are sold for $0.60, $0.75, and $0.85, respectively. Forty percent of the orders are for the small cola, 40% for the medium, and 20% for the large. What is the mean amount charged for cola?

Solution   This is an example of a discrete probability distribution because the outcomes are mutually exclusive. A particular cola cannot be both medium *and* large. A customer can purchase only certain sizes of cola. (It is not possible to purchase a $0.63 cola.)

We determine the mean amount, or expected value, charged by weighting the amount charged by the corresponding percentage of the time each size of cola is purchased. The mean for this probability distribution is $0.71, found by using formula 6–1:

$$\mu = E(X) = \Sigma[X \cdot P(X)]$$
$$= (\$0.60)(0.40) + (\$0.75)(0.40) + (\$0.85)(0.20)$$
$$= \$0.71$$

The calculations above could be shown in a table.

Cola Size	Amount Charged	Probability	X · P(X)
Small	$0.60	0.40	$0.24
Medium	0.75	0.40	0.30
Large	0.85	0.20	0.17
		1.00	E(X) = $0.71

The Variance

We noted in Chapter 3 that the arithmetic mean is a typical value for a distribution. However, as explained in Chapter 4, it does not describe the amount of spread (variation). The variance measures the spread. It allows you to compare two distributions with the same or similar means, but with possibly different amounts of spread.

The formula for the variance (σ^2) is

$$\sigma^2 = \Sigma[(X - \mu)^2 \cdot P(X)] \qquad \text{6–2}$$

To determine the variance

1. subtract the mean from each value,
2. square these differences,
3. multiply each squared difference by its probability, and
4. sum the resulting products to arrive at the variance.

As noted in Chapter 4, the square root of the variance is the standard deviation.

Problem Compute the variance and the standard deviation of the price of cola at the Pizza Palace. Show the details of the calculations and interpret the result.

Solution The information is repeated below.

Size	Amount Charged X	Probability P(X)	$(X - \mu)^2$	$(X - \mu)^2 \cdot P(X)$
Small	$0.60	0.40	$(\$0.60 - \$0.71)^2$	(0.0121)(0.40)
Medium	0.75	0.40	$(\$0.75 - \$0.71)^2$	(0.0016)(0.40)
Large	0.85	0.20	$(\$0.85 - \$0.71)^2$	(0.0196)(0.20)
				0.0094

Using formula 6–2, the variance is 0.0094

$$\sigma^2 = \Sigma\left[(X - \mu)^2 \cdot P(X)\right]$$

$$= (0.0121)(0.40) + (0.0016)(0.40) + (0.0196)(0.20)$$

$$= 0.0094$$

The standard deviation is $0.097, found by $\sqrt{0.0094}$. Both the variance (0.0094) and the standard deviation ($0.097) are measures of variation in the price of cola at the Pizza Palace. Note that the standard deviation is measured in the same units as the original values. Thus, if the Sub Shoppe sells cola and the distribution of its prices has an arithmetic mean of $0.71 and a standard deviation of $0.05, then we know that their colas cost about the same as those at the Pizza Palace, but that there is less difference from customer to customer at the Sub Shoppe.

Self-Review 6–3

The Pizza Palace is located next to the Big Burger. The Big Burger offers four sizes of cola at the following prices:

Size	Price	Probability
Small	$0.50	0.40
Medium	0.60	0.20
Large	0.90	0.20
Extra large	1.10	0.20

a. Compute the mean price.
b. Compute the standard deviation.
c. Compare the means and the standard deviations for colas at the Pizza Palace and the Big Burger.

Exercises

Answers to the even-numbered Exercises are at the back of the book.

1. Compute the mean and the standard deviation of the following probability distribution:

Random Variable	Probability
10	0.60
20	0.30
30	0.10

2. Consider the following probability distribution:

Random Variable	Probability
25	0.10
35	0.25
45	0.40
55	0.20
65	0.05

Compute the mean and the standard deviation.

3. Find the mean (or expected value) and the standard deviation of the number of spots that appear face up when a fair die is rolled one time.

4. The probability distribution for the number of toppings, in addition to cheese, ordered on a pizza is

Number of Items	0	1	2	3	4
Probability	0.3	0.4	0.2	0.06	0.04

Compute the mean and the standard deviation.

5. The Perrysburg City Ambulance Service is requesting additional funds from the city council. In support of its request, the following probability distribution was presented by Dirk Plessner, director of the service:

Number of Trips per Day	Fraction of Days
0	0.10
1	0.40
2	0.30
3	0.10
4	0.10

According to Mr. Plessner, this indicates that on 30% of the days, 2 trips, or ambulance runs, were made. Compute the mean number of trips per day and the standard deviation.

6. Tim Waltzer, principal of Home Street School, holds 3 open houses each school year. Records show the following probabilities that a child's parents (one or both) will attend from 0 to 3 of the open houses:

Number of Open Houses Attended	Probability
0	0.20
1	0.20
2	0.30
3	0.30

Compute the mean and the standard deviation.

7. One hundred tickets are sold for a lottery with 4 prizes: $100, $50, $25, and $10. What is the expected value of a ticket?

8. From experience, an insurance company knows 10% of a particular kind of policy will require payouts of $150,000 each. What would be a "fair" premium to charge for this particular kind of policy? "Fair" is usually interpreted to mean the expected value would be zero, so neither the company nor the policyholder loses any money.

THE BINOMIAL PROBABILITY DISTRIBUTION

In Self-Review 6–1 you were asked to toss a fair coin three times and count the number of "heads." In Chapter 5, we discussed how a tree diagram is useful to show the possible outcomes. Here we can use a tree diagram to show the possible outcomes of the coin tosses, and particularly the number of heads.

First, notice that there are 8 possible outcomes. We assume that the coin is fair (not loaded or fixed in some way) and that each of the 3 tosses is independent of the others. Using the special rule of multiplication, formula 5–5, we determine the probability of each outcome. The probability of three heads is written $P(H_1)P(H_2)P(H_3) = (\frac{1}{2})(\frac{1}{2})(\frac{1}{2}) = \frac{1}{8}$. The probability of each of the other paths through the tree diagram is also $\frac{1}{8}$. Therefore, we say that the outcomes are equally likely. In this case, the tree diagram shows that there are 8 equally likely outcomes. However, we wanted a distribution of the *number of heads*. To find this distribution we use the tree diagram to count the number of occasions on which all 3 coins would be a head. There is 1. There

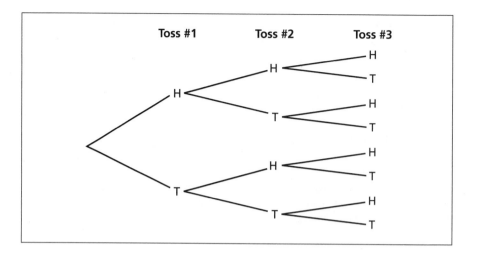

are 3 occasions where there are 2 heads. The complete distribution is as follows:

Number of Heads	Probability
0	1/8
1	3/8
2	3/8
4	1/8

The above situation illustrates one widely used discrete probability distribution—namely, the **binomial distribution.** The following characteristics summarize the binomial distribution:

1. *The binomial distribution is a count of the number of successes.* It is a discrete distribution. *X* can assume only certain integer values—namely, 0, 1, 2, 3, . . . successes. There cannot be, for example, 2½ successes.
2. *Each outcome is classified into one of two mutually exclusive categories.* An outcome is either a "success" or a "failure." For example, a lottery ticket is either a winning number (a success) or not a winning number (a failure). The outcomes are mutually exclusive, meaning that a ticket cannot be both a success and a failure at the same time. We do not record the amount of winnings, only whether we won or lost.
3. *Each trial is independent.* This means that the outcome of one trial does not affect the outcome of any other trial. It is common to use the word *trial* when discussing the binomial probability distribution. Usually, the total number of trials is the same as the size of the sample (*n*).
4. *The probability of success (π) on each trial is the same from trial to trial.* For a lottery ticket, the probability of having a winning number on one ticket is the same as the probability of having a winning number on another ticket. The Greek lowercase π (pi) is often used in statistics to represent the

proportion of successes in a population. It should not be confused with the mathematical constant 3.14159.

Self-Review 6–4

The probability that expectant parents will have a girl is 1/2, or 0.50. Likewise, the probability that they will have a boy is 0.50. The probability of having 0, 1, 2, 3, 4, and 5 girls in a family of 5 children is

Number of Girls X	Probability P(X)
0	0.0312
1	0.1562
2	0.3125
3	0.3125
4	0.1562
5	0.0312

Explain why this distribution qualifies as a binomial probability distribution.

REAL STAT

The *Annals of Internal Medicine* reports that a blood test on a normal person has a 5% chance of being read as abnormal by a laboratory computer. A 12-test profile of a healthy person will therefore appear normal only 54% of the time!

To construct a binomial probability distribution, we must know (1) the total number of trials and (2) the probability of success on each trial. A problem will illustrate the details.

Constructing a Binomial Probability Distribution

Problem   Recall from the Chapter Problem that Lewis Gibson estimates 80% of the voters in Conewango County favor some type of income tax reform in Pennsylvania. To make sure, he randomly selects 6 voters from his county. Construct a binomial probability distribution for the probability that exactly 0, 1, 2, 3, 4, 5, and 6 out of the 6 sampled voters will favor state income tax reform.

Solution   The number of trials is 6 and the probability of success (will vote for tax reform) is 0.80. To illustrate, let us first compute the probability that exactly 5 of the 6 voters selected in the sample will favor tax reform. If F represents a person who favors tax reform and O represents someone who opposes it, one outcome is that the first voter contacted opposes tax reform and the rest favor it. In symbols, this outcome could be written O, F, F, F, F, F. The voter preferences are presumed independent, so the probability of this particular joint occurrence is the product of the individual probabilities. Hence, the likelihood of finding 1 voter who is opposed to tax reform followed by 5 who favor tax reform is

$$(0.2)(0.8)(0.8)(0.8)(0.8)(0.8) = 0.0655$$

However, the problem does not require that the first voter be the one who is opposed to tax reform. The voter opposing tax reform could be any 1 of the 6 people sampled. Any of the 6 outcomes listed below results in exactly 5 people favoring tax reform and 1 opposing it.

Order of Occurrence	Probability of Occurrence
O, F, F, F, F, F	$(0.2)(0.8)(0.8)(0.8)(0.8)(0.8) = 0.0655$
F, O, F, F, F, F	$(0.8)(0.2)(0.8)(0.8)(0.8)(0.8) = 0.0655$
F, F, O, F, F, F	$(0.8)(0.8)(0.2)(0.8)(0.8)(0.8) = 0.0655$
F, F, F, O, F, F	$(0.8)(0.8)(0.8)(0.2)(0.8)(0.8) = 0.0655$
F, F, F, F, O, F	$(0.8)(0.8)(0.8)(0.8)(0.2)(0.8) = 0.0655$
F, F, F, F, F, O	$(0.8)(0.8)(0.8)(0.8)(0.8)(0.2) = \underline{0.0655}$
	0.3930

Therefore, the probability that exactly 5 of the 6 voters sampled will favor tax reform is equal to the sum of the 6 possibilities, or 0.3930.

The probabilities of the other outcomes—0, 1, 2, 3, 4, and 6 of the 6 sampled voters favoring tax reform—could be computed in a similar fashion. A more direct method, however, is to use the **binomial formula**

$$P(X) = \frac{n!}{X!(n - X)!}\pi^X(1 - \pi)^{n-X} \qquad \text{6-3}$$

where

n is the total number of trials. It is 6 in this problem.
π is the probability of success on each trial. It is 0.80.
X is the number of observed successes. Here it is the number favoring reform.

Substituting these values in formula 6–3, the probability that exactly 5 will favor tax reform is

$$P(X) = \frac{n!}{X!(n - X)!}\pi^X(1 - \pi)^{n-X}$$

$$P(5) = \frac{6!}{5!(6 - 5)!}(0.80)^5(1 - 0.80)^{6-5}$$

Recall that 6! (6 factorial) means $6 \times 5 \times 4 \times 3 \times 2 \times 1$. When the same number is included in both the numerator and the denominator, it can be canceled. Continuing:

$$P(5) = \frac{6 \cdot \cancel{5} \cdot \cancel{4} \cdot \cancel{3} \cdot \cancel{2} \cdot \cancel{1}}{\cancel{5} \cdot \cancel{4} \cdot \cancel{3} \cdot \cancel{2} \cdot \cancel{1} \cdot \cancel{1}}(0.80)^5(0.20)^1$$

$$= (6)(0.80)^5(0.20)^1$$

$$= 0.3930$$

This is the same probability obtained earlier. The remaining probabilities for 0, 1, 2, . . . were computed in similar fashion and are shown in Table 6–3.

Problems Let X follow a binomial probability distribution. For each X, compute its probability.

TABLE 6–3	Binomial Probability Distribution for an *n* of 6 and a π of 0.80

Number in Favor of Tax Reform X	Probability of Occurrence $P(X)$
0	0.000
1	0.002
2	0.015
3	0.082
4	0.246
5	0.393
6	0.262
Total	1.000

a. $n = 4, \pi = 0.20, X = 2$
b. $n = 7, \pi = 0.70, X = 5$
c. $n = 5, \pi = 0.60, X = 3$

Solutions

a.   0.154
b.   0.318
c.   0.346

- -

A fire chief estimates that the probability an arsonist will be arrested is 30%. What is the probability that exactly 3 arsonists will be arrested in the next 5 deliberately set fires?

Self-Review 6–5

- -

Exercises

9. For a binomial probability distribution with $n = 6$ and $\pi = 0.4$, what is the probability $X = 4$?

10. For a binomial probability distribution with $n = 8$ and $\pi = 0.3$, what is the probability $X = 6$?

11. An insurance representative has appointments with 4 prospective clients tomorrow. From past experience, she knows that the probability of making a sale on any appointment is 0.20. What is the probability that she will sell a policy to 3 of the 4 prospective clients?

12. It is estimated that 40% of those 65 or older did some type of volunteer work last year. If 5 individuals are randomly selected from this age group, what is the probability that exactly 3 did some type of volunteer work last year?

13. Sixty percent of the applicants to Gator Tech are accepted. What is the probability that exactly four of the next five applicants will be accepted?

14. One-fourth of a certain breed of rabbits are born with long hair. What is the probability that in a litter of five rabbits, none has long hair?

Using Binomial Tables

Another way to find the probabilities needed to construct a binomial distribution is to use tables that give the probabilities for various values of n and π. Such a table is given in Appendix A; a small portion of it, for the case where $n = 6$, is duplicated in Table 6–4.

TABLE 6–4	Binomial Probability Distribution, $n = 6$										
					$n = 6$ Probability						
X	.05	.1	.2	.3	.4	.5	.6	.7	.8	.9	.95
0	.735	.531	.262	.118	.047	.016	.004	.001	.000	.000	.000
1	.232	.354	.393	.303	.187	.094	.037	.010	.002	.000	.000
2	.031	.098	.246	.324	.311	.234	.138	.060	.015	.001	.000
3	.002	.015	.082	.185	.276	.313	.276	.185	.082	.015	.002
4	.000	.001	.015	.060	.138	.234	.311	.324	.246	.098	.031
5	.000	.000	.002	.010	.037	.094	.187	.303	.393	.354	.232
6	.000	.000	.000	.001	.004	.016	.047	.118	.262	.531	.735

In the previous problem involving a sample of 6 voters, a π of 0.80 was used to illustrate the use of formula 6–3. We find the probability of exactly 5 out of 6 voters being in favor of tax reform by using the following procedure, which relies on Appendix A:

1. Find the section where the sample size or the number of trials is given under the first heading n. In this case, it is 6.
2. Within that block, locate the row labeled by the number of successes under the column headed X. It is 5.
3. Move across the row until you reach the column headed by the desired probability, or value of π. It is 0.80.
4. The desired probability is where the row and column of interest intersect. In Table 6–4, for $n = 6$, the value 0.393 is located at the intersection of row 5 and column 8. This result agrees with the earlier calculation.

Problems   Use Appendix A to find each of the following probabilities of X:

a. $n = 6, \pi = 0.20, X = 1$
b. $n = 10, \pi = 0.40, X = 3$
c. $n = 7, \pi = 0.90, X = 6$

Solutions

a.   .393
b.   .215
c.   .372

Self-Review 6–6

In Self-Review 6–5, the fire chief estimated the probability of an arsonist being arrested at 0.3. The question asked was, What is the probability that, for the next 5 arson-related fires, 3 arsonists will be arrested? This probability was computed to be 0.132. Using Appendix A, complete the binomial probability distribution for this problem.

Problem   From Table 6–4, what is the probability that five or more voters favor tax reform? What is the probability that fewer than six favor tax reform?

Solution   We can answer the first question by using the special rule of addition, formula 5–1, which tells us to add the probabilities for 5 and 6: $P(5) + P(6) = 0.393 + 0.262 = 0.655$.

We determine the probability of fewer than 6 either by applying the special rule of addition, formula 5–1: $P(0) + P(1) + P(2) + P(3) + P(4) + P(5) = 0.000 + 0.002 + 0.015 + 0.082 + 0.246 + 0.393 = 0.738$; or by applying the complement rule, formula 5–6: $P(X < 6) = 1 - P(6) = 1 - 0.262 = 0.738$.

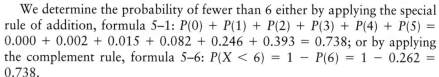

Self-Review 6–7

The binomial probability distribution developed for arsonists in Self-Review 6–6 is

X	P(X)
0	0.168
1	0.360
2	0.309
3	0.132
4	0.028
5	0.002

a. What is the probability that three or more arsonists will be arrested for the next five fires?
b. What is the probability that one or more arsonists will be arrested for the next five fires?

We can also solve the Chapter Problem on tax reform using a statistical software package, such as MINITAB, or a spreadsheet, such as Excel. The MINITAB procedure, for example, is to enter the number of possible successes in a column.

The mouse commands are

Calc ▶ Probability Distribution ▶ Binomial

In the dialog box, set the number of trials as 6 and the probability of a success as .8; then click **OK.** The output is shown on the next page.

MINITAB - Untitled Worksheet - [Data]

File   Edit   Manip   Calc   Stat   Window   Graph   Editor   Help

	C1	C2	C3	C4	C5	C6
↓	X					
1	0					
2	1					
3	2					
4	3					
5	4					
6	5					
7	6					
8	■					
9						

Binomial Distribution   ☒

C1   X

  ⊙ **P**robability
  ○ **C**umulative probability
  ○ **I**nverse cumulative probability

Number of trials: | 6 |
Pro**b**ability of success: | .8 |

⊙ Input **c**olumn: | c1 |
  Optional storage: | |

○ Input co**n**stant: | |
  Optional sto**r**age: | |

[Select]

[?] PDF   [**O**K]   [Cancel]

```
MTB > PDF c1;
SUBC>   Binomial 6 .8.
        K            P( X = K)
      0.00            0.0001
      1.00            0.0015
      2.00            0.0154
      3.00            0.0819
      4.00            0.2458
      5.00            0.3932
      6.00            0.2621
```

The output is the same as Table 6–3, except that there is an additional decimal place.

Appendix A is limited in that it gives probabilities for an *n* of 1 to 20 and π values of 0.05, 0.10, . . . , 0.90, and 0.95. There are two methods for finding probabilities. First use formula 6–3. The second method is to use a statistical software package to generate the probabilities, given particular values for *n* and π. The following is the MINITAB output for an *n* of 22 and a π value of 0.237.

```
MTB > PDF c1;
SUBC>   Binomial 22 .237.
         K          P( X = K)
       0.00           0.0026
       1.00           0.0178
       2.00           0.0580
       3.00           0.1202
       4.00           0.1773
       5.00           0.1982
       6.00           0.1745
       7.00           0.1239
       8.00           0.0721
       9.00           0.0349
      10.00           0.0141
      11.00           0.0048
      12.00           0.0014
      13.00           0.0003
      14.00           0.0001
      15.00           0.0000
```

Several other points should be made regarding the binomial distribution.

1. If *n* remains the same, but π becomes closer to 0.50, the shape of the binomial probability distribution becomes more symmetrical. See Figure 6–2 for an illustration when *n* = 10.

2. If π, the probability of a success, remains the same, but *n* becomes larger and larger, the shape of the distribution also becomes more symmetrical. Figure 6–3 shows various values of *n* with π = .10.

3. The mean (μ) and the variance (σ^2) of a binomial distribution can be computed as follows:

$$\mu = n\pi \qquad\qquad \boxed{6\text{–}4}$$

$$\sigma^2 = n\pi(1 - \pi) \qquad\qquad \boxed{6\text{–}5}$$

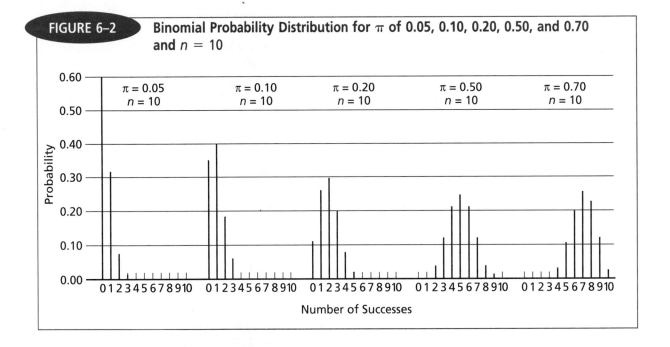

FIGURE 6–2   **Binomial Probability Distribution for π of 0.05, 0.10, 0.20, 0.50, and 0.70 and $n = 10$**

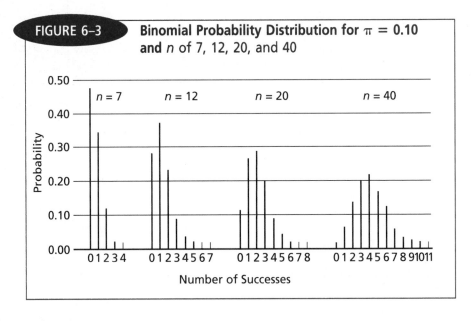

FIGURE 6–3   **Binomial Probability Distribution for $\pi = 0.10$ and n of 7, 12, 20, and 40**

For the Chapter Problem regarding the number of voters favoring tax re-form, where $n = 6$ and $\pi = 0.80$, the mean and variance are computed as follows:

$$\mu = n\pi = 6(0.80) = 4.80$$

$$\sigma^2 = n\pi(1 - \pi) = 6(0.80)(1 - 0.80) = 0.96$$

The mean and variance can also be calculated by formulas 6–1 and 6–2, respectively.

Number in Favor X	P(X)	X · P(X)	$(X - \mu)^2 \cdot P(X)$
0	0.000	0	0
1	0.002	0.002	0.02888
2	0.015	0.030	0.11760
3	0.082	0.246	0.26568
4	0.246	0.984	0.15744
5	0.393	1.965	0.01572
6	0.262	1.572	0.37728
		4.799	0.96260

Notice that the values computed for μ and σ^2 in the above table are the same as those found by formulas 6–4 and 6–5, except for rounding.

How do we interpret these values since in a particular sample there must be a whole number of voters in favor? If we took many samples of 6 voters, then we would find that the mean number favoring reform would be near 4.8. Moreover, most of the samples would have 4, 5, or 6 people in favor.

Exercises

15. Let X follow a binomial probability distribution. Find each of the following probabilities using the binomial tables:
   a. $n = 4$, $\pi = 0.20$, $X = 2$
   b. $n = 7$, $\pi = 0.70$, $X = 5$
   c. $n = 5$, $\pi = 0.60$, $X = 3$

16. Let X follow a binomial probability distribution. Find each of the following probabilities using the binomial tables:
   a. $n = 5$, $\pi = 0.10$, $X = 2$
   b. $n = 10$, $\pi = 0.30$, $X = 4$
   c. $n = 8$, $\pi = 0.40$, $X = 6$

17. The instructor in Political Science 101 gives a weekly five-question multiple-choice quiz. For each question, there are five choices, but only one of them is the correct answer. Suppose a student did not attend class or read the text assignments (a common occurrence). He did, however, take the weekly quiz.
   a. Using the binomial table in Appendix A, develop a binomial probability distribution for the number of correct answers.

   b. The instructor had announced that students who score three or more correct out of five pass the quiz. What is the probability that a student who neither attended class nor read the assignments will pass the test?

18. As noted earlier, half of the newborn babies are girls. For families with five children, what is the probability
   a. of having three girls and two boys?
   b. that all the children are girls?
   c. of having at least one girl?

19. Thirty percent of all automobiles sold in the United States are foreign-made. Four new automobiles are randomly selected.
   a. Referring to Appendix A, what is the probability that none of the four is foreign-made?
   b. Construct a binomial probability distribution showing the probabilities of 0, 1, 2, 3, and 4 out of 4 being foreign-made.
   c. What is the probability that at least one is foreign made?

20. Forty percent of the residents of North Bugenfield are opposed to the widening of Berdan Avenue. In a random sample of ten residents, what is the probability that the number opposed to the widening is

a. exactly five?
b. at least five?
c. between seven and nine, inclusive?
d. fewer than five?

THE POISSON PROBABILITY DISTRIBUTION

Another important discrete probability distribution is the **Poisson distribution.** The Poisson distribution counts the number of successes in a fixed interval of time or within a specified region. Following are some examples where the Poisson distribution could be applied:

- The number of calls arriving per hour at the Nawash Fire Department
- The number of murders reported in a month in Detroit
- The number of orders received by a mail order firm per day
- The number of raisins in a slice of raisin bread

The "fixed interval of time" is an hour in the case of calls to the Nawash Fire Department and a month in the case of reported murders in Detroit. In the example of the slice of raisin bread, we are speaking of a "region." The distribution of the number of raisins in a "thin" slice will be different than in a "thick" slice.

To apply the Poisson distribution, two conditions must be met:

1. The number of successes that occurs in any interval is independent of those that occur in other nonoverlapping intervals.
2. The probability of a success in an interval is proportional to the size of the interval.

A *success* refers to the event of interest, such as the Nawash Fire Department receiving a call. An *interval* refers to the period of time involved, such as an hour. In short, the two important traits of the Poisson distribution are independence and proportionality.

How do these conditions relate to the Nawash Fire Department? If the department receives 3 calls between 8 A.M. and 9 A.M., that outcome does not have any effect on the number they will receive between 10 A.M. and 11 A.M. The probability of a call in the 2-hour interval between 8 A.M. and 10 A.M. is twice that of a call in the 1-hour interval between 3 P.M. and 4 P.M., for example.

The Poisson distribution is described mathematically by the formula

$$P(X) = \frac{\mu^X e^{-\mu}}{X!}$$

6–6

where

μ is the mean number of successes.
e is the mathematical constant 2.7183.

X   is the number of successes in the interval.

$P(X)$ is the probability of X successes in an interval.

The mean of the Poisson distribution is μ, and the standard deviation is $\sqrt{\mu}$. It is positively skewed.

Problem   The number of cars arriving at a gasoline station follows a Poisson distribution, with a mean of 3 cars every 10 minutes. What is the probability that exactly 2 cars will arrive in the next 10 minutes?

Solution   To evaluate this form, we enter 2 for X and 3 for μ in the formula:

$$P(X) = \frac{3^2 e^{-3}}{2!} = \frac{9(.0498)}{2} = 0.224$$

We find that the probability exactly 2 cars will arrive in the next 10 minutes is 0.224.

Figure 6–4 is a graph of the Poisson distribution when $\mu = 3$. Notice that the distribution is positively skewed. For smaller values of μ, the positive skew is more evident. As μ increases, the distribution becomes more nearly symmetrical.

The Poisson distribution can be used to approximate the binomial distribution when the probability of a success (π) is small and the number of trials (n) is very large.

Problem   Records show that the probability a tire will suffer a blowout in the first year is 0.005. If 100 tires are used for a year, what is the probability of exactly 2 blowouts?

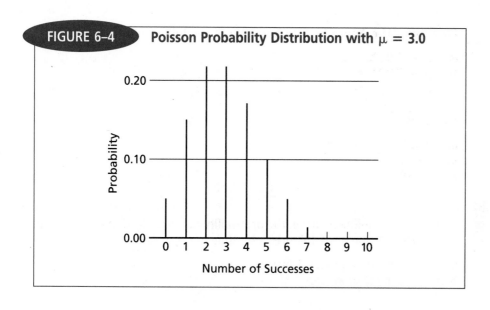

FIGURE 6–4   **Poisson Probability Distribution with $\mu = 3.0$**

Solution   Note that the binomial requirements are satisfied. That is, there is a constant probability of success (0.005), a fixed number of trials (100), only 2 outcomes (the tire blows out or it does not), and independent trials (if the 5th tire blows out, that does not mean that the 18th will). The mean number of blowouts in 100 trials is 0.5, found by formula 6–4, where $\mu = n\pi = 100(0.005)$. The probability of exactly two blowouts is

$$P(2) = \frac{(0.5)^2 \, e^{-.5}}{2!} = \frac{(.25)(.6065)}{2} = 0.0758$$

Appendix B can also be used to obtain the probabilities for selected values of μ. To use Appendix B for this example, find the column headed 0.5. Move down that column to the row where $X = 2$, and read the probability. It is 0.0758, the same as computed previously. This is also given in Table 6–5.

TABLE 6–5   Poisson Probability Distribution

					μ				
X	0.1	0.2	0.3	0.4	0.5	0.6	0.7	0.8	0.9
0	0.9048	0.8187	0.7408	0.6703	0.6065	0.5488	0.4966	0.4493	0.4066
1	0.0905	0.1637	0.2222	0.2681	0.3033	0.3293	0.3476	0.3595	0.3659
2	0.0045	0.0164	0.0333	0.0536	0.0758	0.0988	0.1217	0.1438	0.1647
3	0.0002	0.0011	0.0033	0.0072	0.0126	0.0198	0.0284	0.0383	0.0494
4		0.0001	0.0003	0.0007	0.0016	0.0030	0.0050	0.0077	0.0111
5					0.0002	0.0004	0.0007	0.0012	0.0020
6							0.0001	0.0002	0.0003

Problems   Use Appendix B to find each of the following probabilities:

a. $\mu = 2.0, X = 6$
b. $\mu = 7.0, X = 9$
c. $\mu = 0.9, X = 1$

Solutions

a.   .0120
b.   .1014
c.   .3659

Self-Review 6–8   In the Problem-Solution above, the probability of a tire blowing out in the first year is 0.005. For a set of 100 tires, use Appendix B to find

a. the probability of exactly 1 blowout.
b. the probability of at least 1 blowout.

Exercises

21. If X follows a Poisson probability distribution with $\mu = 0.8$, what is the probability that $X = 1$.
22. If X follows a Poisson probability distribution with $\mu = 0.7$, what is the probability that $X = 3$.
23. The mean number of accidents in a shoe factory is 0.10 per day. What is the probability that during a randomly selected day
    a. there will be no accidents?
    b. there will be exactly 1 accident?
    c. there will be at least 1 accident?
24. Five percent of the graduates of the Modern Method Driving School fail to obtain a driver's license on their first attempt. If there are 60 graduates this month, what is the probability that they will all pass?

25. Lloyd's of London has determined that each year 1 in 1000 people is injured while driving a stunt car for a movie. If Lloyd's insures 400 stunt drivers, what is the probability that in a particular year
    a. none of their policyholders will be injured?
    b. at least 2 of their policyholders will be injured?
    c. no more than 2 of their policyholders will be injured?
26. The mean number of arrests for shoplifting is 0.20 per week at Hidleman's Department Store. What is the probability that in a particular week
    a. there is 1 arrest?
    b. there are 2 or more arrests?

CHAPTER OUTLINE

I. A random variable is a numerical value determined by the outcome of an experiment.
II. A probability distribution is a listing of all possible outcomes of an experiment and the probability associated with each outcome.
   A. A discrete probability distribution can assume only certain values.
      1. The sum of the possible outcomes is 1.
      2. The probability of a particular outcome is between 0 and 1, inclusive.
      3. The outcomes are mutually exclusive.
   B. A continuous distribution can assume an infinite number of values within a specified range.
III. The mean and the variance of a discrete probability distribution are computed as follows:
   A. The mean:

$$\mu = \Sigma[X \cdot P(X)] \qquad \boxed{6\text{--}1}$$

   B. The variance:

$$\sigma^2 = \Sigma(X - \mu)^2 \cdot P(X) \qquad \boxed{6\text{--}2}$$

IV. The binomial distribution is characterized by the following traits:
   A. The distribution is the result of counting the number of successes in a fixed number of trials.
   B. Each outcome is classified into one of two mutually exclusive categories.
   C. The probability of a success remains the same from trial to trial.
   D. Each trial is independent.
V. The binomial distribution is determined from the following equation:

$$P(X) = \frac{n!}{X!(n - X)!}\pi^X(1 - \pi)^{n-X} \qquad \boxed{6\text{--}3}$$

A. The mean of the binomial:

$$\mu = n\pi \qquad \boxed{6\text{-}4}$$

B. The variance of the binomial:

$$\sigma^2 = n\pi(1 - \pi) \qquad \boxed{6\text{-}5}$$

VI. The Poisson distribution counts the number of successes in a fixed interval of time or within a specified region.

A. It is found from the following equation:

$$P(X) = \frac{\mu^X e^{-\mu}}{X!} \qquad \boxed{6\text{-}6}$$

B. The mean and the variance of the Poisson distribution are both equal to μ.

Exercises

27. Sam sells used cars. Last month he sold 1 car on 30% of the days, 2 cars on 40% of the days, 3 cars on 20% of the days, and no cars on 10% of the days. What is the mean number of cars sold per day? What is the standard deviation?

28. The Clegg Truck Stop offers free refills to those ordering coffee. Bill Clegg, the owner, gathered the following information on the number of refills:

Refills	Percentage
1	40.0
2	30.0
3	20.0
4	10.0

Compute the mean and the standard deviation of the number of refills.

29. The Spoon Appliance Store ran an advertisement stating "Come save an additional $10–$25–$50–$100 off the already low price of our TVs. . . . Bust a balloon with a money-saving coupon inside and have that amount taken off of your purchase price." There is a total of 50 balloons with 35, 10, 4, and 1 containing coupons for $10, $25, $50, and $100, respectively. What is the expected savings for each customer? What is the standard deviation?

30. The Utah Department of Natural Resources reports that the mean number of fish caught per hour in the Conewango River is 0.40. What is the probability that, if Victor Anderson fishes for an hour, he will catch 3 or 4 fish? At least 1 fish?

31. The suicide prevention unit in a particular city estimates that 20% of the callers are serious about taking their lives. If on a particular day the unit received 10 calls, what is the probability that none of the callers was serious? What is the probability that at least 2 were serious?

32. It is estimated that 70% of the law school graduates in a particular state pass the state bar examination on the first try. What is the probability that in a group of 12 students, nobody passes? At least 1 student passes? More than half pass?

33. Past records indicate that 30% of the students enrolling at a particular university graduate within 5 years of their entrance. In a group of 14 newly enrolled students, what is the probability that fewer than half will graduate within the 5 years? More than 8 students will graduate?

34. The school band members are selling boxes of candy door to door in an effort to raise money to purchase new band uniforms. Mr. Sprague, the band director, has found the following probability distribution for the number of boxes sold per house:

Number of Boxes Sold X	Probability P(X)
0	0.50
1	0.30
2	0.20

a. How many boxes does a student expect to sell at a particular house?

b. What is the standard deviation of the distribution?

c. What is the probability that a band member will not sell any candy at two consecutive houses?

35. Ten percent of the students in Professor Green's biology class earn the grade of A. In a sample of seven students who have taken this class, compute the probability that the number of A grades is

a. exactly two.

b. at least one.

c. none.

d. seven.

e. How many students out of the seven would you expect to get an A?

36. The Tameron Apartments are a rather small complex. The owner is studying the number of unoccupied apartments and has gathered the following information:

Number of Vacant Units	Probability
0	0.30
1	0.30
2	0.20
3	0.20

Compute the mean and the standard deviation for the number of vacant units.

37. A new pilot TV show has been created. The network president estimates that 60% of the viewing population will like the new show. If the pilot is shown to 15 people and the president is correct, what is the probability that 10 or more of the 15 will indicate they liked it? How many of the 15 should the network president expect to like the show?

38. List the characteristics of the binomial distribution.

39. A poplar tree, if less than 3 feet high and transplanted in the spring, has a 40% chance of survival. If 6 such trees are transplanted, what is the probability that exactly 5 will survive? What is the probability that 2, 3, 4, or 5 will survive? Compute the mean and the variance.

40. A particular type of birth control device is effective 90% of the time when used as directed. If a sample of 15 couples uses the device, what is the probability that none of the devices will fail?

41. If a baseball player with a batting average of .300 comes to bat 5 times in a game, what is the probability he will get

a. exactly 2 hits?

b. fewer than 2 hits?

c. at least 1 hit?

d. Compute the mean and the variance of the number of hits.

42. It is estimated that AT&T has a 70% share of the long-distance telephone market. Fifteen long-distance calls are selected at random.

a. What is the expected number of AT&T calls?

b. What is the probability that 10 are AT&T calls?

c. What is the probability that 10 or less are AT&T calls?

43. Researchers have defined leading a "sedentary lifestyle" as getting less than 20 minutes of exercise 3 times a week. It is claimed that 60% of all Americans lead a sedentary lifestyle. If you randomly select 8 people, what is the probability that

a. exactly 6 lead a sedentary lifestyle?

b. none of them leads a sedentary lifestyle?

c. at least 5 lead a sedentary lifestyle?

44. Binge drinking (consuming 5 or more alcoholic drinks on a single occasion) is a habit of 20% of the national population. If you go to a party with 6 people, what is the probability that

a. none of them engages in binge drinking?

b. exactly 1 does?

c. more than 1 does?

45. National studies reveal that 40% of adult drivers do not use seatbelts. If 9 adult drivers are stopped at random on the Florida Turnpike, what is the probability that
    a. exactly 3 are not wearing seatbelts?
    b. exactly 3 are wearing seatbelts?
    c. at least 4 are not wearing seatbelts?
    d. at least 4 are wearing seatbelts?

46. Twenty percent of all marriages fail in the first year. What is the probability that the marriages of 4 or fewer of the 15 couples married in Findlay, Ohio last weekend will fail within a year?

47. Thirty percent of all elderly people have a flu shot each year. Suppose Dr. Lowe has 12 elderly patients. What is the probability that
    a. none of them receives a flu shot?
    b. at least 1 receives a flu shot?
    c. most of them receive a flu shot?

48. A large department store finds that 1% of its accounts receivable will not be collected. A sample of 100 accounts is selected for audit.
    a. What is the probability that all accounts will be collected?
    b. What is the probability that at least 1 will not be collected?

49. Customers arrive at the drive-in windows of Bank One at a rate of two per minute.
    a. What is the probability that exactly two customers arrive in a particular minute?
    b. What is the probability of no arrivals in a particular minute?
    c. What is the mean and variance number of arrivals?

50. Health records indicate that 70% of all first-graders are fully immunized by the time they reach age 6. A class of 14 first-grade students is being studied. What is the probability that more than 10 are fully immunized?

51. A bank processes 10 personal loan applications per day. The probability that a particular application will be denied is 0.30. What is the probability that more than two will be denied on the same day?

52. A certain medicine is 70% effective; that is, out of every 100 patients who take it, 70 are cured. A group of 12 patients is given the medicine. Find the probability that

a. 8 are cured.
b. fewer than 5 are cured.
c. 10 or more are cured.

53. It is known that 30% of all high-school graduates go to a 4-year college. If a very small graduating class of 20 is considered, find the probability that
    a. more than 8 go to a 4-year college.
    b. fewer than 6 go to a 4-year college.
    c. more than 9 go to a 4-year college.

54. A telephone salesperson expects to make a sale on 8% of her calls. If 50 calls are placed, what is the probability that
    a. 2 will result in a sale?
    b. at least 1 will result in a sale?

55. A recent study showed that 60% of airline flights from Chicago to Atlanta are on time. A sample of 12 flights is selected.
    a. How many flights would you expect to be on time?
    b. What is the likelihood that more than 7 flights are on time?
    c. What is the likelihood that exactly 7 flights are on time?

56. For the hours 4 P.M.–10 P.M., an average of 5 people arrive per hour needing treatment at St. Luke's Hospital Emergency Room.
    a. What is the likelihood that no people will arrive at a particular hour?
    b. What is the likelihood that less than 5 will arrive at a particular hour?
    c. What is the likelihood that less than 5 will arrive in each of 2 consecutive hours?

57. From past experience, 30% of the students at Aloha University graduate within 5 years. In a freshman writing class, there are 15 students.
    a. How many of these students would you expect to graduate in 5 years? What is the standard deviation?
    b. What is the probability that at least 4 of the students will graduate?
    c. What is the probability that less than 4 will graduate?

58. A student accuses an instructor of being chauvinistic and of harassing female students. As evidence, the accuser points to the fact that the instructor has called on a male student on only four of the last ten occasions. Thirty percent of

the large class are women. What is the probability that the result cited might have occurred by chance alone?

59. Describe the differences between the binomial probability distribution and the Poisson probability distribution.

60. In a certain manufacturing process, 5% of the products do not meet specification. A group of 20 parts is selected. Determine the probability that exactly 1 part is defective using both the binomial and the Poisson probability distributions.

DATA EXERCISE

61. Refer to the real estate data set, which reports information on homes sold in Alabama during 1996.
   a. Create a probability distribution for the number of bedrooms. Compute the mean and the standard deviation of this distribution.
   b. Develop the probability distribution of the number of bathrooms. Compute the mean and the standard deviation of this distribution.

CHAPTER ACHIEVEMENT TEST

The answers are at the back of the book.

MULTIPLE-CHOICE QUESTIONS

Select the response that best answers each of the questions.

1. A listing of all outcomes of an experiment and their corresponding probabilities is called a
   a. probability distribution.
   b. frequency distribution.
   c. random variable.
   d. subjective probability.

2. In every probability distribution, the probabilities must
   a. be increasing.
   b. add up to one.
   c. be constant.
   d. be decreasing.

3. Which of the following is *not* required for a binomial distribution?
   a. Only 2 outcomes
   b. At least 50 observations
   c. Independent trials
   d. A constant probability of success

4. The binomial distribution is symmetric when
   a. n is small.
   b. $n\pi$ is 0.5.
   c. π is 0.5.
   d. π is large.

5. A binomial distribution must have
   a. the same probability of success from trial to trial.
   b. a value of π equal to 1.34.
   c. 3 or more possible outcomes.
   d. None of the above

6. In a Poisson distribution, the
   a. variance equals the median.
   b. standard deviation equals the variance.
   c. mean equals the variance.
   d. standard deviation equals the mean.

7. A Poisson distribution
   a. is negatively skewed.
   b. is positively skewed.
   c. is symmetric.
   d. None of the above
8. Within a specific range, a continuous random variable may assume
   a. an infinite number of values.
   b. only certain values.
   c. a countable number of values.
   d. None of the above

9. If a binomial distribution has $n = 10$ and $\pi = 0.4$, then the mean is
   a. 2.4.
   b. 4.0.
   c. 10.0.
   d. None of the above
10. Which are discrete distributions?
   a. The binomial
   b. The Poisson
   c. Both the binomial and the Poisson
   d. Neither the binomial nor the Poisson

COMPUTATION PROBLEMS

11. Suppose 90% of Toro lawn mowers will start on the first pull. Eight Toro mowers are chosen and tried once. Find the following probabilities:
   a. Exactly six will start.
   b. At least six will start.
   c. Three, four, or five will start.
12. The Consumer Protection Division of Casino City receives a complaint on 1 out of 50 calls. If 150 calls were received today, determine the probability that
   a. no complaints were received.
   b. exactly 2 complaints were received.
   c. fewer than 4 complaints were received.
   d. at least 1 complaint was received.

13. The Pizza Palace recorded the number of extra toppings added to each pizza. Compute the mean and the standard deviation for the number of toppings.

Number of Toppings	Percentage of Purchases
0	5
1	15
2	30
3	40
4	10

ANSWERS TO SELF-REVIEW PROBLEMS

6-1 a. In any order, the complete list is

H	H	H
H	H	T
H	T	H
H	T	T
T	H	H
T	T	H
T	H	T
T	T	T

b.

Number of Tails X	Probability of X
0	1/8 = 0.125
1	3/8 = 0.375
2	3/8 = 0.375
3	1/8 = 0.125
Total	8/8 1.000

c.

6-2 **a.** Discrete
b. Discrete
c. Continuous
d. Discrete
e. Discrete
f. Continuous
g. Continuous

6-3 **a.** μ = \$0.50(0.40) + \$0.60(0.20)
+ \$0.90(0.20) + \$1.10(0.20)
= \$0.72

b. σ^2 = $(0.50 - 0.72)^2$ (0.40)
+ $(0.60 - 0.72)^2(0.20)$
+ $(0.90 - 0.72)^2(0.20)$
+ $(1.10 - 0.72)^2(0.20)$
= 0.0576

$\sigma = \sqrt{0.0576} = 0.24$

c. The cola prices are nearly the same (\$0.71 or \$0.72), but there is more variation at Big Burger (\$0.24 versus \$0.097).

6-4 **1.** Outcomes are mutually exclusive. A baby cannot be a girl and a boy at the same time.

2. There are only two possible outcomes: boy and girl.

3. Trials are independent. If a boy is born, this does not affect the sex of the next child.

4. Probabilities remain the same for each birth: 0.50 and 0.50.

6-5 $P(3) = \dfrac{5!}{3!2!}(0.3)^3(0.7)^2$

$= (10)(0.027)(0.49)$

$= 0.132$

6-6 $n = 5, \pi = 0.3$

X	P(X)
0	0.168
1	0.360
2	0.309
3	0.132
4	0.028
5	0.002

6-7 **a.** 0.162, found by $P(3) + P(4) + P(5) = 0.132 + 0.028 + 0.002$

b. 0.832, found by either $1 - P(0) = 1 - 0.168$ or $P(1) + P(2) + P(3) + P(4) + P(5)$ (slight discrepancy due to rounding)

6-8 **a.** $P(1) = 0.3033$

b. $P(X \geq 1) = 1 - P(0) = 1 - 0.6065 = 0.3935$

CHAPTER 7

The Normal Probability Distribution

OBJECTIVES

When you have completed this chapter, you will be able to

- list the characteristics of a normal probability distribution;
- calculate a standardized value or *z*-value;
- calculate the probability that an observation will occur between two particular points of interest on a normal distribution;
- determine a point beyond which a specific percentage of the observations will occur or a point beyond which there is a specific probability of occurrence on a normal distribution.

CHAPTER PROBLEM   How Fast Was I Going, Officer?

The Federal Highway Administration regulates federal funding for roadway construction and improvements. To encourage individual states to observe national standards for speed and safety, it withholds funds from those states that do not make serious efforts to enforce speed limits. In connection with a certification study, the Ohio Department of Transportation conducted a traffic survey on a section of I-75 near Lima. The mean vehicle speed was 66.7 miles per hour, with a standard deviation of 3.5 miles per hour. What percentage of the vehicles traveled at the 65-mile-per-hour speed limit? What percentage traveled more than 70 miles per hour?

INTRODUCTION

In Chapter 2, raw data were organized into a frequency distribution. Then frequency polygons and other graphs were constructed to describe the location and shape of the frequency distribution. In Chapters 3 and 4, we introduced various measures that describe the central tendency and dispersion in the data. The material in these three chapters is referred to as "descriptive statistics." With Chapter 5, we began our exploration of the idea of probability, which measures the likelihood that some uncertain event will occur. The concept of a probability distribution was developed in Chapter 6. A probability distribution extends the notion of a frequency distribution to include a description of some experiment and of all of its possible outcomes. Chapter 6 described two important *discrete* distributions—the binomial and the Poisson.

In this chapter, we will examine a very important *continuous* probability distribution known as the **normal probability distribution.** Its importance is due to the fact that in practice, experimental results very often seem to follow this mounded, "bell-shaped" pattern. The normal probability distribution is also important because most of the sampling methods developed later in this book are based on it. Incidentally, this distribution came to be known as the "normal" probability distribution because around 1800 absolutely every set of data had to follow it or else statisticians would think something was wrong—not "normal"—with the data. The normal distribution requires that the data be of at least interval scale. The data in Table 7–1 and Figure 7–1 are examples of interval-level measurements. Other types of data that frequently array themselves into a bell-shaped pattern are

- weights of 7-year-old girls
- heights of adult males
- IQ scores
- blood-pressure readings
- scores on the first statistics test

TABLE 7–1	Distribution of Ages of Surgical Patients	
	Age of Patient	**Percentage of Total**
	0 up to 10	9
	10 up to 20	10
	20 up to 30	19
	30 up to 40	20
	40 up to 50	17
	50 up to 60	14
	60 up to 70	11

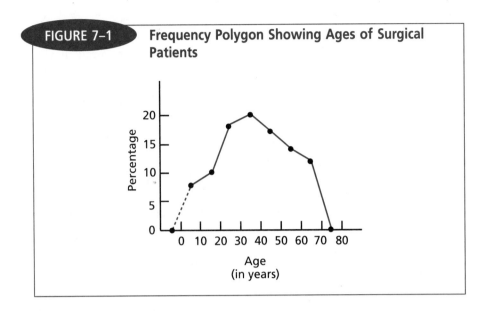

FIGURE 7–1   **Frequency Polygon Showing Ages of Surgical Patients**

CHARACTERISTICS OF A NORMAL PROBABILITY DISTRIBUTION

The normal probability distribution has the following characteristics:

1. The graph of the normal probability distribution has a single peak at the center of the distribution. The mean, the median, and the mode—which in a normal distribution are equal—are all located at the peak. Therefore,

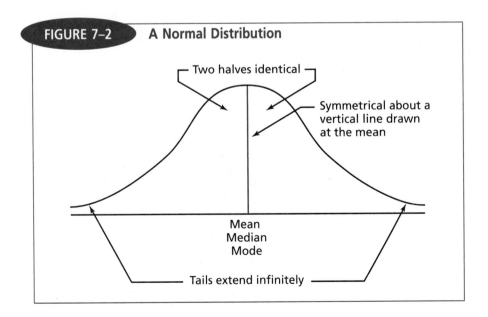

FIGURE 7–2   **A Normal Distribution**

exactly one-half, or 50%, of the area is to the left of the center of the distribution, and exactly one-half of the area is to the right of it.

2. A normal probability distribution is *symmetrical* about its mean. If you were to "fold" the probability distribution along its central value, the two halves would be identical.

3. The normal curve falls off smoothly in a bell shape, and the two tails of the probability distribution extend indefinitely in either direction. In theory, the curve never actually touches the X axis (see Figure 7–2).

THE "FAMILY" OF NORMAL DISTRIBUTIONS

There is not just 1 normal probability distribution. Rather, there is an entire "family" of related normal probability distributions. If you are studying the annual incomes of a group of male employees, that probability distribution might follow a normal distribution, with a mean of $32,394 and a standard deviation of $842. The annual incomes of a group of female employees could follow another normal distribution, with a mean of $27,652 and a standard deviation of $797. For each value of a mean and a standard deviation, there is a particular normal probability distribution. When either the mean or the standard deviation changes, a new normal probability distribution is formed.

Figure 7–3 illustrates three normal probability distributions. Each of them has a mean of 30, but a different standard deviation. Figure 7–4 shows a set of normal probability distributions that have different means, but the same standard deviation of 2.

Figure 7–5 portrays two normal probability distributions for which both the means and the standard deviations are different. Although each of the normal probability distributions displayed in Figures 7–3, 7–4, and 7–5 differs

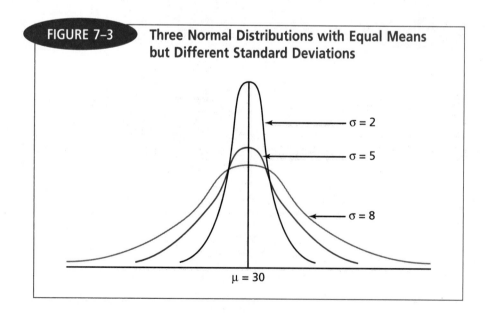

FIGURE 7–3   **Three Normal Distributions with Equal Means but Different Standard Deviations**

σ = 2

σ = 5

σ = 8

μ = 30

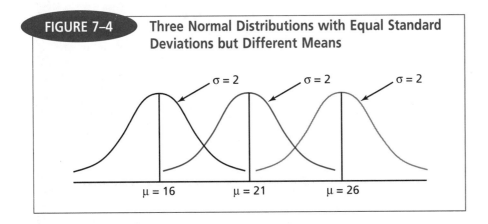

FIGURE 7–4   **Three Normal Distributions with Equal Standard Deviations but Different Means**

$\sigma = 2$ $\sigma = 2$ $\sigma = 2$

$\mu = 16$ $\mu = 21$ $\mu = 26$

somewhat in appearance, each is still a member of the "family" of normal probability distributions.

In dealing with normal probability distributions, we use 3 relationships extensively:

1. About 68% of the distribution is within 1 standard deviation of the mean.
2. About 95% of the observations are within 2 standard deviations of the mean.
3. Virtually all of the area is within 3 standard deviations of the mean.

For example, if a normal probability distribution has a mean of 20 and a standard deviation of 4, then

1. about 68% of the values are between 16 and 24, found by $20 \pm 1(4)$.
2. about 95% of the values are between 12 and 28, found by $20 \pm 2(4)$.
3. virtually all of the values are between 8 and 32, found by $20 \pm 3(4)$.

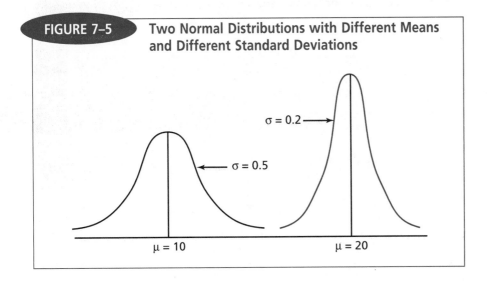

FIGURE 7–5   **Two Normal Distributions with Different Means and Different Standard Deviations**

$\sigma = 0.2$

$\sigma = 0.5$

$\mu = 10$ $\mu = 20$

Problem   The daily temperature for a city in Hawaii is normally distributed, and the mean is 21°C. The standard deviation is 2.5°C. What percentage of the temperatures are within 1, 2, and 3 standard deviations of the mean?

Solution

1.   About 68% of the mean daily temperatures are between 18.5° and 23.5°C, found by $\mu \pm 1(\sigma) = 21 \pm 1(2.5)$.
2.   About 95% of the mean daily temperatures are between 16° and 26°C, found by $\mu \pm 2(\sigma) = 21 \pm 2(2.5)$.
3.   Virtually all of the mean daily temperatures are between 13.5° and 28.5°C, found by $\mu \pm 3(\sigma) = 21 \pm 3(2.5)$.

Self-Review 7–1

Answers to all Self-Review problems are at the end of the chapter.
   A test to measure anxiety is given to males between the ages of 17 and 19. The mean score is computed to be 50 and the standard deviation 6. About 95% of the males have an anxiety score between what 2 values?

THE *STANDARD* NORMAL PROBABILITY DISTRIBUTION

REAL STAT

An individual's skills depend on a combination of hereditary and environmental factors. This suggests that many skills follow the standard normal distribution. For example, scores on the Scholastic Aptitude Test (SAT) are normally distributed, with a mean of 1000 and a standard deviation of 140.

As noted, there are many normal probability distributions—one for each pair of values for a mean and a standard deviation. While this makes the normal probability distribution very versatile in describing many different real-world situations, it would be very awkward to provide tables of areas for each such normal probability distribution. An efficient method for overcoming this difficulty is available. This method calls for *standardizing* the distribution. To find the area between a value of interest (X) and the mean (μ), we first compute the difference between the value (X) and the mean (μ); then we express that difference in units of standard deviation. In other words, we compute the value

$$z = \frac{X - \mu}{\sigma}$$

7–1

where

z is the *standardized* value, or z-value.
X is any observation of interest.
μ is the mean of the normal distribution.
σ is the standard deviation of the normal distribution.

Finally, we find the desired area under the curve, or the probability, by referring to a table whose entry corresponds to the calculated value of z. How the probability is calculated will be discussed shortly.

The value of z actually follows a normal probability distribution with a mean of zero and a standard deviation of one unit. This probability distribution is known as the **standard normal probability distribution.** Thus, we can convert *any* normal distribution to the standard normal distribution by using formula 7–1.

Problem   The ages of patients admitted to the coronary care unit of a hospital are normally distributed with a mean of 60 years and a standard deviation of 12 years. What is the computed z-value (standardized value) for a patient

a.   aged 78?
b.   aged 45?

Solution   The z-values are computed as follows using formula 7–1:

a. $X = 78$ Computing z: $z = \dfrac{X - \mu}{\sigma}$

$$= \dfrac{78 - 60}{12}$$

$$= 1.5$$

b. $X = 45$ Computing z: $z = \dfrac{X - \mu}{\sigma}$

$$= \dfrac{45 - 60}{12}$$

$$= -1.25$$

Note that a z-value merely transforms a selected value to a deviation from the mean expressed in standard deviation units. The locations of the two ages (78 and 45) are shown in Figure 7–6.

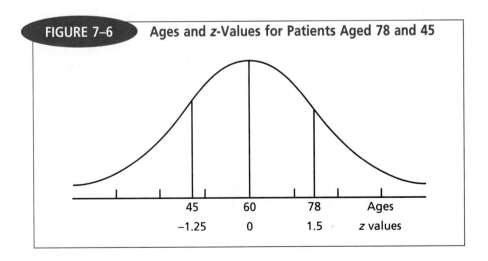

FIGURE 7–6   **Ages and *z*-Values for Patients Aged 78 and 45**

Problems   A normal distribution has a mean of 50 and a standard deviation of 8. Compute the z-value when

a. $X = 56$.
b. $X = 43$.
c. $X = 68$.

Solutions

a.   0.75
b.   −0.875
c.   2.25

Self-Review 7–2	Referring to the ages of patients, what is the z-value ($\mu = 60$ years, $\sigma = 12$ years) for a patient

a. aged 68?
b. aged 33?

Exercises

Answers to the even-numbered Exercises are at the back of the book.

1. A normal distribution has a mean of 30 and a standard deviation of 5. Compute the z-value when $X = 37$.

2. A normal distribution has a mean of 25 and a standard deviation of 8. Compute the z-value when $X = 17$.

3. The monthly food expenditures of families of 5 on welfare were studied. The mean amount spent was \$125 and the standard deviation \$20. Assuming the monthly expenditures are normally distributed,
   a. standardize the expenditure of \$105. That is, convert the expenditure of \$105 to a z-value.
   b. standardize the expenditure of \$145.
   c. What percentage of the welfare families will spend between \$105 and \$145 a month?

4. A mathematics instructor studied the lengths of time required for students to complete the final examination. She found that the mean time was 90 minutes and the standard deviation 10 minutes. If the lengths of time are normally distributed,
   a. 95% of the lengths of time will fall between what 2 times?
   b. Virtually all of the students will complete the final examination between what 2 times?

5. The life of an automatic dryer at a commercial laundry is normally distributed with a mean of 8.5 years and a standard deviation of 0.75 years.
   a. Standardize the value of 8.0 years.
   b. About what percentage of the dryers will last between 7.0 and 10.0 years?

6. A dexterity test is given to young children. The mean score is 20 and the standard deviation is 5.
   a. Standardize a score of 12.
   b. About what percentage of the scores are between 5 and 35?

Comparing Scores on Different Scales

Numbers that are of different scales or are in different units can be compared if we convert them to z-values. This can be explained best by an illustration.

Problem   A person with a mental disability scores 84 on a special anxiety test. The scores for this test are normally distributed with a mean of 80 and a standard deviation of 8. He also scores 28 on a mechanical aptitude test designed especially for persons with disabilities. The scores for this test are normally distributed with a mean of 20 and a standard deviation of 6. Transform the scores to z-values in order to compare the performances of this person on the two tests.

Solution   The z-value for the anxiety score is 0.5, found by formula 7–1:

$$z = \frac{X - \mu}{\sigma} = \frac{84 - 80}{8} = 0.5$$

The z-value for the mechanical aptitude score is 1.33, found as follows:

$$z = \frac{X - \mu}{\sigma} = \frac{28 - 20}{6} = 1.33$$

Interpretation: This person's performance on the anxiety test is slightly above average. This means that relative to other persons with disabilities who took the test, this person's score is 0.5 standard deviation above the arithmetic mean. His performance on the mechanical aptitude test is well above the mean. That is, relative to other persons with disabilities taking the test, this person's score is 1.33 standard deviations above the mean.

It should be noted that a negative z-value would indicate a below-average performance:

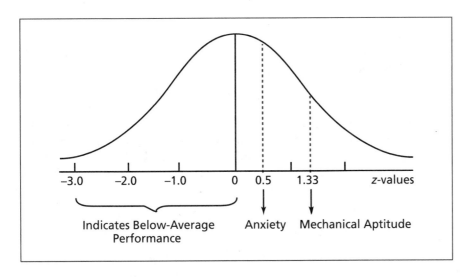

The mean age of prisoners in a state prison is 40 years and the standard deviation 10 years. The ages are normally distributed. The scores on a test measuring the social consciousness of the prisoners are also normally distributed with a mean of 500 and a standard deviation of 100.

A prisoner, age 38, scored 750 on the test. Compare his relative position within the 2 distributions.

Self-Review 7–3

Exercises

7. Martha and Frank Clark are newlyweds. Both are under 25 years of age, and both are college graduates. Their combined income is $65,000 a year. The mean combined income of all newlyweds who are under 25 and are college graduates is $62,000. The standard deviation of that distribution is $5000, and it is normal. Comment on the relative position of the Clarks' income.

8. The hourly wages of 2 people working in the trades are to be compared. Neil Holzmann, a carpenter, earns $12.00 per hour. Joe Bevilacqua, a plumber, earns $14.00 per hour. A survey of both trades in the same city reveals the following information (assume both distributions are normal):

	Plumber	Carpenter
Mean	$15.00	$10.00
Standard deviation	1.75	1.25

Compare the relative positions of the two tradesmen within their respective trades.

9. According to a trade publication, the mean flying time from Detroit to Atlanta is 2 hours, with a standard deviation of 0.15 hours. Yesterday's 10 A.M. flight on Crash Airlines took 2.10 hours, and the 2 P.M. flight took 1.90 hours. The evening flight at 7 P.M. took 2.25 hours. Comment on the relative positions of the 3 flights.

10. The mean time spent in a grocery store is 10 minutes, with a standard deviation of 1.5 minutes. Mrs. Delwhiler shopped this morning and spent 9 minutes in the store. Mrs. Stevens went this afternoon and spent 12 minutes. Comment on the relative amounts of time spent shopping.

Finding Probabilities Between Two Selected Values

Another use of z-values is to determine what percentage of a group of observations will be located *between* two values, or to determine the probability that an observation will occur between two values. Again, we will use a problem to illustrate.

Problem   A research study revealed that the amounts spent by persons seeking a seat on the city council in medium-sized cities are normally distributed, with a mean of $6000 and a standard deviation of $1000. The question to be explored is: What percentage of the candidates seeking office spend between $6000 and $7250?

Solution   The z-value corresponding to $7250 is 1.25, found by formula 7–1:

$$z = \frac{X - \mu}{\sigma} = \frac{\$7250 - \$6000}{\$1000} = 1.25$$

The actual amounts spent, and their corresponding z-values, are depicted on the following graph. The upper scale shows the actual amounts spent and the lower scale shows the z-values. Note that the mean of $6000 on the scale of dollars corresponds to a value of 0 on the z scale.

With that information, we can now determine what percentage of candidates spend between $6000 and $7250. Percentages derived from z-values

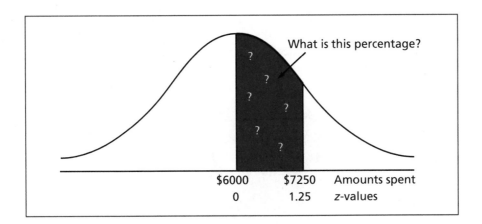

have already been computed and organized into standard tables (see Appendix C, Normal Probability Distribution). The values found in such tables are actually areas under the standard normal curve, or probabilities. A portion of that appendix is shown in Table 7–2.

To find the percentage under the normal curve corresponding to a z-value of 1.25, first go down the left column of the table to a z of 1.2. Then move horizontally to the column headed 5, and read the probability. It is 0.3944. To convert to a percentage, we multiply the number in the table by 100, or, more simply, we move the decimal point 2 places to the right. Converted to a percentage, then, 0.3944 is 39.44%. Interpreting this outcome, we find that 39.44% of the candidates for city council in medium-sized cities spend between $6000 and $7250 on their campaigns.

Recall that one of the characteristics of a normal distribution is its symmetry—that is, the left half of the curve is identical to the right half.

TABLE 7–2	A Portion of a *z*-Table (The Normal Probability Distribution, Appendix C)								
	Second Decimal Place of *z*								
z	0	1	2	3	4	5	6	7	8	9
:	:	:	:	:	:	:	:	:	:	:
1.0	.3413	.3438	.3461	.3485	.3508	.3531	.3554	.3577	.3599	.3621
1.1	.3643	.3665	.3686	.3708	.3729	.3749	.3770	.3790	.3810	.3830
1.2	.3849	.3869	.3888	.3907	.3925	.3944	.3962	.3980	.3997	.4015
1.3	.4032	.4049	.4066	.4082	.4099	.4115	.4131	.4147	.4162	.4177
1.4	.4192	.4207	.4222	.4236	.4251	.4265	.4279	.4292	.4306	.4319
:	:	:	:	:	:	:	:	:	:	:

Thus, the percentage of total observations between the z-values of 0 and +1.00 (34.13%) is the same as the percentage between 0 and −1.00 (also 34.13%). Shown diagrammatically:

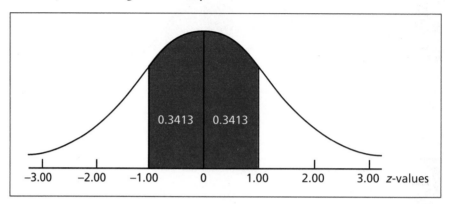

What percentage of the z-values are between −1.00 and +1.00? The answer: 0.3413 + 0.3413 = 0.6826 = 68.26%.

Self-Review 7–4

The weights of boxes of Crunchy breakfast cereal are normally distributed with a mean of 450 grams and a standard deviation of 2 grams. What percentage of the boxes weigh between

a. 450 and 454 grams?
b. 447.3 and 450 grams?

Problem   To illustrate further the procedure for finding the probability between 2 selected values, suppose Cardy Halzey decides to run for city council. What is the likelihood (probability) that she will spend between $5000 and $8000? (Recall that the mean expenditure for all candidates is $6000 and the standard deviation is $1000.)

Solution   The problem can be divided into two parts:

1. What is the probability of a campaign expenditure between $5000 and $6000 ($6000 is the mean)? Using formula 7–1, the z-value is −1.00:

$$z = \frac{X - \mu}{\sigma} = \frac{\$5000 - \$6000}{\$1000} = -1.00$$

The probability for −1.00 (from Appendix C) is 0.3413.

2. What is the probability of a campaign expenditure between $6000 (the mean) and $8000?

$$z = \frac{X - \mu}{\sigma} = \frac{\$8000 - \$6000}{\$1000} = 2.00$$

The probability for 2.00 (from Appendix C) is 0.4772.

Adding 0.3413 and 0.4772 gives 0.8185. Thus, the probability that Halzey will spend between $5000 and $8000 is 0.8185. The various components of this problem are shown in the following diagram:

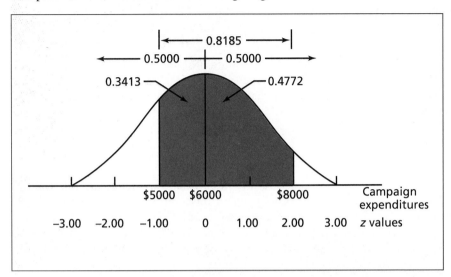

In the previous section, we computed the percentage of observations occurring between two values, or—stated a different way—the probability that a particular observation would occur between two points. In this section, the process is reversed. We will be interested in the percentage of the observations above or below a given point.

Determining Percentage Above or Below a Selected Value

Problem   What percentage of the candidates spend $8000 *or more* on their campaigns for city council?

Solution   Note in the previous diagram that 50% (0.5000) of the candidates spend $6000 or more. The decimal representing the amounts spent between $6000 and $8000 is 0.4772. Subtracting 0.4772 from 0.5000 gives 0.0228, or 2.28%. This is the percentage of candidates who spend $8000 or more on the election.

Problems   A normal distribution has a mean of 50 and a standard deviation of 8. What is the probability that a selected value is

a.   between 50 and 62?
b.   between 44 and 50?
c.   between 44 and 62?
d.   more than 60?

Solutions

a.   0.4332
b.   0.2734
c.   0.7066
d.   0.1056

Problem Recall from the Chapter Problem the Ohio Department of Transportation study. It was found that the mean speed of vehicles on a section of I-75 near Lima was 66.7 miles per hour with a standard deviation of 3.5 miles per hour. What percentage of the vehicles exceeded the 65-miles-per-hour speed limit? What percentage exceeded 70 miles per hour?

Solution First, we assume the distribution of speeds is normal with a mean of 66.7 miles per hour and a standard deviation of 3.5 miles per hour. To find the proportion of vehicles that exceeded 65 miles per hour, we use formula 7–1:

$$z = \frac{X - \mu}{\sigma} = \frac{65.0 - 66.7}{3.5} = -0.49$$

Next, we refer to Appendix C to find the z-value corresponding to 0.49. It is 0.1879. To determine the proportion of observations that exceeded −0.49, we add 0.1879 and 0.5000. The result is 0.6879. We therefore conclude that 0.6879 of the observations exceeded a z-value of −0.49. In other words, about 69% of the vehicles traveled faster than the posted speed limit of 65 miles per hour.

To find the proportion of vehicles that exceeded 70 miles per hour, we compute the z-value associated with 70. It is 0.94, found by (70 − 66.7)/3.5. Appendix C gives us the probability between 0.0 and 0.94 as 0.3264, but we are interested in the probability that z is greater than 0.94. This is found by 0.5000 − 0.3264 = 0.1736. That is, 17.36% of the vehicles traveled more than 70 miles per hour.

REAL STAT

Many processes, such as filling soda bottles and canning fruit, are normally distributed. Manufacturers must guard against both over- and underfilling. If they put in too much, they will give their product away. If they put in too little, the customer will feel cheated. "Control" charts, with limits drawn three standard deviations above and below the mean, are routinely used to monitor these kind of production processes.

Self-Review 7–5

The grade point averages at La Siesta University are normally distributed with a mean of 2.5 and a standard deviation of 0.75.

a. What percentage of the students have a grade point average between 2.35 and 2.5?
b. What percentage of the students have a grade point average between 2.45 and 2.63?
c. What percentage of the students have a grade point average of 3.5 or above?

Exercises

11. A normal distribution has a mean of 30 and a standard deviation of 5. What is the probability that a selected value is between 30 and 37?

12. A normal distribution has a mean of 25 and a standard deviation of 8. What is the probability that a selected value is less than 17?

13. The nicotine contents of a certain brand of king-sized cigarettes are normally distributed with a mean of 2.0 mg and a standard deviation of 0.25 mg. What is the probability that a cigarette has a nicotine content
a. of 1.6 mg or less?
b. between 1.6 and 2.1 mg?
c. of 2.1 mg or more?

14. The mean yearly amount of sap collected for making maple syrup is 10 gallons per tree. The distribution of the amounts collected per tree is normal with a standard deviation of 2.0 gallons.

a. What is the probability that a particular tree produces 13.0 gallons of sap or more a year?
b. What is the probability that a particular tree produces between 9.0 and 12.5 gallons of sap a year?

15. A cola-dispensing machine dispenses 7.00 ounces of cola per cup. The standard deviation is 0.10 ounce. What is the probability that the machine will dispense

a. between 6.8 and 7.00 ounces of cola?
b. between 6.8 and 7.15 ounces of cola?
c. more than 7.15 ounces of cola?

16. The mean starting salary for last year's accounting graduates is $28,000 with a standard deviation of $1500. What percentage of the graduates

a. earned between $26,000 and $27,000?
b. earned less than $25,000?

Determining *X* Values for a Given Probability

In our previous work on the normal curve, we computed the probability of an observation occurring between two values. For example, we determined the probability that Cardy Halzey would spend between $6000 and $8000 on her campaign. Now the process is changed. We are given an area beyond X and are asked to determine X.

Problem Return once more to the expenditures of the candidates for city council where $\mu = \$6000$ and $\sigma = \$1000$. Ten percent of the candidates spent what amount or more on their campaigns? The components of this problem are shown graphically:

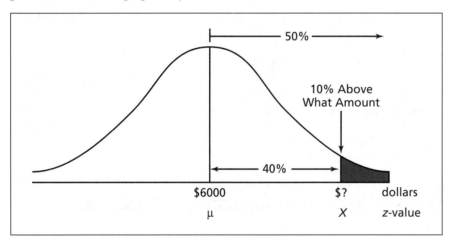

Solution As noted, 50% of the expenses are $6000 or more, and 10% are some unknown amount or more. The unknown amount is designated by X. Logically, 40% of the campaign expenditures are between $6000, or μ, and X. We obtain the z-value corresponding to 0.4000 by searching in Appendix C. The closest probability is 0.3997. This corresponds to a z-value of 1.28.

Inserting the z-value of 1.28 and solving for X using formula 7–1, we find

$$z = \frac{X - \mu}{\sigma}$$

$$1.28 = \frac{X - \$6000}{\$1000}$$

Then

$$\$1280 = X - \$6000$$

$$X = \$7280$$

Thus, about 10% of the candidates spend $7280 or more on their bids for election to the city council.

Self-Review 7–6

From Self-Review 7–5: The mean grade point average of students at La Siesta University was 2.5 and the standard deviation 0.75. The top 3% among students are to be given special recognition. What grade point average (or above) does a student need in order to receive special recognition?

Exercises

17. A normal distribution has a mean of 30 and a standard deviation of 5. Fifteen percent of all the values will be above what value?

18. A normal distribution has a mean of 25 and a standard deviation of 8. The middle 40% of the distribution will fall between what 2 values?

19. The heights of adult males follow a normal distribution with a mean of 70 inches and a standard deviation of 2.6 inches. How tall should a doorway be so that 98% of all men can pass through it without having to stoop?

20. A certain brand of passenger-car tire has a mean tread life of 40,000 miles. The tread lives are normally distributed with a standard deviation of 3000 miles. Five percent of the tires will lose all their tread before they are driven what distance?

21. The times taken by applicants for data entry positions to type a standard passage were normally distributed with a mean of 90 seconds and a standard deviation of 12 seconds. If it is company policy to consider only the top 10% of the applicants, what is the cutoff point between those considered by the company and those not considered?

22. A machine is set to fill milk bottles up to 1.0 liters. Most of the bottles contain 1.0 liters, but some of the bottles are slightly underweight or overweight. The fills are normally distributed, with a standard deviation of 0.05 liter. Find the value below which the smallest 10% of the fills occurs.

THE NORMAL APPROXIMATION TO THE BINOMIAL

Recall from Chapter 6 that the binomial distribution is a discrete probability distribution. It is characterized by π, the probability of a success, and n, the number of trials. (Figure 6–3 on page 198 provides a visual refresher on the effect of n.)

The normal distribution may be used to estimate binomial probabilities. As a general rule of thumb, when $n\pi$ and $n(1 - \pi)$ are both at least 5, the normal probability distribution is a very good approximation for the binomial distribution and is easier to compute. An example will help to illustrate.

Problem   The city's legal affairs director reports that based on past experience, 60% of automobiles reported stolen are recovered and returned to their owners. In a month in which 40 automobiles are stolen, what is the probability that 28 or more will be recovered and returned to their owners?

Solution   For this binomial probability distribution, the total number of trials is 40, and the probability of success on each trial is 0.60. The normal probability distribution may be used to approximate the binomial distribution because both $n\pi$ and $n(1 - \pi)$ exceed 5.

$$n\pi = 40(0.60) = 24$$

$$n(1 - \pi) = 40(1 - 0.60) = 16$$

The steps needed to determine the probability that 28 or more (actually 27.5, as we will see shortly) cars will be recovered and returned are as follows:

Step 1. Compute the mean and the variance of the binomial distribution. They are computed as follows, using formulas 6–4 and 6–5 (μ is used as the designation of the mean, σ^2 as the designation for the variance):

$$\mu = n\pi$$

$$\sigma^2 = n\pi(1 - \pi)$$

The mean is 24 and the variance 9.6:

$$\mu = 40(0.60) = 24$$

$$\sigma^2 = (40)(0.60)(0.40) = 9.6$$

Step 2. Determine the standard normal value (z-value) corresponding to 28. A discrete distribution can be only a number of distinct or separate values. Between these values, there are "gaps" that have no probability. A continuous distribution, on the other hand, takes on any value in a range. The situation is similar to "rounding," where 6.7 is rounded up to 7 and 7.3 is rounded down to 7. In fact, all numbers between 6.5 and 7.5 are *rounded* to 7.

To adjust for a difference between a binomial (discrete) and a normal (continuous) distribution, consider the binomial probability as the area over an interval centered at the discrete value and extending one-half of a unit in both directions. This is called a **correction for continuity.** In this example, the exact binomial probability of recovering 28 cars is approximated by the normal area between 27.5 and 28.5. We find the probability of 28 or more cars returned, then, by computing the probability of 27.5 or more. The standard deviation of the distribution is found by $\sqrt{\sigma^2}$. The square root of the variance of 9.6 is 3.10. Computing z using formula 7–1 gives

$$z = \frac{X - \mu}{\sigma} = \frac{27.5 - 24}{3.10} = 1.13$$

Step 3. Find the probability of a z-value of 1.13 or greater occurring. We determine the area, or probability, between 0 and 1.13 standard normal deviates by referring to Appendix C. Go down the left column to 1.1 and then read the probability headed by the column marked 3. The probability is 0.3708. Thus, the area beyond 1.13 standard normal deviates is 0.1292, found by 0.5000 − 0.3708. Shown graphically:

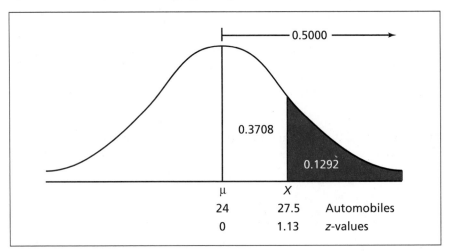

n =	40	= Number of trials					
pi =	0.6	= Probability of "success" on individual trials					
		X is the total number of successes					
x	P(X = x)	P(X <= x)	P(X > x)	x	P(X = x)	P(X <= x)	P(X > x)
0	0.0000	0.0000	1.0000	21	0.0792	0.2089	0.7911
1	0.0000	0.0000	1.0000	22	0.1026	0.3115	0.6885
2	0.0000	0.0000	1.0000	23	0.1204	0.4319	0.5681
3	0.0000	0.0000	1.0000	24	0.1279	0.5598	0.4402
4	0.0000	0.0000	1.0000	25	0.1228	0.6826	0.3174
5	0.0000	0.0000	1.0000	26	0.1063	0.7888	0.2112
6	0.0000	0.0000	1.0000	27	0.0827	0.8715	0.1285
7	0.0000	0.0000	1.0000	28	0.0576	0.9291	0.0709
8	0.0000	0.0000	1.0000	29	0.0357	0.9648	0.0352
9	0.0000	0.0000	1.0000	30	0.0196	0.9844	0.0156
10	0.0000	0.0000	1.0000	31	0.0095	0.9939	0.0061
11	0.0000	0.0000	1.0000	32	0.0040	0.9979	0.0021
12	0.0001	0.0001	0.9999	33	0.0015	0.9994	0.0006
13	0.0003	0.0004	0.9996	34	0.0005	0.9999	0.0001
14	0.0008	0.0012	0.9988	35	0.0001	1.0000	0.0000
15	0.0021	0.0034	0.9966	36	0.0000	1.0000	0.0000
16	0.0050	0.0083	0.9917	37	0.0000	1.0000	0.0000
17	0.0106	0.0189	0.9811	38	0.0000	1.0000	0.0000
18	0.0203	0.0392	0.9608	39	0.0000	1.0000	0.0000
19	0.0352	0.0744	0.9256	40	0.0000	1.0000	0.0000
20	0.0554	0.1298	0.8702				

We therefore determine that the probability is 0.1292 that in a month in which 40 automobiles are stolen, 28 or more will be recovered and returned. To put it another way, 28 or more automobiles will be recovered and returned in 12.92% of the months in which 40 automobiles are stolen.

A spreadsheet, such as Excel, can generate the complete binomial distribution. The various outcomes and corresponding probabilities for the case where $n = 40$ and $\pi = 0.6$ are shown at the bottom of the previous page. The first column lists x, the number of successes. The second column shows the probability of exactly x successes in the 40 trials. For example, the likelihood of 12 successes in 40 trials is .0001. The next column shows the cumulative probability of x or fewer successes in 40 trials. For example, the probability of 18 or fewer successes in 40 trials is .0392. The fourth column gives the likelihood of finding a value larger than x. So the probability of *more than* 18 successes is .9608. This is simply the complement of the previous column. The four columns to the right correspond to the second part of the distribution. We have done this simply to conserve space. In the previous problem, we were interested in the probability of 28 or more cars being returned in a month when 40 cars were stolen. This is the same as having more than 27 cars returned. So, consulting row 27, we see a probability of .1285. This outcome is very close to .1292, which is the value we determined using the normal approximation to the binomial distribution.

Here is a histogram of a binomial distribution generated by the Excel spreadsheet package:

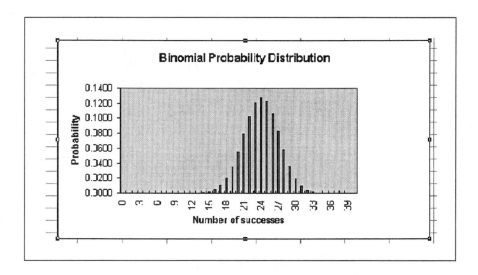

Problems X is distributed as a binomial probability distribution with $n = 30$ and $\pi = 0.30$. Compute each of the following:

a. The mean and the variance of the random variable
b. The probability that X is greater than 12
c. The probability that X is greater than 5

Solutions

a. $\mu = 9$, $\sigma^2 = 6.3$

b. $z = (12.5 - 9.0)/\sqrt{6.3} = 1.39$; probability is 0.0823, found by 0.5000 − 0.4177.

c. $z = (5.5 - 9.0)/\sqrt{6.3} = -1.39$; probability is 0.9177, found by 0.5000 + 0.4177.

Self-Review 7–7

The instructor in Geology 115 gives only a final examination. It consists of 100 multiple-choice questions with 5 possible answers for each. The instructor announces that at least 30 correct answers will be required to pass the course.

Assume you are enrolled in the course, but have never attended class or read any of the assignments. You decide, however, to take the final examination.

a. Can the normal approximation to the binomial be used? Why?
b. Determine the mean and the standard deviation of the number correct.
c. Calculate the probability that you will pass Geology 115.
d. Portray the probabilities and other parts of the problem graphically.

Exercises

23. Find the mean and the variance of a binomial probability distribution with $n = 40$ and $\pi = 0.2$.

24. Find the mean and the variance of a binomial probability distribution with $n = 50$ and $\pi = 0.3$.

25. The Wayward Inn, a 300-room resort hotel, experiences an 85% occupancy rate, on the average, in January. Use the normal approximation to the binomial to find
    a. the probability that at least 260 rooms are occupied in January.
    b. the probability that fewer than 240 rooms are occupied in January.

26. An airline manager estimated that, on the average, 8% of the passengers flying across the Atlantic experience some airsickness. What is the probability that on a transatlantic flight of 150 passengers, at least 5 will experience some airsickness?

27. Suppose 1 out of 10 people defaults on his or her car loan. Last month the Penn Bank approved 50 car loans. What is the probability that at least 1 borrower will default?

28. A mail order company specializes in selling men's slacks. The probability that a customer will respond to a mailing with an order is 0.05. Today's mailing includes 500 people. What is the probability that fewer than 20 people will respond with an order?

THE NORMAL APPROXIMATION TO THE POISSON DISTRIBUTION

The normal distribution was used to approximate the binomial distribution. In a similar fashion, the normal distribution is used to approximate a Poisson distribution. The approximation improves as the mean (μ) increases. A widely followed guideline is to use the normal approximation when the Poisson mean exceeds three.

Problem   The number of people using the emergency room of Addison Community Hospital during a day shift approximates a Poisson distribution with a mean of three people. Compute the probability that between two and five people will arrive during the day shift today. Use both the exact Poisson probabilities and the normal approximation to the Poisson distribution. Compare the results.

Solution   First, to find the probability that between 2 and 5 people will arrive, using the Poisson distribution where $\mu = 3$, we add the probabilities of 2, 3, 4, and 5. From the Poisson table (Appendix B), we find that

$$P(2 \leq X \leq 5) = P(2) + P(3) + P(4) + P(5)$$
$$= 0.2240 + 0.2240 + 0.1680 + 0.1008 = 0.7168$$

The probability is 0.7168. As an alternative, we determine an area between 1.5 and 5.5 for a normal probability distribution with a mean and variance of 3. From our discussion of the Poisson distribution in Chapter 6, recall that the variance of a Poisson distribution equals its mean. In this problem, we need the standard deviation, so $\sqrt{3} = 1.732$. Substituting 1.732 into formula 7–1 for z, we find that

$$z = \frac{X - \mu}{\sigma} \qquad\qquad z = \frac{X - \mu}{\sigma}$$
$$= \frac{1.5 - 3.0}{1.732} \qquad\qquad = \frac{5.5 - 3.0}{1.732}$$
$$= -0.87 \qquad\qquad = 1.44$$

Thus, the standardized values for 1.5 and 5.5 are -0.87 and 1.44.

The area under the normal curve between the mean of 3.0 and X of 1.5, represented by a z of -0.87, is 0.3078 (from Appendix C). The area between 3.0 and 5.5 is represented by the z of 1.44, which is 0.4251. Adding the probabilities from the normal approximation, $0.3078 + 0.4251$, we obtain a total of 0.7329:

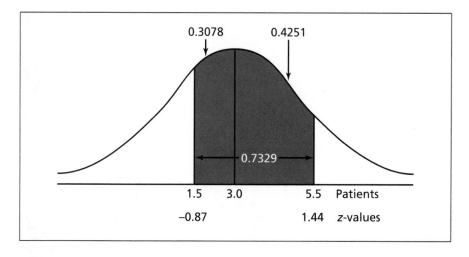

Note that the difference between the probability using the exact Poisson distribution (0.7168) and the probability using the normal approximation (0.7329) is only 0.0161. Again, this emphasizes that under certain circumstances the normal approximation can be used to approximate the Poisson.

Problems   The mean of a Poisson probability distribution is 9. Use the normal approximation to the Poisson to compute the probability (don't forget the correction for continuity) that the value is

a.   greater than 12.
b.   between 7 and 12.
c.   less than 7.

Solutions

a. $z = (12.5 - 9)/3 = 1.17$; probability is 0.1210, found by $0.5000 - 0.3790$
b. $z = (65 - 9.0)/3 = -0.83$ and $z = 1.17$; probability is 0.6757
c.   0.2033, found by $0.5000 - 0.2967$

Self-Review 7–8   A pair of fair dice are rolled 360 times. What is the probability that "snake eyes," a 1 on each die, will appear between 5 and 10 times, inclusively? Remember that the two dice are independent; hence, the probability of "snake eyes" on a single roll is $(1/6)(1/6) = 1/36$.

Exercises

29. The mean of a Poisson probability distribution is five. Use the normal approximation to the Poisson to compute the probability that the value is
a. less than two.
b. between two and ten.
c. less than seven.
30. Sam's Carpet Shop receives an average of four orders per day. The numbers of orders approximate a Poisson probability distribution. Use both the Poisson and the normal distribution to approximate the Poisson to find the probability that on a given day
a. two orders are received.
b. more than seven orders are received.
31. The numbers of hits on a web site approximate a Poisson probability distribution. The mean number of hits per quarter hour is nine. Compute the probability that fewer than seven hits are received in a particular quarter hour.

CHAPTER OUTLINE

I. The normal distribution is a continuous probability distribution with the following characteristics:
   A. It is bell-shaped, and the mean and the median are equal.
   B. It is asymptotic, meaning the curve approaches, but never touches, the X axis.
   C. It is completely described by the mean and the standard deviation.
   D. There is a family of normal distributions, each time the mean or the standard deviation changes, a new distribution is created.
II. The standard normal distribution is a particular normal distribution.

A. It has a mean of zero and a standard deviation of one.

B. Any normal distribution can be converted to the standard normal distribution by the following formula:

$$z = \frac{X - \mu}{\sigma} \qquad \boxed{7\text{--}1}$$

C. By standardizing a normal distribution, we report the distance from the mean in units of the standard deviation.

III. The normal distribution can be used to approximate the binomial distribution under certain conditions.

A. $n\pi$ and $n(1 - \pi)$ must both be greater than 5.

B. The four binomial conditions must be met.

1. The distribution results from a count of "successes" in n trials.

2. There are only two possible outcomes on each trial.

3. π remains the same from trial to trial.

4. The trials are independent.

C. The mean and the variance are computed from formulas 6–4 and 6–5.

D. The continuity correction factor of .5 is used to adjust for using a discrete distribution is being approximated by a continuous one.

IV. The normal distribution can also be used to approximate a Poisson distribution.

A. The mean of the distribution should be at least three.

B. The variance of the distribution is equal to the mean.

C. The continuity correction should be used when calculating probabilities.

- -

Exercises

32. An applicant for a position with the Norton Corporation earned scores of 60 on the Sales Aptitude Test, 150 on the Personnel Human Factors Evaluation Test, and 90 on the Financial Management Test. The scores of each of these tests approximate a normal distribution. The means and the standard deviations of the tests are

Test	Mean	Standard Deviation
Sales Aptitude Test	50	7
Personnel Human Factors Evaluation Test	120	25
Financial Management Test	85	5

a. Compute a z-value for each of the applicant's test scores.

b. On which of the tests did the applicant do the best relative to the entire group?

c. Regarding the Sales Aptitude Test, what percentage scored higher than the applicant?

d. Based on the applicant's performance on the 3 tests, to what area would you assign her—sales, personnel, or finance? Why?

33. If 150 seats are reserved for an airline flight and in the past an average of 7% of the reservations have not shown up, what is the probability a plane with 140 seats will be overbooked?

34. The area a painter covers with 1 gallon of paint is normally distributed with a mean of 400 square feet and a standard deviation of 60 square feet. If the manufacturer specifies that 1 gallon should cover between 375 and 450 square feet, what percentage of the time will the painter exceed the upper limit of the manufacturer's specification? What percentage of the time will he be within the manufacturer's limit?

35. Patients arrive at the emergency room of St. Charles Hospital at the rate of four per hour. Assume that the arrival distribution approximates the Poisson distribution. What is the probability of no arrivals during a given hour? Use the normal approximation to the Poisson and compare your results.

36. It is estimated that 80% of all household plants are overwatered by their owners. In a group of

50 plants, what is the probability that more than 35 were overwatered? That up to 45 were overwatered?

37. The mean life of socks used by the Army is 60 days with a standard deviation of 12 days. Assume the lives of the socks are normally distributed. If 1 million pairs are issued, how many would need replacement after 50 days? After 70 days?

38. The number of sandwiches sold daily by the deli bar in the student union is normally distributed with a mean of 215 and a standard deviation of 20. On what percentage of days does the deli bar sell more than 200 sandwiches? Fewer than how many sandwiches are sold 10% of the days?

39. According to recent newspaper reports, the typical player in the National Football League plays for 3.2 years. If the careers are normally distributed and the standard deviation is 1.80 years, compute the probability that a randomly selected first-year player will play more than five years.

40. Refer to Exercise 39. Assume the lengths of the careers follow a Poisson distribution, with a mean of 3.2 years. Compute the probability that the player will play more than 5 years.

41. The numbers of hours per week college students devote to study are normally distributed with a mean of 30 hours and a standard deviation of 8 hours.
    a. What percentage of students will study less than 20 hours?
    b. What percentage will study more than 35 hours?
    c. Out of a class of 200 students, how many will study between 25 and 35 hours?

42. Memorial Hall is used as a site for both student-sponsored concerts and intercollegiate basketball games. Attendance figures for both the concerts and the basketball games are normally distributed. The means and the standard deviations are:

	Concerts	Basketball
Mean attendance	8600	7200
Standard deviation	560	600

a. What percentage of the basketball games have an attendance of 8000 or more?

b. Tickets to concerts are priced so that, if 8000 tickets are sold, expenses are covered. If fewer than 8000 are sold, the students lose money, and if more than 8000 are sold, they make a profit. What percentage of the time will the students lose money?

43. A juice dispenser is set to fill cups with a mean of 7.5 ounces of fruit juice. The standard deviation of the process is 0.3 ounces.
    a. If 8-ounce cups are used, what percentage of them will overflow?
    b. What percentage of the cups will have less than 6.8 ounces of juice in them?

44. Southern Airways is studying its service from Chicago to Atlanta. Historical data show that the mean number of passengers per flight is 235.6 and the standard deviation of the normal distribution is 36.3 passengers.
    a. What is the probability that a particular flight will carry more than 260 passengers?
    b. What is the probability that a particular flight will have fewer than 180 passengers?
    c. What is the probability that a particular flight will have between 240 and 250 passengers?
    d. What is the probability that a particular flight will have fewer than 240 passengers?

45. The owner of a fast-food restaurant keeps records of the daily hamburger demands, which are normally distributed with a mean of 260 pounds and a standard deviation of 20 pounds.
    a. On what percentage of days will the owner need more than 310 pounds of hamburger?
    b. The owner does not want to run out of meat more than 1% of days. How many pounds should she order every day?

46. A researcher reports that the mean heart rate of rats is 120 beats per minute and that 45% of all rats tested had heart rates in the range 120 to 140. Assume these rates are normally distributed.
    a. What standard deviation is implied by these data? (Hint: Use the formula for z to compute the standard deviation.)
    b. What percentage of the animals have heart rates in the range 100 to 120?
    c. What percentage of the rats have heart rates in excess of 150?
    d. What are the standard scores corresponding to 120 and 140?

47. If an elm tree is less than 3 feet high and is transplanted in the spring, it has a 40% chance of survival. If 50 such trees are transplanted, what is the probability that 25 or more will survive? What is the probability that between 18 and 23 will survive?

48. A new drug is developed to treat a certain disease. It is found to be effective in 90% of patients. If the drug is administered to 80 patients having the disease, what is the probability that it is effective in at least 70 cases?

49. A particular type of birth control device is effective 90% of the time when used correctly. If the device is employed 300 times, how many times would you expect the device to fail? What is the probability that it would fail 35 or more times?

50. The numbers of daily admissions at Riverside Hospital followed a normal distribution with a mean of 39.52 and a standard deviation of 6.29.
   a. What is the standardized z-value corresponding to 50 admissions?
   b. What is the standardized z-value corresponding to 25 admissions?
   c. What percentage of days had more than 50 admissions?
   d. What percentage of days had fewer than 25 admissions?
   e. On the busiest 10% of the days, what was the minimum number of admissions?

51. Suppose the U.S. Postal Service claimed that 80% of letters mailed in New York City destined for Los Angeles are delivered within 3 working days. To verify this claim, you mail 200 letters from New York to various destinations in the Los Angeles area.
   a. What is the probability that more than 150 of the letters will be delivered within 3 working days?
   b. Fewer than 148?
   c. Between 150 and 160?
   d. The probability is 10% that what number or more will be delivered within a minimum of 3 working days?

52. Cars arrive at a car wash at a mean rate of 12 per hour. Compute the probability that more than 15 cars will arrive today between 3 P.M. and 4 P.M.

53. During rush hour, accidents occur, according to a Poisson distribution, at an average rate of 2.5 per hour. Compute the probability that no accidents will occur today between 4 P.M. and 5 P.M.

54. The First National Bank of Sylvania recently opened an automatic teller machine (ATM) at the Southview Mall. The numbers of customers arriving per hour at this ATM follow a Poisson distribution with a mean of 15. Use the normal approximation to the Poisson to estimate the probability that more than 12 customers will use the machine between 2 P.M. and 3 P.M. this afternoon.

55. The high-school grade point averages of students applying to Brownlee University are normally distributed with a mean of 2.80 and a standard deviation of 0.50. If a high-school grade point average of 3.00 is required for admission to Brownlee, what percentage of the students applying meet the requirement?

56. The law firm of Tybo and Associates is concerned with the length of time prospective clients spend on hold when calling one of the firm's attorneys. A study revealed the waiting times to be normally distributed with a mean of 80 seconds and a standard deviation of 12 seconds. What percentage of the clients wait at least a minute?

57. Hannah Simpson, a commuter student at Lourdes College, drives her jeep to class each day. She finds that the driving times are normally distributed with a mean of 20 minutes and a standard deviation of 3.6 minutes. Explain to Hannah why theoretically the probability is 0 that she takes exactly 19.0 minutes to get to class. Using the correction for continuity, develop an estimate of the probability of a particular trip taking 19 minutes.

58. The Heritage Family Restaurants is studying the length of time customers spend in their establishments. The times are normally distributed with a mean of 18 minutes and a standard deviation of 5.5 minutes.
   a. What percentage of the customers spend less than 30 minutes in a given restaurant?
   b. What is the probability of a customer spending exactly 30 minutes in the restaurant?
   c. How would you estimate the probability of a customer spending 30 minutes in the restaurant?

59. According to a recent study by the American Medical Association, 40% of first-year medical students are women. The same study reports that 5.8% are black. Suppose a group of 60 medical students is selected at random.
  a. What is the likelihood that at least 18 are women?
  b. What is the likelihood that more than 5 of the students selected to be in the study are black? (Hint: Does this meet the binomial conditions?)
60. A recent study of home sales in suburban Williston revealed that the mean selling price was $107,800 with a standard deviation of $20,000. The mean number of days on the market was 43 with a standard deviation of 14.5 days. After 50 days on the market, Jeff and Mandy Hall's home sold for $89,550. Assuming that both distributions are approximately normal, determine the percentage of homes that sold for less than the Halls' and the percentage that were on the market for a shorter period of time.
61. Theresa's Tax Service specializes in the preparation of federal tax returns. A recent audit of her returns indicates that an error was made on 6% of the returns her service prepared last year.

Assuming that rate continues this year and 100 returns are prepared, what is the probability that her service will make 8 or more errors? Determine the probability three ways—normal approximation to the binomial, normal approximation to the Poisson, and the Poisson distribution—and compare the results.
62. A racing expert from the Dickey-Bend International Speedway reports that the mean length of time required to complete a routine pit stop is 15 seconds. The times are normally distributed with a standard deviation of 1 second.
  a. The probability is 0.8 that a pit stop will take at least how many seconds?
  b. The probability is 0.9 that a pit stop will take at most how many seconds?
63. A management consultant is studying the daily work habits of senior-level executives. The consultant finds that a mean of 2.50 hours per day is spent performing tasks that could be handled by subordinates. The distribution of hours is approximately normal. It is also found that 10% of the executives spend more than 3.25 hours on such tasks. Estimate the standard deviation of the distribution.

DATA EXERCISES

64. Refer to the real estate data set, which reports information on homes sold in Alabama during 1996.
  a. The mean list price of the 50 homes is $192,180 with a standard deviation of $8250. Use the normal distribution to estimate the number of homes that would list for more than $200,000. Compare this with the actual number. Is the normal distribution a good approximation of the actual results? Why?
  b. The mean area of the 50 homes is 2214.6 square feet with a standard deviation of 435 square feet. Use the normal distribution to estimate the number of homes with more than 2100 square feet. Compare this with the actual number. Is the normal distribution a good approximation of the actual results? Why?

65. Refer to the school data set, which has information on 94 school districts.
  a. The mean amount spent on instruction is $2725 with a standard deviation of $1095. Use the normal distribution to estimate the percentage of districts that spend more than $3000 on instruction. Compare this estimate with the actual proportion. Does the normal distribution appear accurate in this case? Explain.
  b. The mean number of students per district is 2134 with a standard deviation of 3895. Use the normal distribution to estimate the percentage of districts with more than 2000 students enrolled. Compare this estimate with the actual proportion. Does the normal distribution appear accurate in this case? Explain.

CHAPTER ACHIEVEMENT TEST

The answers are at the back of the book.

MULTIPLE-CHOICE QUESTIONS

Select the response that best answers each of the questions.

1. The normal distribution is a
   a. continuous distribution.
   b. discrete distribution.
   c. subjective distribution.
   d. historical distribution.
2. For the normal distribution, which of the following statements is *not* always true?
   a. It is symmetric.
   b. It is unimodal.
   c. The mean, the median, and the mode are equal.
   d. It requires a continuity correction.
3. What percentage of a normal distribution is within 2 standard deviations of the mean? (Select the closest answer.)
   a. 50%
   b. 95%
   c. 75%
   d. 100%
4. Which of the following statements is true for a normal distribution?
   a. It has two parameters: a mean and a standard deviation.
   b. Half of the values are greater than the mean.
   c. It goes to infinity in both directions.
   d. All of the above
5. A normal distribution is standardized by the formula
   a. $n\pi$.
   b. $n\pi(1 - \pi)$.
   c. $(X - \mu)/\sigma$.
   d. None of the above

6. The normal distribution is very close to the binomial distribution when both $n\pi$ and $n(1 - \pi)$ are greater than
   a. 1.
   b. 5.
   c. 30.
   d. The normal distribution is never very close to the binomial distribution.
7. A standardized z-value
   a. expresses distance from the mean in units of the standard deviation.
   b. applies only to the Poisson distribution.
   c. must be positive.
   d. All of the above
8. The probability that a standard normal deviate (z-value) is greater than 1 is
   a. 0.1587.
   b. 0.3413.
   c. 0.8413.
   d. None of the above
9. If we use the standard normal distribution, the probability of obtaining a z-value between 1 and 2 is
   a. 0.0228.
   b. 0.1359.
   c. 0.3413.
   d. 0.4772.
10. The area under the standard normal distribution between -1.5 and 1.5 is
   a. 0.1338.
   b. 0.1668.
   c. 0.4332.
   d. None of the above

COMPUTATION PROBLEMS

11. The lengths of time bank customers must wait for a teller are normally distributed, with a mean of 3 minutes and a standard deviation of 1 minute.
  a. What proportion of bank customers waits between 3 and 4.5 minutes?
  b. What percentage wait more than 4 minutes?
  c. What proportion waits between 2 and 3.5 minutes?
  d. What percentage wait less than 1 minute?
  e. Ninety percent of the customers spend less than what amount of time waiting for a teller?

12. Forty percent of a population have blood type O. Find the following probabilities for a random sample of 80 people:
  a. More than 25 have blood type O.
  b. More than 40 have blood type O.
  c. Fewer than 35 have blood type O.
  d. Between 30 and 36 have blood type O.

13. An average of six customers arrives at the Commodore II Barber Shop every half hour. Compute the following probabilities:
  a. Ten or more customers arrive within a half hour.
  b. More than four customers arrive within a half hour.

. . . .

ANSWERS TO SELF-REVIEW PROBLEMS

7-1   38 and 62: $\mu \pm 2(\sigma)$
   $= 50 \pm 2(6)$
   $= 50 \pm 12$

7-2 a. $\dfrac{68 - 60}{12} = \dfrac{8}{12} = 0.67$

  b. $\dfrac{33 - 60}{12} = \dfrac{-27}{12}$
   $= -2.25$

7-3   Age: The z-value is -0.2, slightly below average age, found by $(38 - 40)/10$.
   Social consciousness: The z-value is 2.50, indicating well above average, found by $(750 - 500)/100$.

7-4 a. $z = \dfrac{X - \mu}{\sigma}$
   $= \dfrac{454 - 450}{2}$
   $= 2.00$
   47.72% from Appendix C

  b. $\dfrac{447.3 - 450.0}{2} = -1.35$
   41.15%

7-5 a. 7.93%
   $z = \dfrac{2.35 - 2.50}{0.75}$
   $= -0.20$
   Refer to Appendix C for 0.0793.

b. 9.54%
   $z = \dfrac{2.45 - 2.50}{0.75}$
   $= -0.07$
   $z = \dfrac{2.63 - 2.50}{0.75}$
   $= 0.17$
   The probabilities from Appendix C are $0.0279 + 0.0675 = 0.0954$.

  c. 9.18%
   $z = \dfrac{3.5 - 2.5}{0.75} = 1.33$
   Then
   $0.5000 - 0.4082 = 0.0918 = 9.18\%$.

7-6   A grade point average of 3.91
   $1.88 = \dfrac{X - 2.50}{0.75}$
   $1.88(0.75) = X - 2.50$
   $1.41 = X - 2.50$
   $X = 3.91$

7-7 a. Yes. $n = 100$, $\pi = 1/5 = 0.20$. Both $n\pi$ and $n(1 - \pi)$ are greater than 5.
   $n\pi = 100(0.20) = 20$
   $n(1 - \pi) = 100(1 - 0.20) = 80$

  b. $\mu = n\pi(100)(0.20) = 20$

$$\sigma^2 = n\pi(1 - \pi)$$
$$= (100)(0.20)(0.80)$$
$$= 16$$
$$\sigma = \sqrt{16} = 4$$

c. $z = \dfrac{X - \mu}{\sigma} = \dfrac{29.5 - 20}{4} = 2.38$

From Appendix C, the probability for z of 2.38 = 0.4913. Then $0.5000 - 0.4913 = 0.0087$. Your chance of passing is less than 1%.

d.

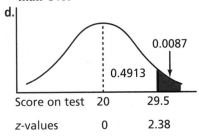

Score on test	20	29.5
z-values	0	2.38

7-8 The mean and variance are 10 [$\mu = n\pi = (360)(1/36) = 10$], so the standard deviation is $\sqrt{10}$, or 3.16. The discrete values from 5 to 10, inclusive, are included in the continuous range from 4.5 to 10.5. When 4.5 and 10.5 are standardized, we have
$$z = (4.5 - 10)/3.16 \text{ and}$$
$$z = (10.5 - 10)/3.16$$
which equal -1.74 and 0.16, respectively. So the normal probability of rolling a "snake eyes" between 5 and 10 times is $0.4591 + 0.0636 = 0.5227$.

Unit Review

In the last three chapters, you were introduced to the fundamental concepts of probability and to discrete and continuous probability distributions. The various methods for determining a probability, the rules for combining several probabilities, and three different probability distributions were described, discussed, and illustrated.

KEY CONCEPTS

1. A **probability** is a number that expresses the likelihood that a particular event will occur. There are three types, or definitions, of probability:
   a. **Classical probability:** Each of the possible outcomes is equally likely. If there are n outcomes, the probability of a particular outcome is $1/n$.
   b. **Empirical probability:** The total number of times the event has occurred in the past is divided by the total number of observations.
   c. **Subjective:** The assignment of probability is based on whatever information is available—personal opinions, hunches, and so on.
2. The fundamental rules of probability are the rule of addition, the rule of multiplication and the complement rule.
   a. **Rule of addition:** If two events, A and B, are mutually exclusive, the probability that one or the other of the events will occur is

   $$P(A \text{ and } B) = P(A) + P(B)$$

   This is called the **special rule of addition.** If the events are *not* mutually exclusive, the probability that one or the other will occur is

   $$P(A \text{ and } B) = P(A) + P(B) - P(A \text{ and } B)$$

   where $P(A \text{ and } B)$ is the probability of the joint occurrence of the two events. This is called the **general rule of addition.**
   b. **Rule of multiplication:** If two events, A and B, are unrelated (independent), the probability of their joint occurrence is the product of the two probabilities:

   $$P(A \text{ and } B) = P(A) \cdot P(B)$$

This is called the **special rule of multiplication.** If the two events are related (not independent), the probability of their joint occurrence is

$$P(A \text{ and } B) = P(A) \cdot P(B|A)$$

where $P(B|A)$ refers to the probability that event B occurs, given that A has already happened. This is called the **general rule of multiplication.**
Complement rule: The probability of the event happening is equal to 1 minus the probability of the event *not* happening.
3. A **probability distribution** is a listing of the outcomes of an experiment that may occur and the corresponding probability associated with each of the outcomes.
4. A probability distribution has two main features:
   a. The likelihood of a particular outcome must be between 0 and 1.0.
   b. The sum of all possible mutually exclusive outcomes must be 1.0.
5. There are two types of **probability distributions**—discrete and continuous. A **discrete probability distribution** can assume only certain distinct values and is usually the result of counting. A **continuous probability distribution** may assume an infinite number of values within a given range.
6. For a discrete probability distribution, the mean is computed by $\mu = \Sigma[X \cdot P(X)]$ and the variance by $\sigma^2 = \Sigma[(X - \mu)^2 \cdot P(X)]$.
7. The **binomial distribution** is an example of a discrete distribution, where π is the probability of a success and n is the number of trials. The mean

is found by $n\pi$ and the variance by $n\pi(1 - \pi)$. A binomial distribution has the following characteristics:

  a. Each outcome is classified in one of two mutually exclusive categories.

  b. Each trial is independent.

  c. The probability of a success remains the same from trial to trial.

  d. It results from counting the number of successes in the total number of trials.

8. The **Poisson distribution** is another discrete distribution. It has the same characteristics as the binomial, but, in addition, n is usually large and π is small. The mean, μ, is also computed by $n\pi$. This distribution depends on only one value—namely, μ.

9. The **normal distribution** is an example of a continuous probability distribution. It has the following main characteristics:

  a. It has a single peak.

  b. It is symmetrical about the mean.

  c. It is bell-shaped, and the two tails extend indefinitely in both directions.

  d. There is a family of normal distributions—a different one for each different mean and standard deviation.

10. The normal distribution is used to approximate the binomial if $n\pi$ and $n(1 - \pi)$ both exceed 5.0.

11. The normal distribution is used to approximate the Poisson if the mean is at least 3.0.

KEY TERMS

Probability	Joint probability	Discrete probability distribution
Experiment	Conditional probability	Binomial distribution
Outcome	General rule of multiplication	Binomial formula
Event	Special rule of multiplication	Poisson probability distribution
Classical probability	Independent events	Continuous probability distribution
Empirical probability	Complement rule	
Subjective probability	Permutations	Expected value
Special rule of addition	Combinations	z-value
General rule of addition	Probability distribution	Standard normal distribution
Mutually exclusive events	Random variable	Correction for continuity

KEY SYMBOLS

$P(A)$ The probability event A will occur.

$P(B|A)$ The conditional probability event B will occur, given that event A has already happened.

$_nP_r$ The number of permutations of n objects taken r at a time.

$_nC_r$ The number of combinations of n objects taken r at a time.

z The value of the standard normal distribution.

CASE STUDY   Will Sampling Save Sara Time?

An item such as an increase in taxes, a recall of elected officials, or an expansion of public services can be placed on the ballot if a required number of valid signatures are collected on a petition. Unfortunately, many people will sign a petition even though they are not registered to vote in that particular district, or they will sign the petition more than once.

Sara Ferguson, the elections auditor, must certify the validity of these signatures after the petition is officially presented. Not surprisingly, her staff is overloaded, so she is considering using statistical methods to validate the pages of 200 signatures instead of validating each individual signature. At a recent professional meeting, she found that in some other communities, election officials were checking only 5 signatures on each page and rejecting the entire page if 2 or more of the 5 signatures were invalid. Some people were concerned that 5 may not be enough to check. They claimed you should check 10 signatures and reject the page if 3 or more are invalid.

In order to investigate these methods Sara asks her staff to pull the results from the last election and sample 30 pages. It happens that they select 14 pages from the Avondale district, 9 pages from the Midway district, and 7 pages from the Kinston district. Each page has 200 signatures, and the data below show the number of invalid signatures on each.

Avondale	Midway	Kinston
9	19	38
14	22	39
11	23	41
8	14	39
14	22	41
6	17	39
10	15	39
13	20	
8	18	
8		
9		
12		
7		
13		

Use the data to evaluate the two proposals Sara has heard about. Calculate the probability of rejecting a page under each of the approaches. Would you get about the same results as you would by examining every single signature? Offer a plan of your own, and discuss how it might be better or worse than one of Sara's two plans.

CHAPTER 8

Sampling Methods and Sampling Distributions

OBJECTIVES

When you have completed this chapter, you will be able to

- design a statistical study;
- list and describe three methods of collecting data;
- describe the various types of probability sampling;
- describe what is meant by sampling error;
- develop and describe the sampling distribution of the sample means.

CHAPTER PROBLEM   Should It Be "Last Hired, First Fired"?

The mayor of Las Palmas is reviewing all city services in an effort to cut the growing budget deficit. It may be necessary to reduce the number of personnel in the Department of Social Services from 6 to 4. The lengths of service for each of the 6 employees are 4, 3, 3, 2, 4, and 3 years. If seniority is the determining factor, which 2 workers should be let go?

INTRODUCTION

What do the following three problems have in common? (1) The president of the United States wants to know the proportion of voters in Texas who will vote to reelect him in November, (2) Revlon wants to know how many women will buy newly developed Siren Red lipstick, and (3) the National Organization for Women (NOW) wishes to know what percentage of all banks in the United States have at least one female director.

These problems have one thing in common: The information is very difficult to obtain. It would be almost impossible (and very expensive) for the president to contact every potential voter in Texas by November. Likewise, it would be impossible and, again, very expensive for Revlon to contact all women in the world (or even just in the United States) and ask each one to try a complimentary stick of Siren Red. NOW would find it difficult to contact every bank in the United States.

The usual, less costly way to gather this kind of information is to take a **sample.** As defined in Chapter 1, a sample is a smaller group selected from the population of interest. The objective of studying the smaller group is to obtain information about the whole **population.** The president may hire a polling service, which in turn may sample 2000 Texans on their political preference in the forthcoming election. NOW might select, say, 50 banks at random and determine what percentage have at least 1 female director. If 15 out of the 50, or 30%, have at least 1 female director, NOW might reasonably conclude that about 30% of *all* banks in the United States have at least 1 female director.

These three situations illustrate how information obtained from a sample can be used to say something about the entire population. This is the process called **statistical inference.** Recall from Chapter 1 that statistical inference is the process of reasoning from specific instances or data to general conclusions about the entire group or population. In this chapter and those that follow, you will be introduced to statistical techniques that are based on **probability sampling.**

If probability sampling is not used, sample results may not be representative of the entire population. In such cases, it is said that the results are **biased.** To illustrate, Revlon might contact 400 women in New York City about Siren Red mainly because the group's location is convenient to Revlon's New York office. The results, however, might not be representative of all women in the United States; the color red may suggest "warmth" to women in New Mexico and "aggressiveness" to those in New York, or vice versa.

In this chapter, we will first discuss several scientific methods of design and selection of a sample and situations in which each method might be used. Then the sampling distribution of a widely used statistic—namely, the sample mean—will be examined.

---

probability sampling
Each member of the population of interest has a known likelihood of being included in the sample.

DESIGNING THE SAMPLE SURVEY OR EXPERIMENT

Nowhere in statistics is the expression "well begun is half done" more applicable than to the design of a sample survey or experiment. If a study is not carefully planned and tested, any calculations that follow from it may be useless. Further, collecting data and conducting a controlled experiment are expensive and time consuming, so it is important to begin with clearly defined objectives and procedures. There are four questions we can ask ourselves to help us plan a sample survey or experiment.

WHAT DO WE WISH TO FIND OUT?   Answer this question precisely! Now is the time to clearly focus the study by avoiding poorly worded goals and emotionally biased statements. Consider the following question sent to all office managers during a national survey: "If modern office equipment was installed, how much would productivity improve in the office?" What exactly is meant by "productivity in the office"? Could managers calculate an exact figure? If so, would it be in terms of dollars or percentages? To some, "modern office equipment" may be a fax machine or a single computer; to others, it may refer to a network of computers or even furniture.

Self-Review 8-1

Answers to all Self-Review problems are at the end of the chapter.

The following question appeared in a survey on smoking, alcohol, and drug abuse given to a group of elementary school pupils: "How often in the last month did you use an inhalant?" Evaluate this question. Is it reasonable or misleading? Why?

WHAT POPULATION ARE WE INTERESTED IN STUDYING?   "All residents of the United States" is a different group than "all eligible voters" or "all registered voters." Since conclusions about one group may not apply to others, it is useful to list every member of the population being studied without duplication. This list is called the **sample frame**. An alphabetical listing of all students enrolled at Kankakee Community College this semester is an example of a sample frame.

sample frame A complete list of the population without duplication.

HOW WILL THE DATA BE COLLECTED?   The procedure for selecting the actual sample is called the **sample design**. This is the step when cost is considered. There are three basic methods for collecting data: mail surveys, personal interviews, and telephone interviews.

Mail surveys are relatively cheap and easy to administer, but a typical response rate is less than 25%. A recent survey sent to 500 MBA graduates resulted in only 89 responses, many of which were unusable! On the other hand, a higher response rate usually occurs when there is an issue involving the respondents. For example, a female student sent a survey to 62 women who were their school's directors of intercollegiate athletics. She received re-

sample design The procedure for selecting the sample.

sponses from 37, well above the typical rate. The difference was probably due to the respondents' interest in women's issues.

Probably the most widely reported and erroneous sampling blunder ever made was committed as part of a mail survey. In 1936, *Literary Digest* published the results of a survey that predicted Alf Landon would defeat Franklin D. Roosevelt in the upcoming presidential election. Among other flaws, this survey had what is termed "self-selection bias"—respondents were supposed to voluntarily send back sample ballots with their "votes" on them. Out of 10 million sample ballots sent out, however, only 2.3 million were returned. Those who returned the ballots clearly did not represent the voting public.

Only people with strong opinions bother to return questionnaires or wait to speak on a call-in radio or TV show. One recent variation of this phenomenon has been to prompt viewers to call, for a charge, one number if they support a particular position and another if they oppose it. For example, baseball fans were asked to call 1–900–283–4545 if they supported interleague play, and 1–900–283–4546 if they did not. Do you think a cross-section of baseball fans bothered to pay 50 cents and voice their opinions? While very committed to causes, those who write letters to editors or call *Larry King Live* usually do not represent the population at large.

Personal interviews are expensive, but get fairly thorough results. In a situation where the interviewer can see the answer, instead of asking the question, the results tend to be more accurate. For example, if the researcher needs to know whether the respondent lives in an apartment, the interview setting will answer the question. Of course, respondents will try to please the interviewer or will be otherwise influenced by the situation. To offset this, for questions where the respondent may give answers he or she feels are "right," rather than true, "check" questions should be included later in the interview. This helps to verify the consistency of the respondent's answers. If not included, the questions may be poorly worded or ones the person may not be qualified to answer. In either case, they should be dropped from the study.

Telephone interviews are a common compromise between mail surveys and personal interviews. Most Gallup polls and *USA Today* surveys are done by phone. With a carefully constructed sample, useful results can be obtained by contacting as few as 1200 people. Here are some recent examples:

- While 89% of Americans could identify Shakespeare, only 47% could identify Freud.
- Sixty-nine percent felt the Postal Service was doing a good job, and 64% approved of the work done by the Defense Department, but only 45% gave the Justice Department good marks.
- Only 61% of American adults are married, down from a peak of 74% in 1960, reflecting the growing social acceptability of remaining unmarried and the increasing financial independence of women.

Unlisted phone numbers were a problem for pollsters until the use of "random-digit dialing." With this procedure, the first three digits of a telephone number (called an exchange) are identified, and random numbers are generated for the last four digits. However, as many as 10% of American households do not have telephones! This could clearly lead to a biased sample if those households are never contacted.

HOW WILL THE RESPONSES BE RECORDED AND PROCESSED?   Trade-offs between effort and accuracy must be made. If, for example, age is a variable, tallying respondents' ages is very tedious, but grouping people into wide ranges, such as 10–20 years, leads to less accuracy. Techniques that can improve studies include "pretesting" questions on a small group to clear up language and using "follow-up" procedures to encourage those who initially do not respond to respond. Pretesting is sometimes called a "pilot study." Including return postage and offering a copy of the study's results are good ways to encourage cooperation.

Thus far, we have only discussed ideas and language related to a **sample survey**. It is sometimes possible, however, to include *every* member of the population of interest in a sample. In that special case, we have conducted a **census**.

In a study on absenteeism, for example, if a company has only 200 employees, it is reasonable to include all of them, rather than taking a sample. All things being equal, a census is better than a sample because there is no sampling error. However, sometimes less experienced personnel are used for census work and errors result. This occurred in the 1990 national census and led to several lawsuits. In one case, the city of Detroit claimed that much of its inner-city population was not counted, resulting in the loss of federal funding.

In some situations, we can do more than simply record information. If we modify some aspect of the group under study and then see the result, we are conducting an "experiment." In Chapter 5, we defined an experiment as the observation of some activity or the act of taking some type of measurement. In most experiments, two groups are formed. In the **control group**, no real change is introduced. The other group is the **experimental group**.

Consider, for example, a study to determine the effect of alcohol on the driving ability of young adults. A driving simulator reports the number of errors made during a 5-minute interval. A sample of 50 young adults is obtained, each of whom is randomly assigned to either the control group or the experimental group. Each member of the control group takes the driving simulator test and the mean number of errors for the group is determined. Each member of the experimental group drinks 2 12-ounce cans of beer in a half-hour period before taking the driving simulator test to determine the mean number of errors for that group. Note that the control group receives no modification or "treatment." However, the experimental group does receive a treatment—consuming 2 cans of beer in a half-hour period. The effect of the modification is measured by the difference between the two groups.

As another illustration, we might form work teams in a study on absenteeism. For half of the employees, no change is made in the work environment; for the other half, employees are allowed to work in an area where music of their choice is played. Then, using employee records, we can see if there is any difference in attendance between the two groups.

As with surveys, it is important to eliminate possible bias. It is common practice to match the two groups as closely as possible with respect to age,

sample survey A study in which the way a treatment has already affected the experimental objects is observed.

census An observation of the entire population.

control group A set of experimental objects that *does not* undergo change or receive a treatment during a test.

experimental group A set of experimental objects that *does* undergo change or receive a treatment during a test.

sex, health condition, type of work, and so on so that these differences cancel each other out. Another way to avoid biasing the results in a direction one thinks they "ought" to go is to conduct "double blind" tests, in which the person doing the experiment does not know which group is the control group.

Self-Review 8–2	In a *USA Today*/CNN/Gallup poll of 602 registered voters taken March 22, 1992, President Bush led Governor Clinton 52% to 43%. A July 16, 1992, poll by the same organizations showed Clinton leading 56% to 33%. How do you explain the difference in the 2 surveys? Was the sample too small or biased in some fashion? Did some of the conditions change? Do you recall Ross Perot?

Exercises

Answers to the even-numbered Exercises are at the back of the book.

1. Identify the "population" for each of the following samples:
   a. A sample of 5000 readers of *Sports Illustrated*
   b. A sample of 400 college students living in southern California
   c. A sample of 30 public high school students in New Jersey

2. You roll a die 5 times with the following results: 5, 3, 2, 3, and 6. What is the population under study? What do you wish to learn about it? What data would you collect?

METHODS OF PROBABILITY SAMPLING

There are several types of probability sampling. We will study simple random sampling, systematic random sampling, and stratified random sampling. Each of these types of sampling has a similar goal: *to allow chance to determine the items or persons that make up the sample.* While each sample outcome may not be predictable when taken alone, groups of samples are quite predictable. The population of interest can consist either of items, such as all the cassette decks produced by Pioneer during the past month, or of persons, such as all the registered voters in Precinct 9. Other examples of a population are all the banks in the United States, all the fish in a pond, and all the students now attending Yale University.

Random Sampling

random sample A sample chosen so that each member of the population has the same chance of being selected for the sample.

The most widely used method of probability sampling is **simple random sampling** or, as it is often called, "random sampling." A random sample does not just happen, nor should just any collection of objects carelessly be called a random sample. The selection of a random sample must be carefully planned. One method we can use is similar to a lottery. First, we write the name of each member of the population, or an identifying number—such as a Social Security number—on a slip of paper. Then the slips of paper are thoroughly mixed. Finally, the desired number of slips of paper is drawn.

As an illustration, let us look at NOW's interest in the percentage of all banks having at least 1 female director. The group could obtain a list of banks in the

United States and assign a number to each of them. If number 121 were drawn first, that bank (say, the First National Bank of Arizona) would be contacted. This procedure would be repeated until the desired sample size had been selected. It is very important that the population of interest be precisely defined. NOW, for example, would have to decide whether branch banks with their own boards of directors should be included on the list; a member of the population should be listed only once. Figure 8–1 depicts the sampling process.

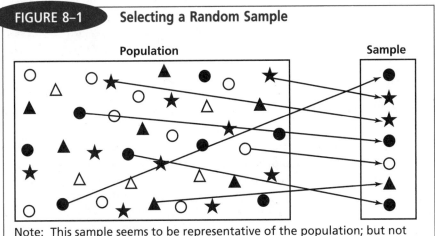

FIGURE 8–1   **Selecting a Random Sample**

Note: This sample seems to be representative of the population; but not perfectly because none of the five open triangles △ was selected to be in the sample.

An easier way to select a random sample employs a **table of random digits.** A portion of such a table, found in Appendix H, is

71529	51996	99289	44268	42759	72434	54402
11776	17395	61317	63290	17067	18408	08992
82437	75248	23715	61194	62175	11149	44793
14997	08398	37662	90175	65331	02562	38020
55317	50018	64380	49047	57111	41641	25427
47422	53721	11419	38616	72171	21523	80967
09540	89442	52381	35035	15884	64273	96028

The table is generated in a random fashion; that is, each of the ten digits has the same chance of being included. Hence, chance determines the outcome of the selection process, and bias does not enter the procedure.

The entire population from which the sample will be drawn is arranged in some systematic fashion (perhaps alphabetically); next, each item is assigned a number. Let us assume our research involves the response of psychiatric patients at Palm Hospital to a new drug. The population consists of 70 psychiatric patients who receive the drug. An identification number is assigned to each patient, starting with 00 and ending with 69. First, a starting point in the table is randomly selected. You could close your eyes and place a pencil down on the page. Suppose, for example, that number were 14 (see the table below).

71529	51996	99289	44268	42759	72434	54402
11776	17395	61317	63290	17067	18408	08992
82437	75248	23715	61194	62175	11149	44793
→[14]997	→[08]398	→[37]662	90175	65331	02562	38020
55317	50018	64380	49047	57111	41641	25427
47422	53721	11419	38616	72171	21523	80967
09540	89442	52381	35035	15884	64273	96028

Starting point   Second patient   Third patient

The patient identified by number 14 becomes part of the sample. To select the next patient in the sample, you can move in any direction you choose. Perhaps you could look at the second hand of a clock and move in the direction in which it is pointing. Let's move horizontally to the right. The first 2 digits in the next column are 08, so patient 08 is also part of the sample. The next patient in the sample is number 37. The following random number is 90. Since no patient is assigned that number, it is omitted. This procedure is continued until we have the desired number of patients for the sample.

Self-Review 8–3

Here is a class roll for a beginning course in sociology. Three students are to be randomly selected to investigate and report on a new community program for the mentally retarded. Suppose you had written the numbers 1 through 38 on slips of paper and then had randomly selected numbers 29, 5, and 11. Which students would be included in the sample?

Winter Quarter   **Preliminary Class Roster**

SS 101   03   INTRO TO SOC SC
2:00 P.M.   3:40 P.M.   MW   UH 422   W Marchal

Name	Rank	Name	Rank
1. August, Nancy M.	FR	20. McFarlin, Ireatha	FR
2. Benner, Robert A.	JR	21. Meinke, Denise M.	JR
3. Brenner, Susan M.	SO	22. Morrison, David D.	JR
4. Clark, Richard C.	FR	23. Navarre, Garry G.	JR
5. Cowan, Timothy J.	JR	24. Oyer, David	SO
6. Cross, Jill M.	SO	25. Pastor, Virginia Marie	SO
7. Daschner, John H.	SO	26. Pickens, Mitch A.	SO
8. Figliomeni, Michael A.	JR	27. Price, Doug C.	SO
9. Grady, Walter P.	JR	28. Rawson, Jeryl L.	FR

10.	Heinrichs, James M.	SO	29.	Rista, Vicki A.	SO
11.	House, James D.	JR	30.	Schmidt, Randy F.	SO
12.	James, Phyllis E.	JR	31.	Sherman, Mike J.	SO
13.	Kimmel, Kurt D.	SO	32.	Shull, Karen A.	SO
14.	Lach, Jerry William	JR	33.	Snow, Sue A.	FR
15.	Lehman, Tim J.	SO	34.	Straub, Jeff J.	SO
16.	Lenz, Matthew H.	SO	35.	Turco, Greg W.	SO
17.	Martin, Diane M.	SO	36.	Von Hertsenberg, Kevin	JR
18.	Mason, Craig D.	SO	37.	Wagner, Holly S.	SO
19.	McCullough, Randy N.	SO	38.	Yamada, Jay A.	JR

. . . .

Exercises

3. The following is a list of the McDonald's Restaurants in a large city. Four locations are to be randomly selected and inspected for cleanliness, safety, customer convenience, and other features. The 28 locations have been coded from 00 to 27. Also noted is whether the location has a play area (P) or not (N).

Number	Location	Play Area	Number	Location	Play Area
00	1560 E Alexis Rd	P	14	5855 Lewis	N
01	835 Lime City Rd	P	15	90 Main	N
02	343 New Towne Square Dr	N	16	567 E Manhattan	N
03	10471 Fremont Pke	P	17	4948 Monroe St	P
04	6555 Airport Hwy	N	18	3345 Monroe St	P
05	4225 Airport Hwy	N	19	2908 Navarre	P
06	5810 W Alexis	N	20	805 N Reynolds	P
07	1736 Broadway	N	21	3138 Secor	P
08	2259 S Byrne	P	22	853 Southwyck Shp Cntr	P
09	3158 Cherry	P	23	3350 W Stern Rd	N
10	1016 Conant	P	24	3740 N Summit	P
11	3240 Dorr St	N	25	1205 W Sylvania Av	P
12	3015 N Holland	P	26	2325 Woodville Rd	P
13	2112 W Laskey	P	27	22201 Woodville Rd	N

a. Suppose the random numbers 11, 17, 61, 03, 93, and 22 are obtained from Appendix H. Which locations would be selected?

b. Using Appendix H, select a random sample of 5 locations.

4. The table at the top of the next page is a list of family practice physicians. Three physicians are to be randomly selected and contacted regarding their fees. The 39 physicians have been coded from 00 to 38. Also noted is whether they are in practice by themselves (S), have a single partner (P), or are in a group practice (G).

Random Number	Physician	Type of Practice	Random Number	Physician	Type of Practice
00	R. E. Scherbarth, M.D.	S	20	Gregory Yost, M.D.	P
01	Crystal R. Goveia, M.D.	P	21	J. Christian Zona, M.D.	P
02	Mark D. Hillard, M.D.	P	22	Larry Johnson, M.D.	P
03	Jeanine S. Huttner, M.D.	P	23	Sanford Kimmel, M.D.	P
04	Francis Aona, M.D.	P	24	Harry Mayhew, M.D.	S
05	Janet Arrowsmith, M.D.	P	25	Leroy Rodgers, M.D.	S
06	David DeFrance, M.D.	S	26	Thomas Tafelski, M.D.	S
07	Judith Furlong, M.D.	S	27	Mark Zilkoski, M.D.	G
08	Leslie Jackson, M.D.	G	28	Ken Bertka, M.D.	G
09	Paul Langenkamp, M.D.	S	29	Mark DeMichiei, M.D.	G
10	Philip Lepkowski, M.D.	S	30	John Eggert, M.D.	P
11	Wendy Martin, M.D.	S	31	Jeanne Fiorito, M.D.	P
12	Denny Mauricio, M.D.	P	32	Michael Fitzpatrick, M.D.	P
13	Hasmukh Parmar, M.D.	P	33	Charles Holt, D.O.	P
14	Ricardo Pena, M.D.	P	34	Richard Koby, M.D.	P
15	David Reames, M.D.	P	35	John Meier, M.D.	P
16	Ronald Reynolds, M.D.	G	36	Douglas Smucker, M.D.	S
17	Mark Steinmetz, M.D.	G	37	David Weldy, M.D.	P
18	Geza Torok, M.D.	S	38	Cheryl Zaborowski, M.D.	P
19	Mark Young, M.D.	P			

a. If the random numbers 54, 08, 44, 38, and 25 are obtained, which physicians would be contacted?

b. Select a sample of 4 physicians using Appendix H.

Systematic Random Sampling

> **systematic random sample** The members of the population are arranged in some fashion. (They may be numbered 1, 2, 3, . . . , listed alphabetically; or ordered by some other method.) A random starting point is selected. Then every *k*th element is chosen for the sample.

If the population is large, such as all the 14,212 students enrolled at Grand Valley State University, located near Grand Rapids, Michigan, it may be more efficient to use systematic random sampling. To use systematic random sampling we begin with a complete list, or sample from, the population. That is, we need a list of all 14,212 students currently enrolled at Grand Valley State. Next, we decide on the approximate sample size. Suppose in this case we decide that about 140 students will be enough. We pick a random starting point—say, 28—and select the corresponding student. We then add 100 and select the 128th student, and so on. The last student selected would be the one corresponding to 14,128 and our sample would consist of 142 students.

The results of a systematic sample will be just as representative as those from a simple random sample. It should not be used, however, if there is any possibility of bias in the ordered list. For example, if you were doing a study on absenteeism, it would be unwise to take a systematic sample of every seventh day. The results would be unduly affected by the starting day. If a Monday was selected as the starting day, then all the other days selected would also be Mondays, and it is well known that absences are higher on Mondays.

Self-Review 8–4

Refer to Self-Review 8–3. Suppose this sample is to consist of every ninth student enrolled in the class after a starting student has been randomly selected from students numbered 1 through 9. Suppose this starting point is the fourth student. Which students will be in the sample?

Exercises

5. Refer to Exercise 3 on page 253. A sample is to consist of every seventh location. The number 02 is selected as the starting point. Which restaurants will be contacted?

6. Refer to Exercise 4 on pages 253–254. A sample is to consist of every fifth physician. A random starting point of physician number 03 is selected. Which physicians will be included in the sample?

Stratified Random Sampling

In planning some types of surveys, it is desirable that the sample be representative not just of the population as a whole, but also of certain subdivisions or groups within it. For example, Revlon might want to ensure that women in each of 10 regions of the United States are included in their research project to determine the market potential of Siren Red lipstick. To accomplish this goal, we would divide the country into 10 geographical regions and randomly select a sample of women within each region. Each woman selected would be asked to try Siren Red and report her reaction. Dividing the country into regions is called **stratifying the population.** Other traits commonly used to form strata are age, income level, and political party affiliation. The manner in which the sample is gathered may be either nonproportional or proportional to the total number of members in each stratum.

We form a **stratified random sample** by identifying the natural subgroupings in the population and selecting an independent random sample from each stratum or subgroup. Obviously, it is important that we define the strata carefully to ensure that each member belongs to only one subgroup.

The advantage of stratified sampling is that one member of a subgroup is usually quite similar to the other members of that subgroup while being quite different from the members of other subgroups. If, for example, the research project involved surveying executives on the role of government in business, the population could be stratified into bankers, executives of large firms, executives of small firms, and so on. Bankers tend to think alike about the role of government in business, but their opinions might differ drastically from those of small business executives. Unless the subgroups differ significantly from each other, nothing is gained by stratification. For example, if our research were concerned with the attitude of college students about compulsory military training, a stratification of the population (all college students) into those from the Far West, the East, and so on would not seem justified unless we believed that geographic location affects attitude toward military service.

stratified random sample After the population of interest is divided into logical strata, a sample is drawn from each stratum or subgroup.

| Self-Review 8–5 | Refer to Self-Review 8–3. Separate the population into three strata: freshmen (FR), sophomores (SO), and juniors (JR). Suppose it has been decided that one freshman, five sophomores, and four juniors will constitute the sample. Randomly select the required number from each stratum. |

Exercises

7. Refer to Exercise 3. Suppose the sample is to consist of four locations, three with play areas and one without. Select a sample accordingly.

8. Refer to Exercise 4. Suppose a sample is to consist of three physicians in group practice (G), two in solo practice (S), and two with a single partner (P). Select a sample accordingly.

THE SAMPLING ERROR

The previous discussion of various scientific sampling methods emphasized the importance of trying to choose a sample in such a way that every member of the population has a known chance of being selected. In other words, the sample should be representative of the population. It would be unreasonable, however, to expect the sample characteristics to match the population *exactly*. The mean of the sample might be different from the population mean by *chance alone*. The standard deviation of the sample will probably be different from the population standard deviation. We can therefore expect some difference between the **sample statistics** (such as the mean and the standard deviation) and the corresponding population values, known as **parameters**. This difference is known as the **sampling error**.

sampling error The difference between the value of the population parameter and its corresponding sample statistic.

The idea of sampling error can be illustrated with a very simple example. Suppose your 5 grades (the population) to date in this course are 69, 86, 82, 70, and 98. A sample of two grades is selected at random from this population of grades to estimate your mean grade. They are, let us say, 70 and 86. The mean of this sample is 78. The mean of another sample of two grades (69 and 98) is 83.5. The mean of all 5 grades (the population) is 81. Notice that sampling errors of -3 and 2.5 are made in estimating the population mean.

Given this potential sampling error, how can political polls, for example, make accurate predictions about the behavior of the voting population based only on sample results? How can a quality control inspector in a manufacturing plant make a decision about the quality of a product after inspecting a sample of only ten parts? We will answer such questions by developing a sampling distribution of the sample means.

THE SAMPLING DISTRIBUTION OF THE SAMPLE MEAN

As noted in the previous section, the means of samples of a specified size selected from a population vary somewhat from sample to sample. When all the sample means that might occur are organized into a probability distribution, the resulting distribution is called the **sampling distribution of the sample mean.** The problem that follows illustrates how such a distribution is constructed.

sampling distribution of the sample mean The probability distribution of the sample means of all possible samples of a given size selected from a population.

Problem Recall from the Chapter Problem that the mayor of Las Palmas is concerned about the city's loss of revenue. To make up for the loss, she is reducing the size of the Department of Social Services from 6 to 4 employees. Table 8–1 lists the 6 employees and their years of service. Note that these data constitute a population and that the mean of this population is 3.16667 years. Suppose the mayor decides, because the lengths of service vary from only 2 to 4 years, that it would be more equitable to randomly select the four workers who are to remain, rather than dismissing the 2 with the fewest years of service. How many different groups of 4 workers are there? How would the sampling distribution of the mean lengths of service appear? What are the possible mean lengths of service for the department when there are 4 employees?

TABLE 8–1	Lengths of Service of Six Social Workers	
Social Worker	Length of Service (in years)	
Don	4	The population mean, μ, is 3.16667:
Gary	3	
Sue	3	$\mu = \dfrac{4 + 3 + 3 + 2 + 4 + 3}{6}$
Bob	2	
Kirk	4	$= 3.16667$
Sandra	3	

Solution The population consists of 6 employees; 4 are to be randomly selected to remain. We can use the idea of a combination, discussed in connection with the binomial distribution in Chapter 6, to determine the total number of possibilities. Let N be the size of the population—in this case, 6—and n the size of the sample, which is 4. So the number of possible samples (groups of remaining employees) is

$$_{N}C_{n} = \frac{N!}{(N - n)! \; n!} = \frac{6!}{(6 - 4)! \; 4!} = 15$$

There are 15 different possible combinations of 4 employees who will remain in the Social Services Department after the cutback. These possible outcomes are listed on the next page, along with the length of service for each retained employee and the mean for each sample (group of employees).

REAL STAT

Is discrimination taking a bite out of your paycheck? Before answering, consider the findings reported in the *Personnel Journal* (December 1995). These findings indicated that good-looking men and women earn about 5% more than average lookers, who in turn earn about 5% more than their plain counterparts. This is true for plain men and even more true for plain women, and it is true in a wide range of occupations from construction work to auto-repair to telemarketing, where it would seem looks should not matter. In another study, lawyers were rated on their appearance using a scale from 1 to 5, and then their earnings were compared. The ugly result: a 12 to 15% pay differential from the highest to the lowest.

	Sample Names			Length of Service (in years)	Sample Mean $\bar{X}$
Don	Gary	Sue	Bob	4, 3, 3, 2	12/4 = 3.00
Don	Gary	Sue	Kirk	4, 3, 3, 4	14/4 = 3.50
Don	Gary	Sue	Sandra	4, 3, 3, 3	13/4 = 3.25
Don	Gary	Bob	Kirk	4, 3, 2, 4	13/4 = 3.25
Don	Gary	Bob	Sandra	4, 3, 2, 3	12/4 = 3.00
Don	Gary	Kirk	Sandra	4, 3, 4, 3	14/4 = 3.50
Don	Sue	Bob	Kirk	4, 3, 2, 4	13/4 = 3.25
Don	Sue	Bob	Sandra	4, 3, 2, 3	12/4 = 3.00
Don	Sue	Kirk	Sandra	4, 3, 4, 3	14/4 = 3.50
Don	Bob	Kirk	Sandra	4, 2, 4, 3	13/4 = 3.25
Gary	Sue	Bob	Kirk	3, 3, 2, 4	12/4 = 3.00
Gary	Sue	Bob	Sandra	3, 3, 2, 3	11/4 = 2.75
Gary	Sue	Kirk	Sandra	3, 3, 4, 3	13/4 = 3.25
Gary	Bob	Kirk	Sandra	3, 2, 4, 3	12/4 = 3.00
Sue	Bob	Kirk	Sandra	3, 2, 4, 3	12/4 = 3.00

The sample means for the retained employees are presented in the form of a probability distribution (see Table 8–2). Logically, it is called the sampling distribution of the sample mean.

The distribution of the sample mean in Table 8–2 gives the probabilities of the sample mean for all possible samples of size 4 taken from the population of 6 social workers. The distribution of the sample mean in Table 8–2 is portrayed graphically in Figure 8–2. The top chart depicts the sampling distribution and the bottom chart the population values.

TABLE 8–2	Sampling Distribution of the Sample Mean	
Sample Mean	**Frequency**	**Probability**
2.75	1	1/15 = 0.0667
3.00	6	6/15 = 0.4000
3.25	5	5/15 = 0.3333
3.50	3	3/15 = 0.2000
	15	1.0000

Note these predictable patterns from Figure 8–2.

1. The mean of the sampling distribution and the mean of the population are equal (3.16667, in this case):

$$\frac{3.00 + 3.50 + 3.25 + \cdots + 3.00}{15} = \frac{47.50}{15} = 3.16667$$

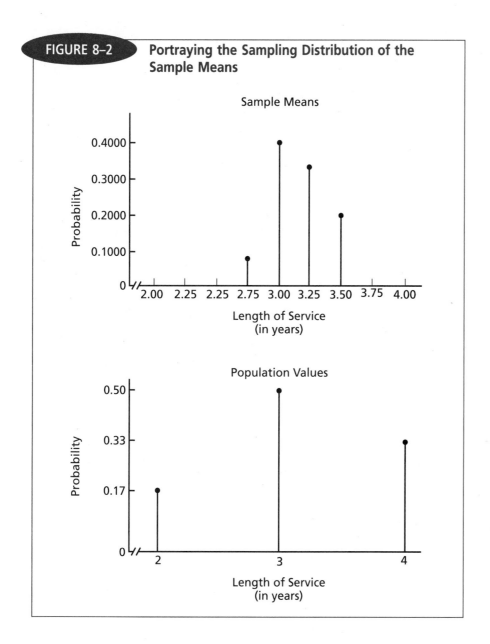

FIGURE 8–2   **Portraying the Sampling Distribution of the Sample Means**

The mean of the sample means is the same as the population mean computed in Table 8–1. This is not a coincidence! The mean of the sample means will always equal the population mean.

2. The spread in the distribution of the sample means is smaller than the spread in the population values; the sample means range from 2.75 to 3.50, while the population ranges from a low of 2 to a high of 4. In fact, the standard deviation of the distribution of the sample means will always equal the population standard deviation divided by the square root of the sample

size. Thus, the formula for finding the standard deviation of the distribution of the sample means is $\sigma/\sqrt{n}$. Notice that, as we increase the size of the sample, the spread of the distribution of the sample means becomes smaller.

3. The shape of the sampling distribution of the means and the shape of the frequency distribution of the population values are different. The distribution of sample means tends to be bell-shaped and to approximate the normal probability distribution.

In summary, we took random samples from a population and for each sample calculated a sample statistic (the mean length of service). Because each possible sample has a known chance of selection, the probability that the mean length of service will be 2.75 years, 3.0 years, and so on can be determined. The distribution of these mean lengths of service is aptly called the sampling distribution of the sample mean.

Even though in practice we see only one particular random sample, in theory any sample could arise. Consequently, we view the sampling process as repeated sampling of the statistic from its sampling distribution. This sampling distribution is then employed to measure how reasonable or likely a particular outcome might be.

Self-Review 8–6

A population consists of five prisoners in cell block M. The length of time each has spent in prison is shown below.

Name	Years
Dow	5
Smith	3
Artz	6
Kim	2
Batt	4

a. Compute the mean length of imprisonment for the population.
b. Select all possible samples of two prisoners from the population. Compute the mean of each sample.
c. Does the mean of the sample means equal the population mean?
d. Give the sampling distribution of the means.
e. Plot the sampling distribution of the means and the population.
f. Does the sampling distribution tend to be bell-shaped, and does it begin to approximate a normal distribution?
g. Cite evidence to show that there is less spread in the sampling distribution compared with the population values.
h. Is the population normally or nonnormally distributed?

Exercises

9. A police training class consists of 4 recruits. They have had 2, 3, 4, and 4 years of education beyond high school.
   a. Determine the population mean.
   b. List all possible samples of 2 recruits.
   c. Compute the mean of each sample. (Hint: There should be 6 samples.)
   d. Develop a sampling distribution of the means.
   e. Portray the sampling distribution of the means and the population values in a histogram.
   f. Compare the two distributions in terms of means, shapes, and ranges.

10. A population consists of the hourly wages of 6 employees. The wages are $10, $4, $12, $11, $9, and $8.
   a. Determine the population mean.
   b. A sample of 4 wages is to be selected at random from the population. List all the possible samples of 4 wages. Then compute the mean of each sample.
   c. Develop a distribution of the sample mean, and compute the mean.
   d. Portray the distribution of the sample mean in the form of a histogram. Immediately below the histogram, show the probability distribution for the population values.
   e. Draw conclusions regarding the 2 means (the population mean and the mean of the sample means). Also, make an observation with respect to the spread of the 2 probability distributions.

11. Five children are in a preschool play group. Their heights are 39, 38, 36, 39, and 38 inches.

   a. Determine the mean height of this population of 5 children.
   b. Compute the standard deviation of their heights.
   c. List all possible samples of 3 children.
   d. Compute the mean of each sample found in part c.
   e. Draw histograms of the original population and of the sampling distribution of the means on the same scale.
   f. Calculate the mean and the standard deviation of the sampling distribution of the means.

12. Last month the Greenville Volunteer Fire Department answered five calls. The length of time, in minutes, from when the call was received until they arrived at the fire is reported below.

   | 18 | 18 | 12 | 18 | 14 |

   a. Determine the population mean.
   b. How many samples of size two are possible from this population?
   c. List the possible samples of size two, and determine the mean of each.
   d. Determine the mean of the distribution of the sample means. How does it compare to the population mean?
   e. Organize the sample means into a frequency distribution. How does the standard deviation of the distribution of the sample means compare to that of the population standard deviation?

CHAPTER OUTLINE

I. In a probability sample, each member of the population has a chance of being selected for the sample.
   A. If probability sampling is not used, the results may be biased.

B. When developing a sample, keep these questions in mind.
   1. What exactly do you wish to find out?
   2. What population are we studying?
   3. How will the data be collected?

a. A mail survey is inexpensive, but the response rate is usually low.

b. Personal interviews are expensive, but the results are usually good.

c. A telephone interview is a compromise between the two above and is used most often.

4. How will the response be recorded or processed?

C. In a census, we include every member of the population.

D. An experiment is the observation of some activity or the act of taking a measurement.

1. In the control group, the experimental objects do not undergo any change.

2. In the experimental group, the objects receive a treatment or undergo a change.

II. There are several types of probability samples.

A. In a simple random sample, every member of the population has the same chance of being selected for the sample.

B. In a systematic random sample, a random starting point is selected, and then every *k*th element is selected for the sample.

C. In a stratified random sample, the population is divided into strata, or groups, and a random sample is selected from each stratum.

III. The sampling error is the difference between the population parameter and the corresponding sample statistic.

IV. A sampling distribution of the sample means is a listing of all sample means and the probability of each outcome.

A. The mean of all the sample means is equal to the population mean.

B. There is less spread or variation in the distribution of the sample means than in the population.

Exercises

13. A study of hotel accommodations in a metropolitan area reveals the existence of 30 such facilities. The city's convention and visitors bureau is surveying charges per day for single-occupancy rooms. The daily rates for this statistical population are

$25	$35	$22	$25	$30	$24
25	20	25	24	28	24
28	25	27	25	35	25
30	25	17	25	21	18
16	24	21	21	13	19

a. Using the random numbers in Appendix H, draw a simple random sample of 6 daily rates from this population.

b. Select a systematic random sample by randomly choosing a starting point among the first 5 hotels and then including every fifth observation.

c. If the last 10 hotels on the list are all "cut-rate" hotels, describe how you could select a sample of 4 regular hotels and 2 "cut-rate" hotels.

14. Since 1935, the Social Security Administration has issued numbers, such as 123–45–6789, to 305 million people.

a. Describe how you would select a random sample of 20 individuals from the agency's files.

b. The leading digit indicates which of the 10 service centers issued the number. Describe how you would create a sample that includes exactly 2 individuals from each service center.

15. You are studying voters' reactions in your state to a piece of statewide legislation. Describe how you would take a sample of these voters.

16. You wish to study the birth weights of newborn infants in your city. How would you go about obtaining the sample?

17. If you wished to estimate the percentage of total working time a secretary spends on specific tasks—typing, answering the telephone, making copies, and so forth—how would you sample his or her activities?

18. A certain variety of flower grows to only three different heights: 2 inches, 4 inches, and 6 inches. If each of these heights is equally likely, draw a histogram of the probability distribution of this population. Find the mean and the standard deviation. List all possible samples of 2 flowers that could be drawn from this population, and calculate the corresponding sample average. Draw a histogram of the probability distribution of sample averages. Find its mean and its standard deviation.

19. Suppose you have borrowed money from your parents on four occasions in the amounts below. They have decided that you should repay only two of the four loans. The four amounts are written on slips of paper, and you select two.

$100 $100 $160 $200

a. What is the mean amount of the population of loans? What is the standard deviation?
b. How many different combinations are there of loans you will not need to repay?
c. Determine the mean amount of the possible samples of loans not to be repaid.
d. Compare the dispersion of the sample means with the dispersion of the population. Which has the larger amount of dispersion?

20. A company manufactured 6 TV sets on a given minute, and these TV sets were carefully inspected and classified as "good" or "defective". The results of the inspection are as follows:

Good	Good	Defective	Defective
Good	Good		

(In later chapters, we will discuss proportion, which is the fraction of a population or sample that has a particular characteristic. In this population, there are 4 good and 2 defective sets, so the proportion of the sets that are good in the population is .6667, found by 4/6. We could think of this in terms of a mean by assigning a 1 for the good sets and a 0 for the defective sets and finding the mean of the number 1, 1, 0, 0, 1, and 1.)

a. What proportion of these sets is good?
b. How many samples of size 2 are possible?
c. Find the mean of the samples.
d. Compare the mean of the samples with the mean of the population.

21. A friend of yours, who will gamble on virtually anything, points out that every dollar bill printed by the federal government has an 8-digit serial number on it, such as J86581144B. He further claims that he can guess the arithmetic mean of the 8 digits on any dollar bill in your pocket. He tells you, before you pull a bill out of your pocket, that the mean of the 8 digits will be between 4 and 6. He challenges you to empty your pockets and give him every bill with a mean between 4 and 6. As incentive, he offers to match any bill whose arithmetic mean does not fall between 4 and 6 and to sweeten the payoff with an extra quarter for every such bill. Would you accept this challenge? Why or why not?

DATA EXERCISES

22. Refer to the real estate data set, which reports information on homes sold in Alabama during 1995. Assume these data are a population. Use Appendix H to select 5 random samples of 10 homes. Compute the mean number of bedrooms in these 5 samples. Compare the mean of the population (3.58 bedrooms) with the sample means. Compare the shape of the population with the shape of the sampling distribution. Write a brief report comparing your findings with what you expected to find.

23. Refer to the school data set, which refers to the 94 school districts in northwest Ohio. Assume these data are a population. Use Appendix H to select 5 random samples of 10 school districts, and determine the mean salary of each of these 5 samples. Compare the mean of the sample means with the mean salary of all 94 school districts. (The mean teacher's salary of all 94 districts is $33,181, the standard deviation is $3549, and the average salaries range from $26,125 to $43,256.) Write a brief report summarizing your findings.

CHAPTER ACHIEVEMENT TEST

The answers are at the back of the book.

MULTIPLE-CHOICE QUESTIONS

Select the response that best answers each of the questions.

1. A list of every member of the population without duplication is called a
   a. sample frame.
   b. sample design.
   c. survey.
   d. census.
2. In a study of poverty, all families receiving welfare benefits are selected. This group represents the
   a. sample design.
   b. census.
   c. survey.
   d. population of interest.
3. A "100%" survey could also be called a(n)
   a. sample frame.
   b. sample design.
   c. census.
   d. unbiased study.
4. In general, the least expensive way to obtain a sample is
   a. an experiment.
   b. a mail survey.
   c. a personal interview.
   d. a telephone survey.
5. The most accurate survey results are generally achieved by
   a. a controlled experiment.
   b. a telephone survey.
   c. personal interviews.
   d. a mail survey.
6. Each item in a population has a chance of being selected. Sampling done under these conditions is called
   a. systematic sampling.
   b. convenience sampling.
   c. confidence sampling.
   d. probability sampling.
7. Each item in a population has the *same* chance to be selected for a sample. What is this process called?
   a. Nonprobability sampling
   b. Random sampling
   c. Judgment sampling
   d. None of the above
8. If you select every tenth item on a list, which method of sampling are you using?
   a. Stratified sampling
   b. Random sampling
   c. Systematic sampling
   d. None of the above
9. The difference between the population mean (μ) and a sample mean ($\overline{X}$) is the
   a. interval estimate.
   b. point estimate.
   c. sampling error.
   d. standard error.
10. When all possible samples are selected and their means found, the mean of these sample means
    a. is smaller than the population mean.
    b. is larger than the population mean.
    c. is equal to the population mean.
    d. cannot be predicted.

COMPUTATION PROBLEMS

11. Listed below are the 10 state-supported universities in Ohio:

Bowling Green	Ohio State
Miami	Akron
Toledo	Youngstown
Kent State	Wright State
Ohio	Cincinnati

   a. Use the following sequence of random numbers to select a sample, without replacement, of 3 universities: 090268.

   b. Select a systematic sample where the starting point is the second university and every third university after that is selected.

12. A population of 5 geriatric patients indicates that the number of care-giving relatives are 6, 4, 7, 3, and 2.

   a. Compute the mean and the standard deviation of this population.

   b. Select all possible samples of 3 patients from this population, and compute the mean of each sample.

   c. Draw a histogram of the sampling distribution of means found in part b.

   d. Compute the mean and the standard deviation of the sampling distribution of the means.

   e. Compare the location and spread in the original population and the sampling distribution of the sample means.

· ·

ANSWERS TO SELF-REVIEW PROBLEMS

8-1 The researcher thought it was reasonable. Use an "inhalant" meant the concentration and breathing of fumes from glue, liquid "white out," or similar products that can yield a "high." Unfortunately, younger students may not understand the term or, even worse, may be using an inhalant under a doctor's direction. They may give a literally correct response that, in fact, misleads the researcher.

8-2 The sample was probably valid. During the time between the two surveys, Ross Perot abruptly decided not to run! So his withdrawal dramatically changed the race.

8-3
29	Vicki Rista
5	Timothy Cowan
11	James House

8-4
4	Richard Clark
13	Kurt Kimmel
22	David Morrison
31	Mike Sherman

8-5 There are many possibilities. One of them is

FR	5	Snow
SO	9	Mason
	3	Daschner
	11	Oyer
	12	Pastor
	8	Martin
JR	7	Lach
	10	Navarre
	4	Grady
	3	Figliomeni

No doubt the composition of your sample will be different.

8-6 a. $20/5 = 4.0$

b.

	$\bar{X}$
Dow, Smith	4.0
Dow, Artz	5.5
Dow, Kim	3.5
Dow, Batt	4.5
Smith, Artz	4.5
Smith, Kim	2.5
Smith, Batt	3.5
Artz, Kim	4.0
Artz, Batt	5.0
Kim, Batt	3.0

c. Yes, $40/10 = 4.0$; same as $\mu = 20/5 = 4.0$

d.

X	f	Probability
2.5	1	1/10 = 0.1000
3.0	1	1/10 = 0.1000
3.5	2	2/10 = 0.2000
4.0	2	2/10 = 0.2000
4.5	2	2/10 = 0.2000
5.0	1	1/10 = 0.1000
5.5	1	1/10 = 0.1000
	10	1.0000

e.

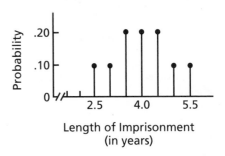

f. Yes; yes

g. The range for the sampling distribution is $5.5 - 2.5 = 3.0$. The range for the population is $6 - 2 = 4.0$.

h. Nonnormally

CHAPTER 9

The Central Limit Theorem and Confidence Intervals

When you have completed this chapter, you will be able to

- describe the central limit theorem;
- develop and describe the sampling distribution of sample means;
- develop and describe the sampling distribution of sample proportions;
- compute confidence intervals for means and proportions;
- understand the need for the finite population correction factor.

CHAPTER PROBLEM

Bill Malrooney began his printing business 20 years ago. It has grown over the years, and he now has 40 employees. He is faced with some major decisions regarding health care for these employees. Before making a final decision on what health care plan to purchase, he decides to form a committee of randomly selected employees. The committee will be charged with the responsibility to study the issues carefully and make a recommendation as to what plan best fits the employees' needs. If he randomly selects this committee, what will be the mean years of experience on the committee? How does the shape of the distribution of years of experience of all employees compare with the distribution of the various sample means?

INTRODUCTION

central limit theorem If
all samples of a fixed size
are selected from any
population, the sampling
distribution of the sam-
ple mean is approxi-
mately a normal distribu-
tion. This approximation
improves with larger
samples.

In this chapter, we will examine one of the most important theorems in statistics—the **central limit theorem.** Its application to the sampling distribution
of the sample mean, introduced in Chapter 8, allows us to use the normal
probability distribution to create confidence intervals for the population mean.

THE CENTRAL LIMIT THEOREM

The central limit theorem states that for large random samples, the shape of
the sampling distribution of the sample mean is close to a normal probability
distribution. The approximation is more accurate for large samples than for
small samples. This is one of the most useful conclusions in statistics. We can
reason about the sample mean with absolutely no information about the shape
of the original distribution from which the sample was taken. In other words,
the central limit theorem is true for all distributions.

If the population has a normal probability distribution, then the sampling
distribution will also be normally distributed. If the population distribution is
symmetrical (but not normal), you will see the results of the central limit
theorem emerge with samples as small as 10. On the other hand, if you start
with a distribution that is skewed or has thick tails, it may require samples of at
least 30 to observe the normality feature. Most statisticians consider a sample of
30 or more "large enough" for the central limit theorem to be employed.

The idea that a population that is not normal will converge to normality
is illustrated by Figures 9–1, 9–2, and 9–3. We will discuss this example in
more detail shortly, but Figure 9–1 is a graph of a discrete probability distribution that is positively skewed. There are many possible samples of 5 that
might be selected from this population. Suppose we randomly select 10 samples of 5 each and compute the mean of each sample. These results are shown
in Figure 9–2. Notice that the shape of the distribution of the sample mean
has changed from the original population even though we only selected 10 of
the many possible samples. To put it another way, we selected 10 random
samples of 5 each from a population that is positively skewed and found the
distribution of the sample mean has changed from the shape of the population.
As we take more samples, we will find the distribution of sample mean will approach the normal distribution. Figure 9–3 is a histogram that shows the results
of 30 random samples of 5 observations from the same population. Observe
the clear trend toward the normal distribution. This is the basis of the central
limit theorem. The following example will underscore this condition.

Problem   The lengths of service (rounded to the nearest year) of the 40 employees currently on the Malrooney Printing, Inc., payroll are as follows:

11	4	18	2	1	2	0	2	2	4
3	4	1	2	2	3	3	19	8	3
7	1	0	2	7	0	4	5	1	14
16	8	9	1	1	2	5	10	2	3

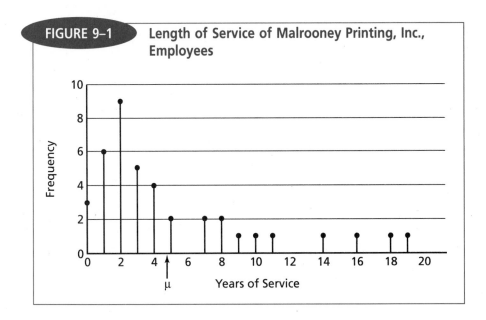

FIGURE 9–1   **Length of Service of Malrooney Printing, Inc., Employees**

This information is depicted in Figure 9–1. Observe that the distribution of lengths of service is positively skewed. A few employees have worked at Malrooney Printing for some time. Specifically, 6 employees have been with the company 10 years or more. However, due to growth in the business, the number of employees has increased in the last few years. Of the 40 employees, 18 have been with the company 2 years or less.

Now let's consider one of Bill Malrooney's problems. He must make decisions regarding health care. He would like to form a committee of five employees to look into the health care question and suggest what type of insurance would be most appropriate for the majority of workers. How should he select the committee? If he selects the committee randomly, what might he expect in terms of mean length of service for those on the committee?

Solution   To begin Bill writes the length of service for each of the 40 employees on a piece of paper and puts them into an old baseball cap. Next, he shuffles the pieces of paper around and randomly selects 5 slips of paper. The lengths of service for these 5 employees are 4, 1, 0, 14, and 9 years. So the mean length of service for these 5 employees is 5.6 years. How does that compare with the population mean? At this point, Bill does not know the population mean, but the number of employees in the population is only 40, so he decides to calculate the mean length of service for all his employees. It is 4.8 years, found by adding the lengths of service for all the employees and dividing the total by 40. That is, $\mu = (11 + 4 + 18 + \cdots + 2 + 3)/40 = 192/40 = 4.8$. The difference between the sample mean, $\overline{X}$, and the population mean, μ, is called **sampling error.** In other words, the difference of 0.8 years between the population mean of 4.8 and the sample mean of 5.6 is the sampling error. It is due to chance. If Bill selected these 5 employees to constitute

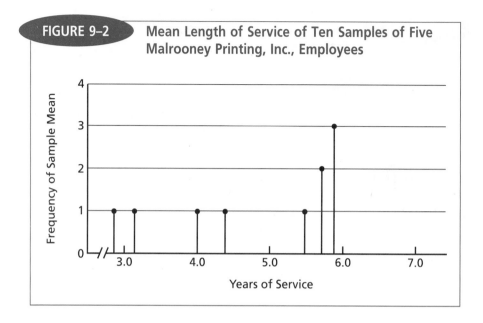

FIGURE 9–2   Mean Length of Service of Ten Samples of Five Malrooney Printing, Inc., Employees

the committee, their mean length of service would be somewhat more than the population mean.

What would happen if Bill put the 5 pieces of paper back into the baseball cap and selected another sample? Would you expect the mean of this second sample to be exactly the same as that of the previous sample? Suppose he finds the lengths of service in this sample to be 8, 3, 1, 1, and 14 years. The second sample mean is therefore 5.4 years. Now suppose Bill repeats the sampling process 8 more times. The result of selecting 10 samples of 5 employees each is shown in Figure 9–2. Notice the difference in the shape of the population and the distribution of the sample mean. The population of the lengths of service for the employees (Figure 9–1) is positively skewed, but the distribution of the 10 sample means does not reflect the same positive skew.

Table 9–1 shows the result of selecting 30 different samples of 5 employees each and computing their sample means. These sample means are then organized into a frequency polygon (Figure 9–3). Compare the shape of this frequency polygon to that of the population of employees in Figure 9–1. You should observe two important features:

1. The shape of the distribution of the 30 sample means is different from that of the population. In Figure 9–1, the distribution of the lengths of service of all employees is positively skewed. However, the sampling distribution of the sample mean, Figure 9–3, is more nearly a normal distribution. This is evidence of the central limit theorem at work.
2. There is less dispersion in the sampling distribution of sample mean than in the population distribution. In the population distribution, the lengths of service ranged from 0 year to 19 years. In the sampling distribution of sample means, the sample means ranged from 2.2 years to 9.2 years.

TABLE 9–1	Random Samples and Sample Means of Malrooney Printing, Inc., Employees					
Sample Number	Sample Data (years of service)					Sample Mean $\overline{X}$
1	4	1	0	14	9	5.6
2	8	3	1	1	14	5.4
3	2	4	2	4	2	2.8
4	11	1	5	2	3	4.4
5	2	1	7	3	3	3.2
6	11	2	10	1	4	5.6
7	4	3	11	2	9	5.8
8	8	3	14	2	2	5.8
9	1	7	8	2	2	4.0
10	14	1	2	10	2	5.8
11	8	2	18	5	0	6.6
12	3	1	4	2	7	3.4
13	0	4	3	3	1	2.2
14	11	4	9	2	8	6.8
15	7	1	2	5	1	3.2
16	2	2	10	11	0	5.0
17	4	2	3	8	1	3.6
18	0	0	4	3	5	2.4
19	1	4	2	3	1	2.2
20	2	7	0	2	3	2.8
21	5	16	2	4	11	7.6
22	9	3	0	2	8	4.4
23	5	1	2	10	0	3.6
24	2	1	2	0	8	2.6
25	19	4	3	3	1	6.0
26	0	4	9	11	8	6.4
27	4	9	4	3	2	4.4
28	2	5	2	7	2	3.6
29	18	8	1	11	8	9.2
30	14	16	0	2	3	7.0

We can also compare the mean of the sample mean to the population mean. The mean of the 30 samples reported in Table 9–1 is 4.7133 years: $\mu_{\overline{X}} = (5.6 + 5.4 + \cdots + 9.2 + 7.0)/30$. We use the symbol $\mu_{\overline{X}}$ to represent the mean of the sample means. The subscript reminds us that the distribution is of sample means. It is read as "mu sub X bar." We observe that the mean of the 30 sample means, 4.7133 years, is very close to the population mean of 4.80 years.

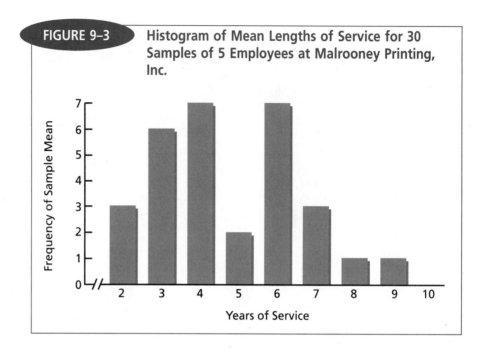

FIGURE 9–3 Histogram of Mean Lengths of Service for 30 Samples of 5 Employees at Malrooney Printing, Inc.

What can we conclude from this example? The central limit theorem indicates that, regardless of the shape of the population, the sampling distribution of the sample mean will approximate the normal distribution. The more samples we select and the larger these samples are, the stronger the convergence is. The Malrooney Printing, Inc., example is empirical evidence of how the central limit theorem works. We began this illustration with a positively skewed population (Figure 9–1). When we selected a small number of samples and looked at that distribution of sample means, we could observe a difference between the shape of the population and that of the distribution of sample means (compare Figures 9–1 and 9–2). When we increase the number of samples from 10 to 30, we begin to see the normality feature. The shape of the distribution of the 30 sample means reported in Figure 9–3 is clearly moving toward a normal distribution.

You will note that the central limit theorem (reread the definition) does not say anything about the dispersion of the sample mean or about a comparison of the mean of the sample means to the mean of the population. However, in our empirical study, we did observe that there is less dispersion in the sample means than in the population by comparing the range of the population and the range of the sample means. Compare the range in Figure 9–1 (0–19) with the range in Figure 9–3 (2.2–9.2). We also observed that the mean of the 30 sample means is very close to the population mean. It can be shown that, if the dispersion in the population is defined by σ, the dispersion in the sample means is $\sigma/\sqrt{n}$, where n is the size of the sample. From this relationship, you can see that, as the sample size increases, the dispersion of the sample means decreases. It can also be demonstrated that, if we compute all the sample means, the mean of the population is exactly equal to the mean of all the sample means. To put it another way, the mean of all the sample means is equal to the population mean.

The empirical example just completed should give you some idea of the importance of the central limit theorem to statistics. It is worthwhile, however, to consider another example. In this second example, we will again see the central limit theorem at work, but this time in a more classical setting. This is how the mathematicians of the 17th century viewed the problem.

Problem   Suppose we have a fair die, roll it twice, and look at the sum of the numbers of spots. If the first roll came up a three and the second a four, we are interested in the total—i.e., seven. What is the shape of the population of the numbers of spots? What are the possible outcomes to the experiment? What is the shape of the sampling distribution of the sums of the numbers of spots when a single die is rolled twice? When it is rolled three times?

Solution   This is actually a sampling situation. The population is a uniform distribution, with each of the whole number values from 1 through 6 having an equal likelihood of occurrence. The following table and graph show the various outcomes in the population and their corresponding probabilities:

Possible Outcomes	Probability
1	1/6 = 0.1667
2	1/6 = 0.1667
3	1/6 = 0.1667
4	1/6 = 0.1667
5	1/6 = 0.1667
6	1/6 = 0.1667

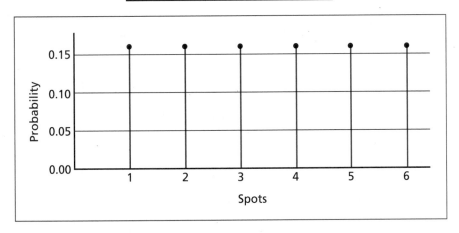

Now, if we roll the die twice, the total numbers of spots appearing can be summarized as follows:

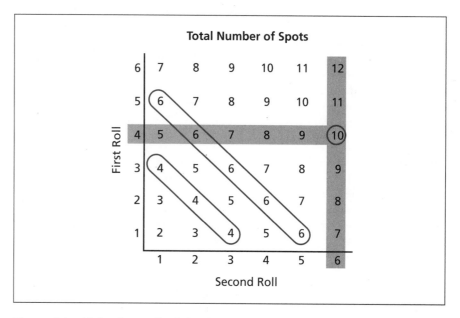

To explain, if the first roll of the die resulted in a 4 and the second roll a 6, the total is 10. This value is circled in the above table. Next, we need to develop a distribution for the total numbers of spots appearing. From the table, there are 36 possible outcomes, and on 1 occasion, the total number of spots appearing is 2, and on 1 occasion, the total is 12. There are 3 occasions when the total is 4, 5 occasions when the total is 6, and so on. By now, you have probably noticed that we can find the number of outcomes that are the same by looking at the diagonal from upper left to lower right. The total numbers of spots and the probability of each are summarized in the following table and chart:

Possible Outcomes	Number of Times Appearing	Probability
2	1	1/36 = 0.0278
3	2	2/36 = 0.0556
4	3	3/36 = 0.0833
5	4	4/36 = 0.1111
6	5	5/36 = 0.1389
7	6	6/36 = 0.1667
8	5	5/36 = 0.1389
9	4	4/36 = 0.1111
10	3	3/36 = 0.0833
11	2	2/36 = 0.0556
12	1	1/36 = 0.0278

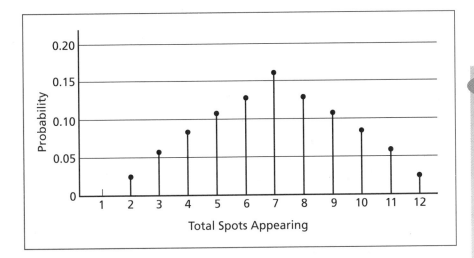

Notice the change in the shape of the distribution of the sums from that of the distribution of the original population. We began with a population that was uniform for the discrete whole numbers 1 through 6. When we rolled the die twice and looked at the total numbers of spots, the distribution changed to a triangular-shaped distribution. Incidentally, we could have used the average numbers of spots on the 2 rolls and obtained the same shape of the distribution. Again, this is the effect of the central limit theorem, which states that, regardless of the shape the population distribution, the distribution of the sample means will approach normal as we increase the size and number of samples.

Without reporting the details, suppose we roll the same die 3 times and report the distribution of the sum of the numbers of spots. First, we know that the smallest possible sum is 3 and the largest possible sum 18. The sums of 10 and 11 will each have a probability of occurrence of 0.1250, which is the highest probability. Figure 9–4 depicts the probability distribution for the sums of 3 rolls as well as of both 1 and 2 rolls. Again observe the clear effects of the central limit theorem. When we move from 1 to 2 and then to 3 rolls of a die, the shape of the distribution changes and moves toward the bell-shaped normal probability distribution. This very interesting result intrigued the mathematicians of the 17th century and has resulted in many modern applications.

Self-Review 9–1

Answers to all Self-Review problems are at the end of the chapter.

Refer to the Malrooney Printing, Inc., data on page 268. Select ten random samples of five employees each. Compute the mean of each sample, and plot the sample means on a chart similar to Figure 9–2. To find the employees to include in the samples, use the methods described in Chapter 8 and the Table of Random Numbers (Appendix H).

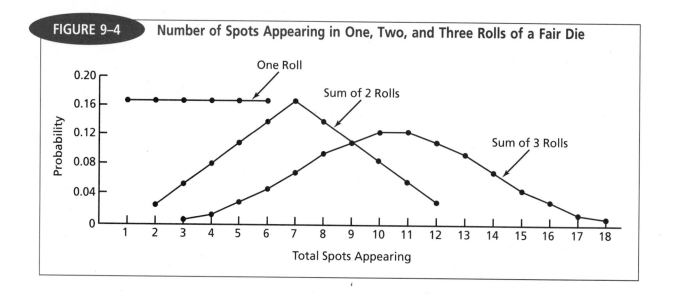

FIGURE 9–4 **Number of Spots Appearing in One, Two, and Three Rolls of a Fair Die**

Exercises

Answers to the even-numbered Exercises are at the back of the book.

1. Identify any population you might like to study, such as a set of telephone numbers, student identification numbers, or the size or weight of potato chips.
   a. Select several random samples of two items from your population. Calculate their means. Repeat the sampling procedure enough times to draw a reasonably accurate histogram or frequency polygon of the sample means.
   b. Repeat this process with a somewhat larger sample—say, four items.
   c. Repeat this process with an even larger sample, and again draw a histogram or frequency polygon of the sample means.
2. Explain in your own words the features of the central limit theorem.

CONFIDENCE INTERVALS FOR MEANS

confidence interval A range of values constructed from sample data so that a parameter occurs within that range at a preselected probability. The preselected probability is termed the "level of confidence."

The information just developed about the shape of the sampling distribution of the sample mean ($\overline{X}$) allows us to locate an interval that has a high probability of containing the population mean (μ). For reasonably large samples, we can state the following:

1. Ninety-five percent of the sample means selected from a population will lie within 1.96 standard deviations of the population mean (μ).
2. Ninety-nine percent of the sample means will lie within 2.58 standard deviations of the population mean.

Intervals computed in this fashion are called the **95% confidence intervals** and the **99% confidence intervals**.

How are the values of 1.96 and 2.58 obtained? The 95% and 99% refer to the approximate percentage of the time that similarly constructed intervals will include the parameter that is being estimated. The 95%, for example, refers to the middle 95% of the observations. Therefore, the remaining 5% is equally divided between the two tails. See the following diagram. The central limit theorem states that the distribution of the sample mean will be approximately normal, so Appendix C may be used to find the appropriate z-values. Locate 0.4750 in the body of the table, and then read the corresponding row and column value. It is 1.96; that is, the probability of being in the interval between $z = 0$ and $z = 1.96$ is 0.4750. Likewise, the probability of being in the interval between -1.96 and 0 is 0.4750. When we combine these two probabilities, the probability of being in the interval -1.96 to 1.96 is 0.95. The z-value corresponding to 0.99 is determined in a similar way.

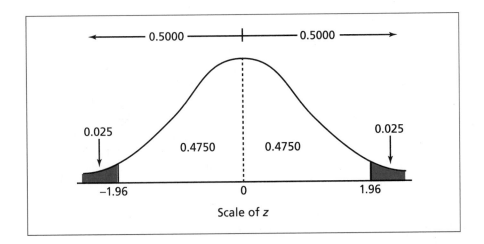

How do you construct the 95% confidence interval? To illustrate, assume your research involves the annual starting salaries of liberal arts graduates with master's degrees. You have computed the means of the samples to be $32,500 and the standard deviation of the sample means to be $200. The 95% confidence interval is from $32,108 to $32,892, found by $32,500 ± 1.96($200). If 100 samples of the same size were selected from the population of interest and the corresponding 100 confidence intervals determined, we could expect to find the population mean in about 95 out of the 100 confidence intervals.

THE STANDARD ERROR OF THE SAMPLE MEAN ($\sigma_{\bar{X}}$)

In the previous section, the standard deviation of the sampling distribution was given as $200. It is called the **standard error of the sample mean** and denoted by the symbol $\sigma_{\bar{X}}$, read "sigma sub X bar," often shortened to the **standard error**.

standard error of the sample mean The standard deviation of the sampling distribution of the sample mean.

The standard error is a measure of the variability of the sampling distribution. It is computed by

$$\sigma_{\overline{X}} = \frac{\sigma}{\sqrt{n}} \qquad \boxed{9\text{--}1}$$

where

$\sigma_{\overline{X}}$ is the symbol for the standard error of the mean.
σ is the population standard deviation.
n is the sample size.

In most real-world situations, the population standard deviation is not known, so we replace it with the sample standard deviation; that is, we replace σ with s. We then change formula 9–1 as follows:

$$s_{\overline{X}} = \frac{s}{\sqrt{n}} \qquad \boxed{9\text{--}2}$$

The size of the standard error is affected by two values. The first is the standard deviation. If the standard deviation is large, then the standard error will also be large. However, the standard error is also affected by the sample size. As the sample size is increased, the standard error decreases, indicating that there is less variability in the distribution of sample means. This conclusion is logical because an estimate made with a large sample should be more precise than one made from a small sample.

When the sample size (n) is at least 30, it is generally accepted that the central limit theorem will ensure a normal distribution of the sample means. This is a very important consideration. If we can depend on the fact that the sample means are normally distributed, we can use the standard normal distribution—that is, z—in our calculations.

Point and Interval Estimates

The data presented in the Problem-Solution regarding Malrooney Printing, Inc. (see p. 268), are a population because they report the lengths of service for all 40 of the company employees. The mean of this population is easy to compute. However, in most cases, the very thing we are trying to estimate is the population parameter. In the Malrooney Printing example, that is the mean length of service of all the employees. This parameter is usually unknown in practice, and we are trying to find its value. The single number with which we estimate a population parameter is called the **point estimate**.

A sample mean, $\overline{X}$, is a point estimate of the population mean, μ. For example, the staff at Best Buy, Inc., wanted to estimate the mean age of buyers of stereo equipment. They selected a random sample of 50 recent purchasers, determined the age of each purchaser, and computed the mean of the buyers in this sample. The mean of this sample is a point estimate of the mean of the population.

. . . .

point estimate The value, computed from a sample, that is used to estimate the population parameter.

However, a point estimate tells only part of the story. While we expect the point estimate to be close to the population parameter, we would like some method to measure how close it is. The **interval estimate** serves this purpose.

For example, we estimate the mean yearly income of a group of successful citrus fruit growers to be $35,000. The range of that estimate might be from $34,000 to $36,000. We can describe how confident we are that the population parameter is in that interval by making a probability statement. The resulting confidence interval is an interval estimate of the population parameter. Confidence levels such as 95% and 99% are often used to indicate the degree of belief or credibility to be placed on a particular interval estimate of a population parameter. We might say, for example, that we are 90% sure that the mean yearly income is between $34,000 and $36,000.

> **interval estimate** A range of values within which we have some confidence that the population parameter lies.

A confidence interval for the population mean is constructed by

Constructing Confidence Intervals

$$\overline{X} \pm z \frac{s}{\sqrt{n}}$$

9–3

where $\overline{X}$ is the sample mean, s is the sample standard deviation, z is the standard normal value corresponding to the desired level of confidence, and n is the sample size. The 95% confidence interval is computed by

$$\overline{X} \pm 1.96 \frac{s}{\sqrt{n}}$$

The 99% confidence interval for the mean is computed by

$$\overline{X} \pm 2.58 \frac{s}{\sqrt{n}}$$

Problem   Construct a 95% confidence interval for the mean hourly wage of apprentice geologists employed by the top 5 oil companies. For a sample of 50 apprentice geologists, $\overline{X} = \$14.75$ and $s = \$3$.

Solution   The standard error of the sample means is estimated to be $0.42. Using formula 9–2:

$$s_{\overline{X}} = \frac{s}{\sqrt{n}} = \frac{\$3}{\sqrt{50}} = \$0.42$$

Thus, the 95% confidence interval for μ goes from $13.93 to $15.57, found by formula 9–3:

$$\$14.75 \pm 1.96(\$0.42)$$

$$\$14.75 \pm \$0.82$$

If we repeat the sampling process for the 50 apprentice geologists, would we expect to see a sample mean of $14.75 each time? Probably not! The central limit theorem tells us that, if the *population* mean of *all* apprentice geologists' hourly wages is $15.00 and we repeat the process of sampling and computing the corresponding 95% confidence intervals, a table similar to Table 9–2 will result.

TABLE 9–2	Twenty Samples ($n = 50$) and Confidence Intervals		
Sample Number	Sample Mean $\overline{X}$	Standard Deviation s	95% Confidence Interval
1	$14.99	$3.01	$14.13–15.84
2	15.61	3.42	14.64–16.58
3	15.12	3.25	14.19–16.04
⋮	⋮	⋮	⋮
18	15.01	2.92	14.18–15.84
19	15.92	2.99	15.07–16.77
20	15.34	3.25	14.41–16.26

Note from Table 9–2 that the population mean value of $15.00 per hour is included in every interval except sample number 19. Therefore, 19 out of 20, or 95%, of the intervals include the population mean value. Figure 9–5 portrays this concept graphically.

Problem   Researchers surveyed 160 Hispanic families in the Dallas area with respect to the proportion of bilingual families, average annual income, mean number and age of children, and so on. "What is the age of your youngest child?" was one of the questions asked. The sample mean was computed to be 6.7 years and the sample standard deviation 2.5 years. Construct a 99% confidence interval for the mean age of the youngest child in the population of Hispanic families in the Dallas area.

Solution   To determine the 99% confidence interval, we need to know the standard error of the mean. It is approximately 0.2 year, found by using formula 9–2:

$$s_{\overline{x}} = \frac{s}{\sqrt{n}} = \frac{2.5}{\sqrt{160}} = 0.2 \text{ year}$$

Recall that 99% of the normal distribution is within 2.58 standard deviations of the population mean. Hence, the 99% confidence interval is

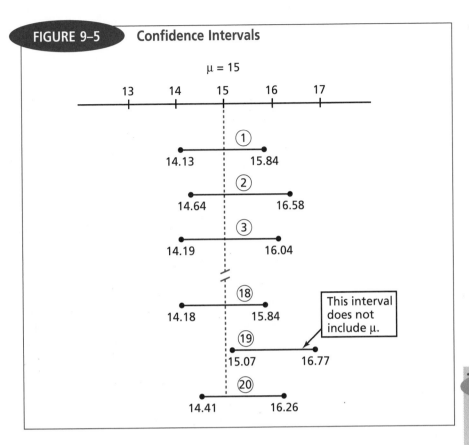

FIGURE 9–5   **Confidence Intervals**

$6.7 \pm 2.58(0.2)$

6.7 ± 0.5

6.2 years up to 7.2 years

Thus, we are 99% confident that the population mean age is between 6.2 years and 7.2 years. In other words, our maximum error is 0.5 year, which is one-half of the width of the interval.

In summary, the researcher selects the desired degree of confidence (such as 95% or 99%) and then proceeds to construct an interval corresponding to that percentage. Note that only one confidence interval is calculated from a particular sample. The confidence interval either includes the population mean, μ, or it does not. Thus, probability statements can be made only before the sample is taken.

We should be careful when interpreting a confidence interval. The level of confidence does *not* specify the probability that a population mean is included in the interval, but rather the percentage of times that similarly constructed intervals could be expected to include the population mean. Referring to the problem above, we can say that about 99% of the similarly constructed confidence intervals for the youngest Hispanic child would bracket the population mean. We cannot say that the probability is 0.99 that the population mean is in that interval.

Self-Review 9–2	a. A sample of 900 registered voters in the state was surveyed about age, political party, and so on. The mean age of the sample was computed to be 42 years and the standard deviation 12 years. What is the 95% confidence interval for the population mean, μ?
	b. A sample of 312 grades on a mathematics test given nationwide reveals that the sample mean is 560 and the sample standard deviation 120. What is the 99% confidence interval for the population mean?

Exercises

3. Define the term *sampling error*.

4. What is the difference between a standard deviation and the standard error of the mean?

5. Find the values of z corresponding to the following levels of confidence:
  a. 80 percent level of confidence.
  b. 96 percent level of confidence.
  c. 98 percent level of confidence.

6. A sample of 45 truck drivers was given the Army General Classification Test. The sample mean is 96.2, with a sample standard deviation of 9.7. Construct a 95% confidence interval for the population mean score of all truck drivers.

7. A study is made of how long inner-city families have lived at their current addresses. A random sample of 40 families revealed a mean of 35 months, with a sample standard deviation of 6.3 months. Construct a 99% confidence interval for the mean time that inner-city families have lived at their current addresses.

8. The Central Ohio Paper Company wants to estimate the mean time required for a new machine to produce a ream of paper, wrap it, and put it in a box ready for shipment. A random sample of 36 reams required a mean machine time of 1.5 minutes. Assuming a standard deviation of 0.3 minute, construct an interval estimate with a confidence level of 95%.

9. For a sample of 80 children enrolled in an elementary school in rural Kansas, the mean distance they were bussed is 17.52 miles (one way), with a sample standard deviation of 4.32 miles. What is the 85 percent confidence interval for the population mean distance they are bussed?

10. A sample of 400 fuses had a mean breaking point (in amperes of current) of 7.5, with a sample standard deviation of 1.0. Construct a 95% confidence interval for the population mean breaking point.

11. A study of the time required between landing clearance and physical touchdown for 49 airplanes at City Airport shows that the sample mean is 280 seconds, with a standard deviation of 21 seconds. Find a 99% confidence interval for the population mean landing time.

12. A sample of 100 high school students revealed that the mean time spent working at an outside job each week is 10.7 hours, with a standard deviation of 11.6 hours. What is the 99% confidence level for the mean number of hours that the population of high school students spent working each week?

13. Sixty-four teenagers who received their driver's licenses during the last month were given a test for reaction times. The mean is 8 seconds, with a standard deviation of 2 seconds. Find a 95% confidence interval for the mean reaction time of the entire population.

THE STANDARD ERROR OF THE SAMPLE PROPORTION (σ_p)

Sometimes you are interested not in a mean value, but simply in the fraction or proportion of a population that has some characteristic. For example, you might wish to determine what proportion of the U.S. population owns a video cassette recorder. The symbol π (Greek letter *pi*) is used to show that fraction of the population that has the characteristic. (Do not confuse this π with the mathematical constant.) If 73% of the U.S. population have a video cassette recorder, then $\pi = 0.73$. The symbol p, on the other hand, is used to indicate a *sample* proportion.

Strictly speaking, the distribution of the number of items in a sample (n) that has a given trait follows a binomial distribution. However, n is frequently large, so we may use the normal approximation to the binomial. This approximation was discussed in Chapter 7. The common rule for employing this approximation is that $n\pi$ and $n(1 - \pi)$ are at least five. Recall that for the binomial, $\sigma = \sqrt{n\pi(1 - \pi)}$. The **standard error of the proportion** is the standard deviation of the sampling distribution of the sample proportion. The symbol σ_p is read as *sigma sub p*. It is found by:

> **proportion** A fraction, ratio, percentage, or probability that indicates what part of a sample or population has a particular trait.

$$\sigma_p = \sqrt{\frac{\pi(1 - \pi)}{n}}$$

9–4

where π is the population proportion that has the trait and n is the size of the sample.

Problem Twenty-five thousand people live in a wealthy Chicago suburb. A travel agent wishes to estimate the proportion of people who have never visited Hawaii. Unknown to the travel agent, 75% have never been there. If the agent selects a random sample of 100, what is the standard error of this sample proportion?

Solution The standard error (σ_p) is found by using formula 9–4:

$$\sigma_p = \sqrt{\frac{\pi(1 - \pi)}{n}} = \sqrt{\frac{(0.75)(0.25)}{100}} = 0.0433$$

CONFIDENCE INTERVALS FOR PROPORTIONS

In general, a confidence interval for a proportion is calculated as follows:

$$p \pm z\sqrt{\frac{p(1 - p)}{n}}$$

9–5

where p is the sample proportion, z is the standard normal value corresponding to the desired level of confidence, and n is the number in the sample.

The 95% and 99% confidence intervals for a sample proportion are estimated by

$$p \pm 1.96\sqrt{\frac{p(1-p)}{n}}$$

and

$$p \pm 2.58\sqrt{\frac{p(1-p)}{n}}$$

respectively.

Problem Recall the survey of 160 Hispanic families in the Dallas area. One of the questions asked was: "Are you fluent in both English and Spanish?" Ninety-six responded "yes." The sample proportion is 96/160, or 0.60. Construct a 95% confidence interval for the proportion of bilingual Hispanic families in the Dallas area. Is it reasonable to conclude that more than 50 percent of the families are bilingual?

Solution First, find the standard error of the sample proportion by

$$\sqrt{\frac{(0.6)(0.4)}{160}} = 0.04$$

Because 95% of a normal distribution is less than 1.96 standard deviations away from the mean, the confidence interval is determined using formula 9–5:

$$0.6 \pm 1.96(0.04)$$

$$0.6 \pm 0.08$$

$$0.52 \text{ up to } 0.68$$

About 95% of similarly constructed intervals will contain the population proportion.

Is it reasonable to conclude that more than 50 percent of the Hispanic population is bilingual? In this case, the confidence interval runs from 0.52 up to 0.68. The lower limit of the confidence interval is larger than 0.50. Hence, we conclude that if all members of the Hispanic population were contacted, more than half would be bilingual. To put it another way, because .50 is *not* included in and is below the confidence interval, we decide that more than half the population is bilingual.

The Greenville Lodge inspected 50 of its air conditioners. Ten were not operating. Develop a 95% confidence interval for the proportion of air conditioners not operating.

Exercises

14. In an "exit" poll of 2200 randomly selected voters conducted in Precinct 29, only 250 said they voted for Governor Long.
  a. Find the standard error of the proportion.
  b. What is a 95% confidence interval for the population proportion that voted for Governor Long?
15. A manufacturing process produced 15 defective parts in a sample of 100.
  a. Determine the standard error of the sample proportion.
  b. Develop a 90% confidence interval for the proportion defective in the population.
16. A sample of 250 urban workers revealed that 25 work with a video display terminal (VDT). What is the 95% confidence interval for the proportion of all urban workers who work with a VDT?
17. A sample of 500 purchasers of a new Mercedes-Benz shows that 325 order two-door models. What is the 99% confidence interval for the population proportion who order two-door models?
18. Of 250 college freshmen polled, 150 said they exercise regularly.
  a. Find a 95% confidence interval for the population proportion of freshmen who exercise regularly.

  b. Find a 99% confidence interval for the same population proportion.
19. Of 40 patients in a survey of current wheelchair users, 32 judged a new, medically approved wheelchair to be an improvement over a standard wheelchair.
  a. Find a 95% confidence interval for the proportion of the population already using wheelchairs for whom the new wheelchair would be an improvement.
  b. Suppose 320 out of 400 wheelchair users judged the new chair to be an improvement. Find a 95% confidence interval for the proportion in the population.
20. An audit of 840 randomly selected tax returns shows that 30% require payment of additional taxes. Find the 95% and 99% confidence intervals for the proportion of all tax returns requiring additional payments.
21. In a national magazine, a poll reported that 36% of the 980 voters surveyed approved of immigration quotas. The article also reported a "margin of error" of plus or minus 3 percentage points. What confidence interval was the pollster using?

THE FINITE POPULATION CORRECTION FACTOR

If you are sampling without replacement from a small population, the standard error is reduced. In other words, if a sample is a relatively large fraction of the population, it yields a better estimate of the population parameter. If a sample (n) is equal to the population's size (N)—that is, $n = N$—we would not expect any error.

Does this make sense? It should because if we know the entire population of values, we will make *no error* when estimating the mean or proportion! At the other extreme, when the population size (N) is quite large compared to the sample size, there is no need for the adjustment to the standard error. (We could never catch and weigh all the fish in Lake Erie because there are so many fish. Therefore, n could not equal N.)

The standard error of the mean or the standard error of the proportion is corrected by the term

$$\sqrt{\frac{N - n}{N - 1}}$$

9–6

where N is the number in the population and n is the number in the sample.

This is called the **finite population correction factor.** The standard error of the sample mean is computed by the following formula, which combines formulas 9–1 and 9–6:

$$\sigma_{\overline{X}} = \frac{\sigma}{\sqrt{n}}\sqrt{\frac{N - n}{N - 1}}$$

The standard error of the sample proportion is

$$\sigma_p = \sqrt{\frac{\pi(1 - \pi)}{n}}\sqrt{\frac{N - n}{N - 1}}$$

9–7

Statisticians usually ignore the correction factor when the sample is less than 5% of the population.

Problem In the small town of Oxford, there are 200 families. A poll of 30 families revealed a mean annual church contribution of $470, with a standard deviation of $130. Construct a 95% confidence interval for the mean annual contribution.

Solution Note that the sample constitutes more than 5% of the population ($30/200 = 15\%$), and the sampling is done without replacement (no family is included twice). The 95% confidence interval is constructed as follows, using formulas 9–3 and 9–6:

$$\overline{X} \pm z \cdot \frac{s}{\sqrt{n}}\left(\sqrt{\frac{N - n}{N - 1}}\right)$$

$$\$470 \pm 1.96\left(\frac{\$130}{\sqrt{30}}\right)\left(\sqrt{\frac{200 - 30}{200 - 1}}\right)$$

$$= \$470 \pm \$46.519(0.924)$$

$$= \$470 \pm \$43$$

$$= \$427 \text{ and } \$513$$

The same study of church contributions in Oxford revealed that 12 of the 30 persons sampled attended church regularly. Construct a 95% confidence interval for the proportion attending church regularly.

Exercises

22. Under what conditions is the finite population correction factor used?

23. Suppose a population consists of 10,000 items and we select a sample of 100. What is the value of the finite populaton correction factor? What is the value of the finite population correction factor if the sample is 1000?

24. A hospital employs 250 nurses. A sample of 50 nurses revealed that 30 were graduates of a diploma school. Develop a 90% confidence interval for the proportion of diploma graduates at the hospital.

25. In River City, a total of 300 traffic citations was issued. A sample of 40 of these tickets showed that the mean amount of the citation was $29

and the standard deviation $5. Construct a 98% confidence interval for the population mean.

26. There are 257 pilots in a secret military training program. In a study of the use of hypnosis to enhance performance, ratings for a sample of 32 pilots led to a mean rating increase of 8.4, with a standard deviation of 2.1. Construct the 95% confidence interval for the population mean.

27. Five hundred students take physical education classes at Wilson High. In a sample of 50 of these students, 10 were able to run 2 miles in less than 14 minutes. Use this sample data to construct the 99% confidence interval for the population proportion of students who can run 2 miles in less than 14 minutes.

CHOOSING AN APPROPRIATE SAMPLE SIZE

A question that usually arises when designing a statistical study is: "How many items should be in the sample?" If a sample is too large, money is wasted collecting the data. Similarly, if the sample is too small, the resulting conclusions will be uncertain. The correct sample size depends on three factors:

1. the level of confidence desired,
2. the margin of error the researcher will tolerate, and
3. the variability in the population being studied.

You, the researcher, select the level of confidence. As noted in the previous section, confidence levels of 95 percent and 99 percent are the ones most often selected. A 95 percent level of confidence corresponds to a z-value of ± 1.96, and a 99 percent level of confidence corresponds to a z-value of ± 2.58.

The maximum allowable error, designated as E, is the amount that is added to and subtracted from the sample mean to determine the end points of the confidence interval. It is the amount of error the researcher is willing to tolerate. It is also one-half the width of the corresponding confidence interval. A small allowable error will require a large sample, and a large allowable error will permit a smaller sample.

The third factor in determining the size of a sample is the population standard deviation. If the population is widely dispersed, a large sample is required. On the other hand, if the population is concentrated (homogeneous), the required sample size will be smaller. However, finding an estimate for the population standard deviation often is difficult. Here are three suggestions.

Use the *comparable study* approach when there is an estimate of the dispersion available from another study. Suppose we want to estimate the number of hours worked per week by librarians. Perhaps information from certain state or federal agencies who regularly sample the workforce might be useful to provide an estimate of the standard deviation. If a standard deviation observed in a previous study is thought to be reliable, it can be used in the current study to help provide an approximate sample size.

If no value from previous experience is available, a *range-based approximation* might be appropriate. To use this approach we need to know or have an estimate of the largest and smallest values in the population. Recall from Chapter 4, where we described the Empirical Rule, that virtually all the observations could be expected to be within ± 3 standard deviations of the mean, assuming that the distribution was approximately bell-shaped—that is, normal. So the distance between the largest and the smallest values is 6σ. We could estimate the standard deviation as about one-sixth of the range. For example, suppose the director of operations of a bank wants an estimate of the number of checks written per month by senior citizens. She believes that the distribution is approximately normal and the minimum number of checks written per month is 2 and the most is 50. The range of the number of checks written per month is 48, found by 50 − 2. The estimate of the standard deviation then would be 8 checks per month (48/6).

The third approach to estimating the standard deviation is to conduct a *pilot study*. This is probably the most common method. Suppose we want an estimate of the number of hours per week worked by students enrolled in the College of Business at the University of Toledo. To determine the validity of our questionnaire we test it on a small sample of students. From this small sample, we compute the standard deviation of the number of hours worked and use this value to determine the appropriate sample size.

We can express this interaction among these three factors and the sample size in the following formula:

$$z = \frac{E}{s / \sqrt{n}}$$

Solving this equation for *n*, we obtain the required sample size:

$$n = \left[\frac{z \cdot s}{E} \right]^2 \qquad \text{9–8}$$

where

E is the maximum allowable error.
s is the estimate of the population standard deviation.

n is the size of the sample.
z is the standard normal value corresponding to the desired level of confidence.

Because the result of this computation is not always a whole number, the usual conservative practice is to round up any fractional result. For example, 72.1 would be rounded up to 73.

Problem A study is to be conducted on the mean salary of mayors of cities with populations of fewer than 100,000. The error in estimating the mean is to be less than $100, and a confidence level of 95% is desired. Suppose the standard deviation of the population is estimated to be $1000. What is the required sample size?

Solution The allowable error, E is $100. The value of z for a 95% level of confidence is 1.96. When we substitute the values into formula 9–8, the required sample size is determined to be

$$n = \left[\frac{(1.96)(\$1000)}{\$100}\right]^2 = (19.6)^2 = 385$$

Thus, a sample of 385 is required. If a higher level of confidence were desired—say, 99%—then a larger sample would also be required:

$$n = \left[\frac{(2.58)(\$1000)}{\$100}\right]^2 = (25.8)^2 = 666$$

When determining the sample size for a proportion, we use the following formula:

$$n = p(1 - p)\left(\frac{z}{E}\right)^2$$

9–9

The value of p can be estimated from a pilot study. If no estimate is available, the value of 0.50 is used. Why? Because the term $p(1 - p)$ can never be larger than when $p = 0.50$. For example, if $p = 0.30$, then $p(1 - p) = 0.30(0.70) = 0.21$, but when $p = 0.50$, $p(1 - p) = 0.50(0.50) = 0.25$.

Problem We are planning a survey to find the proportion of cities that have private garbage collectors. We want a maximum error (E) of 0.10 and a 90% confidence level, and we tentatively estimate the proportion at 0.5. What is the required sample size?

Solution Using formula 9–9:

$$n = (.50)(.50)\left[\frac{1.65}{.10}\right]^2 = 68.0625$$

We need to contact a minimum of 69 cities.

Self-Review 9–5   We are conducting a study designed to estimate the mean number of hours worked per week by suburban housewives. A pilot study revealed that the population standard deviation is 2.7 hours. How large a sample should be selected to be 95% confident that the sample mean differs from the population mean by at most 0.2 hour?

Exercises

28. State in your own words the three factors that determine the size of a sample.
29. State in your own words three methods to estimate the standard deviation to be used to find the sample size.
30. Show that the required sample size for a particular margin of error and standard deviation will be larger for the 99% level of confidence than for the 95% level of confidence.
31. Show with some values you select that, as we decrease the allowable error, we increase the required sample size.
32. A company wishes to estimate the mean starting salaries for security personnel in its manufacturing operations. From a previous study, company staff estimate that the standard deviation is $2.50. How large a sample should they select to be 95% confident that the sample mean dif-

fers from the population mean by at most $0.50?
33. A meat packer is investigating the marked weight shown on links of summer sausage. A pilot study showed a mean weight of 11.8 pounds per link and a standard deviation of 0.7 pound. How many links should the packer sample to be 95% confident that the sample mean differs from the population mean by at most 0.2 pound?
34. The manager of a motel is studying the number of customers who stay more than 1 night. He would like to estimate the proportion within 0.05, with a 95% level of confidence. He estimates the proportion to be 0.30. How large a sample is required?
35. Refer to Exercise 34, but assume that no estimate is available regarding the population proportion. How large a sample is required?

CHAPTER OUTLINE

I. The central limit theorem is fundamental to sampling.
  A. No matter what the shape of the population, the sampling distribution of the sample mean will approximate the normal distribution when the sample is large.
  B. The central limit theorem allows us to use the z distribution.
II. A point estimate is a single value used to estimate a population parameter.
III. An interval estimate is a range of values within which the population parameter is expected to occur.

A. The confidence interval for a population mean is determined by

$$\overline{X} \pm z\frac{s}{\sqrt{n}} \qquad \boxed{9\text{–}3}$$

1. The number of observations in the sample is n.
2. The variability in the population is usually estimated by the sample standard deviation, s.

3. The level of confidence is used to determine the z-value.

B. The confidence interval for a population proportion is determined by

$$p \pm z \sqrt{\frac{p(1 - p)}{n}} \qquad \text{9-5}$$

1. The number of observations in the sample is n.
2. The value of p is found by dividing the number of successes (X) in the sample by the number of observations (n).
3. The level of confidence is used to determine z.

C. The finite population correction factor is applied if n/N is more than 0.05. The correction factor is

$$\sqrt{\frac{N - n}{N - 1}} \qquad \text{9-6}$$

IV. The required size of a sample can be determined for both a mean and a proportion.

A. The formula for finding the size of the sample for a mean is

$$n = \left[\frac{z \cdot s}{E}\right]^2 \qquad \text{9-8}$$

1. The desired level of confidence determines z.
2. The maximum allowable error is E.
3. The variation in the sample is designated by s.

B. The formula for finding the size of the sample for a proportion is

$$n = p(1 - p)\left(\frac{z}{E}\right)^2 \qquad \text{9-9}$$

1. The desired level of confidence determines z.
2. The maximum allowable error is E.
3. The tentative estimate of the population proportion is p. If no value is available, use 0.5.

Exercises

36. A study dealing with divorced couples gathered data on the length of time from marriage to separation. A random sample of 100 divorced couples had an average length of marriage of 5.9 years, with a sample standard deviation of 2.0 years. Construct a 99% confidence interval for the mean length of time from marriage to separation for the population of divorced couples.

37. Refer to Exercise 36. Suppose this study dealt with only the couples who were divorced in a particular city. Assume the population of divorced couples was 500. Compute the confidence interval.

38. A random sample of 80 terms of sentence for rape (first offense) showed a mean equal to 3.9 years, with a standard deviation of 1.8 years. Construct a 95% confidence interval for the mean term of imprisonment for all rape sentences.

39. Refer to Exercise 38. Assume the study referred to only 1 prison, in which there were 250 rape cases (first offense). Compute the confidence interval.

40. Sick-leave reords obtained from a random sample of 200 social workers showed a mean number of days of sick leave equal to 25.6 last year. If the sample standard deviation is 5.1, construct a 90% confidence interval for the mean number of days of sick leave for all social workers last year.

41. The American Restaurant Association collected information on the numbers of meals eaten outside the home per week by young married couples. A survey of 60 couples showed the mean number of meals eaten outside the home was 2.76 per week, with a standard deviation of 0.75

meals. Construct a 97% confidence interval for the population mean.

42. Suppose the National Collegiate Athletic Association (NCAA) reported that the mean number of hours spent per week on coaching and recruiting by college football assistant coaches during the season is 70. A random sample of 50 assistant coaches showed the mean to be 68.6 hours, with a standard deviation of 8.2 hours.
    a. Using the sample data, construct a 99% confidence interval for the population mean.
    b. Does the 99% confidence interval include the value suggested by the NCAA? Does this cast doubt on or reinforce the statement by the NCAA?
    c. Without doing any calculation, would changing the level of confidence from 99% to 95% increase or decrease the width of the confidence interval? What value would change in the calculation?

43. The Human Relations Department of Electronics, Inc., would like to include a dental plan as part of the benefits package. The question is: "How much does a typical employee and his or her family spend per year on dental expenses?" A sample of 45 employees showed the mean amount spent last year was $1820, with a standard deviation of $660.
    a. Construct a 95% confidence interval for the population mean.
    b. The information from part a was given to the president of Electronics, Inc. He indicated he could afford $1700 of employee dental expenses per year. Is it possible that the population mean could be $1700? Justify your answer.

44. In a poll to estimate presidential popularity, each person in a random sample of 1000 was asked to select one of the following statements:
    1. The president is doing a good job.
    2. The president is doing a poor job.
    3. I have no opinion.
    A total of 560 respondents selected the first statement, indicating they thought the president was doing a good job.
    a. Construct a 95% confidence interval for the proportion of respondents who feel the president is doing a good job.
    b. Based on your interval in part a, is it reason- able to conclude that a *majority* (more than half) of the population believes the president is doing a good job?

45. A political candidate wants to estimate his chances of winning the upcoming election. He wants to estimate—within 3 percentage points, with a 95% level of confidence—the proportion of voters who will vote for him. He has no tentative estimate of the population proportion. How many voters should be contacted?

46. The International Council of Shopping Centers reported that a sample of 30 people spent an average of 69 minutes on each visit to a mall or shopping center. The sample standard deviation was 49 minutes. Construct a 95% confidence interval for the mean shopping time.

47. During the month of May, the Department of Transportation monitored 80 flights on a regional airline. Seventy-four of those flights arrived on time. Use this sample data to construct the 99% confidence interval for the proportion of all flights that arrive on time.

48. Hrivnyak Literacy Action Group is planning a survey to determine the percentage of adults in the United States who are illiterate. The group decided that the maximum allowable error should be 0.02. The group plans to use a 95% confidence level. Hrivnyak tentatively estimates the proportion of illiterate adults at 0.15. What sample size would be required to achieve the group's goal?

49. A survey of 50 new-car dealers showed a sample mean of 49 reports of defects per month, with a sample standard deviation of 7 reports per month. Construct a 99% confidence interval for the population mean number of reports of defects per month. A new-car dealers' association claimed that a dealership should expect 45 reports of defects per month. Is this claim consistent with the survey results? What does this suggest about the claim?

50. Election polls usually list the margin of error as plus or minus 3.0%. Suppose you are interested in conducting your own poll, with a 99% level of confidence. Find the appropriate sample size for the following margins of error: 5%, 3%, 1%, 0%. Now justify why many pollsters select 3% as the margin of error.

DATA EXERCISES

51. Refer to the real estate data, which reports information on the homes sold in Alabama in 1996.
    a. Develop a 95% confidence interval for the mean selling price of the homes.
    b. Develop a 95% confidence interval for the mean distance of the home from the center of the city.
    c. Develop a 95% confidence interval for the proportion of homes with a pool.
    d. Develop a 95% confidence interval for the proportion of homes with an attached garage.
52. Refer to the school data set which refers to the 94 school districts in northwest Ohio. Assume these data are a sample. Select the variable referring to proportion of students that are on welfare. Compute the mean and the standard deviation for this variable, *but do not include Lima, Sandusky, Toledo, and Fostoria* in the calculation. Develop a 99% confidence interval for the mean percentage of students on welfare, based on this information. Are the percentages of students on welfare in the excluded school districts in this interval? Does it appear that the percentages of students on welfare in the omitted school districts are different from the others? Why?

CHAPTER ACHIEVEMENT TEST

The answers are at the back of the book.

MULTIPLE-CHOICE QUESTIONS

Select the response that best answers each of the questions.

1. Which of the following statements is *not* true?
   a. A point estimate is more accurate than an interval estimate.
   b. A point estimate is subject to sampling error.
   c. A point estimate says nothing about the error in estimation.
   d. An interval estimate includes the corresponding point estimate.
2. Which of the following is *not* needed to compute a confidence interval for the mean?
   a. The shape of the sampling distribution
   b. The sample size and population size
   c. The population standard deviation
   d. All of the above are needed.
3. If we have *a* sample of 50 items from a very large population, which of the following is *not* needed to compute a confidence interval for the proportion?
   a. The shape of the sampling distribution
   b. The sample size and population size
   c. The degree of confidence
   d. All of the above are needed.
4. The sample standard deviation based on 15 pieces of data is
   a. always larger than the standard error of the mean.
   b. always smaller than the standard error of the mean.
   c. another name for the standard error of the mean.
   d. None of the above
5. You should use the finite population correction factor whenever
   a. the sample is more than 5% of the population.
   b. you are sampling without replacement.

c. the standard error of the mean is too large.

d. the population is finite.

6. For samples of one item, the sampling distribution of the sample mean is normally distributed if the

a. population is symmetric.

b. standard deviation is known.

c. population is normally distributed.

d. standard deviation is less than the mean.

7. The standard error of a proportion becomes larger as

a. π approaches zero.

b. n grows large.

c. π nears one-half.

d. π is close to one.

8. A 95% confidence interval indicates that 95 out of 100 similarly constructed intervals will include

a. a sampling error.

b. an estimation of the parameter.

c. Both a and b

d. None of the above

9. If the confidence level is lowered from 95% to 90% and everything else remains the same, the required sample size will

a. increase.

b. decrease.

c. not change.

d. There is too little information provided to determine the effect.

10. To find a sample size, the use of the finite population correction factor will

a. affect sampling with replacement.

b. decrease the required sample size.

c. increase the required sample size.

d. have more impact on large populations.

COMPUTATION PROBLEMS

11. To estimate the medical charges for an appendectomy Blue Star Insurance has data from a random sample of 70 patients. The sample mean cost is $510, with a sample standard deviation of $70.

a. Find the standard error of the sample mean.

b. Construct a 95% confidence interval for the population mean cost.

12. A real estate agent records the ages of 50 randomly selected home buyers in her sales area. The mean age is 38 years, with a sample standard deviation of 10 years.

a. What is the standard error of the estimate?

b. Find a 99% confidence interval for the population mean age.

13. Larry Clark is a superstar for the Warren High School basketball team. In the game against Dubois, Larry took 20 shots. We want to estimate his long-term shooting percentage.

a. If he normally makes 70% of his shots, what is the standard error of your estimate?

b. If he makes 13 baskets, what is a 99% confidence interval for his long-term shooting percentage?

14. The police department reports that 375 of 500 randomly selected burglar alarms received at the station were false alarms.

a. What is the standard error of an estimate based on this data?

b. What is a 90% confidence interval for the population proportion of false alarms?

15. A study of the incomes of farming households in a particular state is undertaken. How large a sample is required to estimate the mean income within $200 with a 95% level of confidence? The standard deviation is estimated to be $3000.

ANSWERS TO SELF-REVIEW PROBLEMS

9-1 The answers will vary. Here is one possible solution.

	1	2	3	4	5	6	7	8	9	10
	8	2	2	19	3	4	0	4	1	2
	19	1	14	9	2	5	8	2	14	4
	8	3	4	2	2	4	1	14	4	1
	0	3	2	3	1	2	16	1	2	3
	2	1	7	2	19	18	18	16	3	7
Total	37	10	29	35	27	33	43	37	24	17
$\bar{X}$	7.4	2.0	5.8	7.0	5.4	6.6	8.6	7.4	4.8	3.4

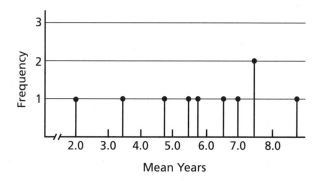

9-2 a. $\sigma_{\bar{X}} = \dfrac{12}{\sqrt{900}} = \dfrac{12}{30} = 0.4$

Then $42 \pm 1.96(0.4) = 42 \pm 0.784 = 41.216$ years and 42.784 years

b. $\sigma_{\bar{X}} = \dfrac{120}{\sqrt{312}} = \dfrac{120}{17.66352} = 6.79$

Then $560 \pm 2.58(6.79) = 560 \pm 17.52 = 542.48$ and 577.52. These would probably be rounded to 542 and 578.

9-3 $p = \dfrac{10}{50} = 0.20$

$0.20 \pm 1.96\left(\sqrt{\dfrac{.2(.8)}{50}}\right)$

0.20 ± 0.11

9-4 $p = \dfrac{12}{30} = 0.40$

$0.40 \pm 1.96\sqrt{\dfrac{(0.40)(0.60)}{30}}\left(\sqrt{\dfrac{200-30}{200-1}}\right)$

0.40 ± 0.162

9-5 $n = \left[\dfrac{(1.96)(2.7)}{0.2}\right]^2$

$n = 701$

CHAPTER 10

Hypothesis Tests: Large-Sample Methods

OBJECTIVES

When you have completed this chapter, you will be able to

- describe the five-step hypothesis-testing procedure;
- distinguish between a one-tailed and a two-tailed statistical test;
- identify and describe possible errors in hypothesis testing;
- conduct a hypothesis test about a population mean;
- conduct a hypothesis test between two population means;
- test a hypothesis about a population proportion;
- test a hypothesis about two population proportions.

CHAPTER PROBLEM   The Luck of the Irish

Are professional money managers worth what they are paid? One New York analyst confessed that he based his stock purchases on the toenail marks his Irish setter left on the financial pages! John Dorfman of the *Wall Street Journal* decided to stage a contest. He challenged four investment professionals to choose portfolios over a 6-month period while he "selected" stocks by throwing darts at an "investment" dartboard. Out of 27 trials, the pros did better than the darts only 15 times. Are the so-called financial wizards just guessing, or does a 56% success rate prove that they know something?

INTRODUCTION

In Chapter 6, we developed a discrete probability distribution to describe the possible outcomes of an experiment. Chapter 7 dealt with the normal probability distribution—a continuous distribution. We noted that it is symmetrical and bell-shaped and that the tails taper off into infinity.

In Chapters 8 and 9, we used both the normal distribution and the notion of a probability distribution to develop a sampling distribution. We also pointed out that it is often too expensive, or impossible, to study an entire population. Instead, we examined a part of the population of interest, called a sample. The purpose of sampling was to learn something about the population. For example, we might want to approximate the mean income or the mean age of the population. We accomplished this by constructing confidence intervals within which the population mean might fall.

This chapter continues our study of the use of the normal distribution in sampling. Instead of building an interval in which a population parameter (such as the mean) is expected to fall, we test the validity of a statement about a population parameter. This is called **hypothesis testing.**

THE GENERAL IDEA OF HYPOTHESIS TESTING

To illustrate the concept, let us first examine a nonstatistical case. It concerns detective Ms. Sharpe as she tries to unmask a mysterious murderer. Upon her arrival at the scene of the crime, Ms. Sharpe observed that the victim had been struck from above by a left-handed person who left a mud print of a size 9 shoe.

What course of action will our detective pursue? Naturally, she will first suspect that the butler performed the foul deed. Then she will examine each piece of evidence in succession to see if it is consistent with the presumption that the butler might be the murderer. If the butler is a short, right-handed man who wears size 11 shoes, it is highly unlikely that he committed the crime, and he will be dismissed as a suspect.

As she considers the butler's innocence or guilt, what goes through the detective's mind? She realizes that she must arrest or release the prime suspect. Either way, the butler may, in fact, be either guilty or innocent. Thus, there are four possibilities that might occur when Ms. Sharpe finally reaches her decision:

1. She can arrest the butler when the butler actually committed the crime—a correct decision.
2. She can release the butler when the butler is innocent—again a correct decision.
3. She can arrest the butler when the butler is actually innocent—an incorrect decision.
4. She can release the butler when the butler is actually guilty—another incorrect decision.

These four possibilities facing our investigator can be summarized as follows:

	Arrest the butler.	Release the butler.
The butler did it.	Correct	Error!
The butler is innocent.	Error!	Correct

If Ms. Sharpe's problem were one involving statistics, the first thing she would do is to set up a **null hypothesis.** The null hypothesis, designated H_0, in the murder case could be

$$H_0: \text{The butler is innocent}$$

null hypothesis A claim about the value of a population parameter.

The null hypothesis is a claim that is established for the purpose of testing. This claim is either rejected or not rejected.

If the evidence is sufficient to reject the null hypothesis, then the **alternate hypothesis** is accepted. In Ms. Sharpe's murder investigation, the alternate hypothesis, designated H_a, is

$$H_a: \text{The butler is not innocent}$$

alternate hypothesis A claim about the population parameter that is accepted if the null hypothesis is rejected.

Note that the procedure is to test the null hypothesis. The alternate hypothesis is accepted if, and only if, the null hypothesis is rejected. The strategy is to make a decision first with respect to the null hypothesis.

After examining all the evidence, the detective might fail to reject the null hypothesis (the butler is innocent) and release him. Or she might reject the null hypothesis and have him arrested.

As noted before, the investigator could make two kinds of mistakes during her investigation. They are called **Type I** and **Type II** errors, respectively.

If, for example, the butler had been arrested as a suspect when he was actually innocent of the murder, a Type I error would occur. But if a null hypothesis is not rejected when it is actually not true, a Type II error occurs. That is, if the butler had been released when he actually did commit the murder, a Type II error would occur. To summarize:

Type I error An error that occurs when a true null hypothesis is rejected.

Action

	Fail to reject H_0	Reject H_0
H_0 is true.		Type I error
H_0 is false.	Type II error	

Type II error An error that occurs when a false null hypothesis is not rejected.

Of course, a Type I error could be avoided if we never reject H_0, the null hypothesis. In the crime problem, if Ms. Sharpe never rejected the H_0 that a suspect was innocent, she would never make the mistake of arresting an innocent man. Clearly, this is a rather extreme way for a crime fighter to avoid a Type I error! It also increases the chances that a Type II error will occur.

A Type II error would be avoided if we were always to reject the null hypothesis. If our private eye were always to reject the null hypothesis and accept the alternate hypothesis, she would never release a murderer. Either extreme position is unrealistic. As will be noted in the following sections, statistical theory deals with "decision rules" based on probability, which attempt to balance these two kinds of potential errors.

We will now leave Ms. Sharpe to her ongoing murder investigation. Many different types of statistical hypothesis-testing problems will be considered in this and in subsequent chapters. As will become evident, however, the thinking and the procedure developed in the nonstatistical murder investigation are very similar to research problems involving statistics. A systematic five-step approach will be applied in solving each problem. A brief discussion of these five steps follows. Later, as each of the tests is presented, the five steps will be explained in greater detail.

STEP 1.   State the **null hypothesis.** Along with the null hypothesis, make a second statement called the **alternate hypothesis.** Accept the alternate hypothesis if you reject the null hypothesis.

STEP 2.   Choose a **level of significance.** Usually, select either the 0.05 or the 0.01 level. The level of significance refers to the probability of making a Type I error.

STEP 3.   Choose a **test statistic.** A test statistic is a quantity calculated from the sample information. Its value will be used in Steps 4 and 5 to arrive at a decision regarding the null hypothesis.

STEP 4.   Set up a **decision rule** based on the level of sigificance chosen in Step 2 and the sampling distribution of the test statistic from Step 3.

STEP 5.   Select **one or more samples.** Then, using the sample results, *compute the value of the test statistic.* In this chapter, the standard normal distribution z will be the only test statistic employed. Finally, use the decision rule in Step 4 to **make a decision**—either to reject the null hypothesis or not to reject the null hypothesis.

As you can see, in statistics we employ a method of "proof by contradiction." Generally, we hope to prove that something is true (the alternate hypothesis) by rejecting the claim in the null hypothesis. If the sample evidence contradicts the null hypothesis, we can reject it and "believe" the alternate hypothesis. On the other hand, if the sample information does not contradict the null hypothesis, we still have not proved that the null hypothesis is true. We have merely found that the null hypothesis cannot be rejected.

A parallel with hypothesis testing is the legal practice, in which a jury finds a defendant either "guilty" or "not guilty," but never declares a defendant innocent. "Not guilty" is the legal equivalent of a null hypothesis. A person is innocent until proven guilty. A prosecutor must present evidence to the jury so that they think that the probability the defendant is innocent is so small as to be unbelievable. Then the jury "rejects the null hypothesis" and says the defendant is "guilty."

A TEST INVOLVING THE POPULATION MEAN (LARGE SAMPLES)

One type of hypothesis-testing problem involves checking whether a reported mean is reasonable. To perform the test, we take a random sample from the population.

Problem   According to the Bureau of the Census, the mean annual income of government employees is $35,000. There is some doubt that this mean is representative of incomes of government employees living in the San Francisco Bay area. Is there sufficient evidence to conclude that the mean annual income of government employees living in the Bay area is different than the national average?

Solution   Even before we start to collect data, the first step is to state the hypothesis to be tested. Recall from the murder investigation that we call this statement the null hypothesis, usually expressed symbolically as H_0. The null hypothesis, in this case, is that the population mean will not be affected—or, stated another way, that the mean in the Bay area is equal to $35,000. The population mean is designated by the Greek lowercase letter μ.

Symbolically, then, the null hypothesis is

$$H_0: \mu = \$35,000$$

The alternate hypothesis—that the mean annual income in the San Francisco Bay area is *not* equal to $35,000—is written

$$H_a: \mu \neq \$35,000$$

Note that the equality condition appears in the null hypothesis. This will *always* be the case.

The second step in testing a hypothesis is to select the **level of significance,** which is the probability of a Type I error. The crucial question in this income problem is when to reject the null hypothesis. The significance level will define more precisely when the sample mean is too far removed from the hypothesized value of $35,000 for the null hypothesis to be plausible. How do you decide what level of significance to select? You should consider the "cost" of being wrong. As an example of the "cost" of being wrong, it is usually a more serious error for meteorologists to predict sunshine and then have it rain than it is for them to predict rain and then have sunshine. So they have a tendency to forecast rain a bit more often than it actually happens. In that way they avoid the bigger cost. The most common levels of significance used in applied work are 0.05 and 0.01, although any value between 0 and 1 is possible. For this illustration, we will select the 0.05 significance level. If rejecting a true hypothesis is relatively serious, then set the significance level quite low. If accepting a false null hypothesis is relatively more serious, then pick a high level of significance.

Suppose, for example, that a test has been devised to detect cancer. The null hypothesis might be that a patient is free of the disease (H_0:no disease).

> **level of significance** The probability of rejecting the null hypothesis when it is true.

A Type I error would reject the null hypothesis when it is true—that is, telling a patient that he has cancer when, in fact, he does not. While such an error would cause much pain and suffering, in a case like this one it is not as serious or costly as a Type II error. Remember, a Type II error means that a false null hypothesis is not rejected. In this example, it means that you tell a patient he is healthy when, in fact, he has cancer. Most physicians would agree that the Type II error here is much more serious than the Type I error; consequently, a high level of significance should be selected for this test—say, 0.10 instead of 0.01.

test statistic A quantity, calculated from the sample information, used as a basis for deciding whether or not to reject the null hypothesis.

The third step is to select the appropriate **test statistic.** Recall from Chapter 9 that according to the central limit theorem, the sampling distribution of the sample mean is approximately normal and the standard deviation of the sampling distribution of means $\sigma_{\overline{X}}$ is $\sigma/\sqrt{n}$. Hence, the following test statistic is appropriate:

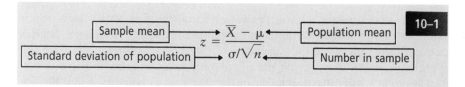

where

$\overline{X}$ is the sample mean.
μ is the population mean which is stated in the null hypothesis.
$\sigma/\sqrt{n}$ is the standard error of the mean.
n is the number in the sample.

Recall that σ is the standard deviation of the population from which the sample was drawn. For this test, we assume either that σ is already known, based on prior studies, or that a good estimate of its value can be obtained from the sample data.

decision rule A statement of the condition or conditions under which the null hypothesis is rejected.

The fourth step is to formulate a **decision rule.** In the study of Bay area wages, the decision rule is an objective statement that will allow us to test the null hypothesis.

The question we are exploring is: "Does the mean annual income of government employees in the Bay area differ significantly from the national average?" The mean in the Bay area could be larger or smaller than the national average for all government employees. The decision rule is designed to accommodate both these possibilities. Thus, the test is called a **two-tailed test.**

As described in Chapter 9, the mean of all possible samples of the same size selected from a population is normally distributed. When the null hypothesis is true, the test statistic z will also be normally distributed. Recall from Chapter 7 that 95% of the area in the distribution is between -1.96 and 1.96. Thus, if the significance level is 0.05, the region of rejection falls to the left of -1.96 and to the right of 1.96—that is, less than -1.96 and greater than 1.96. These two values are called the **critical values.**

critical value A value or values that separate the region of rejection from the remaining values.

So the decision rule is this: Do not reject the null hypothesis if the computed value of z is in the region from -1.96 to 1.96. When that is not the case, the null hypothesis is rejected, and the alternate hypothesis is accepted. The decision rule can be shown in the form of a diagram:

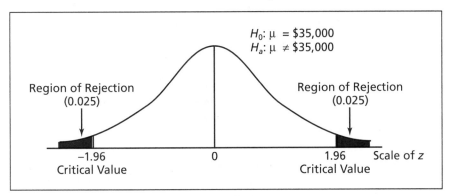

$H_0: \mu = \$35,000$
$H_a: \mu \neq \$35,000$

Region of Rejection
(0.025)

Region of Rejection
(0.025)

−1.96
Critical Value

0

1.96
Critical Value

Scale of z

The final step is to take a sample of the government employees in the Bay area, compute the sample mean $(\overline{X})$, and determine the value of the test statistic z. Based on the computed value of z, the null hypothesis is either rejected or not rejected. For example, suppose the sample consists of 49 employees, the mean of the sample is \$35,600 and the standard deviation of the sample is \$1400. Because the sample is large (greater than 30), the sample standard deviation (s) is used to estimate σ. Using formula 10–1:

$$z = \frac{\overline{X} - \mu}{\sigma/\sqrt{n}} = \frac{\$35,600 - \$35,000}{\$1400/\sqrt{49}} = 3.00$$

The computed value of the test statistic is 3.00. It is in the region beyond 1.96; therefore, the null hypothesis is rejected and the alternate hypothesis accepted. We conclude that the mean income of government employees in the San Francisco Bay area is *not* equal to \$35,000. The difference of \$600 between the sample mean and the hypothesized population mean is too large to have occurred by chance.

When using the hypothesis-testing procedure, we must be careful in drawing our conclusions. We conducted a test to determine if there is a *difference* between the sample mean and the hypothesized population mean. In this case, we concluded that it is unlikely we could select a random sample of 49 government employees with a mean yearly salary of \$35,600 from the population of government employees where the mean salary is actually \$35,000. We must be careful to infer only that there is a difference between the population of government employees and those in San Francisco. The key word here is *difference*. We might be tempted to conclude that, because the sample mean is greater than the population mean, the San Francisco employees earn more. We cannot do that here. Why? We agreed that the null hypothesis is that the mean of the population is equal to \$35,000 $(H_0: \mu = \$35,000)$ and that the alternate hypothesis is that the mean is not equal to \$35,000 $(H_a: \mu \neq \$35,000)$. Further, we stipulated that, if H_0 is rejected, then we will conclude H_a. Because H_a states that the population mean does not equal \$35,000, this is the *only* inference we make. We can only make a conclusion based on the manner in which we establish the null hypothesis and the alternate hypothesis. In a later section, we will describe procedures that would allow us to conclude that San Francisco area government employees earn more than $35,000 or earn less than $35,000.

In general, the correct course of action is to either reject or fail to reject the null hypothesis. By failing to reject the null hypothesis, we take the position that the evidence is not sufficient to rule out—that is, to reject—the null hypothesis. However, in practice, we often think in terms of "accepting" the null hypothesis, rather than "failing to reject" it. While this language is technically not correct, its use is common.

Probability Values (*p*-values) in Hypothesis Testing

> **. . . . .**
> **p-value** The probability of getting a value as extreme as that found in the sample when the null hypothesis is true.

REAL STAT

According to recently published information, the average American consumes 111 pounds of fresh vegetables, including 27.8 pounds of iceberg lettuce, 18.6 pounds of onions, and 15.4 pounds of tomatoes, each year. How do you compare? Could you set up a statistical test to determine if the students in your class are different from the average American?

The five-step hypothesis-testing procedure compares the test statistic to a critical value. A decision is then made whether to reject the null hypothesis. For example, if the critical value is 1.96 and the test statistic is 1.97, the decision is to reject H_0. Likewise, if the computed value of z is 3.76, H_0 is still rejected. No additional information is reported if the value of the test statistic is barely into the rejection region or well into it.

In recent years, more sophisticated computer software has provided information regarding the "strength" of the rejection. Test results are often reported along with **p-values,** or "observed" significance levels. This procedure compares the p-value, or "observed" significance level, with the significance level chosen by the researcher for the test. If the p-value is smaller than the significance level, the null hypothesis is rejected. This procedure not only allows a decision regarding the null hypothesis, but also gives additional insight into the strength of the rejection. A very small p-value—say, 0.0001—indicates a more significant result than a larger p-value of, say, 0.03. In general, for a one-sided test, a p-value is calculated as the probability of observing a value of the test statistic greater than the absolute value of the observed value. For a two-sided test, the p-value is twice the probability for the corresponding one-sided test.

In the statistical test involving the mean annual income of government employees in the Bay area, the null hypothesis is rejected. The value of the test statistic z was computed to be 3.00. The probability of obtaining a z-value of 3.00 or greater is 0.0013, found by $0.5000 - 0.4987$. To compute the p-value, we are concerned with values less than -3.00 as well as those greater than 3.00 because it was a two-tailed test. Hence, the p-value is $2(0.0013) = 0.0026$. This describes the likelihood of observing a value in either tail of the test statistic that is more extreme than the observed test statistic.

The p-value approach not only provides a measure of the significance of the observed statistic, but also lets each reader of the study result pick a significance level he or she is comfortable with in the particular situation.

Self-Review 10–1

Answers to the Self-Review problems are at the end of the chapter.

A recent national survey found that high school students watched an arithmetic mean of 6.8 videos per month. A random sample of 36 college students revealed that the mean number of videos they watched last month was 6.4, with a sample standard deviation of 1.50. Is the number of videos watched per month by college students different from that of high school students? Use the 0.01 significance level.

a. State the null hypothesis and the alternate hypothesis.
b. What is the appropriate test statistic? Write out the formula.
c. Show the decision rule in the form of a diagram.
d. Write out the decision rule.
e. Compute the value of the test statistic.
f. What is your decision regarding the null hypothesis? Interpret this result.
g. Determine the p-value.

. . . .

Exercises

Answers to the even-numbered Exercises are at the back of the book.

1. The following null and alternate hypotheses are given: H_0: $\mu = 10$ and H_a: $\mu \neq 10$. The significance level is 0.03. What are the critical values?
2. The following null and alternate hypotheses are given: H_0: $\mu = 550$ and H_a: $\mu \neq 550$. The significance level is 0.15. What are the critical values?
3. The following information is available:

$$H_0: \mu = 50$$
$$H_a: \mu \neq 50$$

The sample mean is 49, the population standard deviation is 5, and the sample size is 36. Use the 0.05 significance level.
a. State the decision rule.
b. Compute the value of the test statistic.
c. What is your decision regarding the null hypothesis?
d. Determine the p-value.
4. The following information is available:

$$H_0: \mu = 10$$
$$H_a: \mu \neq 10$$

The sample mean is 12, the population standard deviation is 3, and the sample size is 64. Use the 0.05 significance level.
a. State the decision rule.
b. Compute the value of the test statistic.
c. What is your decision regarding the null hypothesis?
d. Determine the p-value.
5. The Public Health Service publishes the *Annual*

Data Tabulations, Continuous Air Monitoring Projects, which recently indicated that a large midwestern city had an annual mean level of sulfur dioxide of 0.12 (concentration in parts per million). To change this concentration, many steel mills and other manufacturers installed antipollution equipment. Plans are to make about 36 random checks during the year to determine if there has been a change in the sulfur dioxide level. The 0.05 level is to be used.
a. State both the null hypothesis and the alternate hypothesis.
b. What is the level of significance?
c. What is the appropriate test statistic? Give its formula.
d. Show the decision rule in the form of a diagram.
e. Thirty-six random checks were made throughout the year. It was found that the sample mean was 0.10 and the sample standard deviation 0.03. At the 0.05 level, does this evidence indicate that there has been a change in the sulfur dioxide level in the city?
f. Compute the p-value.
6. During the past several years, frequent checks were made of the spending patterns of citizens returning from a vacation of 21 days or less to countries in Europe. Results indicated that travelers spent an arithmetic mean of $1010 on items such as souvenirs, meals, film, and gifts. A new survey is to be conducted to determine if there has been a change in the average amount spent. The 0.01 level is to be used.
a. State the null hypothesis and the alternate hypothesis in the forms H_0 and H_a.

b. State the decision rule.

c. A survey of 50 travelers has a sample mean of $1090. The standard deviation of the sample is $300. At the 0.01 level, is there evidence that there has been a change in the mean amount spent abroad, or is the increase of $80 probably due to chance?

d. Compute the *p*-value.

7. A nationwide study of senior citizens revealed that they devote an average of 32 minutes per month to volunteer civic activities. This study also found the distribution of time spent to approximate the normal distribution, with a standard deviation of 4.5 minutes. A sample of 49 former public employees who were receiving pensions from the state of Minnesota showed that they spent a mean of 34 minutes per month on volunteer civic activities. At the 0.01 significance level, is there a difference in the mean amount of time devoted to volunteer activities by former Minnesota public employees?

8. A local manufacturer of doors used in home construction wants to confirm that the mean height of adult males is 70 inches. In a random sample of 121 adult men, the mean height is 72 inches, with a standard deviation of 2 inches. At the 0.05 significance level can we conclude that there has been a change in the mean height?

9. A reading test for 8-year-old children is standardized on a large nationwide basis, so that the mean is 50, with a standard deviation of 10. School authorities in California choose a statewide random sample of 300 8-year-olds to compare reading skills in California with those in the rest of the nation.

a. Formulate the null and alternate hypotheses.

b. For the 0.05 level of significance, what is the decision rule?

c. These children average 51.3 on the test. Compute the test statistic.

d. Is the null hypothesis rejected?

e. Explain the meaning of your results.

10. The daily revenues from public parking fees in the central business district of Corry, Pennsylvania, have a mean of $125 and a standard deviation of $5. The mayor of Corry suggested that, if fees were reduced, downtown shoppers would use public parking more; hence, the daily revenue would increase. After a two-month trial period (44 weekdays), the mean daily revenue was $126. At the 0.05 significance level, is there a difference in the mean daily revenue?

One-Tailed Tests

The earlier example that dealt with the annual incomes of government employees in the San Francisco Bay area required a two-tailed test. Before the sample of 49 employees was taken, there was no knowledge that the mean income would be above or below the national average. For the 0.05 level, there were two rejection regions, one to the right of 1.96, the other to the left of -1.96. (Consult the diagram on p. 303)

Another problem might require the application of a **one-tailed test.** Suppose we suspected that federal employees in the heavily populated Bay area earn *more* than typical government employees. In this case, the only concern is whether the Bay area employees earn significantly *more than* typical government employees. To test our suspicions, we would state in our null hypothesis that the population mean is *equal to or less than* $35,000 or, stated symbolically,

$$H_0: \mu \le \$35,000$$

The alternate hypothesis is that the mean income for Bay area government employees is *greater than* $35,000. It is written

$$H_a: \mu > \$35,000$$

Rejection of the null hypothesis and acceptance of the alternate hypothesis *will under these conditions* allow us to conclude that Bay area government employees earn salaries above the national average.

For the 0.05 level of significance, the critical value is 1.65. We find it by referring to Appendix C again and searching the body of the table for 0.4500, or the value closest to it. The critical z-value is in the margin. Note that, although the null hypothesis contains an inequality ($\mu < \$35,000$), it is the equality condition ($\mu = \$35,000$) that is tested. The decision rule states that the null hypothesis will not be rejected if the computed value of z is equal to or less than the critical value of 1.65. If the computed z is greater than 1.65, the null hypothesis will be rejected and the alternate hypothesis accepted. Shown in diagram:

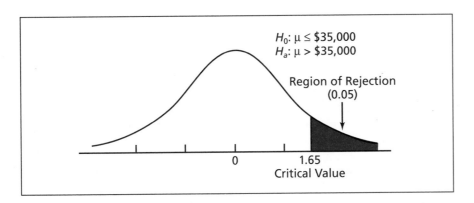

Problem   Hyperactive children are often disruptive in the typical classroom setting because they find it difficult to remain seated for extended periods of time. Baseline data from a very large study show that the typical frequency of "out-of-seat behaviors" was 12.38 per 30-minute period with a standard deviation of 3.52. A treatment known as covert positive reinforcement was applied to a group of 30 hyperactive children. The mean number of "out-of-seat behaviors" was reduced to 11.59 per 30-minute observation period. Using the 0.01 significance level, can we conclude that this decline in "out-of-seat behaviors" is significant? Determine the p-value.

Solution   Use the five-step hypothesis-testing procedure.

Step 1. State the null hypothesis and the alternate hypothesis. The null hypothesis is: The population mean is equal to or greater than 12.38. That is:

$$H_0: \mu \geq 12.38$$

The alternate hypothesis: The mean is less than 12.38. That is:

$$H_a: \mu < 12.38$$

Step 2. State the level of significance. It is 0.01.

Step 3. Give the appropriate test statistic. It is formula 10–1:

$$z = \frac{\overline{X} - \mu}{\sigma/\sqrt{n}}$$

Step 4. State the decision rule. Since there is interest in demonstrating that the covert treatment *lowers* the mean value, a one-tailed test is appropriate. The solution to this problem requires a one-tailed test in the negative direction. So the critical value of z is on the left side of the curve. The critical value for the 0.01 significance level is -2.33 (refer to Appendix C and 0.4900):

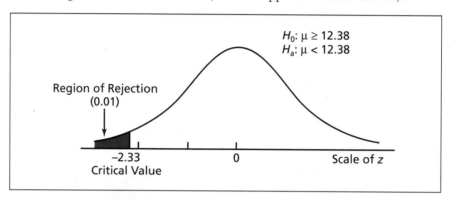

So the decision rule: Reject the null hypothesis if the computed value of z is to the left of -2.33. Otherwise, do not reject the null hypothesis.

Note that the inequality sign in the alternate hypothesis (Step 1) "points" in the negative direction. Thus, the critical region will be in the left tail (the direction in which the inequality is pointing), and the critical value will have a negative sign. If a one-tailed test is employed and the inequality sign in the alternate hypothesis points in the positive direction, the area of rejection will always appear in the positive tail, and the sign of the critical value will be positive.

Step 5. Compute the value of the test statistic and make a decision. Compute z using formula 10–1, and arrive at a decision:

$$z = \frac{\overline{X} - \mu}{\sigma/\sqrt{n}}$$

$$= \frac{11.59 - 12.38}{3.52/\sqrt{30}}$$

$$= -1.23$$

Because the computed z-value of -1.23 is not in the rejection region, we do not reject the null hypothesis. The difference between 11.59 and 12.38 "out-of-seat behaviors" can be attributed to chance.

To determine the p-value we find the probability of a z-value less than -1.23. From Appendix C, the probability of a z-value in the interval between -1.23 and 0 is 0.3907. So the probability of a z-value less than -1.23 is

$0.5000 - 0.3907 = 0.1093$. The p-value is 0.1093. In other words, there is about an 11% chance of obtaining a z-value this large or larger when the null hypothesis is true. Notice that we did not reject the null hypothesis and the p-value was *larger* than the significance level. From a practical standpoint, it cannot be concluded that the covert positive treatment reduced the number of "out-of-seat behaviors" in hyperactive children.

> ### Self-Review 10–2
>
> Experience over a long period of time has shown that, on the average, a mother stayed 2.0 days in the Findlay Children's Hospital after childbirth. The standard deviation was 0.5 day. Hospital administrators, doctors, and other groups decided to make a joint effort to reduce the average time a new mother spends in the hospital. Following their campaign, a sample of the files of 100 mothers revealed that the new mean length of stay was 1.8 days. Are mothers staying in the hospital less time, or could the difference between 2.0 and 1.8 be due to sampling error?
>
> a. What are H_0 and H_a?
> b. Using the 0.05 level, state the decision rule.
> c. Arrive at a decision.

Exercises

11. Suppose we want to conduct a one-tailed test of hypothesis at the 0.10 significance level. Assume the rejection region is in the right tail. What is the critical value?

12. Suppose we want to conduct a one-tailed test of hypothesis at the 0.13 significance level. Assume the rejection region is in the left tail. What is the critical value?

13. A sample of 36 observations is selected. The sample mean is 21 and the sample standard deviation is 5. Conduct the following test of hypothesis using the 0.10 significance level:

$$H_0: \mu \leq 20$$

$$H_a: \mu > 20$$

a. State the decision rule.
b. Compute the value of the test statistic.
c. What is your decision regarding the null hypothesis?
d. Determine the p-value.

14. A sample of 64 observations is selected from a normal population. The sample mean is 215, and the sample standard deviation is 15. Conduct the following test of hypothesis using the

0.04 significance level:

$$H_0: \mu \geq 220$$

$$H_a: \mu < 220$$

a. State the decision rule.
b. Compute the value of the test statistic.
c. What is your decision regarding the null hypothesis?
d. Determine the p-value.

15. A test was constructed to measure the degree of people's alienation. The mean score was 78 and the standard deviation of scores in the population was 16. The test was administered to a sample of 37 Gulf War veterans. Their mean score was 84. At the 0.05 significance level, can we conclude that Gulf War veterans are more alienated than the general population? Use the 0.05 significance level. Compute the p-value and interpret the result.

16. A machine is set to fire 30.00 decigrams of chocolate pellets into a box of cake mix as it moves along the production line. Of course, there is some variation in the weight of the pellets. A sample of 36 boxes of mix revealed that the mean weight of the chocolate pellets was 30.08 decigrams, with a sample standard deviation of 0.50 decigram. Is the increase in the weight of the pellets significant at the 0.05 level? Apply the usual 5 steps to be followed in hypothesis testing. (Hint: This involves a one-tailed test because there is interest in finding out only whether there has been an *increase* in weight.)

17. The board of education of a suburban school district wanted to consider a new academic program funded by the Department of Education. For the school district to be eligible for the federal grant, the arithmetic mean income per household must not exceed $16,000. The board hired a research firm to gather the required data. In its report, the firm indicated that the arithmetic mean income in the district is $17,000. The survey included 75 households and the

standard deviation of the sample was $3000. Use a one-tailed test and the 0.01 level of significance to decide if the board can argue that the larger household income ($17,000) was due to chance.

18. According to informed Pentagon sources, they deploy an average of 90 warheads at each missile site. An international peace-keeping force plans to make 49 inspections at various sites around the country to determine if there has been an increase in the mean number of warheads. They will use the 0.10 level of significance.

    a. State the null and alternate hypotheses.
    b. Give the formula for the appropriate test statistic.
    c. What is the decision rule?
    d. The checks found a sample mean of 92 warheads with a sample standard deviation of 9. What does this evidence indicate?
    e. Compute the *p*-value. Interpret.

A TEST INVOLVING TWO POPULATION MEANS (LARGE SAMPLES)

The previous section dealt with a test of hypothesis involving one large sample (30 or more observations). In this section, we compare two samples to determine if they have the same population mean. The five-step hypothesis-testing procedure is the same, except the formula we use to find the value of z is different.

The idea behind the test is that, if we select random samples from two normal populations, the distribution of the *differences* between the two means is also normal. We can state this more formally:

If a large number of independent random samples are selected from two populations, the differences between the two sample means will be normally distributed. If these differences are divided by the standard error of the difference, the resultant distribution follows the standard normal (z) distribution.

The formula for z is

$$z = \frac{\overline{X}_1 - \overline{X}_2}{\sqrt{\dfrac{s_1^2}{n_1} + \dfrac{s_2^2}{n_2}}}$$

$\overline{X}_1 - \overline{X}_2 \longleftarrow$ difference between two sample means

$\sqrt{\dfrac{s_1^2}{n_1} + \dfrac{s_2^2}{n_2}} \longleftarrow$ standard error of the difference between two sample means

10-2

Problem Each patient at Aloha Memorial Hospital is asked to evaluate the service at the time of their discharge. Recently, there have been several complaints that hospital staff on the surgical wing respond too slowly to the emergency calls of senior citizens. Jack Bartell, the president of the hospital, asked the Quality Assurance (QA) Department to investigate. After studying the problem, the QA Department collected the following information. At the 0.01 significance level, is the response time longer for senior citizens' emergencies?

Patients	Sample Mean	Sample Standard Deviation	Number in Sample
Senior citizens	5.5 minutes	0.40 minute	50
Other patients	5.3 minutes	0.30 minute	100

Solution The first step is to state the null hypothesis and the alternate hypothesis. The null hypothesis is that there is no difference in the mean response times for the 2 groups of patients. In other words, the difference of 0.20 minute between the mean response time for senior citizens and the mean response time for the other patients could be due to chance. The alternate hypothesis is that the mean response time is greater for the senior citizens. We will let μ_s refer to the mean response time for the population of senior citizens and μ_o refer to the mean response time for the population of other patients. The null hypothesis and the alternate hypothesis are stated as follows:

$$H_0: \mu_s \leq \mu_o$$

$$H_a: \mu_s > \mu_o$$

The next step is to determine the decision rule. The 0.01 significance level was given in the problem, and the test statistic follows the standard normal distribution. The decision rule: Reject the null hypothesis if the computed value of z is greater than 2.33. The following chart depicts the decision rule.

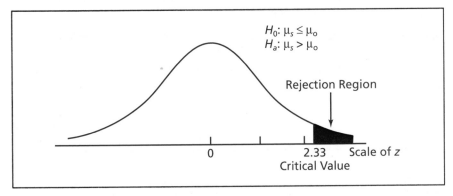

Formula 10–2 is used to compute the value of the test statistic.

$$z = \frac{\overline{X}_s - \overline{X}_o}{\sqrt{\dfrac{s_s^2}{n_s} + \dfrac{s_o^2}{n_o}}} = \frac{5.5 - 5.3}{\sqrt{\dfrac{0.40^2}{50} + \dfrac{0.30^2}{100}}} = \frac{0.20}{0.064} = 3.13$$

The computed value of z is beyond the critical value of 2.33. So the null hypothesis is rejected and the alternate hypothesis accepted. The difference of 0.20 minute between the response time for senior citizens and that for the other patients is too much to have occurred by chance. The QA Department can report to President Bartell that the mean response time is longer for senior citizens than for the other patients.

What is the p-value in this problem? Recall that the p-value is the probability of finding a value of the test statistic as larger or larger when the null hypothesis is true. To calculate the p-value we need to determine the probability of finding a z-value larger than 3.13. From Appendix C, the closest value we have to 3.13 is 3.00. The probability of finding a z-value between 0 and 3.00 is 0.4987. The probability of finding a z-value *greater* than 3.00 is 0.0013, found by $0.5000 - 0.4987$. We can report in this case that the p-value is less than 0.0013.

To properly apply the test involving the difference between two sample means we should meet the following criteria:

1. Both samples should be at least 30. In this problem, one sample was 50 and the other 100, so the sample standard deviations can be used to approximate the population standard deviations.
2. The samples should be from independent populations. This means that the sample of response times for the senior citizens must be unrelated to the sample of response times for the other patients. If Ms. Smith is a senior citizen and her emergency response time is sampled, this does not affect the response time for any of the other patients chosen.

Self-Review 10–3

The admissions officer at a university wants to investigate whether there is any difference between the scores on the mathematics placement test of students who attend classes primarily during the day and those of students who work during the day and attend evening classes.

a. Is this a one-tailed or a two-tailed test? Symbolically, what are the null hypothesis and the alternate hypothesis?
b. The 0.05 level of significance is to be used. State the decision rule.
c. The records of both day and evening students revealed the following:

Student	Mean Score	Standard Deviation	Number in Sample
Day	90	12	40
Evening	94	15	50

Compute z and arrive at a decision.
d. Determine the p-value. Interpret.

Exercises

19. The mean of a sample of 50 observations from one population is 52 and the standard deviation of that sample is 10. The mean of a sample of 40 observations from the second population is 49 with a standard deviation of 12. Conduct the following test of hypothesis, using the 0.04 significance level:

$$H_0: \mu_1 = \mu_2$$

$$H_a: \mu_1 \neq \mu_2$$

a. Is this a one-tailed or a two-tailed test?
b. State the decision rule.
c. Compute the value of the test statistic.
d. What is your decision regarding the null hypothesis?
e. Compute the p-value.

20. The mean of a sample of 65 observations from the first population is 5.67 and the standard deviation of that sample is 0.75. The mean of a sample of 45 observations from the second population is 5.35 with a standard deviation of 0.63. Conduct the following test of hypothesis using the .04 significance level:

$$H_0: \mu_1 \leq \mu_2$$

$$H_a: \mu_1 > \mu_2$$

a. Is this a one-tailed or a two-tailed test?
b. State the decision rule.
c. Compute the value of the test statistic.
d. What is your decision regarding the null hypothesis?
e. Compute the p-value.

21. Two machines fill bottles with a cough syrup. Workers check the performance of each periodically by testing bottles removed at random from the production line. The sample statistics for machine 1 are as follows: The mean weight of the contents of 40 bottles was 202.6 milligrams (mg), and the standard deviation of the sample was 3.3 mg. For machine 2, the mean weight of 50 bottles was 200.0 mg, and the standard deviation was 2.0 mg. Using a two-tailed test and the 0.01 level of significance, test whether there is any difference in the mean performance of the 2 machines.

a. State both the null hypothesis and the alternate hypothesis.
b. State the decision rule.
c. Compute z and arrive at a decision.
d. Determine the p-value. Interpret.

22. You are conducting a study of the annual incomes of probation officers in metropolitan areas of fewer than 100,000 population and in metropolitan areas of greater than 500,000 population. Your sample data are

	Population Less Than 100,000	Population Greater Than 500,000
Sample size	30	60
Sample mean	$54,290	$54,330
Sample standard deviation	$135	$142

Can you conclude that the officers from the areas with larger populations earn more? Use the 0.05 significance level.

23. A sociologist asserts that the mean length of courtship is longer before a second marriage than before a first. She bases this claim on the observation of (1) 80 first marriages in which the mean courtship is 265 days, with a sample standard deviation of 60 days, and (2) 60 second marriages in which the mean is 268 days, with a sample standard deviation of 50 days. Test her assertion by the five-step hypothesis-testing procedure. Use the 0.01 significance level. What is the p-value?

24. Planned Parenthood is investigating the differences between families in the Midwest and those on the Atlantic Coast. A particular study concerns the age of mothers at the birth of their last children. A random sample of 36 women from the Midwest had a sample mean of 32.9 years, with a sample standard deviation of 5.7 years. A second random sample of 49 women from the Atlantic Coast had a mean of 29.6 years, with a sample standard deviation of 5.5 years. Does this represent a significant difference between the 2 geographic regions? Use the 0.05 level of significance. Compute the *p*-value.

25. In an experiment to see if light from ultraviolet sun lamps affects muscle size, 50 pairs of laboratory animals were segregated into 2 groups. One group received ultraviolet light treatments daily for a month. The other did not. A particular muscle on each animal was then weighed.
   a. State the null and alternate hypotheses.
   b. Determine the decision rule for the 0.01 level of significance.
   c. If the mean for the ultraviolet group is 89 mg, with a standard deviation of 9 mg, and the mean for the control group is 57 mg, with a

standard deviation of 7 mg, what is the value of the test statistic?
   d. Make a decision about the null hypothesis.
   e. Compute the *p*-value. Interpret.

26. The burn rates for 2 different types of missile fuel are to be compared. Thirty observations on each will be obtained. You wish to design a test that will use the 0.01 significance level.
   a. What are your hypotheses?
   b. What is your decision rule?
   c. The first sample mean is 20.44 liters per second with a standard deviation of 2.96. The second sample mean is 19.45 with a standard deviation of 2.13. What do you conclude?
   d. Compute the *p*-value and interpret.

A TEST INVOLVING THE POPULATION PROPORTION (LARGE SAMPLES)

We introduced the concept of a proportion in Chapter 8. Now we will explore such questions as these: Will 80% of first offenders placed on probation commit a second crime? Do 20% of inner-city families live on a subsistence income? Is there a difference between the **proportion** of rural voters and that of urban voters who plan to vote for the incumbent governor in the forthcoming election?

Claims are often made that a proportion, or fraction, of a population possesses a certain characteristic. One claim might be that 25% of nonworking adults watched a daytime soap opera yesterday. Another claim might be that 30% of children under age 10 do not brush their teeth daily. A test to verify such claims follows. This test is appropriate when *both* $n\pi$ and $n(1 - \pi)$ are *5 or more*, where π refers to the population proportion. If the data meet this requirement, we can use the normal approximation to the binomial, discussed in Chapter 7. The test statistic, which follows the *z* distribution, is

$$z = \frac{\dfrac{X}{n} - \pi}{\sqrt{\dfrac{\pi(1 - \pi)}{n}}}$$

10–3

where

X is the observed number in the sample possessing the trait.
n is the size of the sample.
π is the population proportion possessing the characteristic.

If we are to employ this test statistic, the items in the sample must be *selected randomly*, and each must have the *same chance of being selected*.

Problem We are investigating the following question: Is the proportion of all inner-city families living on a subsistence income 20%, or 0.20?

Solution Use the five-step hypothesis-testing procedure.

Step 1. State the null hypothesis and alternate hypothesis. In this case, the null hypothesis is that the proportion of all inner-city families living on a subsistence income is 0.20, which is a population parameter. The symbol π (Greek letter *pi*) is used as the symbol for a population proportion. Symbolically:

$$H_0:\pi = 0.20$$

The alternate hypothesis is that the proportion of inner-city families living on a subsistence income is not 0.20. Symbolically:

$$H_a:\pi \neq 0.20$$

Because the alternate hypothesis does not state a direction, the test is two-tailed.

Step 2. Select a level of significance. The significance level determines when the sample measurement is too far away from the hypothesized proportion to be believable. The 0.05 level has been chosen for the inner-city problem.

Step 3. Choose a test statistic. For this problem, a sample of 200 inner-city families will be studied. $n = 200$ and $\pi = 0.20$, so $n\pi$ is $200(0.20) = 40$ and $n(1 - \pi)$ is $200(1 - 0.20) = 160$. Both $n\pi$ and $n(1 - \pi)$ are greater than 5. Therefore, the appropriate test statistic is z, given previously in formula 10–3:

$$z = \frac{\dfrac{X}{n} - \pi}{\sqrt{\dfrac{\pi(1 - \pi)}{n}}}$$

Step 4. Formulate a decision rule. According to the standard normal distribution in Appendix C, the critical value for the 0.05 level of significance is 1.96. We find it by locating 0.4750 in the body of the table and then moving to the margin to read the critical z-value of 1.96. The decision rule states that the null hypothesis will not be rejected if the computed value of z is in the interval between -1.96 and 1.96. Otherwise, it will be rejected and the alternate hypothesis accepted. Shown schematically:

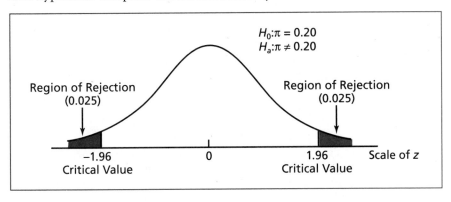

Step 5. Select a sample from the population of all inner-city families, compute z, and then decide whether to reject H_0. A sample of 200 inner-city families revealed that 38 had incomes at the subsistence level. Computing z using formula 10–3, we get

$$z = \frac{\dfrac{X}{n} - \pi}{\sqrt{\dfrac{\pi(1 - \pi)}{n}}}$$

$$= \frac{\dfrac{38}{200} - 0.20}{\sqrt{\dfrac{0.20(1 - 0.20)}{200}}} = -0.35$$

The decision is not to reject the null hypothesis (that $\pi = 0.20$) because -0.35 is between -1.96 and 1.96. Although there is a difference between the sample proportion of 0.19 (found by 38/200) and 0.20, the apparent difference can be attributed to chance (sampling error). A helpful measure of the likelihood of this outcome is the p-value, as described earlier in this chapter. In this case, the p-value is 0.7264, found by $2(0.5000 - 0.1368) = 2(0.3632)$. Note that the p-value is greater than the significance level, indicating that H_0 should not be rejected. This information is summarized in the following chart:

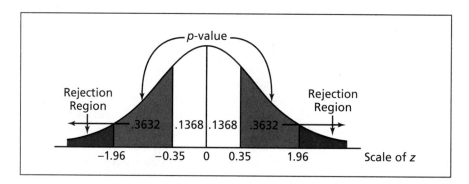

Problem Recall from the Chapter Problem the stock selection contest pitting dart throwing against professional investors. In that competition, the pros did better than the darts 15 times out of 27. Formulate and carry out an appropriate statistical test using this contest data.

Solution Use the five-step hypothesis-testing procedure.

Step 1. In this case, the null hypothesis is that the proportion of contests won by the pros is 0.5. If the pros were just guessing, we would expect them to win half (0.50) of the time and lose half the time. The alternate hypothesis is that the proportion of contests won by the pros is greater than 0.5. This reflects the hope that the pros do have some advantage and suggests that the test is one-tailed.

Step 2. We choose a significance level of 0.10 for this illustration.

Step 3. Since n is 27 and π is 0.5, the normal approximation to the binomial would be appropriate. The test statistic is given in formula 10–3.

Step 4. Consulting the standard normal table in Appendix C, we see that the critical value for a one-sided test with a 0.10 level of significance is 1.28. The decision rule is: Reject the null hypothesis if the test statistic exceeds 1.28.

Step 5. Using formula 10–3, with $X = 15$, we get

$$z = \frac{\dfrac{15}{27} - 0.500}{\sqrt{\dfrac{0.5(1 - 0.5)}{27}}} = \frac{0.556 - 0.500}{0.096} = \frac{0.056}{0.096} = 0.58$$

We fail to reject the null hypothesis because 0.58 is *not* greater than 1.28. Although the professionals won more than half of the contests, the difference between the sample proportion (0.556) and random guessing (0.500) could easily result from sampling error or "dumb luck." The p-value in this case is more than 25%, indicating that the test statistic could easily have happened by chance. There is thus no statistical evidence that the pros will top the dart throwing!

- -

Self-Review 10–4

National figures reveal that the probability that a youthful offender on probation will commit another crime is 0.80. A special rehabilitation program was conducted for youthful offenders. A sample of 100 enrolled in the program showed that 75 subsequently committed another crime. The null hypothesis to be tested was $\pi \geq 0.80$. The alternate hypothesis was $\pi < 0.80$.

a. Based on the way the alternate hypothesis is stated, is this a one-tailed test or a two-tailed test?
b. Using the 0.01 significance level, show the decision rule graphically.
c. Arrive at a decision.

- -

Exercises

27. In the following hypothesis-testing situation:

$$H_0: \pi \leq 0.70$$

$$H_a: \pi > 0.70$$

a sample of 100 observations showed that $x = 73$. Use the 0.01 significance level.

a. State the decision rule.
b. Compute the value of the test statistic.
c. What is your decision regarding the null hypothesis?
d. What is the p-value?

28. In the following hypothesis-testing situation:

$$H_0: \pi = 0.40$$

$$H_a: \pi \neq 0.40$$

a sample of 49 observations showed that $x = 14$. use the 0.10 significance level.
   a. State the decision rule.
   b. Compute the value of the test statistic.
   c. What is your decision regarding the null hypothesis.
   d. What is the p-value?

29. You believe that 90% of the population would continue to work even if they inherited sufficient wealth to live comfortably without working. This assertion is questioned by a social research team. A list of persons who recently inherited more than $400,000 is obtained, and 350 persons are selected from the list. Out of the 350 selected, 308 indicate that they are still working. At the 0.05 level, is this sufficient evidence for the researchers to doubt your claim?

30. The local newspaper conducted a study of the welfare claims in the area. It reported that "at least 20% of the present welfare recipients are ineligible and should not be receiving monthly payments." You are hired by the welfare de-partment to investigate the newspaper's claim. You randomly select 400 recipients from the files and carefully investigate each one. You find that 60 should not be receiving payments because they filed false claims, failed to report that they work, or committed other fraudulent acts. At the 0.01 level of significance, should the newspaper's claim be rejected? Back your decision with statistical evidence.

31. The Joint Monitoring Committee, a federal advisory panel, reports that 28% of adults in the United States are overweight. The Indian Health Commission of Arizona sampled 356 members of a tribe and found that 90 of them are overweight. At the 0.10 level of significance, can you conclude that a smaller proportion of the tribe members are overweight compared with all adults in the United States?

32. Out of 600 addicts admitted in fiscal 1996, 70 patients completed treatment in the California Alcohol Abuse Services Administration programs. By comparison, completion rates at federally funded clinics averaged 16%. Is the difference between the two proportions statistically significant at the 0.05 level? Compute the p-value. Interpret.

A TEST INVOLVING TWO POPULATION PROPORTIONS (LARGE SAMPLES)

Some problems involve testing whether two population proportions are equal. To conduct such a test, we first select a random sample from each population. We assume each sample is large enough that the normal distribution will serve as a good approximation of the binomial distribution. The test statistic is the standard normal distribution. We compute the value of z from the following formula:

$$z = \frac{\dfrac{X_1}{n_1} - \dfrac{X_2}{n_2}}{\sqrt{\bar{p}(1 - \bar{p})\left(\dfrac{1}{n_1} + \dfrac{1}{n_2}\right)}}$$

10–4

where

X_1 is the number possessing the trait in the first sample.
X_2 is the number possessing the trait in the second sample.
n_1 is the number in the first sample.
n_2 is the number in the second sample.
$\overline{p}$ is the proportion possessing the trait in the combined samples. It is referred to as the *pooled estimate* of the proportion:

$$\overline{p} = \frac{\text{Total number of successes}}{\text{Total number of samples}} = \frac{X_1 + X_2}{n_1 + n_2} \qquad \boxed{\textbf{10–5}}$$

Problem   You know that a large group of homosexuals lives in a certain district of a large city. An in-depth study is to be made of this group. One of your objectives is to determine if there is a difference in the respective proportions of white homosexuals to the total white population and of black homosexuals to the total black population living in that district.

Solution   Use the five-step hypothesis-testing procedure.

Step 1. State the null and alternate hypotheses. The null hypothesis is that there is no difference between the two population proportions; that is, the two proportions are equal. Symbolically:

$$H_0 : \pi_1 = \pi_2$$

The alternate hypothesis is that the two population proportions are not equal. Symbolically:

$$H_a : \pi_1 \neq \pi_2$$

The problem does not imply a direction; therefore, a two-tailed test is used.

Step 2. Give the level of significance. You decided on 0.05.

Step 3. The test statistic for problems involving two proportions is z. The formula is

$$z = \frac{\dfrac{X_1}{n_1} - \dfrac{X_2}{n_2}}{\sqrt{\overline{p}(1 - \overline{p})\left(\dfrac{1}{n_1} + \dfrac{1}{n_2}\right)}}$$

Step 4. The decision rule is: Do not reject the null hypothesis if the value of the computed z is between -1.96 and 1.96. Otherwise, reject the null hypothesis and accept the alternate hypothesis. A graphic presentation of the decision rule is

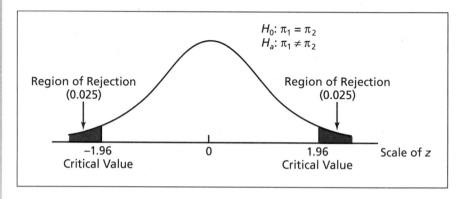

Step 5. Take a sample from the white population living in the district, and determine the number of male homosexuals in the sample. Then take a sample from the black population living in the area, and determine the number of male homosexuals in the sample. The results are

	Number in Sample	Number of Male Homosexuals	Proportion of Male Homosexuals in Population
White	804	575	0.715, found by 575/804
Black	175	111	0.634, found by 111/175

The pooled estimate of the proportion designated by $\bar{p}$ is 0.70, found by using formula 10–5:

$$\bar{p} = \frac{X_1 + X_2}{n_1 + n_2} = \frac{575 + 111}{804 + 175} = \frac{686}{979}$$

$$= 0.70$$

We use formula 10–4 to determine the value of z:

$$z = \frac{\dfrac{X_1}{n_1} - \dfrac{X_2}{n_2}}{\sqrt{\bar{p}(1 - \bar{p})\left(\dfrac{1}{n_1} + \dfrac{1}{n_2}\right)}} = \frac{\dfrac{575}{804} - \dfrac{111}{175}}{\sqrt{0.70(1 - 0.70)\left(\dfrac{1}{804} + \dfrac{1}{175}\right)}}$$

$$= \frac{0.715 - 0.634}{\sqrt{0.001461}} = \frac{0.081}{0.038} = 2.13$$

Because 2.13 is greater than the upper critical value of 1.96, the decision is to reject the null hypothesis. The conclusion is that there is a significant difference between the 2 population proportions. It is unlikely that the difference of 8.1% between the 2 sample proportions is due to chance (sampling error). The *p*-value is 0.0332, found by 2(0.5000 − 0.4834).

The effectiveness of a newly developed allergy-relief capsule is to be compared with that of one that has been on the market for a number of years. A sample of 250 persons using the new capsule revealed that 150 received satisfactory relief. Out of a group of 400 using the older capsule, 232 received satisfactory relief. Using the 0.02 level of significance, test the null hypothesis that the proportion receiving relief from the new capsule is equal to the proportion receiving relief from the old capsule. Use a two-tailed test.

a. State the null and alternate hypotheses using the letters H_0 and H_a.
b. Show the decision rule graphically.
c. Compute z and arrive at a decision. Interpret the result and compute the p-value.

Exercises

33. In the following hypothesis-testing situation:

$$H_0: \pi_1 \leq \pi_2$$

$$H_a: \pi_1 > \pi_2$$

for a sample of 100 observations from the first population, there were 60 successes. For a sample of 150 observations from the second population, there were 80 successes. Use the 0.05 significance level.
a. State the decision rule.
b. Compute the pooled proportion.
c. Compute the value of the test statistic.
d. What is your decision regarding the null hypothesis?
e. Compute the p-value.

34. In the following hypothesis-testing situation:

$$H_0: \pi_1 = \pi_2$$

$$H_a: \pi_1 \neq \pi_2$$

for a sample of 150 observations from the first population, there were 110 successes. For a sample of 200 observations from the second population, there were 170 successes. Use the 0.05 significance level.
a. State the decision rule.
b. Compute the pooled proportion.
c. Compute the value of the test statistic.
d. What is your decision regarding the null hypothesis?
e. Compute the p-value.

35. Subsalicylate bismuth is the active ingredient in several nonprescription drugs for the relief of stomach upsets. Only 14 out of 62 persons sampled who took the bismuth during their 3-week vacations away from home suffered from the gastric and intestinal discomfort that often afflicts travelers. Of 66 persons who did not take bismuth, 40 suffered from stomach upsets. Is the difference in the 2 proportions significant at the 0.05 level? Use a two-tailed test.

36. A major automobile insurance company claims that compared with other drivers, a larger proportion of younger drivers takes high-risk chances when driving. To investigate this claim, you give a driving-simulator test to a sample of young drivers and a sample of other drivers. Out of 100 young drivers tested, 30 took high-risk chances. Out of 200 other drivers tested, 55 took high-risk chances. Do you think there is sufficient evidence to support the insurance company's contention? Use the 0.05 level of significance.

37. Research into the use of physical punishment (spanking) in child rearing yielded the following information: Out of 100 children aged 3 to 9 years, 82 had been spanked by their parents during the previous month. Similarly, 40 out of 60 children aged 10 to 14 years had been spanked. Do parents use spankings significantly less often on older children? Test at the 0.01 level of significance.

38. A noted medical researcher has suggested that a heart attack is less likely to occur among men who actively participate in athletics. A random sample of 300 men is obtained. Of that total, 100 are found to be athletically active. Within this group, 10 had suffered heart attacks; among the 200 athletically inactive men, 25 had suffered heart attacks. Test the hypothesis that the proportion of men who are active and suffered heart attacks is equal to the proportion of men who are not active and suffered heart attacks. Use the 0.05 significance level.

CHAPTER OUTLINE

I. The objective of hypothesis testing is to check the validity of a statement about the value of a population parameter.

II. The procedure used in hypothesis testing is as follows:
 A. State the null hypothesis (H_0) and the alternate hypothesis (H_a).
  1. The null hypothesis is a claim about the value of a population parameter.
  2. The alternate hypothesis is a claim about the population parameter that is accepted if the null hypothesis is rejected.
 B. Select the level of significance.
  1. The level of significance is the probability of rejecting the null hypothesis when it is true.
  2. The 0.10, 0.05, and 0.01 levels of significance are the most common.
 C. Decide on the test statistic.
  1. This is based on characteristics such as the shape of the population and the size of the sample.
  2. The standard normal distribution was used for all tests in this chapter.
 D. State the decision rule.
  1. The decision rule is based on the way the null and alternate hypotheses are stated, the level of significance, and the test statistic selected.
  2. The decision rule defines the conditions when the null hypothesis is rejected.

 E. Select the sample, compute the value of the test statistic, and make a decision regarding the null hypothesis.

III. In hypothesis testing, there are two types of errors.
 A. A Type I error occurs when a true null hypothesis is rejected.
 B. A Type II error occurs when a false null hypothesis is not rejected.

IV. A p-value is the probability of finding a value of the test statistic as extreme as that found in the sample when the null hypothesis is true.

V. Two "large sample" tests of hypothesis were considered for sample means.
 A. The formula to compare a single sample mean to a population mean is

$$z = \frac{\overline{X} - \mu}{\sigma/\sqrt{n}} \qquad \textbf{10-1}$$

 B. The formula to compare two sample means to determine if they are equal is

$$z = \frac{\overline{X}_1 - \overline{X}_2}{\sqrt{\dfrac{s_1^2}{n_1} + \dfrac{s_2^2}{n_2}}} \qquad \textbf{10-2}$$

VI. Two "large sample" tests of hypothesis were considered for sample proportions.

  A. The formula to compare a single sample proportion to a population proportion is

$$z = \frac{\dfrac{X}{n} - \pi}{\sqrt{\dfrac{\pi(1 - \pi)}{n}}} \qquad \boxed{10\text{--}3}$$

  1. X is the number of successes in n observations.
  2. Both πn and $n(1 - \pi)$ should be 5 or more.

  B. The formula to compare two sample proportions to determine if they came from equal populations is

$$z = \frac{\dfrac{X_1}{n_1} - \dfrac{X_2}{n_2}}{\sqrt{\overline{p}(1 - \overline{p})\left(\dfrac{1}{n_1} + \dfrac{1}{n_2}\right)}} \qquad \boxed{10\text{--}4}$$

  where $\overline{p}$ is determined by

$$\overline{p} = \frac{X_1 + X_2}{n_1 + n_2} \qquad \boxed{10\text{--}5}$$

- -

Exercises

39. The Department of Health and Human Services reported that the mean number of years of school completed by adults in the United States was 11.5. The Tennessee Department of Employment Security wants to determine if the mean level of education of its employees is above the national average.

  a. State the null and alternate hypotheses both in words and using H_0 and H_a.
  b. Give the test statistic z.
  c. State the decision rule in words for the 0.05 significance level. Show it in a diagram.
  d. A sample of 100 employees revealed that the mean is 13 years, with a standard deviation of 1.1 years. Compute z and arrive at a decision regarding the null hypothesis.

40. A psychologist wants to evaluate whether windowless schools lead to anxiety in the psychological development of children. A sample of children who are in windowless classrooms is selected and given an anxiety test. Higher scores indicate more anxiety. The same procedure is followed for children in classrooms with windows. The results:

	Windowless Schools	Schools with Windows
Mean	94	90
Standard deviation	8	10
Sample size	100	80

Using the five-step hypothesis-testing procedure, the 0.01 level, and a one-tailed test, arrive at a conclusion regarding the anxiety levels of the 2 groups.

41. A medical researcher contends that lung capacity varies significantly between smokers and nonsmokers. The mean capacity of a sample of 30 nonsmokers is 5.0 liters, with a sample standard deviation of 0.3 liter. For a sample of 40 smokers, the mean lung capacity is 4.5 liters, with a standard deviation of 0.4 liter. At the 0.01 significance level, is there sufficient evidence to conclude that lung capacity is larger among nonsmokers?

42. Patients entering a hospital have complained that it takes at least 30 minutes to fill out the forms required for admittance. As a result of their complaints, the forms and procedure have been revised. A recent analysis of a random sample of 40 incoming patients reveals that the mean time to fill out the forms now is 28.5 minutes, with a standard deviation of 5 minutes. Is there sufficient evidence at the 0.02 level of significance to show that the new system is an improvement?

43. A random sample of 100 freshmen entering Ivy Tech in the fall of 1990 had an average combined ACT score of 23, with a sample standard deviation of 0.80. A sample of 100 similar freshmen in the fall of 1997 showed a mean score of 24, with a sample standard deviation of 0.90. Has there been a significant increase in the test scores over the 7-year period? Use the 0.05 significance level.

44. A group of environmentalists claims that warm water from the Virginia Power nuclear plant at Lake Anna is limiting the growth of striped bass at the lake. Use the 0.01 level of significance.
   a. What are the appropriate null and alternate hypotheses?
   b. Show the test statistic.
   c. State the decision rule.
   d. A study of 50 striped bass at the lake revealed a mean weight of 20 pounds, with a sample standard deviation of 4 pounds. In a similar southeastern reservoir, a sample of 40 striped bass was found to average 40 pounds, with a standard deviation of 8 pounds. Do these data support the environmentalists? What do you conclude?

45. The Municipal Environmental Protection Agency claims there has been a change in the density of the ozone layer over the city. Data from previous monitoring programs indicate that the mean is 384 parts per million (ppm), with a standard deviation of 20 ppm. You want to perform an independent test of the density.
   a. State appropriate null and alternate hypotheses.
   b. If you want the 0.01 significance level, what is your decision rule?
   c. You have collected a random sample of 100 observations, and the mean is 390 ppm. What do you conclude?

46. A soft-drink manufacturing facility turns out "16-ounce" bottles of beverages on a production line that, based on many observations, has a mean filling rate of 16.01 ounces, with a standard deviation of 0.005 ounce. As a newly hired quality control inspector, you decide to draw a sample of 40 bottles for testing.
   a. What are the null and alternate hypotheses?
   b. What is the decision rule for the 0.10 level of significance?
   c. Your sample has a mean of 15.97 ounces. Is that statistically significant?

47. A tradition in a particular fishing village indicates that a certain strain of fish will be found at a depth of 70 meters. You wish to test that lore by using modern equipment to locate the fish and recording the depth in the vicinity of the school.
   a. State your hypotheses.
   b. For the 0.05 level of significance, what is the decision rule?
   c. If you get a mean of 73 meters, with a standard deviation of 2 meters, from a sample of 35 measurements, what do you conclude?

48. A study dealing with comparative nutrition will record the heights of children at age 16 from 2 different countries. One hundred youths will be sampled in the first country and 150 in the second country.
   a. What are the null and alternate hypotheses?
   b. What is the decision rule for the 0.10 level of significance?
   c. The mean from the first country is 52.7 inches, with a standard deviation of 2.5 inches. The mean from the second country is 51.8 inches, with a standard deviation of 2.6 inches. Complete the test of hypothesis.

49. An investigation of 2 kinds of photocopying equipment showed that 60 failures of the first kind of equipment took, on the average, 82.4 minutes to repair, with a standard deviation of 19.4 minutes, while 60 failures of the second kind of equipment took, on the average, 91.6 minutes to repair, with a standard deviation of 18.8 minutes. Test at the 0.01 level of significance whether the difference between these 2 sample means is significant.

50. An EEOC complaint against the Betts Manufacturing Company alleges that men are paid more than women in a particular job category.

a. State the appropriate null and alternate hypotheses.

b. For a test at the 0.02 level of significance, what is the decision rule?

c. Sample data show that 80 men earned, on the average, $353, with a standard deviation of $18, while 80 women earned, on the average, $315, with a standard deviation of $21. State the appropriate conclusion to the complaint.

51. The League of Savings Institutions reported that the mean age of home buyers in 1990 was 35.8 years, with a standard deviation of 14.24 years. You intend to sample 100 home buyers in 1996 and determine their mean age.

a. State appropriate null and alternate hypotheses.

b. For the 0.02 level of significance, what is your decision rule?

c. The mean age of your sample is 37.3. Is that statistically significant?

52. An urban planner claims that nationally 20% of all families renting condominiums move during the year. A random sample of 200 families renting condominiums in a large development revealed that 56 had moved during the past year. Does the evidence demonstrate that this development has a greater proportion of movers than the national average?

53. A drug manufacturer has developed a drug that is said to cure postnatal depression in 85% of cases. A random sample of 150 women who gave birth at the Colorado General Hospital and who used the drug in a 2-year period revealed that 120 of them found it effective. Does this result contradict the manufacturer's claim? Use the 0.02 level of significance.

54. The American Council on Health Policy reports that 65% of citizens in the United States are covered by private health insurance. The staff at the health planning agency of Utah have data from a random sample of 200 residents of Utah. They wish to determine if the Utah proportion is consistent with the nationally reported percentage.

a. Formulate appropriate null and alternate hypotheses.

b. For the 0.10 level of significance, what is the decision rule?

c. In the sample, 120 were covered by private health insurance. Compute the test statistic.

d. What do you conclude?

55. Thirteen percent of the manuscripts submitted to the *New England Journal of Medicine* are accepted for publication. A prominent medical school has submitted 50 manuscripts and claims that its acceptance rate is better than average.

a. What are the null and alternate hypotheses?

b. State the decision rule for a test at the 0.05 level of significance.

c. Ten of the papers are accepted. Compute the test statistic.

d. Complete the statistical test.

56. Genes may play a large part in divorce, according to a recent study. The researchers found that the divorce rate was 45 percent among identical twins and 30 percent among fraternal twins. The study included 722 identical twins and 794 fraternal twins. Identical twins are more genetically alike than fraternal twins. Can we conclude that the divorce rate is higher among identical twins? Use the 0.01 significance level.

57. Citizens for Tax Justice reported that 42 of 250 companies surveyed paid no income tax this year. Last year 130 of the 250 firms legally avoided federal income taxes. Has the proportion of firms that avoid income tax declined, or can the difference be attributed to sampling error?

58. During a baseball game, one team gets 8 hits in 37 at-bats, while another gets 7 hits in 38 at-bats. Is there a statistically significant difference in the proportion of hits for the 2 teams?

59. Past elections indicate that 40% of the central-city votes are needed to pass any amendment that will increase taxes in the entire county. A preelection poll revealed that a tax increase is favored by 300 out of 1000 central-city voters. Is it likely that the difference between the sample proportion (0.30) and the proportion needed (0.40) is due to chance (sampling error)? Use the 0.01 level.

60. Suppose a random sample of 1000 American-born citizens revaled that 200 favor resumption of full diplomatic relations with Cuba. Similarly, 110 out of a sample of 500 foreign-born citizens favor it. Test at the 0.05 level the H_0 that there is no significant difference in the 2 proportions.

61. Past experience at Penta State University indicates that 50% of the students change their major areas of study after the first year of college work. At the end of the past year, a sampling of 100 students revealed that 48 of them had changed their majors at the end of the first year. Has there been a significant decrease in the proportion of students who change their majors, or can the difference between the expected proportion (50%) and the sample proportion (48%) be attributed to chance (sampling error)? Use the 0.05 level of significance.

62. A survey dealing with car pooling in metropolitan Philadelphia found that during a particular October, 83 out of 420 vehicles had more than 1 occupant. Five years later a similar sample of 423 cars showed that 146 had more than 1 occupant. Is this increase significant at the 0.01 level?

63. Preliminary research suggests that the proportion of young couples who use a particular method of birth control is 0.30. A detailed study conducted in Napanee, Indiana, showed a total of 46 users in a random sample of 200 young women. Are these results consistent with the preliminary research? Use the 0.10 level of significance.

64. A sportswriter comments that 50% of college football players suffer damage to their knees. To test this claim, you sample 300 players. If 143 of this group have knee injuries, at the 0.05 significance level can we conclude that less than 50% have knee injuries?

65. The Citizens Action Council claims that over 60% of the citizens of Fowlerville oppose a new highway construction plan. A random sample of 190 residents revealed that 140 were opposed to the construction plan. At the 0.01 level of significance, does this evidence support the claim of the council?

66. To evaluate a new method for treating depression, researchers subjected 70 patients at a clinic to the new method; 60 others were treated by the traditional method. If 46 of the patients in the first group and 28 in the second group showed signs of improvement, what can we conclude at the 0.01 level of significance?

67. In a random sample of 200 Ohio residents, 25 reported allergic reactions during changing weather. In a similar sample of 150 Arizona residents, 10 reported experiencing this type of reaction. Use the 0.02 level of significance. Do the data support the idea that the proportion of persons having allergic reactions is not different in Arizona?

68. Shoplifting has become a national problem. One report indicates that 10% of the customers who enter large discount stores will steal something. The store manager of the Jamesway Discount Store randomly selected 150 customers and observed them from behind a one-way window. Twenty of these shoppers attempted to steal something. Do these data suggest that shoplifting is more of a problem for Jamesway than for the typical large discount store? Use a 0.05 significance level.

DATA EXERCISES

69. Refer to the real estate data set, which reports information on homes sold in Alabama during 1996.
   a. The *Birmingham News* reported that "the typical Alabama home is now selling for less than $190,000." Does this claim appear valid? Write a brief summary of your findings. Use the 0.05 significance level. Be sure to comment on the *p*-value.

   b. The Alabama Chamber of Commerce noted in one of its publications that "because of the good highway system the typical homeowner lives more than 15 miles from the central city." Is this claim valid? Write a brief summary of your findings. Use the 0.05 significance level. Be sure to comment on the *p*-value.

c. Do half of the homes sold have an attached garage, or is the actual proportion less? Use the 0.05 significance level to conduct the test.

d. An important factor in the decision to sell a home is what fraction the selling price is of the list price. At the 0.05 significance level, can we conclude what the mean fraction of the selling price is if the list price is different from 0.975?

70. Refer to the school data set, which includes information on the 94 school districts in northwest Ohio.

a. If we assume the 94 school districts represent a sample, can we conclude that the mean salary in the region is less than $35,000?

b. If we assume the 94 school districts represent a sample, can we conclude that the typical district has more than 5 percent of its population on welfare?

c. If we assume the 94 school districts represent a sample, can we conclude that the typical district has more than 65 percent of its students passing the proficiency exam?

. . . .

CHAPTER ACHIEVEMENT TEST

Answer all the questions. The answers are at the back of the book.

MULTIPLE-CHOICE QUESTIONS

Select the response that best answers each of the questions.

1. A statement about the value of a population parameter is called a
   a. null hypothesis.
   b. sample statistic.
   c. test statistic.
   d. decision rule.

2. The probability of making a Type I error is equal to the
   a. power of a test.
   b. level of significance.
   c. alternate hypothesis.
   d. critical value.

3. The value computed from sample data is called the
   a. decision rule.
   b. critical value.
   c. test statistic.
   d. alternate hypothesis.

4. A Type II error occurs if you
   a. reject a false alternate hypothesis.
   b. accept a false alternate hypothesis.

   c. reject a false null hypothesis.
   d. accept a false null hypothesis.

5. A 1-sample test for a population mean has a critical value of 1.96 and a test statistic of 1.71. The correct statistical conclusion is to
   a. reject the null hypothesis.
   b. reject the alternate hypothesis.
   c. not reject the null hypothesis.
   d. not reject the alternate hypothesis.

6. Tests for proportions require that $n\pi$ and $n(1 - \pi)$ be
   a. equal.
   b. 30 or more.
   c. 5 or more.
   d. 50 or more.

7. The "pooled" estimate of a proportion that results from combining samples will
   a. equal one of the sample proportions.
   b. be greater than either sample proportion.
   c. be between the two sample proportions.
   d. be less than the smaller of the two.

8. To determine if more than 20% of consumers will purchase a new product, a researcher should use a

a. two-tailed test.
b. sample of at least 25 consumers.
c. one-tailed test.
d. 20% level of significance.

9. Tests for proportions are based on which of the following probability distributions?
   a. Binomial approximation of the normal
   b. Poisson distribution
   c. Student's t
   d. Normal approximation of the binomial

10. A sample (as opposed to a population) proportion is found from the
    a. null hypothesis.
    b. sample data.
    c. alternate hypothesis.
    d. decision rule.

COMPUTATION PROBLEMS

11. Coffee cans are filled to a net weight of 32.0 ounces. A random sample of 36 cans revealed a mean weight of 31.8 ounces, with a standard deviation of 0.4 ounce. At the 0.01 significance level, can we conclude that the actual mean weight is different from 32.0 ounces?

12. Kaiser Medical compares the length of hospital stay for a random sample of 40 patients in New York with that for a similar sample of 30 patients in Detroit. Do the data indicate a difference in the mean lengths of a hospital stay in the 2 cities?
    a. State in symbolic form the appropriate null and alternate hypotheses.
    b. At the 7% level of significance, what is the decision rule?
    c. The mean length of stay is 5.21 days in New York, with a standard deviation of 2.3 days. The Detroit average is 4.65 days, with a sample standard deviation of 1.9 days. Show the test statistic and essential calculations.
    d. What do you conclude? State your findings in a sentence.

13. The student newspaper interviewed a random sample of 200 male students and found that 150 were in favor of having a homecoming dance. A random sample of 400 female students revealed that 312 were in favor of the dance. Does the sample evidence show (at the 0.10 significance level) that there is a difference between the proportions of males and of females favoring a dance?

· ·

ANSWERS TO SELF-REVIEW PROBLEMS

10-1 **a.** H_0: $\mu = 6.8$
        H_a: $\mu \neq 6.8$
     b. $z = \dfrac{\overline{X} - \mu}{\sigma/\sqrt{n}}$
     c.

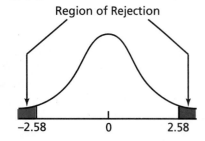

Region of Rejection

−2.58   0   2.58

d. Reject H_0 if $z < -2.58$ or $z > 2.58$.
e. $z = \dfrac{6.4 - 6.8}{1.50/\sqrt{36}} = \dfrac{-0.4}{0.25} = -1.60$
f. H_0 is not rejected. The population mean could be 6.8. The difference between 6.4 and 6.8 could be due to chance.
g. p-value $= 2(0.5000 - 0.4452) = 0.1096$

10-2 **a.** H_0: $\mu \geq 2.0$
        H_a: $\mu < 2.0$
     b. Reject H_0 if the test statistic is to the left of −1.65. Otherwise, do not reject H_0 at the 0.05 level.

c. $z = \dfrac{1.8 - 2.0}{0.50/\sqrt{100}} = -4.00$

Reject H_0. Mothers are staying less time.

10-3 a. Two-tailed

$H_0: \mu_1 = \mu_2$

$H_a: \mu_1 \neq \mu_2$

b. Accept H_0 if computed z is between -1.96 and $+1.96$.

c. $z = \dfrac{90 - 94}{\sqrt{\dfrac{(12)^2}{40} + \dfrac{(15)^2}{50}}} = \dfrac{-4}{2.846} = -1.41$

Do not reject H_0. There is no difference between the two groups.

d. The p-value is 0.1586, found by $2(.5000 - .4207)$. Approximately 16 percent of the time you will get a value as extreme as -1.41 or worse when H_0 is actually true.

10-4 a. One-tailed

b.

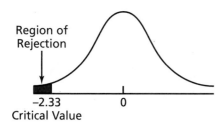

Region of Rejection

−2.33
Critical Value

0

c.

$$z = \dfrac{\dfrac{75}{100} - 0.80}{\sqrt{\dfrac{0.80(1 - 0.80)}{100}}}$$

$$= \dfrac{-0.05}{0.04} = -1.25$$

Do not reject the null hypothesis. The rehabilitation program is not effective.

10-5 a. $H_0: \pi_1 = \pi_2$

$H_a: \pi_1 \neq \pi_2$

b.

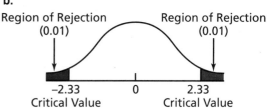

Region of Rejection
(0.01)

Region of Rejection
(0.01)

−2.33
Critical Value

0

2.33
Critical Value

c.

$$z = \dfrac{\dfrac{150}{250} - \dfrac{232}{400}}{\sqrt{0.59(1 - 0.59)\left(\dfrac{1}{250} + \dfrac{1}{400}\right)}}$$

$$= \dfrac{0.60 - 0.58}{\sqrt{0.00157}}$$

$$= 0.50$$

Fail to reject the null hypothesis. The new capsule is no different than the old. The p-value is 0.6170, found by $2(0.5000 - 0.1915)$.

Unit Review

The last three chapters dealt with the fundamental concepts of statistical inference. They described (1) how we can use sampling to develop a range of values within which a population parameter is likely to occur and (2) the procedures for testing hypotheses about population parameters.

KEY CONCEPTS

1. **Sampling** is important because complete knowledge regarding the population is seldom available. The basic purpose of sampling is to provide an estimate about a population parameter based on sample evidence.
2. Sampling is necessary because
   a. it may be impossible to check the entire population;
   b. the cost to study the entire population may be prohibitive;
   c. to contact the entire population would be too time consuming;
   d. the tests may destroy the product.
3. A **confidence interval** is a range of values within which the population parameter is expected to fall for a preselected level of confidence.
4. The **sampling distribution of the mean** is a probability distribution that describes all possible sample means of a given size selected from a population.
5. The **central limit theorem** states that the distribution of the sample mean is approximately normal, regardless of the shape of the population.
6. The **standard error of the sampling distribution of the sample mean** is the standard deviation of the distribution of sample means.
7. The **standard error of the sample proportion** is the standard deviation of the sampling distribution of sample proportions.

8. A **point estimate** is the value computed from a sample that is used to estimate a population parameter. An **interval estimate** is a range of values within which we have some assurance that the population parameter lies.
9. A **proportion** is a fraction, percentage, or ratio that indicates what part of a sample or population has a certain trait.
10. When the sample constitutes more than 5% of the population, the **finite population correction factor** is used. Its purpose is to reduce the standard error because the sample constitutes a large portion of the population. The term $\sqrt{(N - n)/(N - 1)}$ is multiplied by the standard error of the mean or proportion.
11. **Hypothesis testing** is an extension of the concept of interval estimation and offers a strategy for choosing among alternative courses of action. For the purpose of testing, two claims are made about a population parameter. One is called the **null hypothesis** and the other is called the **alternate hypothesis**.
12. Two types of errors may be committed during a test of the null hypothesis. If the null hypothesis is rejected when it is true, a **Type I error** is committed. If the null hypothesis is accepted when it is false, a **Type II error** is committed.
13. The **level of significance** is the probability of rejecting the null hypothesis when it is true. The **test statistic** is a quantity calculated from sample information and used as a basis for deciding whether or not to reject the null hypothesis. The **decision rule** is a statement of the conditions under which the null hypothesis is rejected. A **critical value** or **critical values** separate the rejection region from the remaining values.
14. The *p*-value is the probability (assuming the null hypothesis is true) of getting a value of the test statistic at least as extreme as that found in the sample.

KEY TERMS

Probability sampling
Sample frame
Sample design
Sample survey
Census
Control group
Experimental group
Random sample
Table of random digits
Systematic random sample
Stratified random sample
Sampling error
Sampling distribution of the
sample mean

Sampling distribution of the
sample proportion
Central limit theorem
Confidence interval
Standard error of the sample
mean
Decision rule
Two-tailed test
Critical value
Point estimate
Interval estimate
Standard error of the sample
proportion

Finite population correction
factor
Null hypothesis
Alternate hypothesis
Type I error
Type II error
Level of significance
Test statistic
p-value
One-tailed test
Proportion

KEY SYMBOLS

$\sigma_{\bar{x}}$ The standard error of the mean.

σ_p The standard error of the sample proportion.

E Maximum allowable error.

H_0 The null hypothesis.

H_a The alternate hypothesis.

π The proportion of objects or things in the population that possesses a particular trait.

$\bar{p}$ The pooled estimate of the population proportion based on the two samples.

x The number of successes in a sample of n observation.

CASE STUDY   Excessive Consumer Debt

Banks, financial institutions, and financial planners, as well as psychologists, are concerned with the amount of debt in the United States. In 1996, the amount of credit offered consumers by banks totaled more than $1.02 trillion. The average credit-card holder carries a balance of $3900, and only about one-third of the customers pay their balances at the end of the month. Wendy Lamberg, president of the Genoa Savings Bank, would like to know how customers at her small bank in Tennessee compare with the national statistics given above. Do Genoa

Savings customers, on the average, have larger balances? Do fewer customers pay off their balances at the end of the month? A random sample of 50 Genoa customers with VISA cards serviced by the bank revealed the following data: the customers' balances at the end of last month and whether they paid their entire balances the previous month. Write a brief report to Ms. Lamberg summarizing the experience at Genoa. Be sure to report the hypotheses tested, the p-values, and the conclusions.

Customer	Balance	Paid
1	3880	1
2	3215	0
3	5081	1
4	5019	0
5	4332	1
6	4532	1
7	3378	0
8	4806	1
9	5987	0
10	4167	0
11	2673	0
12	2990	1
13	4430	0
14	3563	0
15	4286	0
16	2829	1
17	4250	1
18	4716	0
19	5315	1
20	5166	0
21	2681	0
22	3366	0
23	4858	1
24	4528	0
25	3732	1
26	5572	1
27	3802	1
28	4011	0
29	4938	0
30	2551	0
31	6984	0
32	4037	1
33	2707	0
34	4343	1
35	5580	0
36	3322	1
37	2454	0
38	4355	1
39	2974	0
40	3101	1
41	3816	1
42	3861	1
43	3562	1
44	4141	0
45	5959	0
46	4687	1
47	5340	0
48	4478	1
49	2983	0
50	5495	1

Hypothesis Tests: Small-Sample Methods

OBJECTIVES

When you have completed this chapter, you will be able to

- describe the *t* distribution;
- conduct a hypothesis test for a population mean when the sample size is less than 30 and the population standard deviation is unknown;
- conduct a hypothesis test for the difference between two population means when either sample is less than 30 and the population standard deviations are unknown;
- conduct a hypothesis test for the difference between paired observations when the sample size is less than 30.

CHAPTER PROBLEM   "We'll Mufflerize in 30 Minutes or Less!"

Tidy Muffler Shops claim on their national radio and TV commercials that they can "mufflerize" any car in 30 minutes or less. The Consumer Protection Agency would like to investigate the validity of this claim. Is the mean time it takes Tidy to install a new muffler more than 30 minutes?

INTRODUCTION

In Chapter 10, we described hypothesis tests from a "large" sample—generally considered by most statisticians to be of size 30 or larger. In many situations, however, there is interest in testing a hypothesis about a normally distributed interval-level variable, but the population standard deviation is not known and the sample size is small—that is, under 30. In such cases, the z statistic is not appropriate because σ is unknown or cannot be accurately estimated based on only a few observations. In its place, we use the **Student's t distribution.** This distribution first appeared in the literature in 1908, when W. S. Gosset, an Irish brewery employee, published a paper about the distribution under the pseudonym "Student." In his paper, Gosset assumed that the small samples were taken from normal populations with unknown standard deviations. Approximate results are also obtained when we sample from nonnormal populations. To begin let's look at the major characteristics of the t distribution.

CHARACTERISTICS OF THE t DISTRIBUTION

The t distribution is similar to the z (standard normal) distribution in some respects, but is quite different in others. It has the following major characteristics:

1. It is a continuous distribution like the standard normal or z distribution used in Chapter 10.
2. It is bell-shaped and symmetrical, again similar to the standard normal distribution. The means of both z and t are zero.
3. There is only one standard normal distribution, z, but there is a "family" of t distributions; that is, each time the size of the sample changes, a new t distribution is created.
4. The t distribution is more spread out at the center (that is, "flatter") than the normal distribution.

Because the t distribution has a greater spread than the z distribution, the critical values of t for a particular significance level are greater in magnitude than the corresponding critical values of z. As the size of the sample increases, the value of t approaches the value of z for a particular significance level. For example, the critical values of a one-tailed test with a significance level of 0.05 are shown in Figure 11–1. The sample sizes of 3 and 15 and the z-value corresponding to a one-tailed test are shown. Note that the t-value for a sample size of 3 is 2.920. For a sample of 15, the t-value is 1.761. Because the first sample size is smaller than the second, its critical value is larger.

FIGURE 11–1 **Critical Values for Corresponding *t* and *z* Distributions**

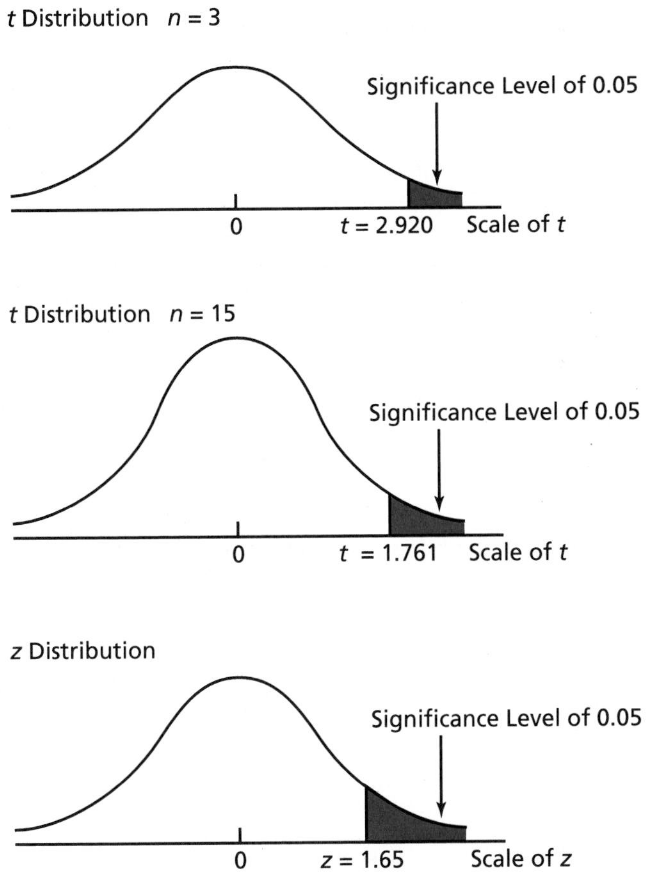

REAL STAT

Do you live to work or work to live? A recent poll of 802 working Americans revealed that, among those who considered their work a career, the mean number of hours worked per day was 8.7 and that among those who considered their work a job, the mean number of hours worked per day was 7.6.

TESTING A HYPOTHESIS ABOUT A POPULATION MEAN

The *t* distribution is used to test a hypothesis about a population mean when the population standard deviation is not known, the sample size is small, and the population is normal. The formula is

$$t = \frac{\overline{X} - \mu}{s / \sqrt{n}} \qquad \textbf{11–1}$$

where

$\overline{X}$ is the sample mean.
μ is the hypothesized population mean.
s is the sample standard deviation.
n is the number of items in the sample.

Note that in the formula for t, the sample standard deviation (s) is used instead of the population standard deviation (σ).

Problem   Recall from the Chapter Problem that Tidy Muffler Shops claim they can "mufflerize" any car in 30 minutes or less. The Consumer Protection Agency decides to investigate this claim. Ten of its fleet of cars, all unmarked and in need of new mufflers, are sent to various Tidy Muffler Shops throughout Denver. The times it took (in minutes) to replace the muffler on each of the 10 vehicles were

| 26 | 32 | 24 | 37 | 28 | 29 | 33 | 31 | 34 | 36 |

Does this evidence indicate that the mean time for Tidy Muffler Shops to replace a muffler is more than 30 minutes, or are they meeting their advertised claim?

Solution   We will use the five-step hypothesis-testing procedure.

Step 1. *The first step is to formulate the null and alternate hypotheses.* The null hypothesis is that the mean time to "mufflerize" a car is 30 minutes or less. The alternate hypothesis is that the mean time is more than 30 minutes. Stated symbolically:

$$H_0 : \mu \leq 30$$

$$H_a : \mu > 30$$

The way the alternate hypothesis is stated dictates a one-tailed test.

Step 2. *Select the level of significance.* It is 0.05.

Step 3. *Choose the appropriate test statistic.* The t distribution is used because the population standard deviation is not known and the sample size is small (10) and we assume a normal population. The value of the test statistic is computed by using formula 11–1.

$$t = \frac{\overline{X} - \mu}{s / \sqrt{n}}$$

Step 4. *State the decision rule.* A critical value separates the region where the null hypothesis is rejected from the region where it is not rejected. Appendix D gives the critical values for Student's t distribution. To use the table, we must determine the number of *degrees of freedom* (df). The number of degrees of freedom is equal to the sample size minus the number of samples. One sample of size 10 is used in this problem. Thus, the degrees of freedom are $n - 1 = 10 - 1 = 9$. Why is 1 subtracted from the sample size? Suppose a sample of 4 measurements is obtained and the values are 4, 6, 8, and 10. The mean of the sample is 7. If any 3 of the values are changed, then the fourth value is automatically fixed so that the mean is still 7. Suppose the first 3 values were changed to 3, 5, and 9. The last number must be 11 for the 4 numbers to have a mean of 7; that is, $(3 + 5 + 9 + 11)/4 = 28/4 = 7$. Because only 3 out of the 4 numbers are free to vary, it is said that there are 3 "degrees of freedom." For this example, the degrees of freedom are determined by $n - 1 = 4 - 1 = 3$ df.

The critical values for Student's t distribution are given in Appendix D. Table 11–1 excerpts a few of these critical values. Note that both two-tailed and one-tailed values are given. To locate the critical value of t, go down the left margin in Table 11–1, labeled DEGREES OF FREEDOM (df), until you locate $n - 1$ degrees of freedom. There are 10 cars being "mufflerized" in the sample, so $10 - 1 = 9$ degrees of freedom. Then move to the right, and read the value given under the ONE-TAILED VALUE column headed by 0.05. The

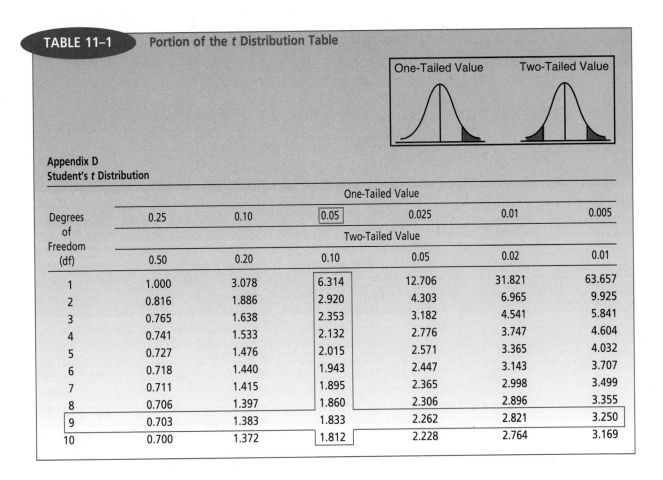

TABLE 11–1   Portion of the t Distribution Table

Appendix D
Student's t Distribution

Degrees of Freedom (df)	One-Tailed Value					
	0.25	0.10	0.05	0.025	0.01	0.005
	Two-Tailed Value					
	0.50	0.20	0.10	0.05	0.02	0.01
1	1.000	3.078	6.314	12.706	31.821	63.657
2	0.816	1.886	2.920	4.303	6.965	9.925
3	0.765	1.638	2.353	3.182	4.541	5.841
4	0.741	1.533	2.132	2.776	3.747	4.604
5	0.727	1.476	2.015	2.571	3.365	4.032
6	0.718	1.440	1.943	2.447	3.143	3.707
7	0.711	1.415	1.895	2.365	2.998	3.499
8	0.706	1.397	1.860	2.306	2.896	3.355
9	0.703	1.383	1.833	2.262	2.821	3.250
10	0.700	1.372	1.812	2.228	2.764	3.169

critical value for 9 df and a one-tailed test is 1.833 for the 0.05 level of significance. Therefore, the decision rule is this: Reject H_0 if the computed value of t is greater than 1.833. Shown schematically:

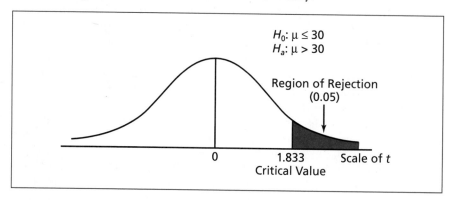

$H_0: \mu \leq 30$
$H_a: \mu > 30$

Region of Rejection
(0.05)

0   1.833   Scale of t
Critical Value

Step 5. *Make the statistical decision.* The mean length of time it took Tidy Muffler to replace the mufflers on the Consumer Protection Agency cars is 31 minutes, found by 310/10. The standard deviation of the sample of 10 is 4.24 minutes, found by applying formula 4–8:

$$s = \sqrt{\frac{\Sigma X^2 - \dfrac{(\Sigma X)^2}{n}}{n - 1}}$$

The essential calculations to compute s are shown in Table 11–2 at the top of the next page.

The computed value of t is 0.746, found by using formula 11–1:

$$t = \frac{\overline{X} - \mu}{s / \sqrt{n}} = \frac{31 - 30}{4.24 / \sqrt{10}} = 0.746$$

Comparing the computed t-value of 0.746 with the critical value of 1.833, we fail to reject the null hypothesis. Although there is a difference between the hypothesized mean (30) and the sample mean (31), it is attributed to sampling error. The Consumer Protection Agency does not have sufficient statistical evidence to disprove Tidy Muffler's claim that its shops are able to "mufflerize" a car in 30 minutes on the average.

In the above Problem, we used the five-step hypothesis-testing procedure to illustrate the t distribution. We can also use the probability, or p-value, approach to do this test. Recall that the p-value is the probability of a value of the test statistic as large or larger than the computed value of t when the null hypothesis is true. Computer software programs, such as MINITAB and Excel, can determine exact p-values. However, the t distribution in Appendix D and Table 11–1 includes only selected critical values, so we usually cannot find the exact p-value. Nevertheless, we can find limits on p-values by scanning the appropriate row of the table based on the degrees of freedom and comparing the computed value of the test statistic with the critical values found in the table.

TABLE 11–2	Essential Calculations for the Sample Standard Deviation	
Time (in minutes)		
X	**X²**	
26	676	
32	1024	
24	576	$s = \sqrt{\dfrac{\Sigma X^2 - \dfrac{(\Sigma X)^2}{n}}{n-1}}$
37	1369	
28	784	
29	841	$= \sqrt{\dfrac{9772 - \dfrac{(310)^2}{10}}{10-1}}$
33	1089	
31	961	
34	1156	$= 4.24$
36	1296	
310	9772	

In the case where $t = 0.746$, we used a one-tailed test with 9 degrees of freedom. Look across the row with 9 degrees of freedom. The value of the test statistic is between 0.703 and 1.383, which correspond to one-tailed significance levels of 0.25 and 0.10, respectively. Refer to the diagram below. So the p-value is more than 0.10, but less than 0.25. The exact p-value is 0.24, calculated by MINITAB (see p. 343). Thus, by using Appendix D, we can get a reasonable estimate of the exact p-value.

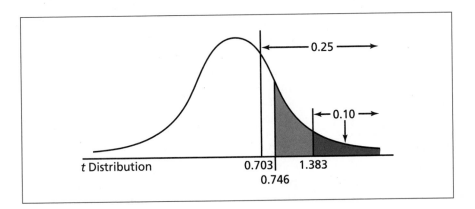

Answers to the Self-Review problems are at the end of the chapter.

Self-Review 11–1

The American Association of Retired Persons reported that typical (average) senior citizens living at high altitudes claim their systolic blood pressures are lower than the arithmetic mean of 160. To test this claim, 16 senior citizens living at high altitudes were selected at random and their blood pressures checked. Their mean systolic pressure was 151; the standard deviation of the sample was 12. Do senior citizens living at high altitudes have significantly *lower* systolic blood pressures? Use the 0.05 level and assume the distribution is normal.

a. State the null hypothesis and alternate hypothesis symbolically.
b. Give the formula for the test statistic.
c. How many degrees of freedom are there?
d. Give the decision rule.
e. Compute t.
f. Arrive at a conclusion.
g. What is the p-value? Interpret it.

Exercises

Answers to the even-numbered Exercises are at the back of the book.

1. Find the critical value of t for the following hypothesis-testing situation when the sample size is 20. Use the 0.05 significance level.

$$H_0: \mu \le 18$$

$$H_a: \mu > 18$$

2. Find the critical value of t for the following hypothesis-testing situation when the sample size is 15. Use the 0.01 significance level.

$$H_0: \mu = 50$$

$$H_a: \mu \ne 50$$

3. A car manufacturer asserts that with the new collapsible bumper system, the mean body repair cost for the damage sustained in a collision impact of 15 miles per hour does not exceed $400. To test the validity of this claim, 6 cars are crashed into a barrier at 15 miles per hour and their repair costs recorded. The mean and the standard deviation are found to be $458 and $48, respectively. At the 0.05 level of significance, do the test data contradict the manufacturer's claim that the repair cost does not exceed $400? Assume a normal population. Find the p-value. Interpret it.

4. It is claimed that a new treatment for prolonging the lives of cancer patients is more effective than the standard one. Records of earlier research show the mean survival period to have been 4.3 years with the standard treatment. The new treatment is administered to a sample of 20 patients, and the durations of their survival are recorded. The sample mean is 4.6 years, and the standard deviation is 1.2 years. Is the claimed effectiveness of the new method supported at the 1% level of significance? Assume a normal population.

5. A manufacturer of strapping tape claims that the tape has a mean breaking strength of 500 pounds per square inch (psi). A random sample of 16 specimens is drawn from a large shipment of tape, and a mean of 480 psi is computed, with the sample standard deviation of 50 psi. Can we conclude from these data that the mean breaking strength for this shipment is less than that claimed by the manufacturer? Use the 0.05 significance level. Assume a normal population.

6. An anthropologist is studying the heights of the adult members of an ancient population. The conventional theory is that the mean height of adult men from the population is 56 inches. A sample of 12 men showed a mean of 53.5 inches, with a standard deviation of 2.0 inches. Can the anthropologist conclude that the mean height is less than 56 inches? Use the 0.05 significance level. Assume a normal population. What is the p-value? Interpret it.

A Computer Example

MINITAB can be used to perform the calculations for a test of hypothesis. The first step is to enter the data. Next, name the variable. In this case, because the times are reported in minutes, the variable is also called Minutes. Using

MINITAB, click on **Stat,** then **Basic Statistics,** and finally **1-Sample t.** We will often show these steps as

Stat ▶ Basic Statistics ▶ 1-Sample t

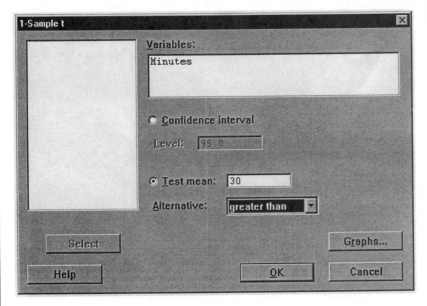

Within the dialog box, select Minutes as the variable by highlighting it with the mouse, then click on the box marked **Select,** next click on the box marked **Test mean,** and finally type "30" in the box to the right to indicate 30 is the hypothesized population mean. Then in the box marked Alternative, select **greater than.** The final step is to click on **OK.** The output is as follows:

```
MTB > set c1
DATA> 26 32 24 37 28 29 33 31 34 36
DATA> end
MTB > name c1 'Minutes'
MTB > TTest 30 'Minutes';
SUBC>   Alternative 1.

T-Test of the Mean

Test of mu = 30.00 vs mu > 30.00

Variable      N      Mean     StDev    SE Mean        T          P
Minutes      10     31.00      4.24       1.34     0.75       0.24
```

The computed value of t is 0.75, the same as we calculated earlier. Included in the output is the p-value. The value of 0.24 is the probability of t being greater than 0.75, with 9 degrees of freedom. The p-value is larger than the significance level of 0.05. So the decision is not to reject the null hypothesis.

A new toy has been developed, and its manufacturer hopes to market it for the coming Christmas season. Before going into full production, the company hand-crafts a large number of toys and sends them to 5 test market areas. The manufacturer plans to start full production if the monthly sales in the test markets average more than $20,000 during the 1-month trial period. The results (in thousands of dollars) were $20, $16, $25, $19, and $24. For the 0.05 level of significance, can it be shown that the mean monthly sales volume is greater than $20,000?

a. State the null and alternate hypotheses.
b. Is this a one-tailed or a two-tailed test?
c. Why is Student's *t* being used and what assumption do we need to make?
d. Calculate the value of the test statistic.
e. What is your decision regarding the null hypothesis?
f. Interpret your results for the toy manufacturer.
g. What is the *p*-value? Interpret it.

Exercises

7. A city health department wishes to determine if the mean bacteria count per unit-volume of water at Siesta Lake Beach is below the safety level of 200. Researchers have collected 10 water samples and have found the bacteria counts per unit-volume to be 175, 180, 215, 188, 194, 207, 211, 195, 198, and 190. Are the data cause for concern? Use the 0.10 level of significance.

8. Sorenson Pharmaceutical has been conducting restricted studies on small groups of people to determine the effectiveness of a measles vaccine. The following measurements are readings on the antibody strength for 5 individuals injected with the vaccine: 1.2, 2.5, 1.9, 3.0, and 2.4. Use the sample data to test at the 0.01 level of signifi-cance the hypothesis that the mean antibody strength for individuals vaccinated with the new drug is more than 1.6.

9. In 5 test runs, a truck operated 8, 10, 7, 9, and 10 miles with 1 gallon of gasoline. At the 0.01 significance level, is this sufficient evidence to show that the truck is operating at a mean rate less than 11.5 miles per gallon? Find the *p*-value.

10. The credit manager of a discount store chain believes the mean age of the stores' customers is less than 40 years. A random sample of 8 customers revealed the following ages: 38, 46, 30, 35, 29, 40, 40, and 46. At the 0.05 significance level, can we conclude that the mean age of customers is less than 40 years?

COMPARING TWO POPULATION MEANS

We are often interested in comparing two sets of data to see if they are from populations with the same mean. For example, AAA wants to know if the mean price of regular unleaded gasoline is the same in Tampa as Atlanta, a statistics instructor wants to know if the mean score on a test is the same in the morning and the afternoon sections, and a plant manager wants to know if the mean number of units produced is the same on the day and the afternoon shifts. We use the *t* distribution to make these kinds of comparisons. We conduct the test by suggesting that the two populations are the same. We assume

1. both populations are normally distributed;
2. their standard deviations are equal, but unknown; and
3. the two samples are unrelated (independent).

The computed t statistic for the 2-sample case with small samples is similar to the 2-sample z statistic in Chapter 10. However, an additional calculation is required. The 2 sample standard deviations (1 from each sample) are "pooled" to form a single estimate of the population standard deviation. This is done to meet the second requirement above: that the two populations have equal standard deviations. Usually, this test is used if either sample has fewer than 30 observations. The formula for t is

$$t = \frac{\overline{X}_1 - \overline{X}_2}{s_p \sqrt{\dfrac{1}{n_1} + \dfrac{1}{n_2}}}$$

11–2

where

$\overline{X}_1$ is the mean of the first sample.
$\overline{X}_2$ is the mean of the second sample.
n_1 is the number in the first sample.
n_2 is the number in the second sample.
s_p is a pooled estimate of the population standard deviation. Its formula is

$$s_p = \sqrt{\frac{(n_1 - 1)s_1^2 + (n_2 - 1)s_2^2}{n_1 + n_2 - 2}}$$

11–3

where

s_1 is the standard deviation of the first sample.
s_2 is the standard deviation of the second sample.

The number of degrees of freedom is equal to the total number of items sampled minus the number of samples. The sample sizes for this test are n_1 and n_2, and there are two samples. Hence, there are $n_1 + n_2 - 2$ degrees of freedom appearing in the denominator of the pooled standard deviation.

Problem   Katy Fit, aerobics instructor at the Anderson Recreation Center, is analyzing the attendance at her classes. She wonders if the mean number of people who come on Monday morning is the same as on Wednesday morning. Some people may be exhausted from their other weekend activities and skip Mondays. Yet others may get busy midweek and be unable to fit exercise into their schedules. Attendance data for 5 randomly selected Mondays and 6 randomly selected Wednesdays are reported in Table 11–3. The sample means are 5 and 6.5. Is there a significant difference in the mean numbers of people who come to class on the 2 days? Test at the 0.01 significance level.

TABLE 11–3	Aerobics Attendance	
	Monday	**Wednesday**
	4	9
	3	6
	5	8
	7	4
	6	7
		5

Solution The null hypothesis is that the means are equal on the two days. The alternate hypothesis is that the means differ between the two days. Stated symbolically:

$$H_0 : \mu_1 = \mu_2$$

$$H_a : \mu_1 \neq \mu_2$$

The alternate hypothesis indicates that this is a two-tailed test. It is reasonable to suppose that both populations are normally distributed with the same standard deviation.

Again the decision rule depends on the number of degrees of freedom. In this case, that number is equal to the combined number of observations in the 2 samples minus the number of samples. This is expressed as $n_1 + n_2 - 2$. For this problem, $n_1 + n_2 - 2 = 5 + 6 - 2 = 9$ degrees of freedom.

The critical value for the 0.01-level, two-tailed test is 3.250 (from Appendix D). The decision rule is that we reject H_0 if $t < -3.250$ or $t > 3.250$. Shown schematically:

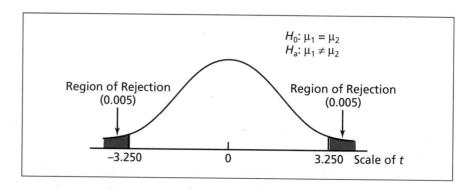

The calculation of Student's t can be accomplished in three stages. First, calculate the standard deviation of each sample. Second, "pool" these standard deviations into a single estimate of the population standard deviation using formula 11–3. Third, calculate t by using formula 11–2.

Stage 1. Calculate the sample standard deviations using these sums and squares:

Monday		Wednesday	
X_1	X_1^2	X_2	X_2^2
4	16	9	81
3	9	6	36
5	25	8	64
7	49	4	16
6	36	7	49
		5	25
25	135	39	271

$$s_1 = \sqrt{\frac{\Sigma X_1^2 - \frac{(\Sigma X_1)^2}{n_1}}{n_1 - 1}}$$

$$= \sqrt{\frac{135 - \frac{(25)^2}{5}}{5 - 1}}$$

$$= 1.58$$

$$s_2 = \sqrt{\frac{\Sigma X_2^2 - \frac{(\Sigma X_2)^2}{n_2}}{n_2 - 1}}$$

$$= \sqrt{\frac{271 - \frac{(39)^2}{6}}{6 - 1}}$$

$$= 1.87$$

Stage 2. Pool the standard deviations by using formula 11–3:

$$s_p = \sqrt{\frac{(n_1 - 1)s_1^2 + (n_2 - 1)s_2^2}{n_1 + n_2 - 2}}$$

$$= \sqrt{\frac{(5 - 1)(1.58)^2 + (6 - 1)(1.87)^2}{5 + 6 - 2}}$$

$$= \sqrt{3.05}$$

$$= 1.746$$

Stage 3. Calculate t using formula 11–2:

$$t = \frac{\overline{X}_1 - \overline{X}_2}{s_p\sqrt{\frac{1}{n_1} + \frac{1}{n_2}}} = \frac{5.0 - 6.5}{1.746\sqrt{\frac{1}{5} + \frac{1}{6}}} = -1.42$$

Because -1.42 falls between -3.250 and 3.250, the null hypothesis is not rejected. There is no statistically significant difference in the mean numbers of people attending class on the 2 days. The difference in the sample means could easily happen by sampling variation.

To find the p-value for this problem, refer to the row with 9 degrees of freedom in Appendix D or Table 11–1. The test statistic 1.42 is greater than 1.383, but smaller than 1.833. For a two-sided test, those values have significance levels of 0.20 and 0.10, so the p-value is between 0.10 and 0.20.

A Computer Example

The MINITAB output for the Katy Fit problem is shown below. To begin we enter the data in columns c1 and c2. The columns are named Monday and Wed-day. (We use the name Wed-day for Wednesday because MINITAB allows only eight letters for any variable name.) Next, we click the following mouse commands:

Stat ▶ Basic Statistics ▶ 2-Sample t

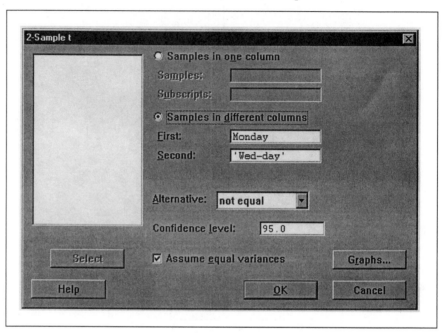

Within the dialog box, we select the samples in different columns option and click on **c1 Monday, Select, c2 Wed-day,** and **Select.** To complete the process we click on the **Assume equal variances** box and then click on **OK.** The output is as follows:

```
MTB > set c1
DATA> 4 3 5 7 6
DATA> end
MTB > set c2
DATA> 9 6 8 4 7 5
DATA> end
MTB > name c1 'Monday' c2 'Wed-day'
MTB > TwoSample 95.0 'Monday' 'Wed-day';
SUBC>    Alternative 0;
SUBC>    Pooled.
```

Two Sample T-Test and Confidence Interval

```
Two sample T for Monday vs Wed-day
            N      Mean     StDev   SE Mean
Monday      5      5.00      1.58      0.71
Wed-day     6      6.50      1.87      0.76

95% CI for mu Monday - mu Wed-day: ( -3.89,  0.89)
T-Test mu Monday = mu Wed-day (vs not =): T= -1.42  P=0.19  DF=  9
Both use Pooled StDev = 1.75
```

The computed value of t, -1.42, is the same as that calculated earlier. The actual p-value is 0.19, which is within the limits 0.10 to 0.20 we defined earlier.

A visual representation of the original data is shown below, where the letter M stands for the attendance on a Monday, W stands for the attendance on a Wednesday, and a □ is placed at the sample mean:

```
            W      W      W      W      W      W
     M      M      M      M      M
   — 3 ———— 4 ———— 5 ———— 6 ———— 7 ———— 8 ———— 9
                   □             □
```

Notice that, even though the sample mean of the Wednesday group (6.5) is higher than that of the Monday group (5), there is such variation from day to day that we cannot safely say the population means are different. When would the null hypothesis be rejected? There are two factors that affect this decision. The null hypothesis might be rejected if the Wednesday group regularly got 2 more people to attend class. Then the sample mean for the Wednesday group would increase to 8.5 because each of the data points is now 2 units higher. The new t statistic would then be -3.31, leading to rejection of the null hypothesis because -3.31 is more extreme than -3.25. Here is a "picture" of the new situation so you can see why the decision would change:

```
                   W      W      W      W      W      W
     M      M      M      M      M
   — 3 ———— 4 ———— 5 ———— 6 ———— 7 ———— 8 ———— 9 ———— 10 ———— 11 —
                   □                    □
```

The null hypothesis also might be rejected if there was less variation within the sample attendance. Then the pooled estimate of the population variance in the denominator of the statistic would be smaller. Consequently, the statistic itself would get larger. Consider, for example, the following possible samples in this low variation case:

```
                          M
                   M      W      W
                   M      W      W
            M      M      W      W
   — 3 ———— 4 ———— 5 ———— 6 ———— 7 ———— 8 ———— 9
                   □             □
```

The sample means (5 and 6.5) are the same as in the original data. However, the variances are now much smaller. The standard deviation for Monday is 0.7071, and the standard deviation for Wednesday is 0.5471, making the "pooled" standard deviation 0.6236. Finally, the value of the t statistic becomes -3.98, which would lead to rejection of the null hypothesis. Repeat the detailed calculation for yourself to confirm this part.

In summary, then, this t statistic is the ratio of the difference between two sample means to the pooled standard deviation. A change in either may alter the statistical decision. These ideas will come up again in Chapter 12, where we compare several means using a single simple statistic that is the ratio of two variances.

Self-Review 11–3

An orthodontist wants to investigate the effectiveness of the interceptive treatment she prescribes for some of her patients. (Interceptive treatment is dental work performed on relatively young patients in hopes of forestalling more extensive treatment later.) She compares the length of time a sample of interceptive patients must wear braces with a random sample of noninterceptive patients. The results (time is in months):

Time Wearing Braces			
Interceptive		Noninterceptive	
Joseph	12	Sally	16
Karen	13	George	22
Pam	11	Enrico	14
Peter	12	Aldine	18
Nickie	14	Jenny	19
Rosa	9		
Kurt	11		

REAL STAT

Supplying soldiers with enough water was a major problem during Operation Desert Storm. The typical American consumes an average of 8 8-oz. glasses of water per day. The typical soldier in Desert Storm consumed an average of 96 8-oz. glasses of water. Is the increase in water consumption significant?

Is there sufficient evidence to indicate that interceptive patients spend less time in braces? Use the 0.05 significance level.

a. State the null and alternate hypotheses. Use the subscript $_1$ to refer to the interceptive group.
b. What assumption is necessary and how many degrees of freedom are there in this problem?
c. Calculate the two sample standard deviations.
d. Compute the pooled estimate of the standard deviation.
e. What is the decision rule?
f. Calculate the value of the test statistic.
g. What is your decision regarding the null hypothesis?
h. Interpret the results.

Exercises

11. Test the following hypotheses:

$$H_0 : \mu_1 = \mu_2$$

$$H_a : \mu_1 \neq \mu_2$$

given the following information: $n_1 = 21, \overline{X}_1 = 51, s_1 = 11, n_2 = 26, \overline{X}_2 = 42, s_2 = 17$.

a. Using the 0.05 significance level, what is the decision rule?
b. Find s_p.
c. Compute the value of t.

d. What is your decision regarding the null hypothesis?

12. Test the following hypotheses:

$$H_0 : \mu_1 \le \mu_2$$

$$H_a : \mu_1 > \mu_2$$

given the following information: $n_1 = 11$, $\overline{X}_1 = 41$, $s_1 = 9$, $n_2 = 13$, $\overline{X}_2 = 32$, $s_2 = 15$.

a. Using the 0.01 significance level, what is the decision rule?

b. Find s_p.

c. Compute the value of t.

d. What is your decision regarding the null hypothesis?

13. The peak oxygen intake per unit of body weight, called the "aerobic capacity," of an individual performing a strenuous activity is a measure of work capacity. For a comparative study, measurements of aerobic capacities are recorded for a group of 20 highland natives from Peru and for a group of 10 from the United States. The following summary statistics were obtained from the data:

	Peruvians	Americans
Mean	46.3	38.5
Standard deviation	5.0	5.8

At the 0.05 level of significance, do the sample data indicate a significant difference in the mean aerobic capacity?

14. To compare the effectiveness of isometric and isotonic exercise methods of abdomen reduction, 20 overweight business executives are included in an experiment. Ten use each type of exercise, and after 10 weeks, the reductions in abdomen measurements are recorded in centimeters:

	Isometric	Isotonic
Mean	2.5	3.1
Standard deviation	0.8	1.0

At the 0.01 level of significance, do these data support the claim that the isotonic method is more effective? Find the p-value.

15. Do cars traveling in the right lane of I-75 travel slower than those in the left lane? The following sample information was obtained. Data are reported in miles per hour.

Right Lane	Left Lane
65	69
70	72
60	68
62	73
68	64
	68

Use the 0.05 significance level.

16. The budget director of a large firm compared the luncheon expense vouchers for executives in the sales department with those for executives in the production department. The following sample information was gathered (rounded to the nearest dollar):

Sales: $17, 23, 26, 30
Production: $19, 15, 14, 17, 20

At the 0.05 significance level, can we conclude that luncheon expenses are less for executives in the production department?

TESTING WITH DEPENDENT OBSERVATIONS

In some experiments, the investigator is concerned with the *difference* in a pair of related observations. In other words, the samples are *dependent* or related. We can identify two types of dependent samples: (1) those characterized by "before" and "after" measurements and (2) those where the subjects have been "matched."

The first type of dependent sample is a measurement, followed by an intervention of some type, followed by another measurement of the same trait.

If, for example, we wished to show the effectiveness of a diet program in terms of weight loss, we would weigh the patients first, then have them follow the diet for a period, and finally weigh them again. The data used in the test would then be the *differences* between their weights before the diet and their weights after the diet.

The second type of dependent sample occurs when the researcher "pairs" subjects in an attempt to eliminate or control for the effects of other variables. If, for example, we wished to compare the intelligence of men and women, we might select married couples and administer an intelligence test to both the husband and the wife. In this way, we would hope to eliminate other effects, such as age, race, or socioeconomic status, and measure only the difference in intelligence between men and women.

The distribution of this population of differences is assumed to be normal, with an unknown standard deviation. The mean of this population of differences is designated μ_d. As pointed out earlier, it is often impossible to study the entire population of differences. Therefore, a sample is selected. The symbol d is used to designate a particular observed difference and $\overline{d}$ the mean difference. The formula for t is

$$t = \frac{\overline{d}}{s_d / \sqrt{n}}$$

11–4

where

$\overline{d}$ is the mean difference between the paired observations.
s_d is the standard deviation of the differences between the paired observations.
n is the number of paired observations.

Problem   ETM Chemical Company is considering the purchase of a new Colton word processor. Colton claims that its new-model word processor increases data entry efficiency. Ten data entry clerks randomly selected from the typing pool are assigned to test the new model. The results are:

Clerk	Proposed Model	Current Model
Woodstock	61	55
LaRoche	60	54
Hayes	56	47
Kootz	63	59
Kinney	56	51
Markas	63	61
Chung	59	57
Jaynes	56	54
Shue	62	63
Keller	61	58

Using the 0.01 level of significance, what would you conclude?

Solution   Before making the change to the new model, the company would like to know that the clerks can type *faster* on the proposed model. The difference, *d*, is $X - Y$, where X is the speed on the proposed model and Y the speed on the current model. The null hypothesis is that the clerks will perform at least as well on the currently used model. Colton hopes the null hypothesis will be rejected. Thus, a one-tailed test is necessary. The null and alternate hypotheses are

$$H_0 : \mu_d \leq 0$$

$$H_a : \mu_d > 0$$

Rejecting H_0 and accepting H_a would indicate that the mean difference is greater than zero. Hence, we would conclude data entry speeds on the proposed model are faster. The sample size, *n*, is equal to the number of paired observations (10). In this example, there are 9 (10 − 1) degrees of freedom. The decision rule is to reject H_0 if the computed value of *t* is to the right of 2.821 using the .01 significance level. (see Appendix D).

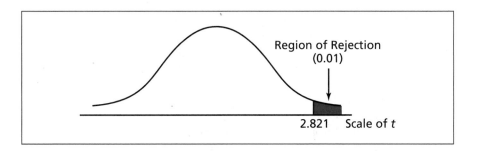

The sample data and the calculations to find *d* and d^2 are given in Table 11–4 on page 354. The mean difference is computed by formula 3–1, substituting *d* for *X*:

$$\overline{d} = \frac{\Sigma d}{n} = \frac{38}{10} = 3.8$$

The $\overline{d}$ of 3.8 indicates that the typical data entry clerk increased his or her speed by 3.8 words per minute. The sample standard deviation is found by using formula 4–8, repeated below. Inserting the sums from Table 11–4, the standard deviation of the differences is 2.82 minutes:

$$s_d = \sqrt{\frac{\Sigma d^2 - \frac{(\Sigma d)^2}{n}}{n-1}} = \sqrt{\frac{216 - \frac{(38)^2}{10}}{10 - 1}} = 2.82$$

The value of the test statistic, *t*, is 4.26, found by using formula 11–4:

$$t = \frac{\overline{d}}{s_d / \sqrt{n}} = \frac{3.8}{2.82 / \sqrt{10}} = \frac{3.8}{0.892} = 4.26$$

TABLE 11–4	Comparative Results of Efficiency			
Clerk	Proposed Model X	Current Model Y	X − Y = d	d^2
Woodstock	61	55	6	36
LaRoche	60	54	6	36
Hayes	56	47	9	81
Kootz	63	59	4	16
Kinney	56	51	5	25
Markas	63	61	2	4
Chung	59	57	2	4
Jaynes	56	54	2	4
Shue	62	63	−1	1
Keller	61	58	3	9
			38	216

Because 4.26 is to the right of 2.821, the null hypothesis is rejected. Since 4.26 > 3.25, the p-value is less than 0.005! The alternate hypothesis ($\mu_d > 0$) is accepted. Therefore, the conclusion is that the clerks perform more effectively on the proposed model.

Note that in Table 11–4, X represents the proposed model and Y the current model. The difference, d, is $X - Y$, where X is expected to be larger than Y. This leads to the positive direction of the alternate hypothesis. (The entire test could have been reversed if d were found by $Y - X$. In that case, the d values would have been negative, and the direction of the alternate hypothesis would have been negative.)

COMPARING DEPENDENT AND INDEPENDENT SAMPLES

Why do we prefer dependent samples over independent samples? The answer is that by using dependent samples, we are able to reduce the variation in the sampling distribution and produce a more powerful statistical test. We will use the ETM Chemical Company Problem/Solution just completed to illustrate.

If we assume the data entry speeds for the proposed model and the current model, from Table 11–4, are independent samples. To compute the value of the test statistic we use formula 11–2. The mean of the 10 data entry speeds on the proposed model is 59.7 words per minute, and the standard deviation is 2.83. For the current model, the mean of the 10 data entry speeds is 55.9 words per minute, and the standard deviation is 4.75. The value of the pooled estimate of the standard deviation is

$$s_p = \sqrt{\frac{(n_1 - 1)s_1^2 + (n_2 - 1)s_2^2}{n_1 + n_2}} = \sqrt{\frac{(10 - 1)2.83^2 + (10 - 1)4.75^2}{10 + 10 - 2}} = 3.91$$

Using formula 11–2, t is 2.173:

$$t = \frac{\overline{X}_1 - \overline{X}_2}{s_p\sqrt{\frac{1}{n_1} + \frac{1}{n_2}}} = \frac{59.7 - 55.9}{3.91\sqrt{\frac{1}{10} + \frac{1}{10}}} = \frac{3.80}{1.749} = 2.173$$

There are 2 samples of 10, so the degrees of freedom are $10 + 10 - 2 = 18$. From Appendix D, using the 0.01 significance level and a one-tailed test, the decision rule is to reject the null hypothesis if the computed value of t is greater than 2.552. But 2.173 is less than 2.552. So the null hypothesis is not rejected. We have *not* proved that the proposed model is more effective than the current model. That is not the same conclusion that we got before!

Why does this happen? The numerator is the same in the paired observations test. However, the denominator is smaller. In the paired test, the denominator is 0.892 (see the calculations on p. 353). In the case of the independent samples, the denominator is 1.749. This accounts for the difference in the t values and the difference in the statistical decisions. The denominator measures the "standard error" of the statistic. When the samples are not paired, two kinds of variation are present: differences among the operators and differences among the machine models. Hayes, for example, is slow on both machines, but Shue is fast on both machines. These data show how different the operators are, but we are really interested only in the difference between the two models of word processors.

The trick is to "pair" the Hayes measurements and the Shue measurements and all the other operators to eliminate the effect of the variation among the operators. The paired test uses only the observation of the difference between the two machines with the *same* operator. Thus, the paired or dependent statistic includes only the variation between machine models, and its standard error is always smaller. That in turn leads to a larger test statistic and a greater chance of rejecting the null hypothesis. So whenever possible you should "pair" the data.

There is a bit of bad news here. In the paired observations test, the degrees of freedom are half of what they are if the samples are not paired. For the ETM example, the degrees of freedom drop from 18 to 9 when the observations are paired. However, in most cases, this is a small price to pay for a better test.

- -

Self-Review 11–4

Advertisements for the Sylph Physical Fitness Center claim that completion of its course will result in a loss of weight. A random sample of recent students revealed the following body weights before and after completion of the course. Assume these populations follow a normal distribution. At the 0.01 level, can it be concluded that the course will result in a significant weight loss?

Name	Before	After
Wellman	155	154
Gersten	228	207
Tamayo	141	147
Miller	162	157
Ringman	211	196
Garbe	185	180
Monk	164	150
Heilbrunn	172	165

a. State the null and alternate hypotheses.
b. What is the critical value of t?
c. Calculate the value of the test statistic.
d. What is the decision regarding the null hypothesis?
e. Interpret the results.

Exercises

17. Test the following hypotheses:

$$H_0: \mu_d = 0$$

$$H_a: \mu_d \neq 0$$

when $n = 20$, $\overline{d} = 2.5$, $s_d = 2.3$, α "significance level" $= 0.05$.
a. What is the decision rule?
b. Find t.
c. Do you reject H_0?
d. What is the p-value?

18. Test the following hypotheses:

$$H_0: \mu_d \leq 0$$

$$H_a: \mu_d > 0$$

when $n = 15$, $\overline{d} = 1.5$, $s_d = 0.9$, $\alpha = 0.01$.
a. What is the decision rule?
b. Find t.
c. Do you reject H_0?
d. What is the p-value?

19. Two methods of memorizing difficult material are being tested. Nine pairs of students are matched according to IQ and background and then assigned to one of the two methods at random. A test is finally given to all the students, with the following results:

	Pair								
	1	2	3	4	5	6	7	8	9
Method A	90	86	72	65	44	52	46	38	43
Method B	85	87	70	62	44	53	42	35	46

Using the 0.05 level of significance, test to determine if there is a difference in the effectiveness of the two methods. What is the p-value?

20. Measurements of the left-handed and right-handed gripping strengths of 10 left-handed persons are recorded:

	Person									
	1	2	3	4	5	6	7	8	9	10
Left hand	140	90	125	130	95	121	85	97	131	110
Right hand	138	87	110	132	96	120	86	90	129	100

Do these data provide evidence, at the 0.01 level of significance, that those tested on average have greater gripping strength in their left hands?

21. A manufacturer of shock absorbers is comparing his model's durability with that of his competitor's models. To conduct the test the manufacturer installed one of his and one of a competitor's shock absorbers on each of 9 cars selected at random. Each of the cars was driven 20,000 miles. Then each shock was measured for strength. The results are shown below.

Car	Manufacturer's Shock	Competitor's Shock
1	10.0	9.6
2	11.7	11.2
3	13.7	13.1
4	9.9	9.4
5	9.8	9.0
6	14.4	14.0
7	15.1	14.6
8	10.6	10.3
9	9.8	9.4

At the 0.05 significance level, do the data provide evidence that the manufacturer's shocks on average last longer?

22. The Austin-Hall Department Store plans to renovate one of its floors, increasing either the men's or the women's clothing department. The final decision will be based on mean sales, and the department that has the greater sales will be enlarged. The last 12 months of sales data (in thousands of dollars) are shown below.

Month	Women's Clothing	Men's Clothing
Jan.	$20.0	$22.5
Feb.	14.0	13.9
Mar.	29.0	24.8
Apr.	30.0	27.5
May	16.0	24.9
June	17.0	27.4
July	20.1	21.4
Aug.	20.5	20.7
Sept.	21.3	20.6
Oct.	22.0	22.8
Nov.	25.7	23.5
Dec.	42.2	44.2

At the 0.05 significance level, can the department store conclude that there is a difference in the sales? Find the p-value.

. . . .

CHAPTER OUTLINE

I. The objective of tests of hypothesis using small samples is to check the validity of quantitative statements.

II. Student's t distribution is the appropriate test statistic for populations with unknown standard deviations.
   A. It is used when two conditions exist:
      1. The sample size is less than 30.
      2. The population or populations are normally or nearly normally distributed.
   B. The Student's t distribution has four major characteristics:

1. It is a continuous distribution.
2. It is bell-shaped and symmetrical.
3. There is a family of t distributions. All have the same mean—zero—but different standard deviations, depending on the sample size.
4. It is more spread out and flatter at the apex of the curve than the standard normal distribution.

III. The formula for a test of hypothesis about a population mean using Student's t distribution is

$$t = \frac{\overline{X} - \mu}{s / \sqrt{n}}$$ **11–1**

with $n - 1$ degrees of freedom, where

$\overline{X}$ is the sample mean.
μ is the hypothesized population mean.
s is the sample standard deviation.
n is the number in the sample.

IV. When using a test of hypothesis involving the difference between two population means, we make three assumptions:

   A. The observations in one sample are independent of those in the other sample and independent of each other.

   B. The two populations are normally distributed.

   C. The two populations have equal variances.

V. The t statistic is similar to the z statistic, but requires additional calculation.

   A. The formula for pooling the variances is

$$s_p = \sqrt{\frac{(n_1 - 1)(s_1^2) + (n_2 - 1)(s_2^2)}{n_1 + n_2 - 2}}$$ **11–3**

where

n_1 is the number of observations in the first sample.

n_2 is the number of observations in the second sample.
s_1^2 is the variance of the first sample.
s_2^2 is the variance of the second sample.

   B. The formula for the test statistic, t, is

$$t = \frac{\overline{X}_1 - \overline{X}_2}{s_p \sqrt{\left(\dfrac{1}{n_1} + \dfrac{1}{n_2}\right)}}$$ **11–2**

VI. If the samples are dependent (paired);

$$t = \frac{\overline{d}}{s_d / \sqrt{n}}$$ **11–4**

where

$$\overline{d} = \frac{\Sigma d}{n} \qquad s_d = \sqrt{\frac{\Sigma d^2 - \dfrac{(\Sigma d)^2}{n}}{n - 1}}$$

VII. A p-value is the likelihood of a value of t as extreme as that computed when the null hypothesis is true.

Exercises

Use the five-step hypothesis-testing procedure in Exercises 23–38.

23. The university library is interested in determining whether the mean number of books checked out per visit has increased. In the past, the mean was 3.0 books per student visit. A random sample of 10 students revealed a mean of 4.1 books, with a standard deviation of 2.0 books. At the 0.05 level of significance, does this information provide sufficient evidence to show that students are now checking out more books per visit?

24. A recent newspaper article claims that the typical American is 20 pounds overweight. To test this claim, 15 randomly selected persons were weighed. They were an average of 18 pounds overweight, with a standard deviation of 5 pounds. At the 0.05 significance level, is there sufficient evidence to reject the newspaper's claim?

25. An association of college textbook publishers recently reported the mean retail cost of its members' books to be $35.00. A group of students lobbying for increased state support to students because of higher education costs has challenged this claim. A random sample of 20 books is selected. Calculations indicate that the mean cost is $35.80, with a standard deviation of $3.80. At the 0.05 level of significance, is there sufficient evidence to reject the claim of the publishers' association? Can the students assert that the mean cost is higher? Find the p-value.

26. Your new car has an EPA rating of 26.0 miles per gallon. The mileage figures actually obtained on 6 trips were 24.3, 25.2, 24.9, 24.8, 25.6, and 25.4. Is there sufficient evidence, at the 0.01 level of significance, to conclude that the car performs below the EPA specifications?

27. A recent newspaper headline indicated that the typical residential home in a certain community is now selling for less than $100,000. A sample of 10 recent transactions revealed the following selling prices (in thousands of dollars): $96.0, $92.0, $95.0, $97.0, $103.0, $105.0, $94.0, $97.0, $96.0, and $95.0. At the 0.01 significance level, can we conclude that the mean selling price of these homes is less than $100,000?

28. The National Weather Service reports that the high temperature on July 1 is as likely to be above 25°C as below. Test this claim at the 0.10 significance level, given the following July 1 temperature readings over the past 16 years: 22, 26, 28, 24, 27, 20, 29, 32, 28, 21, 25, 27, 26, 28, 30, 22. What is the p-value?

29. A toothpaste manufacturer claims that children who brush their teeth with Bianca will have fewer cavities than those who brush with Sparkle. In a carefully supervised study, the numbers of cavities that occurred with each of the 2 brands are compared. Is there sufficient evidence, at the 0.01 level of significance, to support the manufacturer's claim?

	Cavities
Bianca	1, 2, 3, 4, 2, 0, 2
Sparkle	4, 5, 4, 2, 1, 2, 4

30. The following data are the weight gains (in pounds) of babies from birth to age 6 months. All babies in the sample weighed between 7 and 8 pounds at birth. One group of babies was breast-fed, and the other was fed a specific formula. Is there evidence that the weight gains are different between the 2 groups? Use the 0.01 significance level.

Breast-Fed	Formula-Fed
7	9
8	10
6	8
10	6
9	7
8	8
9	

31. Do managers in the retail industry earn less than those in the auto industry? A sample of 15 retail managers shows a mean salary of $47,250, with a standard deviation of $5500. A sample of 10 auto managers shows the mean salary to be $56,720, with a standard deviation of $6200. Use the 0.05 significance level. What is the p-value?

32. A manufacturer of wrist supports for bowlers maintains that use of this support will improve a bowler's average. A sample of 12 bowlers who have league-sanctioned averages of over 150 roll 2 complete games, 1 with the wrist support and 1 without it. Does the evidence support the manufacturer's claim? Assume a significance level of 0.01.

Bowler	With Wrist Band	Without Wrist Band
Clarke	230	217
Redenback	225	198
Simmons	223	208
Pelton	216	222
Griffin	229	223
Farthy	201	214
Hawkins	205	187
Nugent	193	187
Bryan	177	178
Hucklebury	201	195
Baker	178	169
Berry	207	194

33. Six junior executives were sent to a class to improve their verbal skills. To test the quality of the program, the executives were tested before and after taking the class, with the following results:

	Before	After
Levin	18	30
Baker	38	70
Craft	8	20
Denfrey	10	4
Longhi	12	10
Foster	12	20

Do these records indicate a significant improvement in verbal skills at the 0.10 significance level?

34. To measure the effectiveness of his sales training program a car dealer selects at random 8 sales representatives to take the course. The weekly sales volume of each sales representative is shown below. At the 0.05 significance level, can it be concluded that the new program is effective in increasing sales? Find the p-value.

Sales Rep	Gross Sales After the Course (in thousands of dollars)	Gross Sales Before the Course (in thousands of dollars)
1	$14.0	$13.5
2	10.7	11.4
3	12.4	10.7
4	11.1	11.1
5	10.9	9.8
6	10.5	9.6
7	10.8	10.7
8	13.0	11.7

35. Dhondt Doors manufactures garage doors. Mr. Dhondt is considering two sources for the springs used in the doors. The research department of Dhondt Doors conducted strength tests of the two springs, with the following results:

Sevits Manufacturing	Molnar Industries
$n = 20$	$n = 32$
$\overline{X} = 66.0$ kg	$\overline{X} = 68.3$ kg
$s = 2.1$	$s = 1.4$

At the 0.05 significance level, is there a difference in the strength of the two springs?

36. Last year the mean number of new homes started per month in the Perrysburg area was 6.00. Because of good weather, an increased demand for housing, and lower mortgage rates, the Perrysburg Builders Association claims that the number of housing starts has increased this year. The mean number of new homes started for the first 6 months of this year is 7.2, with a standard deviation of 1.05. Assuming the data from the first 6 months to be a sample, can we conclude that the mean number of housing starts has increased? Use the 0.01 significance level.

37. Kazimer's Supermarkets are considering locating a new store in either Whitehouse or Waterville. Before making a final decision, the company has hired a marketing research firm to determine the mean amount spent weekly per family unit on groceries in the two communities.

Whitehouse	Waterville
$n = 12$	$n = 15$
$\overline{X} = \$132$	$\overline{X} = \$140$
$s = \$13$	$s = \$16$

At the 0.05 significance level, can Kazimer's conclude that there is a difference in the mean amounts spent weekly on groceries in the two communities? Find the p-value.

38. Do sellers generally receive their asking prices when selling their homes? The prices at the top of the next page. (in thousands of dollars) were obtained from a sample of 12 homes recently sold in the Pittsburgh, Pennsylvania, area.

Single-Family Home	Asking Price	Selling Price
1	$103.0	$ 99.3
2	127.0	124.3
3	114.9	110.8
4	102.0	102.0
5	84.6	80.0
6	160.5	158.8
7	99.6	95.4
8	173.0	167.3
9	212.5	210.1
10	89.2	86.3
11	99.9	96.8
12	138.0	132.6

At the 0.01 significance level, can we conclude that sellers normally take less than their asking prices?

39. To service her customers more effectively Ms. Dodd, owner of Damschroders, an exclusive women's boutique, wants to know their mean age. A random sample of 15 customers revealed the mean to be 46.3 years, with a standard deviation of 10.6 years. When Ms. Dodd opened her boutique, the manager of the mall where the shop is located had reported the mean age of shoppers to be 40. Can Ms. Dodd conclude that her clients are older? Use the 0.05 significance level.

40. A government testing agency routinely tests various foods to ensure that they meet label requirements. A random sample of 10 1-liter bottles of a soft drink actually contained the following amounts when tested: 0.93, 0.97, 0.96, 1.02, 1.05, 1.01, 1.02, 0.97, 0.98, and 0.97. At the 0.05 significance level, can the testing agency show that the soft-drink manufacturer is underfilling the product?

41. The personnel manager of a large corporation believes that people are now retiring at a later age. A sample of 20 employees who retired in 1970 revealed that their mean age at retirement was 63.7 years, with a standard deviation of 3.2 years. A sample of 15 employees who retired last year revealed the mean to be 66.5 years and the standard deviation 4.3 years. At the 0.01 signif-icance level, can the manager conclude that the mean age at retirement has increased? Find the p-value.

42. The dean of Northern University believes that student grade point averages have increased in recent years. She obtains a sample of 7 student grade point averages for 1990 and 8 for this year. Based on these data, can the dean conclude that grades have increased? Use the 0.01 significance level.

Student Grade Point Averages	
1990	This Year
2.90	2.70
2.85	3.35
2.67	3.60
1.98	2.75
3.20	2.35
2.65	2.90
2.35	2.98
	3.01

43. A sample of eight upper-middle-class families were surveyed to determine their annual medical expenses. The mean amount was $960 with a sample standard deviation of $135. Can we conclude that the mean of the population's medical expenses is greater than $900 per year? Use the 0.05 significance level.

44. A study was designed to determine if drinking affects a driver's reaction time. A random sample of 12 people was given a driving-simulator test. Each person was then asked to drink 2 ounces of whiskey and to repeat the test. The "errors" made in each test were tabulated. They can be found at the top of the next page. At the 0.05 significance level, can it be concluded that people make more errors after having drunk 2 ounces of whiskey? Find the p-value.

Subject	Before Drinking	After Drinking
A	8	9
B	9	12
C	10	13
D	8	14
E	11	15
F	6	11
G	12	12
H	15	14
I	10	13
J	7	12
K	8	13
L	10	19

that Inspector Powell consistently rates the restaurants higher than does Inspector Saner?

	Inspector	
Restaurant	Powell	Saner
Gallion	79	75
Maumee	70	61
Oak Harbor	46	32
Sylvania	55	59
Rossford	65	58
Russell	72	70
Bradford	63	60
Erie	65	65

45. A large university is concerned about salary discrimination on the basis of sex. To investigate, a sample of 11 female instructors is obtained, and their salaries are determined. For each female selected, a male faculty member with similar tenure status, academic rank, discipline, and so on is obtained, and his salary is paired with that of the female instructor. The data (in thousands of dollars) are shown in the table that follows. Can it be concluded that female instructors earn significantly less? Use the 0.05 significance level.

Faculty Pair	Male	Female
1	$46.0	$45.9
2	42.9	43.5
3	43.9	42.7
4	47.8	41.7
5	45.5	44.7
6	44.3	43.0
7	40.7	40.8
8	45.4	44.8
9	46.4	49.5
10	51.0	47.2
11	38.3	38.1

46. All Chicken and Eggs Restaurants are rated for cleanliness and food quality by a corporate quality control department. For a sample of 8 restaurants, the ratings by 2 inspectors are obtained. At the 0.05 significance level, can we conclude

47. A fertilizer-mixing machine is set to give 10 pounds of nitrate for every 100 pounds of fertilizer. Eight 100-pound bags were examined, and the pounds of nitrate were 8, 10, 9, 11, 7, 10, 9, and 10. Is there reason to conclude that the mean is not equal to 10 pounds? Use the 0.01 significance level.

48. A random sample of 12 women not employed outside the home was asked to estimate the selling prices of 2 25-inch color TV sets. Their estimates are:

Homemaker	Model A	Model B
1	$715	$810
2	830	650
3	815	620
4	770	760
5	650	830
6	680	720
7	770	800
8	760	830
9	990	830
10	550	900
11	670	620
12	760	630

Is there sufficient evidence, at the 0.05 level of significance, to claim that homemakers perceive Model B to be more expensive than Model A? Find the *p*-value.

DATA EXERCISES

49. Refer to the real estate data set which reports information on homes sold in Alabama during 1996.
   a. There are 21 homes with fireplaces. At the 0.05 significance level, can we conclude that the mean selling price of these homes is different from $185,000?
   b. Consider the homes with attached garages. Is there a difference in the mean selling price of those homes with a fireplace and those without a fireplace? Use the 0.05 significance level.
50. Refer to the schools data set which has information on 94 school districts.
   a. Select a sample of every fourth school district

in the data set. Is the mean amount spent on instruction different from $3000?
   b. Is there a difference between the mean amounts spent on instruction between the "large" (more than 3000 students) districts and the "small" (fewer than 1000 students) ones?
   c. Is there a difference between the mean amounts spent on instruction between the districts with a "low" (less than 5%) proportion of students on welfare and those with a "high" (more than 10%) proportion on welfare?

CHAPTER ACHIEVEMENT TEST

The answers are at the back of the book.

MULTIPLE-CHOICE QUESTIONS

Select the response that best answers each of the questions.

1. The t test for the difference between the means of two independent samples assumes
   a. the samples were obtained from normal populations.
   b. the population standard deviations are equal.
   c. the samples were obtained from independent populations.
   d. All of the above
2. If we are testing for the *difference* between the means of 2 related samples with $n = 15$, the number of degrees of freedom is equal to
   a. 28.
   b. 30.
   c. 15.
   d. 14.
   e. None of the above

3. The t distribution approaches which distribution as the sample size increases?
   a. Binomial
   b. Normal
   c. Poisson
   d. None of the above
4. A random sample of 10 is selected from a normal population. The population standard deviation is unknown. If a two-tailed test of significance is to be used at the 0.01 significance level, the null hypothesis is not rejected if
   a. z is between -2.58 and 2.58.
   b. t is between -3.250 and 3.250.
   c. t is between -3.169 and 3.169.
   d. t is less than -2.764.
   e. None of the above
5. The two-sample t test and the t test for paired observations will
   a. always yield the same results.

b. always have the same degrees of freedom.

c. always have the same sample sizes.

d. None of the above

For Questions 6–11, use the following information.

A U.S. congressman claims that the mean enrollment in 2-year public institutions of higher learning is fewer than 4500 students. To test this claim, a random sample of 6 schools is selected. The enrollments at the selected schools are shown below.

School	Number of Students (in thousands)
Central CC	3.2
Edison CC	1.7
Lacy State	5.6
Northside Tech	1.4
Shawnee State	1.7
Washington CC	0.7

6. The appropriate null and alternate hypotheses are
  a. $H_0: \overline{X}_1 \geq \overline{X}_2$
      $H_a: \overline{X}_1 < \overline{X}_2$
  b. $H_0: \mu \geq 4.5$
      $H_a: \mu < 4.5$
  c. $H_0: \mu = 4.5$
      $H_a: \mu \neq 4.5$
  d. $H_0: \overline{X} \geq 4.5$
      $H_a: \overline{X} < 4.5$

7. Assuming the population approximates a normal distribution, the test statistic is t because

a. the population standard deviation is not known.

b. the sample size is less than 30.

c. the enrollments are about equal.

d. Both a and b

e. None of the above

8. The population standard deviation must be estimated from the sample information. The best estimate is
  a. 1.77.
  b. 1.62.
  c. 15.75.
  d. 3.15.
  e. None of the above

9. If the congressman assumed the 0.05 significance level, the null hypothesis would be rejected if
  a. t is to the left of -1.476.
  b. t is to the left of -2.015.
  c. t is outside the interval from -2.015 to 2.015.
  d. z is to the left of -1.65.
  e. None of the above

10. The computed value of the test statistic is
  a. $t = -2.93$.
  b. $z = -1.25$.
  c. $t = 2.93$.
  d. None of the above

11. At the 0.05 significance level, which of the following would be a correct conclusion?
  a. Enrollment mean is not fewer than 4500.
  b. Enrollment mean is at least 4500.
  c. Enrollment mean is fewer than 4500.
  d. You cannot determine.
  e. None of the above

COMPUTATION PROBLEMS

Use the five-step hypothesis-testing procedure in problems 12–14.

12. The Food and Drug Administration is conducting tests on a certain drug to determine if it has the undesirable side effect of reducing the body's temperature. It is known that the mean human temperature is 98.6°F. The new drug is administered to 25 patients and the patients' mean temperature drops to 98.3°F, with a standard deviation of 0.64°F. At the 0.05 significance level, is there sufficient reason to conclude that the drug reduces mean body temperature?

13. A home builder claims that the addition of heat pumps will reduce electric bills in all-electric homes. To support his claim, he tests 7 electric bills for the month of January for 2 consecutive years, 1 before the heat pump was installed and 1 after. Is there sufficient evidence to show that heating bills were reduced at the 0.01 level?

Customer	Before	After
Garcia	$180	$160
Huffman	156	164
Johnson	188	172
Palmer	132	130
Kerby	208	200
Beard	196	190
Sauve	190	184

14. The fire chief for Slocum County is evaluating 2 plans for the location of emergency medical equipment. One plan calls for supplies to be kept near the engines most often used by paramedical personnel. The second plan calls for storage near the crew's sleeping quarters. To decide objec-

tively if 1 location is better than the other, the chief tries each of the 2 locations, clocking the time it takes paramedics to collect their equipment under emergency conditions. The results (in seconds) are listed below. Use the 0.05 significance level to determine if there is a significant difference in the time needed for the paramedics to collect their equipment. What is the p-value?

Plan 1	Plan 2
10	15
55	9
30	47
30	3
53	34
	41
	30
	29

ANSWERS TO SELF-REVIEW PROBLEMS

11-1 a. $H_0: \mu \geq 160$
$H_a: \mu < 160$

b. $t = \dfrac{\overline{X} - \mu}{s / \sqrt{n}}$

c. 15, found by $16 - 1$

d. Reject H_0 if computed t falls to the left of -1.753.

e. $t = \dfrac{151 - 160}{12 / \sqrt{16}} = -3$

f. Reject H_0. Senior citizens living in high altitudes do have a lower mean systolic blood pressure.

g. the p-value is less than 0.005. There is very little chance H_0 is true.

11-2 a. $H_0: \mu \leq 20$
$H_a: \mu > 20$

b. One-tailed

c. σ is unknown; the sample is small; and it is reasonable to assume a normal population.

d. $s = \sqrt{\dfrac{2218 - \dfrac{(104)^2}{5}}{5 - 1}}$
$= 3.70$
$t = \dfrac{20.8 - 20.0}{3.70 / \sqrt{5}}$
$= 0.48$

e. Fail to reject H_0 because 0.48 is less than the critical value of 2.132.

f. Although the sample mean of 20.8 is greater than 20, the difference could be due to sampling error. The product should not be marketed.

g. The p-value is greater than 0.25. There is a good chance of finding a mean this large due to sampling variability.

11-3 a. $H_0: \mu_1 \geq \mu_2$
$H_a: \mu_1 < \mu_2$

b. The populations are normal with equal standard deviations. The $df = 10$, found by $7 + 5 - 2$.

c. $s_1 = \sqrt{\dfrac{976 - \dfrac{(82)^2}{7}}{6}} = 1.60$

$s_2 = \sqrt{\dfrac{1621 - \dfrac{(89)^2}{5}}{4}} = 3.03$

d. $s_p = \sqrt{\dfrac{(6)(1.60)^2 + (4)(3.03)^2}{7 + 5 - 2}}$

$s_p = 2.28$

e. Reject H_0 if t is less than (to the left of) -1.812.

f. $t = \dfrac{11.714 - 17.80}{2.28\sqrt{\dfrac{1}{7} + \dfrac{1}{5}}} = -4.56$

g. Reject H_0.

h. Interceptive treatment is effective in reducing the mean time in braces.

11-4 a. $H_0 : \mu_d \leq 0$
   $H_a : \mu_d > 0$

b. $t = 2.998$

c. $t = \dfrac{7.75}{8.60 / \sqrt{8}} = 2.55$

d. H_0 is not rejected.

e. The program cannot be shown to result in a significant mean weight loss.

CHAPTER 12

An Analysis of Variance

OBJECTIVES

When you have completed this chapter, you will be able to

- describe the F distribution;
- conduct a test of hypothesis to determine if two sample variances are equal;
- construct an ANOVA table;
- test for a difference among two or more population means;
- construct a confidence interval for treatment means.

CHAPTER PROBLEM   Leaving the Nest

Dr. Barbara Waite is a clinical psychologist studying the age at which people become psychologically independent of their families. She wants to identify factors that affect the age of independence, such as the number and educational level of parents living in the household, religious affiliation, the number of siblings, and so on. She believes that there are no differences in the mean ages of the different religious groups. Is she correct?

INTRODUCTION

This chapter continues the discussion of hypothesis testing introduced in Chapter 10. Recall that in Chapter 10, we developed the general theory of hypothesis testing and applied it to the means of two normally distributed populations using the z distribution. We also tested the difference between two proportions, again employing the z statistic. In Chapter 11, we analyzed differences between means of two normally distributed populations, but we used Student's t distribution instead of the z statistic.

The basic features of a distribution are its mean and standard deviation. We discussed tests for population means in earlier chapters. It is natural at this point, then, to discuss tests for the standard deviation. In this chapter, we will describe the **F distribution.** This probability distribution is used as the test statistic under two conditions. First, we use it when testing two population variances to determine if they are equal. Second, we use the distribution to simultaneously compare two or more population means. This simultaneous comparison of several population means is called **analysis of variance (ANOVA).** In both situations, the data must be at least interval scale, and the populations must approximate a normal distribution.

THE *F* DISTRIBUTION

In Chapter 11, we described Student's t distribution. The major characteristics of the t distribution were these: It is continuous and symmetrical; both tails of the t distribution approach the X axis, but never reach it; and it is specified by the degrees of freedom, so that there are many t distributions.

This chapter uses another probability distribution as its test statistic—the F distribution. What are the major characteristics of the F distribution?

1. The value of F is not negative. Its values may range from zero with no upper limit.
2. The F distribution is continuous.
3. The F distribution is positively skewed.
4. There is a "family" of F distributions. A particular member of the family is determined by two parameters—namely, a pair of degrees of freedom.

The following graph, in which three F distributions are shown schematically, illustrates the point. There is one F distribution for the combination of 32 degrees of freedom in the numerator and 30 degrees of freedom in the denominator. There is a second, different F distribution for the pair of 20 degrees of freedom in the numerator and 8 degrees of freedom in the denominator. The third F distribution has 8 df in the denominator and 8 df in the numerator.

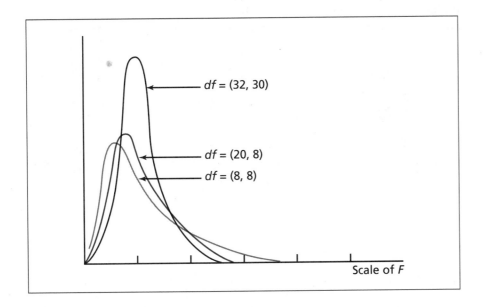

COMPARING TWO POPULATION VARIANCES

The F distribution can be used to test the hypothesis that the variance of one normal population equals the variance of another normal population. This test determines whether one population is more variable than another. For example, a nursing supervisor is interested in whether there is more variation in the temperatures taken with one type of thermometer than with another.

This test also determines whether the assumptions underlying certain statistical tests are valid. Recall that in the independent t test to determine whether two population means are the same, it was necessary to assume that the variances of the two populations are equal. This assumption can be validated by the following test.

In the null hypothesis, the variance of one normal population, σ_1^2, equals the variance of the other normal population, σ_2^2. To conduct this hypothesis test, we obtain a random sample of n_1 observations from one population and a sample of n_2 observations from the second population.

The test statistic is s_1^2/s_2^2, where s_1^2 and s_2^2 are the respective sample variances. If the null hypothesis ($H_0: \sigma_1^2 = \sigma_2^2$) is true, the test statistic is the F distribution with $n_1 - 1$ and $n_2 - 1$ degrees of freedom. For a two-tailed test, the larger sample variance is placed in the numerator. We find the critical value of F by dividing the level of significance by 2, written $\alpha/2$, where the Greek lowercase letter alpha α represents the significance level and then referring to the appropriate number of degrees of freedom in Appendix E.

REAL STAT

Have you ever waited in line for a telephone, and it seemed like the person using the phone talked on and on? Apparently, people tend to spend more time on a public telephone when someone is waiting than when they are alone. In a recent study, researchers measured the length of time 56 shoppers in a mall spent on the phone (1) when they were alone, (2) when a person was using an adjacent telephone, and (3) when a person was using an adjacent phone *and* someone was waiting to use a phone. The study revealed that shoppers talked *less* when alone than when a person was using an adjacent phone or someone was waiting!

Problem A study involves the number of absences per year among union and nonunion workers. Of concern is whether the 2 populations have equal variances. A sample of 16 union workers has a sample standard deviation of 3.0 days. A sample of 10 nonunion workers has a standard deviation of 2.5 days. At the 0.02 significance level, can we conclude that there is a difference in variation between the two groups?

Solution Use the five-step hypothesis-testing procedure.

Step 1. State the null and alternate hypotheses. This test is two-tailed because we are looking for a difference in the variances. We are not testing whether one variance is less than the other, which would be a one-tailed test.

$$H_0 : \sigma_1^2 = \sigma_2^2$$

$$H_a : \sigma_1^2 \neq \sigma_2^2$$

Step 2. The significance level is 0.02.

Step 3. The appropriate test statistic is $s_1^2/s_2^2 = F$.

Step 4. Obtain the decision rule from Appendix E, a portion of which is shown in Table 12–1. Because we are using a two-tailed test, the significance level is 0.01, found by $\alpha/2 = 0.02/2$. There are $n_1 - 1 = 16 - 1 = 15$ degrees of freedom in the numerator and $n_2 - 1 = 10 - 1 = 9$ degrees of freedom in the denominator. To locate the critical value, move horizontally across the top portion of the F table, given in Table 12–1, to 15 degrees of freedom in

TABLE 12–1		Portion of the F Distribution Table $\alpha = .05$			
		Degrees of Freedom in Numerator			
		10	12	15	20	24
	1	6,056	6,106	6,157	6,209	6,235
	2	99.4	99.4	99.4	99.4	99.5
	3	27.2	27.1	26.9	26.7	26.6
	4	14.5	14.4	14.2	14.0	13.9
	5	10.1	9.89	9.72	9.55	9.47
	6	7.87	7.72	7.56	7.40	7.31
Degrees of	7	6.62	6.47	6.31	6.16	6.07
Freedom in	8	5.81	5.67	5.52	5.36	5.28
Denominator	9	5.26	5.11	4.96	4.81	4.73
	10	4.85	4.71	4.56	4.41	4.33
	11	4.54	4.40	4.25	4.10	4.02
	12	4.30	4.16	4.01	3.86	3.78
	13	4.10	3.96	3.82	3.66	3.59
	14	3.94	3.80	3.66	3.51	3.43
	15	3.80	3.67	3.52	3.37	3.29

the numerator. Then move down that column to the critical value opposite 9 degrees of freedom in the denominator. The critical value of F is 4.96, so the decision rule is: If the ratio of the sample variances exceeds 4.96, H_0 is rejected.

Step 5. The computed value of the test statistic is 1.44, found by $(3.0)^2/(2.5)^2$. The null hypothesis that the population variances are equal cannot be rejected. The data do not indicate a difference in the variation of days absent for union and nonunion workers.

For a one-tailed test, the numerator is determined from the statement of the alternate hypothesis. For example, if $H_a : \sigma_2^2 > \sigma_1^2$, the appropriate test statistic is $F = s_2^2/s_1^2$. The critical value of F is obtained for α (not $\alpha/2$) and the given degrees of freedom. The decision rule is: Reject H_0 if the computed statistic F exceeds the critical value.

If the problem involved determining whether union members were absent more irregularly than nonunion members, the test would be one-tailed. Appendix E gives only the 0.05 and 0.01 critical values; therefore, if we use the text, we can select only these significance levels. For other significance levels, use a statistical software package like MINITAB.

. . . .

Self-Review 12–1

Answers to the Self-Review problems are at the end of the chapter.

The Ace Construction Company has 2 front-end loaders and employs 2 workers to operate them. The dirt is scooped from the construction site and deposited in a waiting truck. There seems to be a difference in the volume of dirt scooped up by the 2 operators. A random sample of 9 loads from Operator A showed a standard deviation of 0.90 ton, while a sample of 6 loads from Operator B showed a standard deviation of 0.40 ton. Both samples appear normally distributed. Can we conclude, at the 0.05 significance level, that there is more variation in Operator A's work?

. . . .

Exercises

Answers to the even-numbered Exercises are at the back of the book.

1. Find the critical value of F (using Appendix E) under the following conditions:
   a. $H_0 : \sigma_1^2 = \sigma_2^2$; $H_a : \sigma_1^2 \neq \sigma_2^2$, where $n_1 = 6$ and $n_2 = 10$. Use the 0.02 significance level.
   b. $H_0 : \sigma_1^2 \leq \sigma_2^2$, $H_a : \sigma_1^2 > \sigma_2^2$, where $n_1 = 6$ and $n_2 = 10$. Use the 0.05 significance level.

2. Find the critical value of F (using Appendix E) under the following conditions:
   a. $H_0 : \sigma_1^2 = \sigma_2^2$; $H_a : \sigma_1^2 \neq \sigma_2^2$, where $n_1 = 8$ and $n_2 = 7$. Use the 0.10 significance level.
   b. $H_0 : \sigma_1^2 \leq \sigma_2^2$; $H_a : \sigma_1^2 > \sigma_2^2$, where $n_1 = 4$ and $n_2 = 10$. Use the 0.01 significance level.

3. A study is made concerning hours spent reading the newspaper. A sample of 10 men showed a standard deviation of 4 hours per week. A sample of 8 women revealed a standard deviation of 5.7 hours per week. Can we conclude that there is more variation among the women? Both samples appear normally distributed. Use the 0.05 significance level.

4. The raw materials supplied by 2 vendors are compared. The vendors seem to provide materials that are normally distributed with the same mean, but the variability is a matter of concern. A sample of 16 lots from Vendor A yields a variance of 150, and a sample of 21 lots from Vendor B reveals a variance of 225. Are the population variances equal? Use the 0.10 significance level.

UNDERLYING ASSUMPTIONS FOR ANOVA

The second use of the F distribution is in the **analysis-of-variance (ANOVA)** technique. To use ANOVA, we assume the following conditions:

1. The populations being studied are normally distributed.
2. The populations have equal standard deviations (σ).
3. The samples selected from those populations are independent and random.

When these conditions are met, the F statistic is used (instead of z or t) to test if the *means* of the populations are equal. Whenever the assumptions about the normality of the population distributions and equal standard deviations cannot be met, an analysis-of-variance technique developed by Kruskal and Wallis, to be discussed in Chapter 17, is used.

Statistician Ronald Fisher developed the technique of analysis of variance while working in agriculture. The crops were provided different kinds of fertilizer, called "treatments." He wanted to see if some of these treatments were more effective than others. Of course, in modern statistics, these treatments are called *populations*.

A specific example may clarify the term *treatment* and also illustrate the basis of the ANOVA test. Suppose a farmer wanted to know if the mean yield per acre of wheat differs as a result of the brand of fertilizer used. Three brands—Porter, Anderson, and Scoots—are available. As an experiment, the farmer divides his field into 15 equal plots. The plots are planted in the usual manner, all at the same time, but the farmer randomly puts Porter on 5 plots, Anderson on 5 plots, and Scoots on the remaining 5 plots. The different fertilizers, in this case, are the different treatments. At the end of the growing season, he records the number of bushels of wheat produced by each plot. Here are the results:

> **treatment** A specific source or cause of variation in a set of data.

Porter	Anderson	Scoots
40	72	51
45	71	55
47	68	60
50	75	57
47	66	54

A visual "plot" of the data is shown in Figure 12–1 to help you "see" whether the treatment means are different. What do you think? All five of the Porter readings are lower than any of the others. All five of the Anderson readings are higher than all of the other brands. That consistent pattern within the brands would probably convince most people that the population mean yields are different.

On the other hand, suppose someone made a mistake in recording the data. The first number (40) under Porter should have been under Anderson, and the first number (72) under Anderson should have been under Porter. What

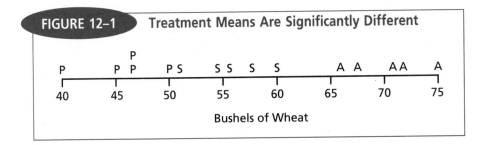

FIGURE 12–1   **Treatment Means Are Significantly Different**

Bushels of Wheat

does this apparently small change do to our evidence? Figure 12–2 is the graph of this situation. Now what do you think?

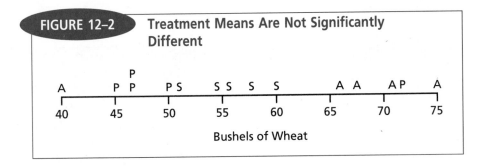

FIGURE 12–2   **Treatment Means Are Not Significantly Different**

Bushels of Wheat

Now things are not so clear. One of the Porter readings is now the second highest, while the other four are still among the five lowest. One of the Anderson readings is now the lowest and quite different from the other four for that brand. In short, the large variation within the brands themselves and the fact that the sample means are closer together now make it very difficult to say whether there is really a difference among the population mean yields of the fertilizers.

The ANOVA statistic computes the variance within the samples and the variance between the sample means and then compares them to make a clear decision whether the population means could be equal.

THE ANOVA TEST

The purpose of this section is to explore some of the reasoning behind ANOVA. First, ANOVA breaks down the total variation into two parts. One part measures the variation between sample means (treatments). The other part measures the variation of the observations from their treatment mean; that is, it measures the variability within each of the sampled populations. Next, we compare this relationship between the two sources of variation by forming the F ratio in the following manner:

$$F = \frac{\text{Estimate of the population variance based on differences between sample means}}{\text{Estimate of the population variance based on variation within samples}}$$

The following problem provides additional insight into the analysis-of-variance technique.

Problem   Recall from the Chapter Problem that Dr. Waite, a clinical psychologist, is studying the age at which people become psychologically independent of their families. One factor that may affect this variable is religious orientation. To test this possibility, she selects a sample of 39 young people and classifies each by religion. The possible categories are Protestant, Catholic, Jewish, and Other. A tally of the religious affiliations of those sampled by the ages at which they became independent of their families is given in Table 12–2.

TABLE 12–2	Comparison of Religious Affiliations and Ages at Which Those Sampled Became Independent		
Religious Affiliation			
Jewish	Catholic	Protestant	Other
22	27	20	18
19	25	18	16
13	22	21	24
19	27	21	19
23	19	16	22
15	23	17	22
16	21	20	24
18	28	18	
20	23	17	
20	25	19	
	27	18	

Dr. Waite's hypothesis is: There is no difference in the mean age at which independence was achieved for the four religious groups.

Solution   Use the five-step hypothesis-testing procedure.

Step 1. The null hypothesis is that the mean age is the same for each faith:

$$H_0 : \mu_1 = \mu_2 = \mu_3 = \mu_4$$

The alternate hypothesis is that the mean ages are not all equal:

$$H_a : \text{Not all means are equal.}$$

Step 2. The significance level is 0.01.

Step 3. The appropriate test is based on the F statistic.

Step 4. Formulate the decision rule. Remember that, in order to arrive at a decision rule, we need to identify the *critical value*. The critical values for the F statistic can be found in Appendix E, a portion of which follows in Table 12–3. You will find the critical values for the 0.05 significance level on the first page of Appendix E and the values for the 0.01 significance level on the second. To use the table, you need to know two numbers: the degrees of freedom in the numerator and the degrees of freedom in the denominator. The degrees of freedom in the numerator refer to the number of treatments, designated k, minus 1. That is, the degrees of freedom are found by $k - 1$. The degrees of freedom in the denominator refer to the total number of observations, designated n, minus the number of treatments, that is, $n - k$. For this problem, there are 4 treatments and a total of 39 observations.

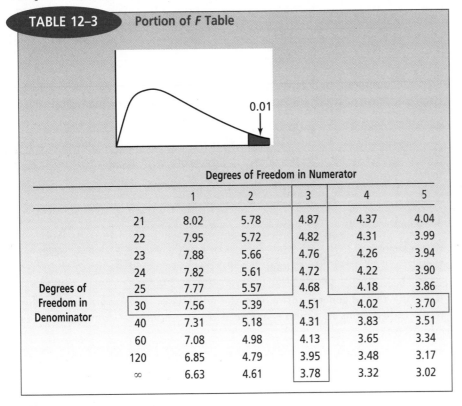

TABLE 12–3   **Portion of F Table**

Degrees of Freedom in Denominator	Degrees of Freedom in Numerator				
	1	2	3	4	5
21	8.02	5.78	4.87	4.37	4.04
22	7.95	5.72	4.82	4.31	3.99
23	7.88	5.66	4.76	4.26	3.94
24	7.82	5.61	4.72	4.22	3.90
25	7.77	5.57	4.68	4.18	3.86
30	7.56	5.39	4.51	4.02	3.70
40	7.31	5.18	4.31	3.83	3.51
60	7.08	4.98	4.13	3.65	3.34
120	6.85	4.79	3.95	3.48	3.17
∞	6.63	4.61	3.78	3.32	3.02

Thus:

Degrees of freedom in numerator $= k - 1 = 4 - 1 = 3$
Degrees of freedom in denominator $= n - k = 39 - 4 = 35$

Refer to Appendix E or Table 12–3 and the 0.01 level of significance. Move horizontally at the top of the table to 3 degrees of freedom in the numerator. Then move down that column to the critical value opposite 30 degrees of freedom. We selected 30 because the value for 35 is not given in the table. Why did we select the smaller value? Selecting the smaller value is more conservative, that is, it will result in the larger value for F. The decision rule is to reject the null hypothesis if the computed value of F is larger than 4.51. Recall that the F distribution is positively skewed. Shown diagrammatically, the decision rule is:

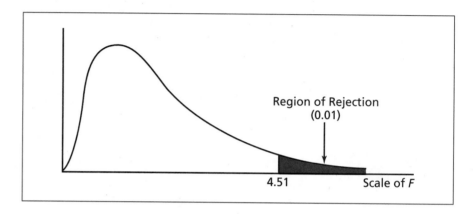

Step 5. Compute F and make a decision. It is often convenient to record the computations for the F statistic in an ANOVA table. The general form is

ANOVA Table				
Source of Variation (1)	Sum of Squares (2)	Degrees of Freedom (3)	Mean Square (4)	F (5)
Between treatments	SST	$k - 1$	$\dfrac{SST}{k - 1}$	$\dfrac{SST/(k - 1)}{SSE/(n - k)}$
Within samples	SSE	$n - k$	$\dfrac{SSE}{n - k}$	
Total	SS total			

Referring to the general format for the analysis of variance, note that three totals, called **sums of squares,** are needed to compute F—namely, SST, SSE, and SS total.

1. SST is the abbreviation for the sum of squares due to the treatment effect and is found by

$$SST = \sum\left(\frac{T_c^2}{n_c}\right) - \frac{(\Sigma X)^2}{n} \qquad \boxed{\textbf{12-1}}$$

where

T_c is the column total for all observations in the treatment.
n_c is the number of observations (sample size) for each respective treatment.

ΣX is the sum of all observations.
k is the number of treatments.
n is the total number of observations.

2. *SSE* is the abbreviation for the sum of squares within samples (error), which is computed by

$$SSE = \Sigma(X^2) - \Sigma\left(\frac{T_c^2}{n_c}\right) \qquad \textbf{12-2}$$

3. *SS* total is the total variation; that is, it is the sum of *SST* and *SSE*.

$$SS \text{ total} = SST + SSE$$

As a check, *SS* total is calculated as follows:

$$SS \text{ total} = \Sigma(X^2) - \frac{(\Sigma X)^2}{n} \qquad \textbf{12-3}$$

To compute *F*:

$$F = \frac{\dfrac{SST}{k-1}}{\dfrac{SSE}{n-k}} \qquad \textbf{12-4}$$

The calculations for *F* are shown in Table 12–4.
The entries for the ANOVA table are computed as follows:

$$SST = \Sigma\left(\frac{T_c^2}{n_c}\right) - \frac{(\Sigma X)^2}{n}$$

$$= \frac{(185)^2}{10} + \frac{(267)^2}{11} + \frac{(205)^2}{11} + \frac{(145)^2}{7} - \frac{(802)^2}{39}$$

$$= 234.93$$

$$SSE = \Sigma(X^2) - \Sigma\left(\frac{T_c^2}{n_c}\right)$$

$$= (22^2 + 19^2 + \cdots + 24^2) - \left[\frac{(185)^2}{10} + \frac{(267)^2}{11} + \frac{(205)^2}{11} + \frac{(145)^2}{7}\right]$$

$$= 256.66$$

TABLE 12–4	Calculations Needed for Computed F				
	Religious Affiliation				
	Jewish	Catholic	Protestant	Other	**Grand Totals**
	Ages (X)				
	22	27	20	18	
	19	25	18	16	
	13	22	21	24	
	19	27	21	19	
	23	19	16	22	
	15	23	17	22	
	16	21	20	24	
	18	28	18		
	20	23	17		
	20	25	19		
		27	18		
Column total (T_c)	185	267	205	145	802 ← ΣX
Sample size (n_c)	10	11	11	7	39 ← n
Squared total (ΣX^2)	3509	6565	3849	3061	16,984 ← $\Sigma(X^2)$

$$SS \text{ total} = SST + SSE$$
$$= 234.93 + 256.66$$
$$= 491.59$$

As a check, use formula 12–3:

$$SS \text{ total} = \Sigma(X^2) - \frac{(\Sigma X)^2}{n}$$

$$= 16,984 - \frac{(802)^2}{39}$$

$$= 491.59$$

Insert these values into an ANOVA table:

Source of Variation	Sum of Squares	Degrees of Freedom	Mean Square
Between treatments	234.93	3	78.31
Within samples	256.66	35	7.33
Total	491.59		

Now compute F by applying formula 12–4:

$$F = \frac{\dfrac{SST}{k-1}}{\dfrac{SSE}{n-k}} = \frac{234.93/3}{256.66/35} = \frac{78.31}{7.33} = 10.68$$

Because the computed F value of 10.68 is greater than the critical F value of 4.51 (determined in Step 4), the null hypothesis is rejected at the 0.01 level. Therefore, the p-value is less than 0.01. This indicates that it is quite unlikely that the differences in the 4 means could have occurred by chance. Thus, we conclude that the population means are not all equal. Apparently, religious affiliation affects the age at which people become independent of their families. To determine which pairs of means differ, additional analysis is required. How to recognize which pairs are significantly different is illustrated in the following example.

A Software Example

MINITAB has several procedures for computing a one-way analysis of variance. The output that follows was obtained with the procedure AOVONE-WAY. To begin we enter the four columns of ages, corresponding to the four religions, and name the variables.

Data			
C1	**C2**	**C3**	**C4**
↓ Jewish	Catholic	Prot.	Other
1 22	27	20	18
2 19	25	18	16
3 13	22	21	24
4 19	27	21	19
5 23	19	16	22
6 15	23	17	22
7 16	21	20	24
8 18	28	18	
9 20	23	17	
10 20	25	19	
11	27	18	

Next, we select

Stat ▶ ANOVA ▶ Oneway (Unstacked)

from within the dialog box, click the left mouse button to highlight all four columns of data, click on **Select,** and then click on **OK.**

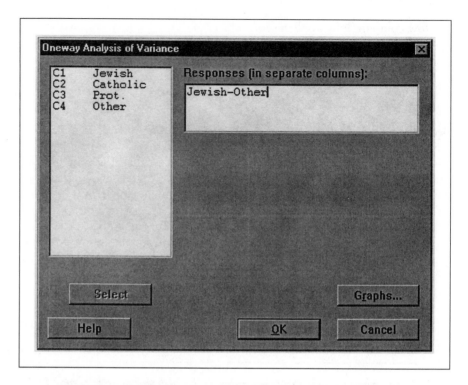

The output, including the data entry information, is as follows:

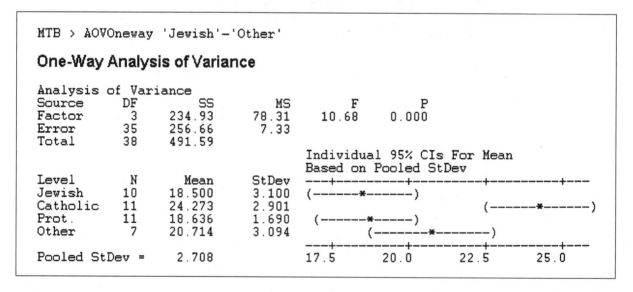

```
MTB > AOVOneway 'Jewish'-'Other'
```

One-Way Analysis of Variance

```
Analysis of Variance
Source    DF        SS        MS        F        P
Factor     3    234.93     78.31    10.68    0.000
Error     35    256.66      7.33
Total     38    491.59
```

```
                                  Individual 95% CIs For Mean
                                  Based on Pooled StDev
Level     N       Mean    StDev   ---+---------+---------+---------+---
Jewish   10     18.500    3.100   (------*------)
Catholic 11     24.273    2.901                            (------*------)
Prot.    11     18.636    1.690     (------*------)
Other     7     20.714    3.094          (--------*--------)
                                  ---+---------+---------+---------+---
Pooled StDev =    2.708          17.5      20.0      22.5      25.0
```

Note that the computer output gives the computed value of F—namely, 10.68. It also reports a p-value of 0.000. MINITAB supplies some additional information, such as the sample size, the mean, and the standard deviation for each of the 4 religious categories. Further, a confidence interval for the mean of each category is given. The endpoints for each interval are designated by parentheses. If H_0 is rejected, we can use these confidence intervals to determine which of the particular pairs of means differ. Note that the intervals for Jewish and Catholic do not overlap. That is, the upper limit for Jewish is about 21 years and the lower limit for Catholic is about 22.5 years. Hence, we can conclude that there is a difference in these groups. Continuing, we conclude that Catholic and Protestant differ, but that Jewish and Protestant, Jewish and Other, Protestant and Other, and Catholic and Other do not differ.

Problem The head nurse at University Hospital has the responsibility of assigning personnel to the emergency room. Present policy calls for the same number of registered nurses to be assigned to all three shifts. The head nurse, however, thinks that the number of emergencies handled may not be the same for each shift. It is decided to use the ANOVA technique to investigate whether the same number of emergencies is handled on each shift.

A random sample of five days from each shift is selected. The results are shown. The "treatment" in this problem is the shift. Note that the sample sizes are equal. ANOVA follows the five-step hypothesis-testing procedure outlined in Chapter 10.

A Second Example

	Number of Emergency Cases Reported per Shift		
	Day	Afternoon	Night
	44	33	39
	53	42	24
	56	15	30
	49	30	27
	38	45	30
Mean	48	33	30

Solution Repeating this table with all necessary calculations, we get

	Day		Afternoon		Night		Grand Totals	
	X	X^2	X	X^2	X	X^2		
	44	1936	33	1089	39	1521		
	53	2809	42	1764	24	576		
	56	3136	15	225	30	900		
	49	2401	30	900	27	729		
	38	1444	45	2025	30	900		
Column total (T_c)	240		165		150		555	$\leftarrow \Sigma X$
Sample size (n_c)	5		5		5		15	$\leftarrow n$
Squared Total (ΣX^2)		11,726		6003		4626	22,355	$\leftarrow \Sigma(X^2)$

Step 1. The null hypothesis is that the mean number of emergencies is the same for each shift; that is, $\mu_1 = \mu_2 = \mu_3$. The alternate hypothesis is that the mean numbers of emergencies are not all equal.

Step 2. The level of significance is 0.05.

Step 3. The distribution can be assumed normal with equal standard deviations. The appropriate statistical test is F.

Step 4. Formulate the decision rule. There are 2 degrees of freedom in the numerator and 12 in the denominator. So we reject H_0 if F exceeds 3.89.

Step 5. The F statistic is determined by formula 12–4:

$$F = \frac{\dfrac{SST}{k-1}}{\dfrac{SSE}{n-k}}$$

where k is the number of treatments and n the total number of items sampled.
Calculate SST (the sum of squares due to the treatment effect):

$$SST = \sum \left(\frac{T_c^2}{n_c} \right) - \frac{(\Sigma X)^2}{n}$$

$$= \frac{(240)^2}{5} + \frac{(165)^2}{5} + \frac{(150)^2}{5} - \frac{(555)^2}{15}$$

$$= 930$$

Next, calculate SSE [the sum of squares within samples (error)]:

$$SSE = \Sigma(X^2) - \sum \left(\frac{T_c^2}{n_c} \right)$$

$$= 22{,}355 - \left(\frac{(240)^2}{5} + \frac{(165)^2}{5} + \frac{(150)^2}{5} \right)$$

$$= 22{,}355 - 21{,}465 = 890$$

We can now determine the F statistic:

$$F = \frac{\dfrac{930}{3-1}}{\dfrac{890}{15-3}}$$

$$= 6.27$$

The decision rule, formulated in Step 4, stated that H_0 is to be rejected if the computed F-value exceeds 3.89. Because 6.27 is greater than 3.89, we reject the null hypothesis at the 0.05 level of significance. To put it another way, the differences in mean numbers of emergencies handled per shift (48, 33, and 30) cannot be attributed to chance. From a practical standpoint, it may be concluded that the mean numbers of emergency cases handled on the 3 shifts are not the same. If the 0.01 significance level were used, the critical value would be 6.93. Since F is between 3.89 and 6.93, the p-value is between 0.01 and 0.05.

Here are the steps to do this ANOVA example with Excel.

1. Type the data into a worksheet. For example, put it in the range a1:c6, with names in the first row.
2. From the Menu bar, select Tools ▶ Data Analysis, highlight Anova: Single Factor, and click **OK.**
3. In the Anova: Single Factor dialog box:
   a. Put the cursor on the *I*nput Range text box, and type "a1:c6" or the appropriate range. Push the Tab key.
   b. Select Grouped by Columns. Push the Tab key.
   c. Check mark the box to select Labels in First Row by pushing the L key. Push the Tab key.
   d. The Alpha text box should display 0.05.
   e. Select the Output Range text box, type "a8" or an appropriate range, and click **OK.**

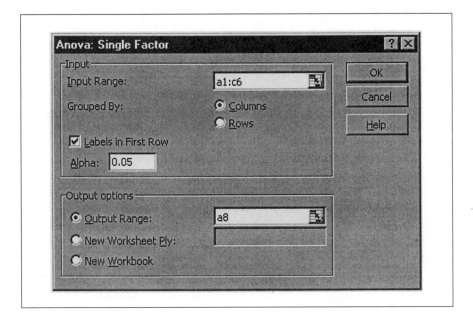

4. You may do the following to make the output easier to read:
   a. Point at a18 or an appropriate range to make it the active cell.
   b. From the Menu bar, select Format ▶ Column ▶ AutoFit Selection.

Here is the Excel software output for this data set:

	A	B	C	D	E	F	G
1	Day	Afternoon	Night				
2	44	33	39				
3	53	42	24				
4	56	15	30				
5	49	30	27				
6	38	45	30				
7	Anova: Single Factor						
8							
9	SUMMARY						
10	Groups	Count	Sum	Average	Variance		
11	Day	5	240	48	51.5		
12	Afternoon	5	165	33	139.5		
13	Night	5	150	30	31.5		
14							
15							
16	ANOVA						
17	Source of Variation	SS	df	MS	F	P-value	F crit
18	Between Groups	930	2	465	6.269663	0.013675	3.88529
19	Within Groups	890	12	74.16667			
20							
21	Total	1820	14				

Self-Review 12–2 Energy shortages have caused many schools to turn down the heat in the classroom. The principal at Penn Street Elementary School is concerned that this may have an effect on achievement. To investigate this question further, students in the fifth-grade mathematics class were randomly assigned to 1 of 3 groups. The 3 groups were then separated and placed in rooms having different temperatures. Each group received televised instruction in long division. At the end of the lesson, the same 10-question examination was given to all 3 groups. The results, scoring the number of correct answers out of 10, were

Temperature		
65°F	72°F	78°F
3	7	4
5	6	6
4	8	5
3	9	7
4	6	6
	8	5
	8	4
		3

Is there any significant difference in the mean scores? Use the 0.05 significance level.

a. What assumptions are necessary.
b. Compute *SST.*
c. Compute *SSE.*
d. Compute *SS* total.
e. Arrange the values in an ANOVA table, and compute *F*.
f. Arrive at a decision.
g. What is the *p*-value?

A final note: In an analysis of variance, the null hypothesis is that *the means are all equal*. The alternate hypothesis is that *at least one of the means is different*. Rejection of the null hypothesis and acceptance of the alternate hypothesis indicate the conclusion that there is a significant difference between at least one pair of means. Rejecting the null hypothesis does not pinpoint which pairs—or how many pairs—of means differ significantly. It only indicates that there is a significant difference between at least one pair of means. Multiple-comparison tests, discussed in more advanced texts, may be used to identify the treatment that differs significantly.

. . . .

Exercises

5. There are five data values observed for each of three treatments in an experiment.
   a. What are the degrees of freedom for the Between treatments estimate?
   b. Find the degrees of freedom for the Within samples estimate.
   c. How many total degrees of freedom are there?
   d. What is the critical value of *F* at the 0.05 level of significance?

6. There are three data values for treatment one, four for treatment two, and six for treatment three in an experiment.

   a. What are the degrees of freedom for the Between treatments estimate?
   b. Find the degrees of freedom for the Within samples estimate.
   c. How many total degrees of freedom are there?
   d. What is the critical value of *F* at the 0.01 level of significance?

7. The quantity of oxygen dissolved in water is a measure of water pollution. Samples are taken at 3 locations in a lake, and the quantities of dissolved oxygen appear at the top of the next page (lower readings indicate greater pollution).

Location	Quantity
Bayview	6.5, 6.4, 6.9
Council Bluffs	6.7, 7.1, 6.9, 7.3
Swan Creek	7.4, 6.9, 7.2

Do these data indicate significant differences in the mean amounts of dissolved oxygen at the 3 locations? Use the 0.05 significance level in determining your answer. Assume the populations are normal with equal standard deviations.

8. There are 3 banks in Warren, Pennsylvania. Customers are randomly selected from each bank, and their waiting times before being served are recorded:

Bank	Waiting Time (in minutes)
Pittsburgh Bank	3.8, 4.5, 5.3
Mountain Trust	6.9, 9.6, 7.3, 8.2
Western National	4.3, 5.1

Do these data indicate significant differences in the mean waiting times at the 3 banks? Use the 0.05 significance level. Find the p-value. Assume the populations are normal with equal standard deviations.

9. Deals-on-Wheels, a manufacturer of mobile homes, is interested in the ages of buyers of each of 5 available floor designs. The president suspects that certain designs tend to appeal to younger buyers more than others. A random sample of 18 records is selected from last year's sales files. The respective ages of the principal buyers from the files are

		Floor Design		
A	B	C	D	E
30	48	54	52	44
32	52	60	50	48
31	45	56	43	50
	38	50	42	

a. State the null and alternate hypotheses.
b. How many degrees of freedom are there in the numerator and the denominator?
c. For the 0.05 level of significance, what is the critical value of F?
d. Compute F.
e. State your decision.
f. Interpret the results.

10. A social researcher wants to evaluate the ethical behavior of attorneys in 4 major regions of the United States. Samples are selected and an index of ethical behavior computed for each attorney in the study. Refer to the data below. At the 0.01 level of significance, are there statistically significant differences in ethical behavior among the 4 regions? Use the five-step hypothesis-testing procedure.

West	South	North Central	East
12	19	34	19
16	20	29	21
12	18	31	17
14	9	19	24
26	22	26	
	19		

. . . .

CHAPTER OUTLINE

I. The F distribution has four major characteristics:
A. It is continuous.
B. Its values cannot be negative.
C. It is positively skewed.
D. There is a family of F distributions. Each time the degrees of freedom in either the numerator or the denominator change, a new distribution is created.

II. The F distribution is used to test whether two sample variances come from the same or equal populations.
A. The sampled populations must be normal.

B. The ratio of the two sample variances is computed and the result compared to the critical value of F.

C. The larger of the two sample variances is placed in the numerator, forcing the ratio to always be greater than 1.00.

III. ANOVA is used to compare two or more treatment means to determine if they come from the same or equal populations. A *treatment* is a source of variation.

   A. Three assumptions underly ANOVA:

     1. The samples are obtained from normal populations.

     2. The populations have equal standard deviations.

     3. The populations are independent.

   B. An ANOVA table is developed, which uses SS total, SST, and SSE. These values are computed as follows, where n is the total number of observations, T_c is a column total, and n_c is the number of observations in each treatment (column).

     1. SS total, the total sum of squares, is

$$SS \text{ total} = \Sigma X^2 - \frac{(\Sigma X)^2}{n} \qquad \boxed{\textbf{12–3}}$$

2. SST, the sum of squares due to treatment effect, is

$$SST = \Sigma\left(\frac{T_c^2}{n_c}\right) - \frac{(\Sigma X)^2}{n} \qquad \boxed{\textbf{12–1}}$$

3. SSE, the sum of squares within samples (error), is

$$SSE = \Sigma(X^2) - \Sigma\left(\frac{T_c^2}{n_c}\right) \qquad \boxed{\textbf{12–2}}$$

4. The F test statistic is

$$\frac{\dfrac{SST}{(k-1)}}{\dfrac{SSE}{(n-k)}} \qquad \boxed{\textbf{12–4}}$$

Exercises

11. Two brands of cigarettes have the same mean nicotine contents. However, there seems to be a difference in the amount of variation in the two brands. A random sample of 13 cigarettes of Brand A showed a standard deviation of 3.5 grams. A random sample of 8 cigarettes of Brand B showed a standard deviation of 2.0 grams. At the 0.05 significance level, can we conclude that there is more variation in Brand A? Find the p-value.

12. Last year's graduates in accounting and sales had the same mean starting salaries. However, the standard deviation was $2000 (sample of 8) among the accountants and $4200 (sample of 7) among the sales majors. At the 0.05 significance level, can we conclude there is more variation among the sales graduates?

13. A physician randomly selects 18 patients among those she is treating for high blood pressure.

These patients are randomly assigned to 3 groups and treated with 3 different drugs, all designed to reduce blood pressure. The amount of reduction (in millimeters of mercury) is shown. At the 0.01 level of significance, is there sufficient evidence to show that the drugs act differently?

Drug A	Drug B	Drug C
10	13	9
10	14	8
9	11	6
10	10	10
7	9	10
6	10	7

14. There are 4 different instructors for the history course Introduction to Western Civilization, taught at Scandia Tech. Following are the numbers of pages of reading assigned by each instructor every week for the first 5 weeks of the course. At the 0.05 level of significance, is there sufficient evidence to show differences in the average lengths of the readings assigned by the 4 instructors? State your conclusion and interpret the results. What is the p-value?

Mr. Barr	Dr. Sedwick	Dr. Reading	Dr. Faust
25	35	30	28
29	20	27	32
30	20	18	33
42	17	19	35
35	30	26	24

15. The merchandising manager for Food Mart grocery stores is analyzing the effect of various placements of candy displays within the company's stores. He decides to conduct an experiment by locating the display in a different area in each of 4 Food Mart outlets. The amounts (in pounds) sold at the various locations each week are recorded. Is there sufficient evidence to indicate that there are differences in sales at the various locations? Use the 5% significance level. State your conclusion and interpret the results.

Back of Store, Near Bread	Top Shelf, Near Cookies	End of Aisle, Near Meat	Third Shelf, Near Soda Pop
76	73	89	96
75	70	82	92
83	81	85	104
87	78	79	89
81	76	80	94

16. A psychologist wants to investigate the effect of social background on the time (in minutes) it takes freshmen to solve a puzzle. A random sample of students from different backgrounds is selected, resulting in the following data. Use the 0.05 level of significance to test the hypothesis that social background has no effect on the time required to solve the puzzle. State your conclusion and interpret the results.

Inner City	Urban	Suburban	Rural
16.5	10.9	18.6	14.2
5.2	5.2	8.1	24.5
12.1	10.8	6.4	14.8
14.3	8.9		24.9
	16.1		5.1

17. A wholesaler is interested in comparing the weights in ounces of grapefruit from Florida, Texas, and California:

Florida	Texas	California
12.6	12.8	16.0
13.8	13.2	15.1
14.0	12.4	13.9
	13.2	14.3
		15.0

a. What are the null and alternate hypotheses?
b. Fill in an ANOVA table.
c. What is the critical value of F, assuming the 0.01 level of significance?
d. What decision should the wholesaler make? Describe your conclusion. What is the p-value?

18. Theft in student rooms is a major problem at a large university. In an effort to reduce the problem, the university conducted an experiment in which 15 dormitories were randomly divided into 3 groups. In the first group of 4 dorms, all students were involved in structured discussion groups about the theft problem. Informal "peer" group meetings were held for the students in 5 other dorms. The 6 dorms in the third group were not subjected to any changes. The number of reported thefts in each dorm after 1 semester is shown by group at the top of the next page.

Cluster A Structured	Cluster B Informal	Cluster C Control
35	18	14
20	10	3
32	21	16
27	14	10
	13	11
		12

Do these data present sufficient evidence to indicate a difference among the 3 methods? Use the 0.01 level of significance. Describe your conclusion.

19. An oncologist—a physician who specializes in the treatment of tumors—has 24 patients with advanced lung cancer. He is aware of 3 treatments, reported in medical journals, that may gain remission for his patients. To assess the effect of the treatments, the doctor randomly assigns patients to each treatment and then keeps careful records on the numbers of days the patients live after treatment starts. At the 0.05 level, can it be concluded that there are any differences in the effects of the treatments?

Laetrile	Chemotherapy	Radiation
75	80	64
88	82	90
62	64	58
97	45	64
62	67	82
81	84	71
93	55	59
	39	66
	60	

20. Applicants for a position as a school bus driver are given a psychological test to determine their abilities to stay calm under difficult conditions. The following table gives the scores of the 13 applicants by age:

Age of Applicant		
25–34	35–44	45–54
48	61	78
67	75	82
58	82	81
56	59	80
	69	

At the 0.05 significance level, do the means of the 3 age groups differ? Find the p-value.

21. Professor Lim, who teaches a large section of Introduction to Marketing, asked students to rate his instruction as excellent, good, fair, or poor. The results are shown below.

Excellent	Good		Fair	Poor
85	80	88	73	71
77	70	75	71	75
74	78	73	70	76
77	72	80	79	81
90	74	82	73	79
94	77	83	76	70
89	79	74	76	79
	78	76	68	82
	82	91	80	
	78		78	

A graduate student collected the ratings and assured the students that Professor Lim would not receive them until after final grades were filed in the Registrar's Office. After the grades were filed, the rating given by each student was matched with the student's score on the final examination. At the 0.05 significance level, are there differences in the mean scores of the four rating groups?

22. An independent testing agency has been hired by a large manufacturer of tires. Specifically, the tire manufacturer would like to know if there are differences in the tread wear of a tire on a variety of road surfaces. To assure uniformity, the testing agency measured the wear on only the right front tire, drove the cars in all types of traffic, and attempted to ensure that the weight was the same in all test vehicles. The data (in millimeters of wear) are shown below.

Type of Road Surface			
Concrete	Composite	Brick	Gravel
8.8	10.1	11.9	13.4
9.6	10.1	11.1	13.0
8.3	10.3	11.0	11.9
9.3	9.8	12.1	12.6
9.1	9.9	12.6	12.7
8.3	10.6	10.9	13.0
8.4	10.8	11.8	
	10.3	12.9	
		12.3	

At the 0.05 significance level, are there differences in the mean amount of wear from the various surfaces?

23. The Plumber's Union has gathered data on the hourly wages of random samples of plumbers in 4 southwestern cities. At the 0.05 significance level, can it be shown that the mean hourly wages differ in the 4 cities?

City			
A	B	C	D
$14.20	$14.40	$16.20	$15.40
15.40	15.20	15.40	15.80
14.80	15.80	16.00	16.60
13.80	16.20	16.00	14.80
	16.00	15.80	
	15.60		

24. A health spa has 2 1-week programs for grossly overweight persons. A client may select either of these. To assess the effectiveness of 1 program over the other, a sample of 300-pound persons was selected and their weight losses (in pounds) recorded. Seven were in Program A and 9 in Program B.

Weight Loss	
Program A	Program B
40	39
38	21
41	29
52	42
27	43
32	28
40	29
	28
	46

Use the 0.05 level of significance to test whether there is a statistically significant difference between the 2 means of the weight-loss programs. What is the p-value? Explain what it indicates.

25. The health spa also has 3 programs designed to lower stress. An incoming group of patients was randomly assigned to the 3 programs. The reductions in stress (in percent) after the week-long programs are shown below.

Stress Reduction		
Program A	Program B	Program C
27	17	22
21	26	31
18	33	16
32	26	18
26		27
		32

Test at the 0.01 level the hypothesis that there are no differences in the effectiveness of the 3 stress-reducing programs.

26. The manager of Sally's Hamburger World suspects that the mean number of customers between 4 P.M. and 7 P.M. differs by day of the week. The data show the numbers of customers during that period for randomly selected days of the week. At the 0.01 significance level, can it be concluded that the number of customers differs by day of the week?

Monday	Tuesday	Wednesday	Thursday	Friday
86	77	69	78	84
96	102	91	77	88
78	54	86	90	94
66	98	74	84	102
100		82	72	96
		78	74	
		84		

27. Suppose the United States has 2 sources for short-range rockets. The rockets built by the first source have a mean target error of 68 feet, with a standard deviation of 18 feet, for a sample of 8 rockets. Those built by the second have a mean target error of 48 feet, with a standard deviation of 12 feet, for a sample of 6 rockets. At the 0.05 significance level, can we conclude that the population variances are the same? If yes, can we conclude that there is more mean error in the first source's rockets at the 0.05 significance level? Find the p-value.

28. An investor is trying to decide between oil stocks and utility stocks. He is concerned that there is more variability in the oil stocks. For a sample of 10 oil stocks, the standard deviation of the dividends paid was $12.51. For a sample of 8 utility stocks, the standard deviation of the dividends paid was $4.56. At the 0.05 significance level, can we conclude there is more variability in the oil stocks?

DATA EXERCISES

29. Refer to the real estate data, which reports information on homes sold in Alabama during 1996.
  a. Are there differences in the mean selling prices of the homes with different numbers of bathrooms? Use the number of bathrooms as a treatment variable, and test the hypothesis to determine if there is no difference in treatment means. Use the 0.05 significance level.
  b. Are there differences in the mean selling prices of the homes with different numbers of bedrooms? Use the number of bedrooms as a treatment variable, and test the hypothesis to determine if there is a difference in treatment means. Use the 0.05 significance level.
30. Refer to the schools data set, which has information on 94 school districts.
  a. Consider the 3 sizes of school district: small (less than 1000 students), medium (between 1000 and 3000 students), and large (more than 3000 students). Are there differences among the 3 in terms of the mean amount spent on instruction? Use the 0.05 significance level.
  b. Now treat the data as samples from 3 populations based on the percentage of students on welfare: "low" (less than 5%), "moderate" (between 5 and 10%, inclusive), and "high" (more than 10%). Compare the mean numbers of these students per district for these 3 samples at the 0.05 significance level.

CHAPTER ACHIEVEMENT TEST

The answers are at the back of the book.

MULTIPLE-CHOICE QUESTIONS

Select the response that best answers each of the questions.

Questions 1–4 refer to the following problem: Mr. Tourtellotte can drive to work along one of three different routes. The following data show the numbers of minutes it takes to make the trip on five different occasions for each route. Is there sufficient evidence to indicate that there are differences in the mean times it takes to drive the three routes?

Via Expressway	Via Downtown	Past the University
33	22	14
35	26	21
34	17	24
32	18	25
38	20	15

1. The number of treatments is
   a. 3.
   b. 2.
   c. 12.
   d. 15.
   e. 14.
2. The number of degrees of freedom for the denominator of the F distribution is
   a. 3.
   b. 2.
   c. 12.
   d. 15.
   e. 14.
3. If the 0.05 significance level is used, the critical value of F is
   a. 19.4.
   b. 3.49.
   c. 3.89.
   d. 6.93.
   e. 1.96.
4. The value of F was computed to be 23.07. The correct conclusion is to
   a. not reject H_0.
   b. reject H_a.
   c. reject H_0 and accept H_a.
   d. find the samples too small to be appropriate.

5. The ANOVA technique was used to compare 3 population means based on samples of size 9 drawn from each population. It was found that $SST = 90$ and $SSE = 200$. At the 0.05 level of significance, the correct conclusion is
   a. $F = 0.50$; do not reject H_0.
   b. $F = 2.22$; do not reject H_0.
   c. $F = 5.40$; reject H_0.
   d. $F = 12$; reject H_0.
   e. that the value of F cannot be determined from the information given.
6. In a particular ANOVA test, the calculated value of F is between zero and the value of the F table. The correct conclusion is to
   a. not reject H_0 and to conclude that the treatment means being tested are not significantly different.
   b. not reject H_0 and to conclude that the treatment means being tested differ significantly.
   c. reject H_0, accept H_a, and conclude that the treatment means are not significantly different.
   d. reject H_0, accept H_a, and conclude that the treatment means are significantly different.
7. If the sample means for the individual treatment groups were identical, the value of the F statistic would be
   a. equal to 1.00.
   b. zero.
   c. infinite.
   d. a negative number.
   e. a number between 0 and 1.00.
8. The F distribution
   a. cannot be negative.
   b. is positively skewed for small samples.
   c. is determined by two parameters.
   d. All of the above
9. Suppose the within-samples (error) estimate of the variance was found to be negative. This means that
   a. the null hypothesis is not rejected.
   b. the null hypothesis is rejected.
   c. we should have used the t distribution.
   d. a mistake was made in arithmetic because the variance cannot be negative.
10. Rejecting H_0 in an ANOVA test indicates that there is a significant difference

a. between at least one pair of means.
b. between all means.

c. between the F values.
d. between SST and SS total.

COMPUTATION PROBLEM

11. A manufacturer of automobiles is testing a new design for brakes. The director of engineering reports that test data are available for both the existing design and two proposed designs. Analysts obtained test results by measuring the stopping distances (in feet) of the cars traveling at a speed of 15 miles per hour:

Existing Design	First Proposal	Second Proposal
5	5	8
7	5	4
6	8	5
	7	9
	6	

a. What are the null and alternate hypotheses?
b. Complete an ANOVA table.
c. What is the critical value of F, assuming the 0.05 level of significance?
d. Can we conclude that there are differences in the mean stopping distances?
e. What is the p-value?
f. What assumption should you make to perform the test?

ANSWERS TO SELF-REVIEW PROBLEMS

12-1 $H_0: \sigma_a^2 \leq \sigma_b^2$
$H_a: \sigma_a^2 > \sigma_b^2$
H_0 is rejected if $F > 4.82$.
$F = (0.90)^2/(0.40)^2 = 5.0625$
H_0 is rejected. There is more variability in Operator A's work.

12-2 a. Assume a normal population with equal variances.

b.

3	7	4
5	6	6
4	8	5
3	9	7
4	6	6
	8	5
	8	4
		3
$\overline{19}$	$\overline{52}$	$\overline{40}$

$\Sigma X = 111$

Then:

$$SST = \frac{(19)^2}{5} + \frac{(52)^2}{7} + \frac{(40)^2}{8} - \frac{(111)^2}{20}$$

$$= 42.44$$

c. $SSE = 3^2 + 5^2 + 4^2 + \cdots + 3^2$

$$- \left[\frac{(19)^2}{5} + \frac{(52)^2}{7} + \frac{(40)^2}{8} \right]$$

$$= 681.000 - 658.49 = 22.51$$

d. $42.44 + 22.51 = 64.95$

e.

Source of Variation	Sum of Squares	df	Mean Square
Between treatments	42.44	2	21.22
Within samples	22.51	17	1.32
Total	64.95		

$$F = \frac{21.22}{1.32} = 16.08$$

f. The critical value of F is 3.59, found by 0.05 level, 2 df in numerator, 17 df in denominator. Since $16.08 > 3.59$, H_0 is rejected. The means are not all equal.

g. Since 16.08 is also greater than 6.11, the p-value is less than 0.01.

Unit Review

In the last two chapters, we examined small-sample methods of hypothesis testing for both means and variances from normal populations. First, we used Student's t distribution as the test statistic when we compared a single sample mean to a population mean. Next, we compared two sample means, again using the t distribution, to determine if they came from populations with the same mean. Then we used the F distribution to test the hypothesis that two populations had the same variance. We also used the F distribution to compare more than two populations to determine if they have the same mean.

. . . .

KEY CONCEPTS

1. **Student's t distribution** is used to test hypotheses about the mean of a single population, the means of two populations, and the difference in related observations. In each of these instances, the population is normal, with an unknown standard deviation. Usually, the sample size is smaller than 30. The t distribution has the following characteristics:
   a. It is a continuous distribution.
   b. It is somewhat bell-shaped and symmetrical.
   c. There is a family of t distributions; that is, each time the sample size changes, the t distribution changes.
   d. As the sample size increases, the t distribution approaches the normal distribution. Often, when $n > 30$, the standard normal distribution is used instead of t because the values of these 2 statistics are close.
   e. The t distribution is more spread out than the normal distribution.
2. When testing the means of more than two populations simultaneously, we use the **analysis of variance (ANOVA).** The test statistic follows the F **distribution,** which has the following characteristics:
   a. It is either zero or positive (it cannot be negative).
   b. It is a continuous distribution.
   c. It is positively skewed.
   d. The F distribution is based on the ratio of two variances. There is a different F distribution each time either of the sample sizes is changed.
3. For the F distribution to be employed to test the difference in more than two populations, the following conditions must be met:
   a. The data must be at least of interval scale.
   b. The populations should be approximately normally distributed.
   c. The population variances should be equal.
   d. The samples must be randomly selected.
4. The F distribution is also used to test the hypothesis that two populations have equal variances.

. . . .

KEY TERMS

Student's t distribution	Dependent samples	Sum of squares error
Degrees of freedom	F distribution	Sum of squares treatment
Pooled variance estimate	Treatment	Analysis of variance (ANOVA)
Independent samples	Sum of squares total	

. . . .
KEY SYMBOLS

t Depending on its usage, t refers to the t test itself, the computed value of t, or the critical value of t.

df Degrees of freedom.

s_p Pooled estimate of the population standard deviation.

s_d Standard deviation of the paired differences.

F The F probability distribution.

k The number of treatments.

n The total number of observations in the sample.

SST Sum of squares treatment (between treatments).

SSE Sum of squares error (within samples).

SS total Sum of squares total.

 **CASE STUDY** . . . . **Medical Center Scheduling**

Jean Dempsey manages the emergency care facility at Grove Bell Medical Center. One of her responsibilities is to staff enough nurses so that incoming patients who call for service can be handled promptly. It certainly would not be appropriate to ask a patient with an emergency to wait a long time because she does not have enough staff. She has recorded the number of calls for service actually received over the last five weeks. She has divided each day into three segments, and the center is not open on Sundays. Are there any statistically significant differences among the various periods that she should know about when doing scheduling?

Calls	Day	Shift
13	Mon	morn
17	Mon	after
8	Mon	night
12	Tue	morn
9	Tue	after
7	Tue	night
17	Wed	morn
16	Wed	after
5	Wed	night
8	Thur	morn
13	Thur	after
7	Thur	night

Calls	Day	Shift
13	Fri	morn
12	Fri	after
8	Fri	night
4	Sat	morn
5	Sat	after
4	Sat	night
11	Mon	morn
9	Mon	after
5	Mon	night
12	Tue	morn
14	Tue	after
6	Tue	night
19	Wed	morn
11	Wed	after
6	Wed	night
10	Thur	morn
4	Thur	after
5	Thur	night
9	Fri	morn
13	Fri	after
7	Fri	night
7	Sat	morn
3	Sat	after
3	Sat	night
13	Mon	morn

Calls	Day	Shift	Calls	Day	Shift
9	Mon	after	8	Thur	after
8	Mon	night	8	Thur	night
8	Tue	morn	14	Fri	morn
10	Tue	after	13	Fri	after
8	Tue	night	5	Fri	night
10	Wed	morn	5	Sat	morn
7	Wed	after	10	Sat	after
3	Wed	night	3	Sat	night
11	Thur	morn	13	Mon	morn
12	Thur	after	13	Mon	after
8	Thur	night	6	Mon	night
15	Fri	morn	12	Tue	morn
8	Fri	after	10	Tue	after
4	Fri	night	9	Tue	night
6	Sat	morn	12	Wed	morn
4	Sat	after	14	Wed	after
4	Sat	night	4	Wed	night
16	Mon	morn	14	Thur	morn
10	Mon	after	17	Thur	after
8	Mon	night	4	Thur	night
11	Tue	morn	14	Fri	morn
6	Tue	after	14	Fri	after
7	Tue	night	4	Fri	night
12	Wed	morn	6	Sat	morn
13	Wed	after	10	Sat	after
6	Wed	night	4	Sat	night
14	Thur	morn			

CHAPTER 13

Correlation Analysis

When you have completed this chapter, you will be able to

- describe the relationship between two variables;
- compute and interpret Pearson's coefficient of correlation;
- compute and interpret the coefficient of determination;
- test the statistical significance of the coefficient of correlation;
- compute and interpret Spearman's coefficient of rank correlation.

CHAPTER PROBLEM   **Our Books Are Priceless!**

Hannah Simpson is the president of student government at North Central State University. For some time, she and her colleagues have been concerned with the increasing cost of textbooks. At a recent meeting, Ms. Simpson decided to form a committee to examine the matter. What factors affect the cost of a textbook? The number of pages? The number of pictures? The number of formulas? Should the student committee look at books in technical areas, such as mathematics and chemistry, to find out if they cost more than, say, sociology and psychology textbooks?

INTRODUCTION

In Chapters 10–12, we dealt with hypothesis tests involving means and proportions. The techniques concentrated on only a *single* feature of the sampled item, such as income. This chapter begins a study of the relationship between *two or more* variables. We may want to determine if there is any relationship between the number of years of education completed by federal government employees and their incomes. Or we may want to explore one of these questions: Does the crime rate in inner cities vary with the unemployment rate in those cities? Is there a relationship between the amount of money spent advertising a product, such as a toothpaste, and its sales? Is there any relationship between the number of hours studied and a student's grade on an examination? Note that in each case, there are two separate characteristics—years of education and income, for example.

correlation analysis The statistical techniques used to determine the strength of the relationship between two variables.

CORRELATION ANALYSIS

The study of the relationship between two variables is called **correlation analysis**. The objective of correlation analysis is to study the strength of the association between two variables. Our attention will focus first on the correlation between two interval-scaled variables and then on the relationship between two ordinal-scaled variables.

The Scatter Diagram

One tool that is very useful for visualizing the relationship between two variables is the **scatter diagram**.

Problem   Recall from the Chapter Problem that a student committee was formed to look into the factors affecting the cost of textbooks. After some discussion within the committee and consultation with book publishers, the committee decided to first consider the relationship between the cost of the text and its length.

scatter diagram A graphic tool that portrays the relationship between two variables.

Solution   The committee selected a sample of six textbooks currently in use at North Central. For each selected text, the committee determined the number of pages and the selling price. The results are reported in Table 13–1.

We call the selling price the dependent variable and the number of pages the independent variable. The dependent variable is thus the variable being estimated, and the independent variable is the variable used as the estimator. It is traditional to put the dependent variable on the vertical axis (Y) and the independent variable on the horizontal axis (X) of the scatter diagram. We plot the paired data for *Introduction to Psychology* ($X = 400$, $Y = 44$) by moving to 400 on the X-axis, then moving vertically to a position opposite 44 on the Y-axis, and placing a dot at that intersection (see the scatter diagram in Figure 13–1). This process is continued for all the books in the sample. The completed scatter diagram is shown in Figure 13–1.

TABLE 13–1	Numbers of Pages and Selling Prices for a Sample of North Central Textbooks		
Title		**Number of Pages**	**Selling Price**
Introduction to Psychology		400	$44
Behavior in Organizations		600	47
Introduction to Criminology		500	48
Principles of Sociology		600	48
Nursing Education		400	43
Introduction to Residential Construction		500	46

FIGURE 13–1   **Scatter Diagram of Numbers of Pages and Selling Prices**

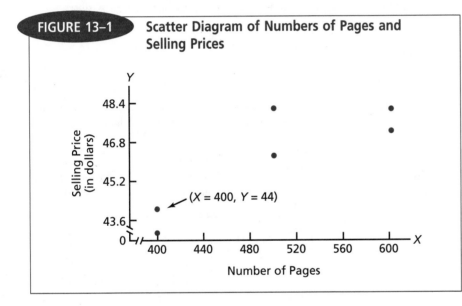

As will be explained shortly, there is indeed a very strong relationship (correlation) between the number of pages in each book and its selling price. A book like *Principles of Sociology*, which is relatively long (600 pages), sells for a higher-than-average price ($48).

Self-Review 13–1

Answers to the Self-Review problems are at the end of the chapter.

An astute college recruiter noted that the enrollment figures at the local campus of the state university seemed to fluctuate with the unemployment rate for the region. To establish whether or not his suspicions were justified the recruiter collected relevant unemployment and enrollment figures for the same time periods. Her findings were

Percentage Unemployed X	State University Campus Enrollment Y
3.4	13,500
3.3	14,500
4.6	15,200
5.1	14,900
4.5	14,700
3.5	14,300
5.7	15,700
7.6	17,100

a. Construct a scatter diagram for the paired data.

b. As unemployment increases, does enrollment appear to increase, decrease, or remain the same?

Exercises

Answers to the even-numbered Exercises are at the back of the book.

1. A Peace Corps specialist in soil management is studying the relationship between the mean temperature and the yield in bushels per acre for a crop. He collected the following information for several regions:

Region	Temperature (in °C) X	Yield (in bushels per acre) Y
1	4	1
2	8	9
3	10	7
4	9	11
5	11	13
6	6	7

a. Construct a scatter diagram for the paired data.

b. As temperature increases, does yield appear to increase, decrease, or remain the same?

2. A human resources trainee believes that there is a relationship between the number of years of employment with the company and the number of days a year an employee is absent from work. She gathers the following information from the company records of six employees picked at random:

Employee	Length of Employment (in years) X	Number of Absences Last Year (in days) Y
Phuong	1	8
Fadale	5	1
Sasser	2	7
Dilone	4	3
Thelin	4	2
Tortelli	3	4

a. Construct a scatter diagram for the paired data.

b. As the years of employment increase, do absences appear to increase, decrease, or remain the same?

3. Kuppenheimer Men's Apparel wants to determine the relationship between the number of suits sold during a day and the number of salespeople. A random sample of the company's files resulted in the paired information listed at the top of the next page.

Number of Salespeople on Duty X	Number of Suits Sold Y
3	7
1	5
2	6
3	6
4	9
5	10
1	4

First Die X	Second Die Y
6	4
4	4
5	1
6	2
1	3

Plot a scatter diagram with the number of salespeople on the X axis and the number of suits sold on the Y axis. Does it appear that more suits are sold when more salespeople are on duty?

4. A pair of dice is rolled five times, and the number of spots appearing face up is noted each time. The results are

Develop a scatter diagram scaling the first die on the X axis and the second die on the Y axis. What comments can you make? Would you expect to find a relationship?

. . . .

THE COEFFICIENT OF CORRELATION

About 1900, Karl Pearson, who made significant contributions to the science of statistics, developed a measure that describes the relationship between two sets of *interval-scaled variables*. It is called the **coefficient of correlation** and is designated by the letter *r*. Pearson's *r* is also known as the product-moment correlation coefficient, partly to distinguish it from other correlation coefficients. It is a valid measure of correlation if the relationship between the variables is *linear*. As illustrated by the scatter diagram in Figure 13–2, if all data points lie on a straight line, the correlation coefficient is either +1.00 or −1.00, depending on the direction of the slope of the line. Coefficients of +1.00 and −1.00 describe *perfect correlation*.

If there is no linear relationship between X and Y, *r* will be zero, and the points on the corresponding scatter diagram will be randomly scattered. These situations and several others are portrayed in Figure 13–3.

The strength and direction of the coefficient of correlation are summarized in the diagram on page 405. Negative numerical values, such as −0.92 and −0.48, signify inverse correlation, whereas positive numerical values, such as +0.83 and +0.46, indicate direct correlation. The closer Pearson's *r* is to ±1.00, the greater the strength of the correlation is. Note, however, that *the strength of the correlation is not dependent on the direction*. Therefore, −0.12 and +0.12 are equal in strength (both weak). Coefficients of +0.94 and −0.94 are also equal in strength (both very strong). The diagram at the top of page 404 summarizes this information.

How is the value of the coefficient of correlation determined? Suppose we draw a vertical line through the mean X-value and a horizontal line

coefficient of correlation
A measure of the strength of the association between two variables.

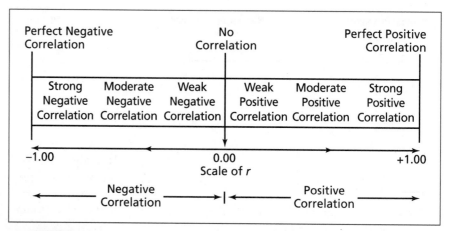

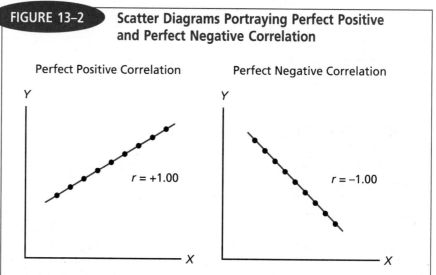

FIGURE 13–2   **Scatter Diagrams Portraying Perfect Positive and Perfect Negative Correlation**

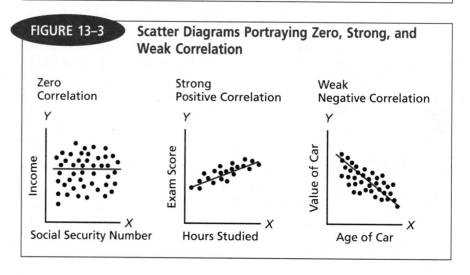

FIGURE 13–3   **Scatter Diagrams Portraying Zero, Strong, and Weak Correlation**

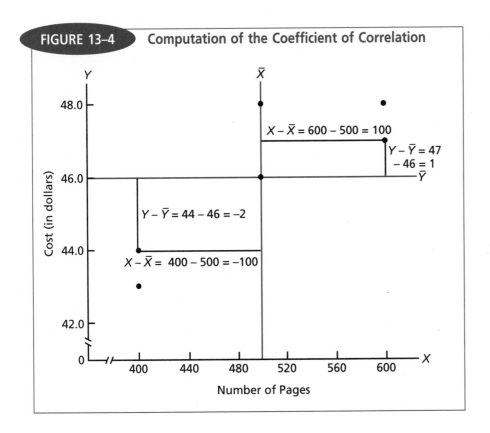

FIGURE 13–4   **Computation of the Coefficient of Correlation**

through the mean Y-value. For example, look at Figure 13–4, where the data on the numbers of pages and the selling prices of the sampled text-books are presented. A vertical line is added through 500 pages, the mean number of pages for the 6 textbooks in the sample. A horizontal line is drawn through $46, the mean cost of the 6 sampled texts. These lines pass through the "center" of the data and divide the scatter diagram into 4 quadrants.

If the 2 variables are directly related, when the selling price is above the mean, the number of pages is as well. These points appear in the upper-right quadrant. Similarly, when the selling price is below the mean, so is the number of pages; these points fall in the lower-left quadrant. The first text, *Introduction to Psychology,* for example, is 100 pages shorter than the mean length and sells for $2 less than the mean selling price. It is therefore located in the lower-left quadrant. The second text, *Behavior in Organizations,* is 100 pages longer than the mean length and sells for $1 more than the mean selling price. It is therefore located in the upper-right quadrant. The deviations from the mean number of pages and from the mean selling price are summarized in Table 13–2 for the entire sample of the 6 text-books.

TABLE 13–2	Deviations from the Means and Their Products				
Title	Number of Pages	Selling Price	$X-\overline{X}$	$Y-\overline{Y}$	$(X-\overline{X})(Y-\overline{Y})$
Introduction to Psychology	400	$44	−100	−2	200
Behavior in Organizations	600	47	100	1	100
Introduction to Criminology	500	48	0	2	0
Principles of Sociology	600	48	100	2	200
Nursing Education	400	43	−100	−3	300
Introduction to Residential Construction	500	46	0	0	0
					800

In both the upper-right and the lower-left quadrants, the product of $(X - \overline{X})(Y - \overline{Y})$ is positive because both of the factors have the same sign. This happens for all the points in the textbook example. The resulting correlation coefficient is a positive value.

If two variables are inversely related, one variable will be above the mean and the other below the mean. Most of the points in this case occur in the upper-left and lower-right quadrants. Now $(X - \overline{X})$ and $(Y - \overline{Y})$ will have opposite signs, so their product is negative. The resulting correlation coefficient is negative.

What happens if there is no linear relationship between two variables? The points on the scatter diagram will appear in all four quadrants. The positive products of $(X - \overline{X})(Y - \overline{Y})$ will be offset by the negative products, and their sum will be near zero. This leads to a correlation coefficient near zero.

Pearson also wanted r to be unaffected by the particular scales being used. For example, if we had used cents instead of dollars in our textbook example, we would want the value of the correlation coefficient to be the same. The correlation coefficient is independent of the scale used if we divide the term $\Sigma(X - \overline{X})(Y - \overline{Y})$ by the sample standard deviations. It is also made independent of the sample size and bounded by the values $+1.00$ and -1.00 if we divide by $(n - 1)$.

This reasoning leads to the following formula:

$$r = \frac{\Sigma(X - \overline{X})(Y - \overline{Y})}{(n - 1)(s_y)(s_x)} \qquad \boxed{13-1}$$

To compute the coefficient of correlation, we use the standard deviations of the sample of 6 page lengths and 6 selling prices. The standard deviation of the length is 89.44 pages, and the standard deviation of the selling price is $2.10. First, formula 4–8 is used to compute the sample variance; then we take the square root to get the sample standard deviation:

$$s_x^2 = \frac{\Sigma X^2 - \frac{(\Sigma X)^2}{n}}{n - 1} = \frac{1{,}540{,}000 - \frac{(3000)^2}{6}}{6 - 1} = 8000$$

$$s_x = \sqrt{8000} = 89.44$$

$$s_y^2 = \frac{\Sigma Y^2 - \frac{(\Sigma Y)^2}{n}}{n - 1} = \frac{12{,}718 - \frac{(276)^2}{6}}{6 - 1} = 4.40$$

$$s_y = \sqrt{4.40} = 2.10$$

We now insert these values into formula 13–1 to determine the coefficient of correlation:

$$r = \frac{\Sigma(X - \overline{X})(Y - \overline{Y})}{(n - 1)(s_y)(s_x)} = \frac{800}{(6 - 1)(89.44)(2.10)} = 0.85$$

The coefficient of correlation, r, can also be computed by the following formula, which is based on the actual values of x and y:

$$r = \frac{n(\Sigma XY) - (\Sigma X)(\Sigma Y)}{\sqrt{\left[n(\Sigma X^2) - (\Sigma X)^2\right]\left[n(\Sigma Y^2) - (\Sigma Y)^2\right]}} \qquad \boxed{13\text{--}2}$$

where

n is the number of paired observations.
ΣX is the sum of the X variable.
ΣY is the sum of the Y variable.
ΣX^2 is the X variable squared and the squares summed.
$(\Sigma X)^2$ is the X variable summed and the sum squared.
ΣY^2 is the Y variable squared and the squares summed.
$(\Sigma Y)^2$ is the Y variable summed and the sum squared.

Problem The data on selling prices and numbers of pages in the textbooks are repeated below.

Title	Number of Pages X	Selling Price Y
Introduction to Psychology	400	$44
Behavior in Organizations	600	47
Introduction to Criminology	500	48
Principles of Sociology	600	48
Nursing Education	400	43
Introduction to Residential Construction	500	46

Determine the coefficient of correlation using formula 13–2.

Solution The totals and the sums of squares needed are in Table 13–3. Computing gives

TABLE 13-3   Calculations Needed to Determine the Coefficient of Correlation

Title	X	Y	XY	X²	Y²
Introduction to Psychology	400	44	17,600	160,000	1936
Behavior in Organizations	600	47	28,200	360,000	2209
Introduction to Criminology	500	48	24,000	250,000	2304
Principles of Sociology	600	48	28,800	360,000	2304
Nursing Education	400	43	17,200	160,000	1849
Introduction to Residential Construction	500	46	23,000	250,000	2116
	3000	276	138,800	1,540,000	12,718

$$r = \frac{n(\Sigma XY) - (\Sigma X)(\Sigma Y)}{\sqrt{[n(\Sigma X^2) - (\Sigma X)^2][n(\Sigma Y^2) - (\Sigma Y)^2]}}$$

$$= \frac{6(138,800) - (3000)(276)}{\sqrt{[6(1,540,000) - (3000)^2][6(12,718) - (276)^2]}}$$

$$= \frac{4800}{\sqrt{(240,000)(132)}} = 0.85$$

Because 0.85 is close to the perfect correlation value of 1.00, the student committee at North Central State can conclude that there is a strong positive relationship between the number of pages in a book and its selling price. To put it another way, as the number of pages increases, so does the price of the book.

Self-Review 13–2

Is there a relationship between the numbers of votes received by candidates for public office and the amounts spent on their campaigns? The following sample information was gathered for a recent election:

Candidate	Amount Spent on Campaign (in thousands of dollars) X	Votes Received (in thousands) Y
Weber	$3	14
Taite	4	7
Spencer	2	5
Lopez	5	12

a. Draw a scatter diagram.
b. Compute the Pearson coefficient of correlation.
c. Interpret the correlation coefficient.

Exercises

5. In Exercise 1, a soil management specialist was studying the relationship between the average temperature (in °C) and the yield in bushels per acre for a certain fall crop. The data are repeated below.

Region	Temperature (in °C) X	Yield (in bushels per acre) Y
1	4	1
2	8	9
3	10	7
4	9	11
5	11	13
6	6	7

Compute the coefficient of correlation, and interpret it.

6. In Exercise 2, a human resources trainee was studying the relationship between the number of years of employment with the company and the number of days absent from work last year. To repeat, the following information was gathered from company records:

Employee	Length of Employment (in years) X	Number of Absences Last Year (in days) Y
Phuong	1	8
Fadale	5	1
Sasser	2	7
Dilone	4	3
Thelin	4	2
Tortelli	3	4

Compute the coefficient of correlation, and interpret it.

7. Is age related to the length of stay of surgical patients in a hospital? We gathered the following sample information to study the relationship:

Age	Numbers of Days in Hospital
40	11
36	9
30	10
27	5
24	12
22	4
20	7

a. Draw a scatter diagram. X is age and Y is days in hospital.
b. Compute the coefficient of correlation.
c. Interpret the results.

8. Is there a relationship between the number of golf courses in the United States and the divorce rate? The following sample information was collected for six regions:

Number of Golf Courses X	Divorce Rate (per 1000 population) Y
28	2.2
38	2.5
47	3.5
54	4.1
62	4.8
66	5.0

a. Draw a scatter diagram.
b. Compute the coefficient of correlation.
c. Interpret the results.

THE COEFFICIENT OF DETERMINATION

coefficient of determination The proportion of the total variation in one variable that is explained by the other variable.

The coefficient of correlation allowed us to make statements such as "The relationship between the two variables is very strong." But how can one measure "very strong"? A measure of association that does have a more precise meaning is the **coefficient of determination.**

This technique results in a proportion, or percentage, that makes it relatively easy to arrive at a precise interpretation. We compute it by squaring the coefficient of correlation. The coefficient of determination may vary from 0 to 1.00 or, converted to a percentage, from 0 to 100%. It is usually represented by r^2.

To illustrate its computation and meaning, let us return to the Chapter Problem, where the relationship between the numbers of pages and the selling prices of a sample of 6 textbooks was considered. The coefficient of correlation, r, is computed to be 0.85. The coefficient of determination, r^2, is 0.72, found by $(0.85)^2$. Thus, the student committee can conclude that the variation in page lengths of the textbooks explains, or accounts for, 72% of the variation in selling price.

The magnitude of the coefficient of determination is smaller than that of the coefficient of correlation. It is a more conservative measure of the relationship between 2 variables, and for this reason, many statisticians prefer it. To put it another way, the coefficient of correlation tends to overstate the association between 2 variables. A correlation coefficient of 0.70, for example, would suggest that there is a fairly strong relationship between 2 variables. Squaring r, however, gives a coefficient of determination of 0.49, which is a somewhat smaller value; that is, $(0.70)^2 = 0.49$. The major weakness of the coefficient of determination is that it does not tell us anything about the direction of the association. Why is this so? When we square the value of r, the result is always a positive number.

Self-Review 13–3

In Self-Review 13–2, you computed the coefficient of correlation between campaign expenses and number of votes received to be 0.43. Determine the coefficient of determination, and interpret is value.

Exercises

9. A study is made of the relationship between the number of passengers on an aircraft (X) and the total weight in pounds of luggage stored in the aircraft's baggage compartment (Y). The coefficient of correlation is computed to be 0.94. Determine the coefficient of determination, and interpret its value.

10. The coefficient of correlation between the number of people on the beach at 4 P.M. and the high temperature for that day at Sunner's Creek is computed to be 0.96. Determine the coefficient of determination, and interpret its value.

11. In a government study of the relationship between the tar content and the nicotine content

(in milligrams) of a cigarette, the coefficient of correlation is 0.43. Determine the coefficient of determination, and interpret its value.

12. At Middletown High, a coefficient of correlation of -0.68 was found between the number of high school activities offered and the number of drug-related suspensions. Determine the coefficient of determination, and interpret its value.

TESTING THE SIGNIFICANCE OF THE COEFFICIENT OF CORRELATION

The student committee at North Central State University selected only 6 text-books for a study of the relationship between selling prices and number of pages. The coefficient of correlation was calculated to be 0.85, indicating a very strong positive association between the two variables. The question arises, however, whether it is possible—due to the small size of the sample—that the correlation in the population is really zero and the apparent relationship is due to chance. The "population" in this case might be *all* textbooks sold at the university.

To decide formally whether the correlation in the population could be zero, we apply the five-step hypothesis-testing procedure used in Chapters 10–12. Assuming the two variables are normally distributed. The statement that the coefficient of correlation in the population is zero becomes our null hypothesis. The alternate hypothesis is that the coefficient of correlation in the population is not zero. The population correlation coefficient is usually represented by the Greek letter rho (ρ). Symbolically, then, the null hypothesis and alternate hypothesis are

$$H_0: \rho = 0$$

$$H_a: \rho \neq 0$$

The appropriate test statistic follows the t distribution, with $n - 2$ degrees of freedom, and has the following formula:

$$t = \frac{r\sqrt{n - 2}}{\sqrt{1 - r^2}}$$

13-3

where

r is the sample coefficient of correlation.
n is the number of items in the sample.

If the computed t-value does not lie in the rejection region, we fail to reject the null hypothesis, and we conclude that there is no correlation between the variables. A computed value of t in the rejection region indicates that there is a correlation in the population between the two variables. The following diagram depicts the situation:

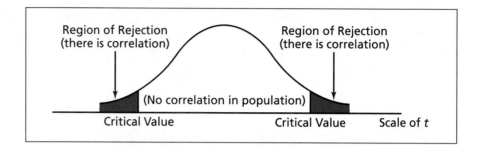

Problem   Dr. Alan Schwissow, a physician specializing in internal medicine, wants to investigate the relationship between systolic blood pressure and the weight of men over age 40. He collected the following information on a sample of 25 patients:

Patient	Pressure	Weight	Patient	Pressure	Weight
1	109	174	13	166	206
2	124	204	14	158	265
3	118	160	15	182	210
4	128	187	16	112	159
5	120	158	17	162	203
6	130	189	18	130	195
7	130	206	19	176	185
8	114	151	20	114	194
9	116	180	21	124	212
10	122	162	22	114	160
11	116	238	23	118	215
12	112	167	24	109	164
			25	148	213

Plot the data. Does there seem to be an association between the systolic pressure and weight? Determine the coefficient of correlation and the coefficient of determination. At the 0.05 significance level, can Dr. Schwissow conclude that there is a positive association between the variables in the population? What is the *p*-value?

Solution   We used the Excel spreadsheet and the "ChartWizard" to develop the scatter diagram at the top of page 413.

The weight is plotted on the *X* axis and systolic blood pressure on the *Y* axis. Based on this chart, there seems to be a direct relationship between the variables; that is, a heavier patient seems to have higher systolic pressure. However, the association does not seem to be too strong. To determine the strength of the association we computed the coefficient of correlation using the Excel spreadsheet. The steps are

Tools ▶ Data Analysis ▶ Correlation

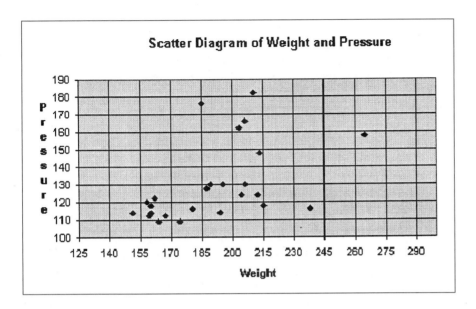

In the dialog box, indicate that the input range is a1:b25, that there are no labels, and that the output range is c1 (or wherever you want to put it).

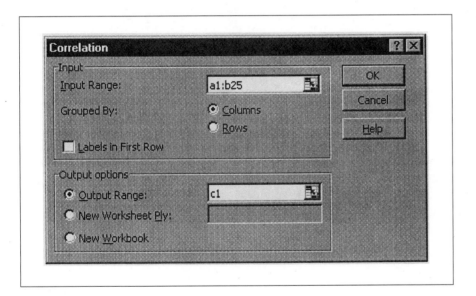

The output is

	Pressure	Weight
Pressure	1	
Weight	0.48463	1

The correlation between systolic blood pressure and weight is 0.48463, or 0.485. Thus, we conclude that there is a moderate positive association between the 2 variables. The coefficient of determination is 0.235, found by $(0.485)^2$. This value indicates that 23.5% of the variation in systolic blood pressure is accounted for by the variation in weight.

Dr. Schwissow would like to show that the amount of correlation is too large to occur by chance. To do this, his first step is to state the null and alternate hypotheses. He would like to show that there is a positive association, so we will conduct the following one-tailed test:

$$H_0 : \rho \leq 0$$

$$H_a : \rho > 0$$

There are 25 patients, so the degrees of freedom are equal to $n - 2 = 25 - 2 = 23$. To locate the critical value use Appendix D. Move down the left-hand margin to the row with 23 degrees of freedom and then move across that row to the column with the 0.05 significance level and a one-tailed test. The value is 1.714. The null hypothesis will be rejected if the computed value of t is greater than 1.714.

To determine the value of the test statistic, we use formula 13–3:

$$t = \frac{r\sqrt{n-2}}{\sqrt{1-r^2}} = \frac{0.485\sqrt{25-2}}{\sqrt{1-(0.485)^2}} = 2.660$$

Because the computed value is greater than the critical value, the null hypothesis is rejected. There is a positive association between systolic blood pressure and weight. The p-value for the test is less than 0.01 because the computed value of t is more than 2.500 (0.01), but less than 2.807 (0.005).

Self-Review 13–4

In Self-Review 13–3, we found that the correlation between number of votes received and campaign expense was 0.43. Using the 0.01 significance level, can we conclude that the correlation in the population is different from zero?

Exercises

13. A major airline selected a random sample of 25 flights and found that the correlation between the number of passengers and the total weight (in pounds) of luggage stored in the luggage compartment is 0.94. Using the 0.05 significance level, can we conclude that there is a positive association between the 2 variables?

14. A sociologist claims that the success of students in college (measured by their GPAs) is related to family income. For a sample of 20 students, the coefficient of correlation is 0.40. Using the 0.01 significance level, can we conclude that there is a positive correlation between the variables?

15. An Environmental Protection Agency study of 12 automobiles revealed a correlation of 0.47 between the engine size and the performance. At the 0.01 significance level, can we conclude that there is a positive association between these variables? What is the *p*-value? Interpret.

16. A study of college soccer games revealed the correlation between the number of shots attempted and the number of goals scored to be 0.21 for a sample of 20 games. Is it reasonable to conclude that there is a positive correlation between the 2 variables? Use the 0.05 significance level. Determine the *p*-value.

SPEARMAN'S COEFFICIENT OF RANK CORRELATION

In previous sections, we computed the coefficient of correlation between two interval- or ratio-scaled variables. In the study of the cost of textbooks at North Central, we measured the association between the cost of the book and the number of pages. Both of these variables used the ratio level of measurement. In the study involving the relationship between weight and blood pressure, again the variables were of the ratio scale. But how do we proceed if one or both of the variables being studied are on the ordinal scale? For example, a mathematics instructor wants to know the relationship between the order of finish on her test and the score earned; that is, do those who finish first also have the highest scores? The scores are interval scale, but the order of finish involves ranked data. Because the order of finish is ordinal (ranked) data, Pearson's coefficient of correlation is not appropriate.

Charles Spearman, a British statistician, introduced a measure of correlation for ordinal-level data. This measure allows us to study the relationship between sets of ranked data. For example, two staff members at the Office of Research at the University of California were asked to rank ten faculty research proposals in the order in which the proposals should be funded. We want to study the association between the ratings of the two staff members; that is, do the two staff members agree on which proposals are most worthy and which are least worthy? Spearman's coefficient of rank correlation, denoted r_s, provides a measure of the association. The coefficient of rank correlation is computed from the following formula:

$$r_s = 1 - \frac{6\Sigma d^2}{n(n^2 - 1)} \qquad \boxed{\text{13–4}}$$

where

d is the difference between the ranks for each pair of observations.
n is the number of paired observations.

Spearman's rank-order correlation coefficient ranges (as does Pearson's *r*) from −1.00 to +1.00, with a +1.00 indicating that the ranks are in perfect agreement. A −1.00 indicates that the variables are inversely related, but in perfect agreement. An r_s of 0 reveals that there is no relationship between the ranks.

Problem   The director of nursing at a city hospital asked two staff nurses to rank ten patients according to the difficulty of care required. A "difficult" assignment would include giving the patient a complete bath and monitoring the intravenous equipment. A less difficult assignment would be to give medication to a patient only at bedtime. Table 13–4 summarizes the ranking of each patient by the two nurses.

TABLE 13–4	Two Nurses' Rankings of Difficulty of Care for Ten Patients	
Patient	Nurse Scott's Ranking	Nurse Palmer's Ranking
Ms. Garcia	1	2
Ms. Hibner	2	1
Ms. Schaefer	3	5
Ms. Gerko	4	3
Ms. Owens	5	7
Mr. Bunt	6	6
Mr. Kolby	7	4
Mr. Cassow	8	10
Mr. Bianchi	9	9
Mr. Loftop	10	8

1. Compute Spearman's rank-order correlation coefficient.
2. At the 0.05 significance level, can we conclude that the ranks are positively correlated in the population?

Solution   The computations needed to determine the rank-order correlation coefficient are in Table 13–5.

Determining Spearman's rank-order correlation coefficient, we get

$$r_s = 1 - \frac{6\Sigma d^2}{n(n^2 - 1)} = 1 - \frac{6(28)}{10(99)} = 1 - 0.170 = 0.83$$

The coefficient of 0.83 indicates a very strong agreement between the 2 nurses with respect to the rankings given the patients.

Now we will determine if the ranks are positively correlated in the population. When the number of paired observations is 10 or more, we can use Student's t to determine the significance of Spearman's correlation coefficient. The sampling distribution follows the t distribution with $n - 2$ degrees of freedom. The computed value of t is found by formula 13–3.

TABLE 13–5	Two Nurses' Rankings and Calculations Needed for the Rank-Order Correlation Coefficient			
Patient	Nurse Scott's Ranking	Nurse Palmer's Ranking	Difference Between Rankings d	Difference Squared d^2
Ms. Garcia	1	2	−1	1
Ms. Hibner	2	1	1	1
Ms. Schaefer	3	5	−2	4
Ms. Gerko	4	3	1	1
Ms. Owens	5	7	−2	4
Mr. Bunt	6	6	0	0
Mr. Kolby	7	4	3	9
Mr. Cassow	8	10	−2	4
Mr. Bianchi	9	9	0	0
Mr. Loftop	10	8	2	4
			0	28

In this example, the null and alternate hypotheses are

H_0: There is no correlation among the ranks in the population.

H_a: There is positive rank correlation in the population.

The decision rule, determined from Appendix D, calls for rejecting the null hypothesis if t is greater than 1.860. This critical value is determined by using 8 df and a one-tailed test at the 0.05 significance level.

The computed value of t is 4.21:

$$t = \frac{r\sqrt{n-2}}{\sqrt{1-r^2}} = \frac{0.83\sqrt{10-2}}{\sqrt{1-(0.83)^2}} = 4.21$$

The null hypothesis is rejected and the alternate is accepted because 4.21 > 1.860. There is positive agreement between the rankings of the 2 nurses as to the more difficult patients in the population. The p-value is less than 0.005, for a one-tailed probability. Recall that a p-value reports the probability of a value of the test statistic this large or larger when the null hypothesis is true.

In the previous Problem/Solution, each nurse *ranked* a patient with respect to difficulty of care. In some problems, we may need to convert data to ranks before computing the rank-order correlation coefficient. For example, 2 executives rated the performance of 6 subordinate executives on a scale of 0 to 100, with 0 representing very undesirable performance and 100 representing outstanding performance (see the Ratings columns in Table 13–6). Because the data are not known to be of interval scale, or perhaps they are not clearly normally distributed, Pearson's r is not appropriate. However, if we convert

TABLE 13–6	Ratings of Subordinates by Two Executives, with Conversion to Rankings			
	Rating		**Ranking**	
Subordinate	Harn's	Kander's	Harn's	Kander's
Glasser	42	55	4	3
Kingman	81	40	1	4
Cardellini	27	10	6	5
O'Hara	60	90	3	1
Compton	36	8	5	6
Wong	76	86	2	2

the executive ratings to rankings, the rank-order correlation coefficient may be used to describe the agreement between the 2 executives' ratings.

Referring to the first column of Table 13–6, note that Harn rated Kingman highest (81). Therefore, Kingman is ranked 1. Wong was rated next highest (76) and is ranked 2, and so on. Kander's ratings are treated the same way; that is, O'Hara (with a 90 rating) is ranked 1, Wong (with an 86 rating) is ranked 2, and so on. (Note that the ratings for each executive were ranked from high to low. However, had each set of ratings been ranked from low to high, the rank-order correlation coefficient would be the same.)

Suppose now that Harn had rated Kingman's and Wong's performances as equal, assigning 83 points to each. How would their respective ranks be determined? Both would be ranked 1.5, which is the average of ranks 1 and 2. Had Kander rated O'Hara, Wong, and Glasser highest (with 91 points each), how would their ranks be determined? These 3 would tie for ranks 1, 2, and 3. The average of the 3 tied ranks would be 2, found by (1 + 2 + 3)/3. Therefore, each of the 3 would be given the rank 2.

Self-Review 13–5

A professional and a blue-collar worker were asked to rank 12 occupations according to the social status they attached to each. A ranking of 1 was assigned to the occupation with the highest status, a 2 to the occupation with the next highest status, and so on. Their rankings are given below.

a. Compute the rank-order coefficient of correlation.
b. Interpret the coefficient.
c. Can we conclude that there is a positive rank correlation at the 0.05 significance level?

Occupation	Rank of Professional	Rank of Blue-Collar Worker
Physician	1	1
Dentist	4	2
Attorney	2	4

Occupation	Rank of Professional	Rank of Blue-Collar Worker
Pharmacist	6	5
Optometrist	12	9
Schoolteacher	8	12
Veterinarian	10	6
College professor	3	3
Engineer	5	7
Accountant	7	8
Health care administrator	9	11
Government administrator	11	10

Exercises

17. There is an opening for police sergeant. Applicants for promotion to sergeant are rated separately by both the police board and the county commissioners. On a scale of 1 to 50, the ratings of 6 applicants are

Applicant	Police Board's Rating	County Commissioners' Rating
Arbuckle	42	40
Cantor	36	40
Silverman	16	21
Trepinski	42	40
Lopez	49	47
Condon	8	9

a. Rank the ratings of the applicants. (Watch the ties.)

b. Compute Spearman's rank-order correlation coefficient.

18. A sales manager and a personnel manager were asked to rank graduates of the company's sales training program in terms of the likelihood of their success on the job. Determine the coefficient of rank correlation.

Trainee	Sales Manager	Personnel Manager
A	3	2
B	1	5
C	8.5	9
D	2	1
E	4	6
F	8.5	12
G	5	3
H	7	4
I	10	11
J	6	7
K	11	8
L	12	10

a. Compute Spearman's rank-order correlation coefficient.

b. Can it be shown statistically that there is positive rank correlation?

CAUTION: REASONING ABOUT CAUSE AND EFFECT.   If a strong correlation, such as 0.85, is found between 2 variables, it is tempting to assume that a change in 1 variable *causes* the other variable to change. This may not be true. For example, it has been discovered that the birth rate goes up when beer consumption goes up. An increase in the consumption of beer, however, does not cause the birth rate to increase. This only indicates that there is a relationship between the 2 variables. Results of this nature are called nonsense, or spurious, correlations.

CHAPTER OUTLINE

I. A scatter diagram is a graphic tool used to portray the relationship between two variables.
  A. The dependent variable is scaled on the Y-axis and is the variable being estimated.
  B. The independent variable is scaled on the X-axis and is the variable used as the estimator.
II. Pearson's coefficient of correlation measures the strength of the linear association between two variables.
  A. Both variables must be at least of the interval scale of measurement.
  B. The coefficient of correlation can range from -1.00 up to 1.00.
  C. If the correlation between 2 variables is 0, there is no linear association between them.
  D. A value of 1.00 indicates perfect positive correlation and -1.00 perfect negative correlation.
  E. A positive sign means there is a direct relationship between the variables, and a negative sign means there is an inverse relationship
  F. It is designated by the letter r and computed using the following equation:

$$r = \frac{n\Sigma XY - \Sigma X\Sigma Y}{\sqrt{\left[n\Sigma X^2 - (\Sigma X)^2\right]\left[n\Sigma Y^2 - (\Sigma Y)^2\right]}} \qquad \boxed{13\text{-}2}$$

  G. The following test statistic is used to determine if the correlation in the population of two normally distributed variables is different from zero.

$$t = \frac{r\sqrt{n - 2}}{\sqrt{1 - r^2}} \qquad \boxed{13\text{-}3}$$

III. The coefficient of determination is the fraction of the variation in one variable that is explained by the other variable.
  A. It ranges from 0 to 1.00.
  B. It is the square of the coefficient of correlation.
IV. Spearman's coefficient of rank correlation is used to measure the association between two ordinal-scaled variables.
  A. It ranges from -1.00 to 1.00, with 0 indicating no association between the ranks and -1.00 and 1.00 perfect association between the ranks.
  B. The sign of r_s indicates the direction of the association.
  C. It is computed using the following formula:

$$r_s = 1 - \frac{6\Sigma d^2}{n(n^2 - 1)} \qquad \boxed{13\text{-}4}$$

  D. Formula 13–3 is used to test if the population coefficient of rank correlation is different from zero.

Exercises

19. The following table shows the percentages of the vote actually received by the candidates for mayor in ten large cities and the corresponding percentages predicted by a national polling organization:

Predicted	Actual Percentage
59	62
47	51
43	42
55	56
57	57
58	63
46	53
43	49
55	58
48	47

a. Plot the data on a scatter diagram.

b. Compute the coefficient of correlation, and interpret.

c. Compute the coefficient of determination.

d. At the 0.01 significance level, can you conclude that the population correlation is different from zero?

20. The following data relate family size to annual food expenditures for a sample of ten families:

Family	Food Expenditures (in hundreds of dollars)	Family Size
1	$19	5
2	20	6
3	15	5
4	6	2
5	11	3
6	17	3
7	18	3
8	15	3
9	13	2
10	16	3

a. Compute the coefficient of correlation.

b. Can we conclude that there is a relationship between food expenditures and family size?

21. A researcher studied the relationship between the years of education beyond high school and the monthly salary of a sample of 27 welfare workers. The correlation was 0.25. At the 0.01 significance level, can we conclude that there is a positive association between the variables?

22. The manufacturer of Cardio Glide exercise equipment wants to study the relationship between the number of months since the glide was purchased and the length of time the equipment was used last week.

Person	Number of Months Owned	Number of Hours Exercised
Rupple	12	4
Hall	2	10
Bennett	6	8
Longnecker	9	5
Phillips	7	5
Massa	2	8
Sass	8	3
Karl	4	8
Malrooney	10	2
Veights	5	5

a. Plot the information on a scatter diagram. Let the number of hours exercised be the dependent variable. Comment on the graph.

b. Determine the coefficient of correlation. Interpret.

c. At the 0.01 significance level, can we conclude that there is a negative association between the variables?

23. A biologist is studying the relationship between the age of a tree (in years) and the amount the tree grew (in centimeters) in a 5-year period. She gathered the information below for a sample of 16 trees:

Tree	1	2	3	4	5	6	7	8	9	10	11	12	13	14	15	16
X Growth	18	18	15	15	14	14	13	13	12	11	11	10	10	9	9	5
Y Height of tree	400	380	300	225	340	290	335	305	340	225	165	255	230	260	150	70

a. Plot the information on a scatter diagram. Comment on the graph.

b. Determine the coefficient of correlation between the growth and the height of the trees.

c. Determine the coefficient of determination, and interpret.

d. At the 0.05 significance level, can we conclude that there is a positive association between the variables? Conduct an appropriate test of hypothesis.

24. Is there a relationship between the mean high temperature in January and the amount of precipitation (in inches) for the same month for cities in the United States? A sample of 18 reporting stations is shown below.

Weather Station	High Temperature	Precipitation
Albany, NY	21	2.4
Atlantic City, NJ	31	3.5
Bismarck, ND	9	0.5
Charleston, SC	48	3.5
Dallas, TX	43	1.8
Duluth, MN	7	1.2
Grand Rapids, MI	22	1.8
Huron, SD	13	0.4
Kansas City, MO	26	1.1
Los Angeles, CA	58	2.9
Milwaukee, WI	19	1.6
Nashville, TN	36	3.6
Omaha, NE	21	0.7
Portland, ME	21	3.5
Richmond, VA	37	3.2
San Diego, CA	57	1.8
Seattle, WA	41	5.4
Washington, DC	31	2.7

a. Compute the coefficient of correlation between the mean January high temperature and the amount of precipitation.

b. At the 0.10 significance level, can we conclude that the correlation is greater than zero?

25. The design department of a large automobile manufacturer developed ten different mock-up models of a mid-sized automobile. The design group is interested in determining the opinions of teenagers and senior citizens with respect to the models. A group of teenagers was asked to rank each model. Based on their rankings, a composite rank for each model was determined. For example, the teenagers ranked mock-up model E the most appealing and model J the next most appealing. The same procedure was followed for the senior citizens. The rankings of the two groups are:

Model	Ranking by Teenagers	Ranking by Senior Citizens
A	6	8
B	4	1
C	8	10
D	3	3
E	1	2
F	10	7
G	7	9
H	5	5
I	9	6
J	2	4

a. Draw a scatter diagram. Place the rankings of the teenagers on the X-axis and the rankings of the senior citizens on the Y-axis.

b. Compute Spearman's rank-order correlation coefficient.

c. Interpret your findings.

26. Professor P. Dant believes that students who complete her examinations in the shortest times receive the highest grades and that those who take the longest to complete them receive the lowest grades. To verify her suspicion, she numbers the midterm examination paper as each student completes it.

Name	Order of Completion	Grade (50 possible)
Gorney	1	48
Gonzales	2	48
McDonald	3	43
Sadowski	4	49
Jackson	5	50

Name	Order of Completion	Grade (50 possible)
Smythe	6	47
Carlson	7	39
Archer	8	30
Namath	9	37
MacFearson	10	35

a. Rank the grades.

b. Draw a scatter diagram. Put the order of completion on the X axis and the grades on the Y axis.

c. Compute the rank-order correlation coefficient.

d. Interpret your findings.

27. There has recently been much debate about the cost of traveling on the Ohio Turnpike and whether or not the tolls should be reduced. As part of a citizens' investigation, the following information on other turnpikes and toll roads was obtained:

Toll Road	Length (in miles)	Cost per Mile (in cents)
Massachusetts Turnpike	123.0	4.15
Pennsylvania (east–west)	358.9	4.10
Pennsylvania (north–south)	111.1	3.74
New Jersey Turnpike	118.0	3.89
Florida Turnpike	265.0	3.75
New York Thruway (Mainline)	390.0	3.10
New York Thruway (Erie)	67.0	3.13
New York Thruway (Saratoga)	24.0	2.97
Indiana Toll Road	156.9	2.96
Maine Turnpike	106.0	2.92
Oklahoma (Turner)	86.0	2.33
Oklahoma (Will Rogers)	88.5	2.26
Ohio Turnpike	241.2	2.03

a. Determine Pearson's coefficient of correlation between the length of the toll road and the cost per mile to travel on the toll road.

b. Determine the coefficient of determination.

c. At the 0.05 significance level, can we conclude that the correlation between the length of the typical toll road and the cost per mile on that toll road is different from zero?

d. Rank the data regarding the length and cost per mile for the turnpikes. Determine Spearman's coefficient of rank correlation. Compare your answer with that in part a. Comment.

28. A real estate agent has sample data on the selling prices and distances from the center of the city for 75 homes. The mean distance from the center of the city is 14.9 miles, with a standard deviation of 4.9 miles. The mean selling price of the homes is $167,000, with a standard deviation of $36,000. The sum of the products of the deviations from the mean is −4821.

a. Determine the coefficient of correlation.

b. At the 0.05 significance level, can we conclude that the distance from the center of the city is inversely related to the selling price?

29. Harvey Maertin is an analyst for the health department for the city of Spokane. As a part of a recent project, he computed the correlation between the appraised home value and the amount spent on medical expenses to be −0.618 for a sample of 11 residents. He was somewhat surprised at the inverse relationship. The sample data in Harvey's study are reported below. Does the relationship seem to be inverse? What help or suggestions can you offer Harvey? You might begin by plotting the data.

Name	Medical Expenses	Home Value
Miller	$5032	$154,672
Hendrix	4508	151,773
Sines	2326	150,995
Kennedy	4147	148,928
Abernathy	2867	150,691
Lennox	3352	150,310
Tusko	3538	147,243
Lutz	2784	151,651
Sell	15,000	145,038
Brigden	3489	151,157
Hewitt	4289	150,635

30. A major highway construction project by the Ohio Department of Transportation will close a portion of Interstate 75 near the Rossford exit in northwestern Ohio this construction season. Ray, the owner of Ray's Marathon, is very concerned about how this will affect his business. The closing of the section of I-75 means that fewer cars will pass in front of his station. Ray assumes that, if fewer cars pass in front of his station, fewer will come in for gasoline. He was recently discussing this situation with 2 of his young employees, Mark, who works as a part-time mechanic, and Sarah, who is a cashier. Both attend the nearby college and are currently enrolled in a statistics class. Mark says that, because the station has been there for many years, most customers will continue to buy gasoline there, even if it means going out of their way a little. Sarah suggests that Ray should gather some information on the number of gallons of gas pumped in a day and that she and Mark will estimate the volume of cars that pass the station. For a sample of 12 days, they collect the following information. Does it appear that there is any relationship between the traffic flow and the number of gallons of gasoline pumped? Conduct an appropriate test of hypothesis.

Day	Traffic Count (in hundreds) X	Total Gallons Sold (in thousands) Y
1	12	330
2	17	332
3	10	368
4	6	172
5	16	347
6	9	322
7	11	298
8	4	313
9	7	360
10	6	324
11	6	257
12	12	187

31. For several large cities, the typical sizes of downtown offices (in mean number of square feet per worker) and the average monthly rental rates (in dollars per square foot) are listed below.

City	Square Footage	Rental Rate
Austin	350	$14.52
Dallas	344	13.10
Houston	327	12.32
Billings	306	12.78
New Orleans	302	10.95
Seattle	300	13.73
San Jose	293	15.47
Atlanta	288	10.62
Portland, OR	277	13.91
Phoenix	275	11.76
Cleveland	272	12.00
Birmingham	268	14.34
Kansas City, MO	268	10.35
Pittsburgh	268	15.30
Jacksonville	266	17.75
Omaha	265	14.13
Denver	262	12.96
Chicago	257	18.07
St. Louis	254	12.61
Minneapolis	251	15.26
San Diego	249	12.31
Miami	246	16.97
Orlando	242	12.11
Philadelphia	229	16.43
Sacramento	223	18.87
Milwaukee	215	14.91
Charlotte	206	14.93
Columbus, OH	202	15.27
Honolulu	201	17.30
Baltimore	191	16.69
Detroit	187	12.78
Indianapolis	182	13.40

a. Develop a scatter diagram.
b. Compute the coefficient of correlation.
c. What is the coefficient of determination?
d. Interpret your findings.
e. Using the 0.05 significance level, test the hypothesis that the correlation in the population is zero.

DATA EXERCISES

32. Refer to the real estate data set, which reports information on homes sold in Alabama during 1996.

  a. Determine the coefficient of correlation between the selling price and the area of each home in square feet. At the 0.05 significance level, can we conclude that there is a positive association between the 2 variables?

  b. Determine the coefficient of correlation between the selling price and the distance from the center of the city of each home. At the 0.05 significance level, can we conclude that there is a negative association between the 2 variables?

  c. Determine the coefficient of correlation between the selling price and the number of bedrooms in each home. At the 0.05 significance level, can we conclude that there is a positive association between the 2 variables?

33. Refer to the school data set in Appendix, which refers to the 94 school districts in northwestern Ohio.

  a. Determine the coefficient of correlation between the percentage of students passing the ninth-grade proficiency test and the number of students in the school system. At the 0.01 significance level, can we conclude that there is an inverse relationship?

  b. Determine the coefficient of correlation between the average teacher salary and the percentage of students passing the ninth-grade proficiency test. How much of the variation in the percentage of students passing the test is accounted for by the average teacher salary? Can we conclude that the correlation is different from zero?

  c. Determine the correlation between the median income and the average teacher salary. At the 0.05 significance level, can we conclude that the correlation is greater than zero.

CHAPTER ACHIEVEMENT TEST

The answers are at the back of the book.

MULTIPLE-CHOICE QUESTIONS

Select the response that best answers each of the questions.

Questions 1–5 are based on the following graphs:

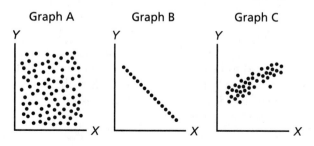

Graph A   Graph B   Graph C

1. These graphs are called
  a. tree diagrams.
  b. shotgun blasts.
  c. scatter diagrams.
  d. frequency polygons.
  e. None of the above

2. Graph A: The coefficient of correlation, if computed, would be approximately
  a. 100.
  b. -1.00.
  c. $+1.00$.
  d. 0.
  e. None of the above

3. Graph B: The coefficient of determination, if computed, would be approximately
   a. −1.00.
   b. 0.50.
   c. 0.
   d. 1.00.
   e. None of the above
4. Graph B: The linear relationship between the two variables
   a. is very weak.
   b. is moderate.
   c. is perfect.
   d. cannot be determined from this graph.
   e. None of the above
5. Graph C: The coefficient of determination, if computed, would be
   a. about 60%.
   b. negative.
   c. about 0.
   d. near infinity.
6. Pearson's product-moment coefficient of correlation for a set of paired observations was computed to be 0.70.
   a. The relationship is inverse.
   b. The coefficient of determination is 0.49.
   c. The correlation in the population is 0.
   d. The correlation in the population is 0.49.
   e. None of the above

7. Pearson's r was computed to be −0.55. Which of the following values of r represents a stronger relationship than −0.55?
   a. 0
   b. +0.50
   c. −0.70
   d. 20.9
   e. None of the above
8. Spearman's rank-order correlation coefficient
   a. can range from −1.0 to 1.0.
   b. is always positive.
   c. is positively skewed.
   d. All of the above
9. The term $\Sigma(X - \overline{X})(Y - \overline{Y})$ affects the
   a. sign of r.
   b. standard deviation of y.
   c. standard deviation of x.
   d. size of the sample.
10. Spearman's rank-order correlation coefficient was computed to be −0.76 for a sample of 15 observations. Which of the following are correct statements?
    a. Something is wrong with the calculation because the rank-order correlation coefficient cannot be negative.
    b. As one ranking increases, so does the other.
    c. As one ranking increases, the other decreases.
    d. None of the above

COMPUTATION PROBLEMS

11. The ages and corresponding prices of five bottles of wine selected at random and sold at a wholesale auction are shown in the following list. The relationship between age and price is to be explored.

Age (in years) X	Price Y
2	$ 5
6	16
10	18
5	9
8	15

a. Portray the paired data in a graph.
b. Compute Pearson's coefficient of correlation.
c. Compute the coefficient of determination.
d. Test the hypothesis that the coefficient of correlation in the population is zero. Use the 0.05 significance level.
e. Interpret your findings.
12. Early in the basketball season, 12 teams appeared to be outstanding. A panel of sportswriters and a panel of college basketball coaches were asked to rank the 12 teams. Their composite rankings were as follows:

Team	Coaches	Sportswriters
Notre Dame	1	1
Indiana	2	5
Duke	3	4
North Carolina	4	6
UCLA	5	2
Louisville	6	3
Ohio State	7	10
Syracuse	8	11
Georgetown	9	7
LSU	10	12
Clemson	11	8
St. John's	12	9

a. Compute the appropriate coefficient of correlation to evaluate the agreement between the sportswriters and coaches with respect to the rankings.
b. Interpret your findings.
c. Can we conclude that there is a significant correlation between the rankings of the coaches and those of the sportswriters at the 0.05 significance level?

ANSWERS TO SELF-REVIEW PROBLEMS

13-1 a.

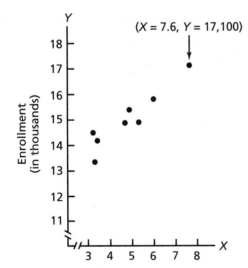

$(X = 7.6, Y = 17,100)$

Percentage of Unemployment

b. As unemployment in the region increases, enrollment at the state university campus also increases.

13-2 a.

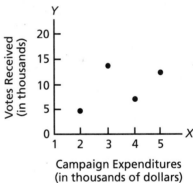

Campaign Expenditures
(in thousands of dollars)

b.
$$r = \frac{4(140) - (14)(38)}{\sqrt{[4(54) - (14)^2][4(414) - (38)^2]}}$$
$$= \frac{560 - 532}{\sqrt{[216 - 196][1656 - 1444]}}$$
$$= \frac{28}{\sqrt{[20][212]}} = \frac{28}{\sqrt{4240}}$$
$$= \frac{28}{65.12} = 0.43$$

c. There is a moderate-to-weak relationship between the two variables.

13-3 The coefficient of determination is 0.18, found by $(0.43)^2$. Only about 18% of the variation in the number of voters is accounted for by the level of campaign expenses.

13-4 $H_0: \rho = 0$
$H_a: \rho \neq 0$
Reject H_0 if $t < -9.925$ or $t > 9.925$.

$$t = \frac{0.43\sqrt{4-2}}{\sqrt{1-(0.43)^2}} = 0.67$$

H_0 is not rejected. The correlation in the population could be zero.

13-5 a.

Rank	Rank	d	d^2
1	1	0	0
4	2	2	4
2	4	−2	4
6	5	1	1
12	9	3	9
8	12	−4	16
10	6	4	16
3	3	0	0
5	7	−2	4
7	8	−1	1
9	11	−2	4
11	10	1	1
		0	60

$$r_s = 1 - \frac{6(60)}{12\left[(12)^2 - 1\right]} = 1 - \frac{360}{1716} = 0.79$$

b. There is quite strong agreement between the two rankings.

c. H_0: No correlation in the ranks.
H_a: Positive correlation in the ranks.
H_0 is rejected if $t > 1.812$.

$$t = \frac{0.79\sqrt{12-2}}{\sqrt{1-(0.79)^2}} = 4.07$$

H_0 is rejected. There is positive correlation in the ranks.

CHAPTER 14

Regression Analysis

OBJECTIVES

When you have completed this chapter, you will be able to

- describe a linear relationship between two variables;
- determine the linear equation using the method of least squares;
- compute the standard error of estimate and explain its use;
- construct a confidence interval and a prediction interval for estimates of the dependent variable.

CHAPTER PROBLEM **Our Books Are Priceless!**

Recall from Chapter 13 that a committee of the student government at North Central State University began a study of the cost of textbooks. They found a strong positive relationship between the number of pages in a textbook and its selling price. They would like to develop an equation from the same sample data that can be used to estimate the selling price of a textbook based on the number of pages.

INTRODUCTION

Chapter 13 dealt with measures of correlation between two variables—Pearson's product-moment correlation coefficient and Spearman's rank-order correlation coefficient. In this chapter, we will continue the study of two related variables. We will develop a mathematical equation that allows us to estimate one variable based on another variable. For example, if a student studied three hours for an examination, it may be possible to estimate his or her score on the examination. Similarly, the sales of a product may be predicted by the level of advertising expenditures. The technique used to make these estimations is called *regression analysis*.

LINEAR REGRESSION

The word *regression* was first used by Sir Francis Galton (1822–1911) in the late 19th century. A study by Galton revealed that the height of the children of tall parents was above average, but seemed to fall back, or regress, toward the mean height of the population. Thus, the general process of predicting one variable (such as a child's height) based on another variable (the parents' heights) became known as *regression*.

The usual first step in studying the relationship between two variables is to plot the paired data in a scatter diagram, discussed in Chapter 13. To illustrate the concept of regression, we will examine the Chapter Problem relating the length of a book to its selling price. The data are repeated in Table 14–1 and plotted in the form of a scatter diagram in Figure 14–1. This time we are interested in *estimating* a selling price based on the number of pages in the text. Recall from Chapter 13 that the variable being predicted (selling price) is the *dependent* variable—designated Y—and the variable used to make the prediction (number of pages) is the *independent* variable—designated X.

The objective is to find a line that—when translated into equation form—can describe the relationship between the two variables. An important

TABLE 14–1	Numbers of Pages and Selling Prices of Selected Books	
Title	Number of Pages	Selling Price
Introduction to Psychology	400	$44
Behavior in Organizations	600	47
Introduction to Criminology	500	48
Principles of Sociology	600	48
Nursing Education	400	43
Introduction to Residential Construction	500	46

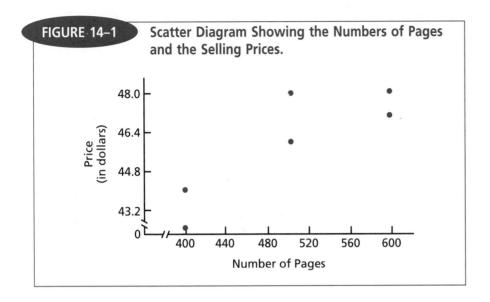

FIGURE 14–1 Scatter Diagram Showing the Numbers of Pages and the Selling Prices.

use of the equation for the line is to approximate a value of the dependent variable based on a selected value of the independent variable. For example, in the Chapter Problem, where we are concerned with the selling price and the number of pages, we may want to approximate the selling price of a book that has 450 pages. The equation for that line is called the **regression equation.**

The linear equation that expresses the relationship between two variables is

$$Y' = a + bX \qquad \boxed{\textbf{14–1}}$$

> **regression equation** A mathematical equation that defines the relationship between two variables.

where

Y' (read "Y prime") is the predicted value of Y for a selected X value. In the example, it is the selling price for a given number of pages.
a is a constant: The value at which the straight line intersects the Y' axis. It is also the value of Y' when $X = 0$.
b is also a constant and is the slope of the line. It is the change in Y' for each change of one (either increase or decrease) in X. In this problem, b is the increase in the selling price for an increase of one page.
X is any value of X that is selected. In this problem, X is the number of pages.

To further illustrate the meanings of a and b, suppose the regression equation in the form of $Y' = a + bX$ for the data in Figure 14–2 had been computed to be $Y' = 20 + 10X$ (in thousands of dollars). When $X = 0$, the line intercepts the Y-axis at $20,000. This is a, which is sometimes called the Y-intercept.

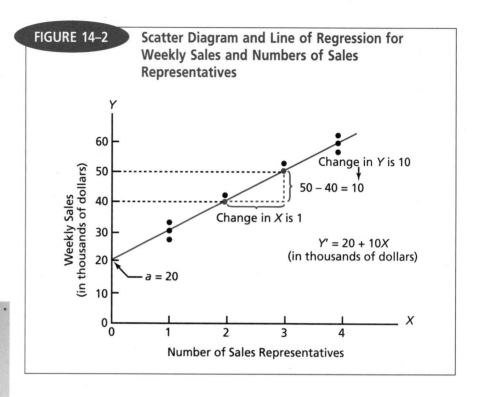

FIGURE 14–2   Scatter Diagram and Line of Regression for Weekly Sales and Numbers of Sales Representatives

Figure 14–2 also shows that weekly sales increase from $40,000 a week when 2 sales representatives call on clients to $50,000 when 3 representatives call. Thus, with each increase of 1 sales representative, weekly sales increase by $10,000. This is *b*.

The regression equation $Y' = 20 + 10X$ (in thousands of dollars) can be used to estimate the average weekly sales if 4 sales representatives are employed, $X = 4$. Then $Y' = 20 + 10(4) = 60$, or $60,000.

Self-Review 14–1

Answers to the Self-Review problems are at the end of the chapter.

Some data dealing with the years of education and the weekly incomes of residents of Precinct 13 have been collected and are presented on the next page. The purpose of the study is to predict income based on education.

a. Interpret the meaning of the point for Don Murray.
b. What is the value of *a* in the regression equation $Y' = a + bX$?
c. What is the value of *b* in the regression equation? What does it indicate?
d. What is the regression equation?
e. Based on the regression equation, estimate the average weekly income for a person with four years of education beyond high school.

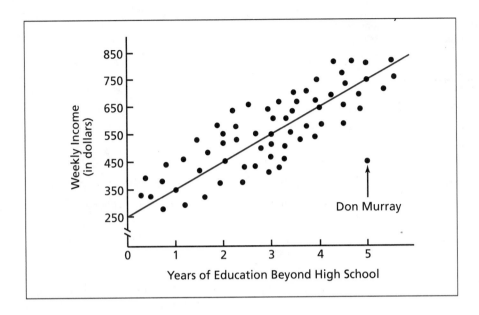

Exercises

Answers to the even-numbered Exercises are at the back of the book.

1. The lengths of confinement in days (X) and the dollar costs (Y) for six patients at St. Matthew's Hospital were tabulated and then plotted on a scatter diagram:

Length of Confinement (in days) X	Cost Y
5	$1110
2	490
12	2500
4	880
7	1530
1	270

a. What is the general form of the regression equation?
b. Based on the scatter diagram, what is a?
c. What is b (the slope) of this regression line?
d. If a patient stayed six days, what would you estimate his or her bill to be?
e. Interpret in words the values of a and b.

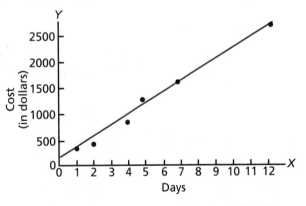

2. A study attempting to relate the number of minority employees (Y) to the number of nonminority employees (X) in eight firms revealed the following:

Firm	Number of Nonminority Employees X	Number of Minority Employees Y
London Mfg.	4628	182
ABC Chemicals	4272	299

Firm	Number of Nonminority Employees X	Number of Minority Employees Y
Cork Floors, Inc.	1663	173
Sanson Industries	691	44
Martin Trucking	5355	1069
Cable Lead	5618	777
Parson Farms	2934	802
Tayo Electric	3650	790

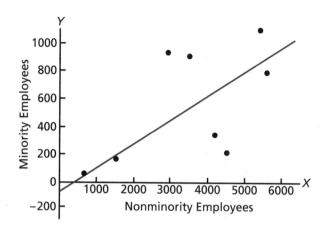

a. What is the general form of the regression equation?

b. Based on the following scatter diagram, what is the Y-intercept?

c. What is b, the slope of the regression line?

d. If a firm employs 3000 nonminority persons, estimate the number of minority employees.

e. Interpret the values of a and b.

3. An equation that relates the annual cost to heat a single-family residence (Y) to its size in square feet of living area (X) is $Y' = -200 + 0.3X$.

a. Estimate the cost of heating a 3500-square-foot home.

b. What is the slope of this line?

c. What is the Y-intercept?

d. Interpret in words the meaning of a and b.

4. The equation $Y' = 39 - 0.002X$ relates the annual birth rate per thousand population (Y) of a country to its median annual income (X).

a. Estimate the birth rate for a country whose median annual income is $7000.

b. What is the Y-intercept of the line of regression?

c. What is the slope of the line?

d. Interpret the meaning of b.

FITTING A LINE TO DATA

How should the line through the plots on a scatter diagram be drawn? We could simply place a ruler on the scatter diagram and draw a line through the middle of the points. Yet if this job of drawing the best-fitting line "right through the middle of the points" were given to three persons, their lines would probably all be different. Consider the Chapter Problem concerned with predicting selling price on the basis of the number of pages in a book. The information for the six books comes from Table 14–1. Figure 14–3 shows how three different people might draw a straight line. Each is different! What is needed is a precise method of arriving at the best-fitting straight line. Then the identical line will be obtained every time, regardless of who does the calculations.

The Least-Squares Principle

least-squares principle A method used to determine the regression equation by minimizing the sum of the squares of the distances between the actual Y values and the predicted Y values.

We can avoid having to rely on personal judgment as a way to find the exact location of the line by employing a method called the **least-squares principle**, which will provide us with the "best-fitting" line. How does it work? Note that in Figure 14–2 (p. 432), the plotted points are not exactly on the straight line. The difference between an actual value (a plotted point) and a predicted value (a corresponding value on the line) can be thought of as the "error" we

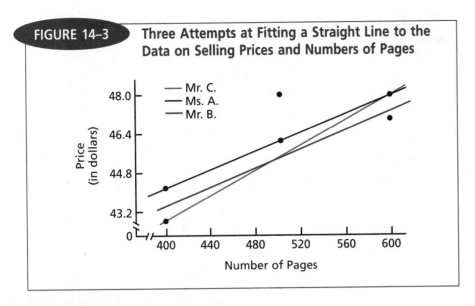

FIGURE 14–3 **Three Attempts at Fitting a Straight Line to the Data on Selling Prices and Numbers of Pages**

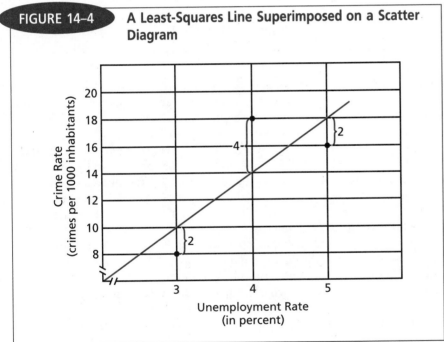

FIGURE 14–4 **A Least-Squares Line Superimposed on a Scatter Diagram**

make when using the regression equation to predict a specific value of the dependent variable. Put another way, this difference, or error, can be considered to be the deviation of the actual value from the predicted value.

By using this principle, we obtain a *unique* regression equation that is "best" in the sense that the squared errors around the regression line are smaller than those around any other line; that is, the sum of the squares of

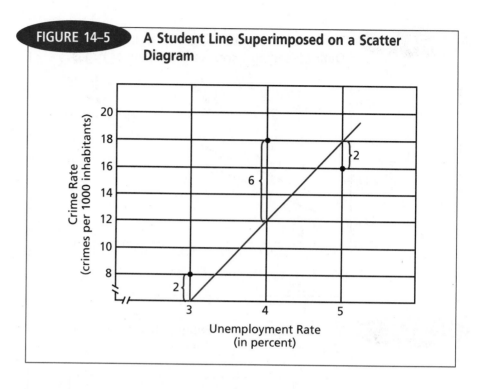

FIGURE 14–5   A Student Line Superimposed on a Scatter Diagram

the deviations between the actual values (Y) and the predicted values (Y') is minimized. The regression line is unique—meaning that there is only one least-squares line, and this line can be determined by anyone doing the calculations.

Figure 14–4 illustrates this concept. The line was determined by the method of least squares. The first plot ($X = 3$, $Y = 8$) deviates vertically by 2 units from the predicted value of 10 on the straight line. The squared deviation is 4, found by 2^2. The squared deviation for $X = 4$, $Y = 18$ is 16, found by $(18 - 14)^2$. The squared deviation for $X = 5$, $Y = 16$ is 4. The sum of the squared deviations is 24, found by $4 + 16 + 4$.

The same crime and unemployment data in Figure 14–4 also appear in Figure 14–5. The line through the data in Figure 14–5, however, is *not* determined by the least-squares method. A beginning student thought the line should be positioned as shown. The sum of the vertical squared deviations from the freehand line is 44, found by $2^2 + 6^2 + 2^2$, which is greater than 24, the sum of the squared deviations around the least-squares line.

Self-Review 14–2

The following scatter diagram shows the number of units sold and the amount spent on advertising for the XYZ Corporation. The line was carelessly drawn by a student.

a. Determine the sum of the squared vertical deviations from the line.
b. What is the dependent variable in this diagram?

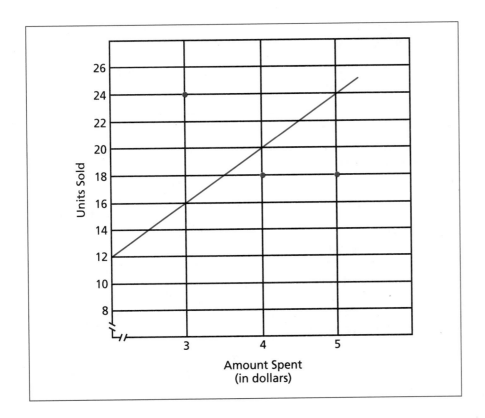

THE METHOD OF LEAST SQUARES

In the previous section, we stated that the "best" equation is the one that minimizes the sum of the squared deviations. Now we will show how this line is determined. The determination is made by the mathematical process of minimization; that is, among all possible lines, the one where $\Sigma(Y - Y')^2$ is the smallest is obtained, and it is unique. The following formulas, based on the least-squares principle, are used to compute the unique values of b and a:

$$b = \frac{n(\Sigma XY) - (\Sigma X)(\Sigma Y)}{n(\Sigma X^2) - (\Sigma X)^2}$$

14–2

$$a = \frac{\Sigma Y - b(\Sigma X)}{n}$$

14–3

where

X is a value of the independent variable.

Y is a value of the dependent variable.
n is the number of items in the sample.

Problem   The data on the lengths of books and their selling prices from the Chapter Problem are repeated below.

Title	Number of Pages X	Selling Price Y
Introduction to Psychology	400	$44
Behavior in Organizations	600	47
Introduction to Criminology	500	48
Principles of Sociology	600	48
Nursing Education	400	43
Introduction to Residential Construction	500	46

Using the method of least squares and formulas 14–2 and 14–3, determine the linear regression equation. Use this equation to estimate the selling prices for these books based on the number of pages in each. Then estimate the selling price for a book with 450 pages.

Solution   The products, squares, and totals needed for the method of least squares are shown in Table 14–2. Solving for b and a using formulas 14–2 and 14–3, we obtain

$$b = \frac{n(\Sigma XY) - (\Sigma X)(\Sigma Y)}{n(\Sigma X^2) - (\Sigma X)^2}$$

$$= \frac{6(138,800) - (3000)(276)}{6(1,540,000) - (3000)^2} = \frac{4800}{240,000} = 0.02$$

$$a = \frac{\Sigma Y - b(\Sigma X)}{n}$$

$$= \frac{276 - 0.02(3000)}{6} = \frac{216}{6} = 36$$

TABLE 14–2   Calculations Needed for Least-Squares Regression Equation

Title	X	Y	XY	X^2	Y^2
Introduction to Psychology	400	$ 44	17,600	160,000	1936
Behavior in Organizations	600	47	28,200	360,000	2209
Introduction to Criminology	500	48	24,000	250,000	2304
Principles of Sociology	600	48	28,800	360,000	2304
Nursing Education	400	43	17,200	160,000	1849
Introduction to Residential Construction	500	46	23,000	250,000	2116
	3000	$276	138,800	1,540,000	12,718

The regression equation is $Y' = 36 + 0.02X$. Thus, the estimated selling price for a book with 450 pages is $45, found by $Y' = a + bX = 36 + 0.02(450) = 36 + 9 = 45$.

DRAWING THE LINE OF REGRESSION

Now that the regression equation has been determined, we can plot the least-squares line of regression on the scatter diagram. The first step is to select an X value—say, 0. When $X = 0$, $Y' = 36$, found by $Y' = 36 + 0.02(0)$. The computations for some other points on the line are as follows:

When X is	Y' is	found by
200	40	$Y' = 36 + 0.02(200)$
400	44	$Y' = 36 + 0.02(400)$
600	48	$Y' = 36 + 0.02(600)$

Figure 14–6 shows the original scatter diagram with the line of regression superimposed on it.

How do we interpret the regression equation? The a value is the point where the equation crosses the Y axis. This is also the point where $X = 0$. In our textbook example, if we substitute $X = 0$, we find that the value of Y' is $36.00 ($Y' = 36 + 0.02X = 36 + 0.02(0) = 36$). The literal interpretation is that a book with no pages can be expected to cost $36.00. The b value of 0.02 tells us the rate of change. That is, for each additional page in the book, the selling price is expected to increase $0.02. Or, to put it another way, we

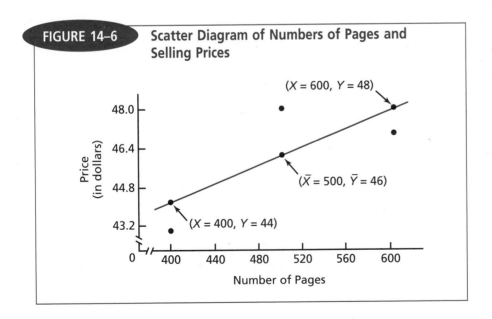

FIGURE 14–6   **Scatter Diagram of Numbers of Pages and Selling Prices**

would expect a 400-page book to cost $2.00 more than a 300-page book, found by 100 pages multiplied by $0.02.

Is it possible for either a, the intercept, or b, the slope, to be negative? Absolutely! If the value of the intercept is negative, this means that the regression equation intersects the Y axis below the origin. See the example below. If the value of b is negative, we conclude that there is an inverse relationship between the variables; that is, as X increases, the value of Y' decreases. We will find many real-world problems where there is an inverse relationship between the dependent variable and the independent variable. In a study of the relationship between the cost to heat a home in the Northeast and the mean monthly temperature, we would find that, as the outside temperature increased, the cost to heat the home would decrease. Therefore, the sign of b in the regression equation is negative.

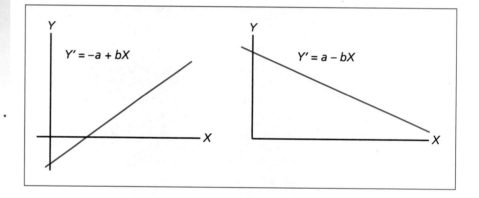

There is interest in determining if the number of votes that will be received in an election can be predicted based on the amount of money spent during the campaign. A tabulation of available data follows:

Candidate	Amount Spent on Campaign (in thousands of dollars) X	Number of Votes Received (in thousands) Y
Weber	$3	14
Taite	4	12
Spence	-2	5
Henry	5	20

a. Draw a scatter diagram.
b. Determine the regression equation using the method of least squares.
c. How many additional votes could a candidate expect for each additional $1000 spent?
(This problem will be continued; save your calculations.)

Exercises

5. The following table lists the populations of five counties (X) and the numbers of physicians (Y) who practice there:

County	Population (in thousands)	Number of Physicians
Lucas	38	21
Wood	76	35
Samborn	36	13
Atlee	62	24
Ohio	32	18

a. Draw a scatter diagram.
b. Find the least-squares regression equation.
c. Interpret the values of a and b.
d. Estimate the number of physicians for a county with a population of 45,000.

6. The numbers of adult movie theaters (X) and the annual numbers of sex-related crimes (Y) reported for five cities are

Number of Adult Movie Theaters	Number of Sex-Related Crimes
15	175
20	220
10	120
12	152
16	181

a. Draw a scatter diagram.
b. Find the least-squares regression equation.
c. Interpret the values of a and b.
d. Estimate the number of sex-related crimes for a city with 18 adult movie theaters.

7. The mean nighttime low temperatures (X) and the scores on an index of collective aggression (Y) are recorded for six urban areas:

Mean Nighttime Low (in °F)	Index of Collective Aggression
73	51
82	63
90	67
60	35
51	23
40	20

a. Draw a scatter diagram.
b. Find the least-squares regression equation.
c. Interpret the values of a and b.
d. Predict the value of the index when the mean nighttime low is 80.

8. Patients of various ages (X) undergoing the same medical treatment suffer asthma attacks of different durations (Y).

Age (in years)	Duration of Attack (in minutes)
30	15
25	28
65	30
50	22
40	24

a. Draw a scatter diagram.
b. Find the least-squares regression equation.
c. Interpret the values of a and b.
d. Predict the duration of an attack if the patient is 42 years old.

THE STANDARD ERROR OF ESTIMATE

If all the data points fall precisely on the line of regression, then our prediction is exact. In the previous chapter, this situation was described as perfect correlation. The coefficient of correlation would be either -1.00 or $+1.00$. The coefficient of correlation, then, is a possible measure of how well the line of regression fits the paired data.

Perfect predictions of Y based solely on X are practically nonexistent in real-world situations. Instead, there is usually some scatter around the regression line. The **standard error of estimate** is a measure of the scatter of the observed Y-values around the Y'-values on the line of regression. The "standard error," as it is usually called, is an indicator of how well the predicted Y'-values compare to the actual Y-values. It is designated by $s_{Y \cdot X}$ and has the formula

$$s_{Y \cdot X} = \sqrt{\frac{\Sigma(Y - Y')^2}{n - 2}}$$

14–4

where the $\Sigma(Y - Y')^2$ component refers to the sum of the squared differences between the actual Y-values and the values predicted for Y, designated Y'. The subscript $_{Y \cdot X}$ reminds us that we are predicting Y based on values of X, and n refers to the size of the sample. The denominator is $n - 2$ because the least squares coefficients, namely a and b, are being estimated from sample data.

The standard error measures variation in a manner similar to that of the standard deviation, except that, instead of summing the squared deviations from the mean as the standard deviation does $[\Sigma(Y - \overline{Y})^2]$, it employs the squared deviations between the actual values of Y and the predicted values of Y'—that is, $\Sigma(Y - Y')^2$. In fact, it may be thought of as a conditional standard deviation of dependent variable Y for a given value of independent variable X.

The table below shows the calculation of the standard error based on the prices and lengths of textbooks.

Title	Number of Pages X	Selling Price Y	Prediction Y'	Error Y − Y'	Squared Error (Y − Y')²
Introduction to Psychology	400	$44	$44	0	0
Behavior in Organizations	600	47	48	−1	1
Introduction to Criminology	500	48	46	2	4
Principles of Sociology	600	48	48	0	0
Nursing Education	400	43	44	−1	1
Introduction to Residential Construction	500	46	46	0	0
					6

The standard error of estimate is $1.22, found by using formula 14–4:

$$s_{Y \cdot X} = \sqrt{\frac{\Sigma(Y - Y')^2}{n - 2}} = \sqrt{\frac{6}{6 - 2}} = 1.22$$

Formula 14–4 requires that the difference between each Y observation and its corresponding Y'-value be determined and then squared. These are very time-consuming and error-prone calculations, especially for large samples. An equivalent, simpler, form is

$$s_{Y \cdot X} = \sqrt{\frac{\Sigma Y^2 - a\Sigma Y - b\Sigma XY}{n - 2}}$$ **14-5**

Problem Recall the Chapter Problem. The selling prices and numbers of pages in the textbooks, as well as the needed calculations for the standard error of estimate, are shown in Table 14–2.

Solution Earlier the regression equation was found to be $Y' = a + bX = 36 + 0.02X$. The sums needed to determine $s_{Y \cdot X}$ are in Table 14–2 on page 438. They are $\Sigma Y = 276$, $\Sigma XY = 138,800$, $\Sigma X^2 = 1,540,000$ and $\Sigma Y^2 = 12,718$. Inserting the appropriate values in formula 14–5 for the standard error of estimate, we get

$$
\begin{aligned}
s_{Y \cdot X} &= \sqrt{\frac{\Sigma Y^2 - a\Sigma Y - b\Sigma XY}{n - 2}} \\[2mm]
&= \sqrt{\frac{12,718 - 36(276) - 0.02(138,800)}{6 - 2}} \\[2mm]
&= \sqrt{1.5} \\[2mm]
&= \$1.22 \text{ (same as calculated before)}
\end{aligned}
$$

Further interpretation of this value will follow shortly.

The problem begun in Self-Review 14–3 is continued here. The data are

Self-Review 14–4

Candidate	Amount Spent on Campaign (in thousands of dollars) X	Number of Votes Received (in thousands) Y	XY	X^2	Y^2
Weber	$ 3	14	42	9	196
Taite	4	12	48	16	144
Spence	2	5	10	4	25
Henry	5	20	100	25	400
	$14	51	200	54	765

This regression equation is $Y' = -2.3 + 4.3X$. Compute the standard error of estimate.

Exercises

9. Refer to Exercise 5. The relationship between the numbers of physicians (Y) and the populations (X) is

$$Y' = 3.259 + 0.388X, \; n = 5, \; \Sigma Y = 111,$$

$$\Sigma XY = 5990, \text{ and } \Sigma Y^2 = 2735$$

Compute the standard error of estimate.

10. Refer to Exercise 7 on the index of aggression and its relationship to the nighttime low temperature in July.

$$Y' = -25.7 + 1.04X, \; n = 6,$$

$$\Sigma Y = 259, \; \Sigma XY = 18,992, \text{ and } \Sigma Y^2 = 13,213$$

What is the standard error of estimate?

11. Sample data on the unemployment rate (X) and the crime rate (Y) are as follows:

X	Y
3	8
4	18
5	16

a. Calculate the regression equation.
b. Find the standard error of estimate.

12. The data from Exercise 1, showing lengths of confinement in days (X) and the hospital costs (Y) at St. Matthew's Hospital, are repeated here:

X	Y
5	$1110
2	490
12	2500
4	880
7	1530
1	270

a. Calculate the regression equation.
b. Find the standard error of estimate.

ASSUMPTIONS ABOUT REGRESSION

The techniques of regression and correlation are based on several assumptions. For the techniques to be properly employed, the following assumptions should be met or should be approximately true:

1. For each value of X, the Y observations are normally distributed around the line of regression.
2. The standard deviation of each of these normal distributions is the same. This common standard deviation is estimated by the standard error of estimate ($s_{Y \cdot X}$).
3. The deviations from the regression line are independent. There is no pattern in the size or in the direction of these deviations.

These assumptions are graphically summarized in Figure 14–7, wherein each of the distributions depicted is normal and has the same standard deviation, estimated by $s_{Y \cdot X}$.

Assuming that the observed Y-values are normally distributed around the line of regression, we can say

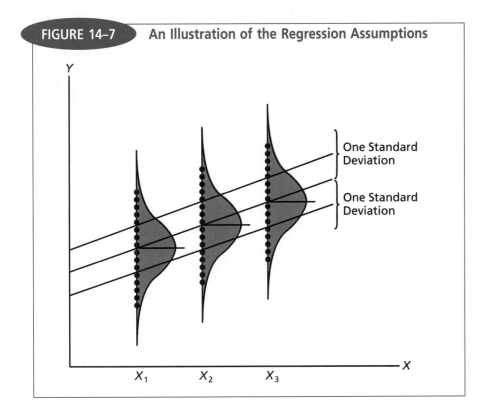

| FIGURE 14–7 | An Illustration of the Regression Assumptions |

$\overline{Y}' \pm 1s_{Y \cdot X}$ encompasses about 68% of the points,
$\overline{Y}' \pm 2s_{Y \cdot X}$ encompasses about 95.5% of the points, and
$\overline{Y}' \pm 3s_{Y \cdot X}$ encompasses virtually all of the points.

What does the standard error of estimate in the Chapter Problem tell us? It is a measure of the accuracy of our prediction. Suppose we were to repeat this experiment with a large number of books and predict their selling prices. Then the difference between our predicted selling price and the actual selling price would be less than 1 standard error of the estimate ($1.22) in 68% of the cases. Ninety-five percent of our predictions would be "off" by no more than 2($1.22) = $2.44. And virtually all of our predictions would be "off" by no more than 3($1.22) = $3.66.

CONFIDENCE AND PREDICTION INTERVALS

Suppose we wanted to calculate a confidence interval for our Chapter Problem. We would consider two factors. One is the size of the sample. The other is the fact that errors of prediction are larger as we move away from the mean value of the independent variable (X).

We will consider two applications. In the first case, the *mean* value of Y is estimated for a given value of X and is referred to as a confidence interval.

In the other case, an *individual* value of Y is estimated for a given value of X and is called a prediction interval.

First, for the confidence interval for the *mean value* of Y, the formula is

$$Y' \pm t(s_{Y \cdot X}) \sqrt{\frac{1}{n} + \frac{(X - \overline{X})^2}{\Sigma X^2 - \dfrac{(\Sigma X)^2}{n}}} \qquad \boxed{14\text{--}6}$$

Second, the formula for the prediction interval for an *individual value* of Y is

$$Y' \pm t(s_{Y \cdot X}) \sqrt{1 + \frac{1}{n} + \frac{(X - \overline{X})^2}{\Sigma X^2 - \dfrac{(\Sigma X)^2}{n}}} \qquad \boxed{14\text{--}7}$$

where

t is the value from Student's t distribution in Appendix D.
X is any selected value of the independent variable.
$\overline{X}$ is the mean of the independent variable (X).
n is the number of paired observations.
Y' is the point estimate of the dependent variable.
$s_{Y \cdot X}$ is the standard error.

Note that there is only a slight difference in the 2 formulas. There is a 1 in the formula for finding the prediction interval. Logically, the inclusion of 1 under the radical widens the confidence interval for the individual. This is reasonable because individuals are more unpredictable than group averages.

Problem Recall the Chapter Problem concerning the selling prices and lengths of textbooks at North Central State University.

Title	X	Y	XY	X²	Y²
Introduction to Psychology	400	$ 44	17,600	160,000	1936
Behavior in Organizations	600	47	28,200	360,000	2209
Introduction to Criminology	500	48	24,000	250,000	2304
Principles of Sociology	600	48	28,800	360,000	2304
Nursing Education	400	43	17,200	160,000	1849
Introduction to Residential Construction	500	46	23,000	250,000	2116
	3000	$276	138,800	1,540,000	12,718

What is the 95% confidence interval for the *mean* selling price for all books with 450 pages?

Solution   Using the regression equation computed previously, the predicted selling price for a book with 450 pages is $45:

$$Y' = 36 + 0.02X$$

$$\doteq 36 + 0.02(450)$$

$$= 45$$

We find the value of t by referring to Student's t distribution in Appendix D. Read down the left column until you locate $n - 2$, or $6 - 2 = 4$ degrees of freedom. Then move horizontally to the t-value under the 0.05 level of significance for a two-tailed test ($t = 2.776$).

Other values needed to set the confidence-interval estimate were computed previously: $s_{Y \cdot X} = \$1.22$ and, from the table above, $n = 6$, $\Sigma X = 3000$, $\Sigma X^2 = 1{,}540{,}000$, and $\overline{X} = 3000/6 = 500$. Inserting these values gives

$$Y' \pm t(s_{Y \cdot X}) \sqrt{\frac{1}{n} + \frac{(X - \overline{X})^2}{\Sigma X^2 - \frac{(\Sigma X)^2}{n}}}$$

$$= \$45 \pm (2.776)(\$1.22) \sqrt{\frac{1}{6} + \frac{(450 - 500)^2}{1{,}540{,}000 - \frac{(3000)^2}{6}}}$$

$$= \$45 \pm 3.4\sqrt{.23}$$

$$= \$45 \pm 1.63$$

$$= \$43.37 \text{ and } \$46.63$$

Interpreting, we find that the mean selling price for a group of books, all of whose lengths are 450 pages, is $45 and the 95% confidence interval is from $43.37 to $46.63.

Problem   Estimate a 95% prediction interval for the selling price of a *specific* book that is 450 pages long. This differs from the previous illustration in that we are concerned about a specific book that is 450 pages long, rather than the mean of all the books that are 450 pages long.

Solution   The formula for a prediction interval for an individual value is

$$Y' \pm t(s_{Y \cdot X}) \sqrt{1 + \frac{1}{n} + \frac{(X - \overline{X})^2}{\Sigma X^2 - \frac{(\Sigma X)^2}{n}}}$$

For a specific book that is 450 pages long:

$$\$45 \pm (2.776)(\$1.22) \sqrt{1 + \frac{1}{6} + \frac{(450 - 500)^2}{1{,}540{,}000 - \frac{(3000)^2}{6}}}$$

$$= \$45 \pm 3.4\sqrt{1.23}$$

$$= \$45 \pm 3.77$$

$$= \$41.23 \text{ and } \$48.77$$

Interpreting, we conclude that for a specific book 450 pages long, the 95% confidence interval for the selling price is between $41.23 and $48.77.

Self-Review 14–5

In Self-Reviews 14–3 and 14–4, which dealt with campaign expenditures and the numbers of votes received, we computed these values:

$$Y' = -2.3 + 4.3X, \; s_{Y \cdot X} = 3.3, \; n = 4, \; \Sigma X^2 = 54, \; \Sigma X = 14, \text{ and } \overline{X} = 3.5$$

a. Determine the 90% confidence interval for the *mean* number of votes gained by all candidates who spent $3800 on their campaigns.
b. Interpret your answer to part a.
c. Compute the 90% prediction interval for Sarah Norbett, whose campaign cost $3800.
d. Interpret your answer to part c.
e. Compare your responses to parts b and d.

Exercises

13. Refer to the data in the table on page 446.
 a. What is the 90% confidence interval for the mean selling price for all books with 600 pages?
 b. What is the 90% prediction interval for the selling price of a specific book that is 600 pages long?
14. Exercise 7 dealt with data relating an aggression index (Y) to the mean nighttime low temperature (X). The equation was $Y' = -25.7 + 1.04X$. The predicted value of the index was 57.5 when the temperature was 80°F. In Exercise 10, the standard error of estimate was found to be 3.6, $n = 6$, $\Sigma X = 396$, and $\Sigma X^2 = 27{,}954$.
 a. Construct a 95% confidence interval for the *mean* value of the aggression index when the temperature stands at 80°F.

b. On a *particular* night, the temperature is 80°F. Find the 95% prediction interval for the aggression index on this evening.
15. Exercise 5 presented some data relating the populations of five counties (X) to the numbers of physicians who practice in those counties (Y). Exercise 9 asked you to compute the standard error of estimate.
 a. Based on these data, construct a 95% confidence interval for the mean number of physicians practicing in counties with a population of 45,000.
 b. Based on these data, construct a 95% prediction interval for the number of physicians practicing in a particular county with a population of 45,000.

16. Refer to the data from St. Matthew's Hospital (in Exercises 1 and 12).

   a. Construct a 90% confidence interval for the *mean* cost of a 6-day stay.

   b. Construct a 90% prediction interval for the cost of an *individual's* 6-day stay.

A COMPREHENSIVE EXAMPLE

In Chapters 13 and 14, we discussed measures to (1) describe the relationship between two variables and (2) estimate the value of one variable based on another. The following problem reviews these topics. As you have observed, the calculations can be both tedious and lengthy. Therefore, we will use a statistical software package.

Problem   Allison Heflen is a social worker who specializes in marital problems among young married couples. She notes that young couples often have difficulty disciplining themselves to use credit wisely. As part of a study for the social services department for the city of Seattle, she collected a sample of 25 young married Seattle-area couples. She interviewed each couple and determined the number of different credit cards they used in the last year and their combined income for the last year. This sample information is reported in Table 14–3.

As part of her report to the social services department, she needs to respond to the following questions:

TABLE 14–3	Number of Credit Cards Used and Combined Income for a Sample of Young Married Couples				
Couple	Credit Cards	Combined Income (in thousands of dollars)	Couple	Credit Cards	Combined Income (in thousands of dollars)
Anderson	3	50.2	Eltimsayh	5	42.4
Jones	4	62.0	Rupple	8	91.0
Whitman	4	64.8	Rao	6	58.7
Bohn	4	68.7	Bhatt	2	60.2
Perry	2	53.9	Deans	8	80.2
Murnen	6	73.7	Saner	3	57.6
Harmon	6	64.3	Niles	2	33.8
Pohlman	5	53.7	Hercher	3	64.2
Kiel	8	67.1	Korosa	6	66.7
Hall	3	49.8	Minitel	7	83.1
Wolfe	4	69.6	Clegg	2	48.0
Jorgensen	4	51.7	Sevits	9	79.3
Williamson	7	83.4			

1. Is there a relationship between combined income and the number of credit cards? Is it strong or weak? Direct or indirect?
2. Can she estimate combined income on the basis of the number of credit cards?

Solution   To better focus her study, Allison developed the following list of goals for her report:

1. Develop a scatter diagram. Based on this scatter diagram, does there appear to be a linear relationship between combined income and number of credit cards?
2. Determine the coefficient of correlation. Is it strong or weak? Is it positive or negative?
3. Determine the coefficient of determination. Interpret.
4. Determine the standard error of estimate. Interpret.
5. Determine the regression equation. Interpret the meaning of the slope and the intercept values. Estimate the combined income for a young couple that used six different credit cards last year.
6. Develop a confidence interval and a prediction interval for couples who used six credit cards last year.

The first step is to develop a scatter diagram. Using MINITAB

Graph ▶ Character Graph ▶ Scatter Plot

Within the dialog box select "Income" as the *X*-variable and "Cards" as the *Y*-variable, then click OK.

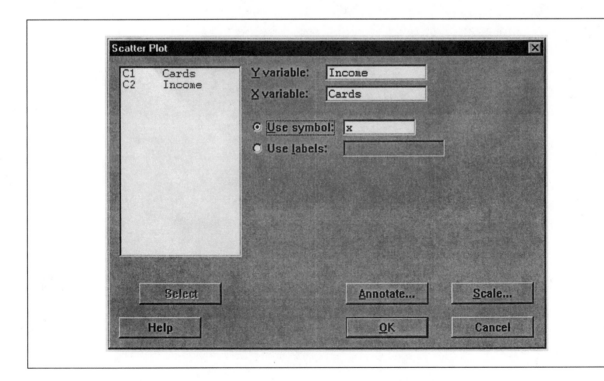

The independent variable is the number of credit cards used last year, which is scaled along the horizontal axis. The dependent variable is the combined income (reported in thousands of dollars), which is scaled along the vertical axis. Note that in the following graph the number "2" means that there are 2 data points at that location. There does appear to be a direct relationship between the two variables. As the number of credit cards increases, so does the combined income.

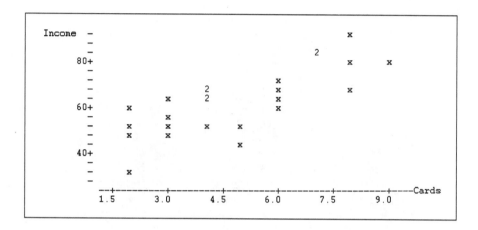

Again using the statistical software, Allison determined the coefficient of correlation. Again using the MINITAB system, the steps are:

Stat ▶ Descriptive Statistics ▶ Correlation

Within the dialog box, select "Cards" and "Income" as the variables, then click OK. MINITAB computes the correlation to be 0.746. The value 0.746 indicates a strong positive association between the variables. Could this association have occurred by chance, or can Allison conclude that the 2 variables are related? To answer this question she conducted the following test of hypothesis:

$$H_0: \rho \le 0$$

$$H_a: \rho > 0$$

The sample included 25 couples, so there are 23 degrees of freedom ($n - 2 = 25 - 2 = 23$). From Appendix D, the critical value for a one-tailed test at the 0.05 significance level is 1.714. The null hypothesis is rejected if t is greater than 1.714. From Chapter 13, Allison used formula 13–3.

$$t = \frac{r\sqrt{n-2}}{\sqrt{1-r^2}} = \frac{0.746\sqrt{25-2}}{\sqrt{1-0.746^2}} = 5.372$$

Because the computed value of 5.372 is greater than the critical value, the null hypothesis is rejected. Allison can conclude that there is a positive association between the number of credit cards used and the combined income.

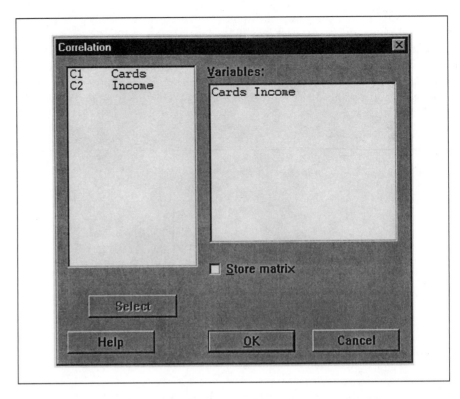

```
MTB > Correlation 'Cards' 'Income'.
```
Correlations (Pearson)

```
Correlation of Cards and Income = 0.746
```

When the number of credit cards used increases, income also increases (and vice versa).

The coefficient of determination is 0.557, found by 0.746^2. This indicates that more than 55% of the variation in the combined incomes of the young couples is accounted for by the number of credit cards used.

The MINITAB statistical software system is used to determine the standard error of estimate and the regression equation. The steps are:

$$\text{Stat} \blacktriangleright \text{Regression} \blacktriangleright \text{Regression}$$

Within the dialog box (shown on the next page), select "Income" as the Response variable and "Cards" as the Predictor variable, then click OK. The results are shown below the dialog box on the next page.

The standard error of estimate is 9.409. This is in thousands of dollars, so about 95% of the values would be within 18.818 thousand dollars of the regression line—that is, 2(9.409). The regression equation is $Y' =$

39.963 + 4.7853X. The *b*-value, which is the slope of the regression line, is 4.7853. This indicates that for each additional credit card used, the typical combined income increases about $4785. The *Y*-intercept, the *a*-value, is at 39.963. This is the point at which the regression line crosses the *Y* axis. This is the estimated income for a couple that did not use any credit cards

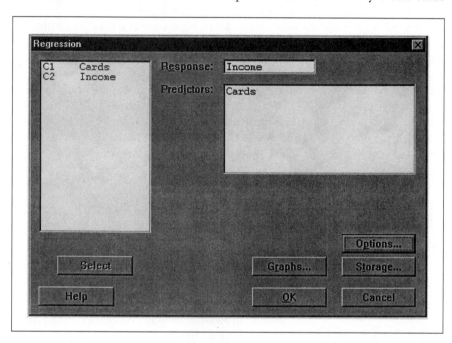

```
The regression equation is
Income = 40.0 + 4.79 Cards

Predictor        Coef       StDev           T         P
Constant       39.963       4.708        8.49     0.000
Cards          4.7853      0.8916        5.37     0.000

S = 9.409      R-Sq = 55.6%      R-Sq(adj) = 53.7%

Analysis of Variance

Source         DF          SS            MS        F         P
Regression      1       2550.1        2550.1    28.80     0.000
Error          23       2036.2          88.5
Total          24       4586.3

Unusual Observations
Obs    Cards    Income       Fit   StDev Fit   Residual   St Resid
 14     5.00     42.40     63.89        1.89     -21.49     -2.33R

R denotes an observation with a large standardized residual

     Fit   StDev Fit       95.0% CI            95.0% PI
   68.67        2.15   (  64.23,   73.12)  (  48.71,   88.64)
```

last year. This value is found by substituting zero in the regression equation for X.

The estimated combined income for a young couple using 6 credit cards is 68.6748, which is $68,675. This value is shown on the last row of the MINITAB output under the heading "Fit." It is also computed as follows: $Y' = 39.963 + 4.7853X = 39.963 + 4.7853(6) = 68.6748$.

To develop confidence intervals and prediction intervals we use formulas 14–6 and 14–7. The MINITAB system prints these values at the bottom of the output. The 95% confidence interval ranges from 64.23 up to 73.12. This means we are 95% "confident" that the mean income for the couples using 6 credit cards is between $64,230 and $73,120. The 95% prediction interval is from 48.71 up to 88.64. So a particular young couple—say, Les and Stephanie Holler—who used 6 credit cards last year could be expected to have earned between $48,710 and $88,640. Notice again that the interval for a particular couple is much wider than that for the mean of all couples using a particular number of credit cards.

CHAPTER OUTLINE

I. In regression analysis, we estimate one variable based on another variable.
   A. The variable being estimated is the dependent variable.
   B. The variable used to make the estimate is the independent variable.
      1. The relationship between the variables must be linear.
      2. Both the independent variable and the dependent variable must be of interval or ratio scale.
      3. The least-squares criterion is used to determine the regression equation.
II. The least-squares regression line is of the form $Y' = a + bX$.
   A. Y' is the estimated value of Y for a selected value of X.
   B. a is the constant or intercept.
      1. It is the value of Y' when $X = 0$.
      2. a is computed using the following equation:

$$a = \frac{\Sigma Y - b\Sigma X}{n} \qquad \boxed{14\text{–}3}$$

C. b is the slope of the line.
   1. It shows the amount of change in Y' for a change of one in X.
   2. A positive value for b indicates a direct relationship between the two variables and a negative value an indirect relationship.
   3. The sign of b and the sign of r, the coefficient of correlation, are always the same.
   4. b is computed using the following equation:

$$b = \frac{n\Sigma XY - \Sigma X\Sigma Y}{n\Sigma X^2 - (\Sigma X)^2} \qquad \boxed{14\text{–}2}$$

D. X is the value of the independent variable.
III. The standard error of estimate measures the variation around the regression line.
   A. It is in the same units as the dependent variable.
   B. It is based on squared deviations from the regression line.
   C. Small values indicate that the points cluster closely about the regression line.

D. It is computed using the following formula:

$$s_{Y \cdot X} = \sqrt{\frac{\Sigma Y^2 - a\Sigma Y - b\Sigma XY}{n - 2}}$$ `14–5`

IV. Inference about linear regression is based on the following distributional assumptions:
  A. For a given value of X, the values of Y are normally distributed about the line of regression.
  B. The standard deviation of each of the distributions is the same for all values of X and is estimated by the standard error of estimate.
  C. The deviations from the regression line are independent, with no pattern to the size or direction.
V. There are two types of interval estimates.
  A. In a confidence interval, the mean value of Y is estimated for a given value of X.
    1. It is computed from the following formula:

$$Y' \pm t s_{Y \cdot X} \sqrt{\frac{1}{n} + \frac{(X - \overline{X})^2}{\Sigma X^2 - \frac{(\Sigma X)^2}{n}}}$$ `14–6`

2. The width of the interval is affected by the level of confidence, the size of the standard error of estimate, and the size of the sample as well as the value of the independent variable.
B. In a prediction interval, the individual value of Y is estimated for a given value of X.
  1. It is computed from the following formula:

$$Y' \pm t s_{Y \cdot X} \sqrt{1 + \frac{1}{n} + \frac{(X - \overline{X})^2}{\Sigma X^2 - \frac{(\Sigma X)^2}{n}}}$$ `14–7`

2. The difference between formulas 14–6 and 14–7 is the 1 under the radical.
  a. The prediction interval will be wider than the confidence interval.
  b. The prediction interval is also based on the level of confidence, the size of the standard error of estimate, the size of the sample, and the value of the independent variable.

- -

Exercises

17. A tanning booth survey has produced the following information on lengths of time in a booth and results. Each person surveyed was asked how long his or her average stay lasted. Then the employees scored each tan on a scale of 1 to 10. The data:

Name	Tanning Time (in minutes)	Score
Jill	30	7
Jodi	20	6
Mark	40	9
Ted	15	5
Tami	25	7

a. Find the least-squares regression equation.
b. Interpret the values of a and b.
c. Predict the score that a customer who spends 50 minutes in the tanning booth will receive.
18. Data show the following relationship between a man's height and shoe size:

Height (in inches)	Shoe Size
76	12
72	11
65	8
78	13
82	16
67	9

a. Calculate the regression equation. Shoe size is the dependent variable.

b. Find the standard error of estimate.

19. The numbers of requests for information (X) and actual enrollments (Y) for the past six years at Nordam University are

Number of Requests for Information (in thousands)	Actual Enrollment (in thousands)
3.0	3.3
3.5	4.1
4.2	5.6
4.8	5.2
5.0	5.9
5.1	5.5

a. Draw a scatter diagram.

b. Determine the least-squares regression equation.

c. Calculate the standard error of estimate.

d. Compute a 90% confidence interval for the mean enrollment when the requests for information number 4500.

20. A student of criminology, interested in predicting the age at incarceration (Y) using the age at first police contact (X), collected the following data:

Age at First Contact X	Age at Incarceration Y
11	21
17	20
13	20
12	19
15	18
10	23
12	20

a. Draw a scatter diagram.

b. Find the regression equation by the least-squares method.

c. Compute the standard error of estimate.

d. Estimate the age at incarceration for an individual who at age 14 had his first contact with police. Develop a 95% prediction interval for your estimate.

21. A study is undertaken regarding public accounting firms. Of particular interest is the relationship between the number of years a firm has been in business and the number of accountants employed by the firm. A sample of eight firms is selected. The results are as follows:

Firm	Number of Years	Number of Accountants
A & G Accounting	12	36
Kramer Associates	18	72
Whitte and Whitte	27	160
Englekay and Jerdonic	5	13
Huffman and Clark	7	18
Grasser and Lewis	22	68
M & M Accountants	15	40
Bookkeeping, Inc.	10	33

a. Draw a scatter diagram. Let number of years be the independent variable. Does there appear to be a relationship between years and numbers of accountants?

b. Determine the regression equation.

c. Compute the standard error of estimate.

d. Estimate the number of accountants in a 10-year-old firm.

e. Develop a 95% confidence interval for all 10-year-old firms.

22. The lengths of time children are exposed to a common virus (X) and the percentages of the group (Y) who contract the disease are

Weeks X	Percentage Y
2	0.1
3	0.3
3	0.5
4	0.8
4	1.2
6	1.8
7	2.5
8	3.4

a. Draw a scatter diagram of the data.
b. Compute the least-squares line of regression.
c. Find the standard error of estimate.
d. Make a prediction about the percentage of a group who will contract the virus after a 5-week exposure.
e. Determine a 99% prediction interval for the percentage of a group exposed 5 weeks that contract a virus.

23. The lengths of various magazine articles (X) and individual reading time (Y) were recorded:

Length of Article (in pages)	Reading Time (in minutes)
5	13
6	15
6	15
7	18
4	10
3	8
7	18
12	25

a. Draw a scatter diagram.
b. Calculate the regression line.
c. What is the standard error of estimate?
d. Predict the average reading time for a particular 10-page article, and construct a 95% prediction interval about your estimate.

24. A statistics instructor wishes to study the relationship between the numbers of cuts (absences) students take and their final course grades. The data obtained are

Number of Absences	Grade
1	98
2	90
4	83
3	88
5	71
2	85
4	76
3	81
6	71

a. Draw a scatter diagram, and determine the regression equation.
b. Compute the standard error of estimate.
c. Develop a 90% confidence interval for all students with 3 absences.
d. If a particular student, Julia Grayon, has 3 absences, what is the 90% prediction interval for her score?

25. The Kansas Department of Transportation is investigating the relationship between the number of bidders on a highway construction project and the winning bidder's price. Does the number of bidders decrease the price of the project? A sample of 20 recent construction projects revealed the following information:

Project	Number of Bidders	Winning Bid (in millions of dollars)
1	5	$ 8.7
2	4	9.5
3	3	8.8
4	1	10.9
5	6	7.1
6	7	7.4
7	3	9.3
8	5	7.4
9	6	6.8
10	5	8.7
11	3	8.9
12	6	6.2
13	9	6.3
14	4	9.0
15	6	6.5
16	3	9.3
17	2	9.7
18	5	6.3
19	4	9.2
20	4	8.0

a. Determine the regression equation.
b. Note that the sign of b is negative. Does this surprise you? Comment.
c. How much does the cost of a project decline for each additional bidder?
d. Determine a 95% confidence interval for all highway projects with 5 bidders.

e. Determine a 95% prediction interval for the project of building a bridge across the Cone-wango River. There are 5 bidders for the project.

26. The Yuppie Inn is an exclusive restaurant that accepts reservations for Saturday night only. The owner would like to estimate the number of meals to be served on the basis of the number of reservations. The following is a sample of recent experience:

Number of Reservations	Number of Meals Served
48	79
56	69
65	98
67	82
65	89
63	91
76	95
46	94
92	91
84	115
52	70
90	114
46	86
32	70
100	110
80	91
60	96
90	110
81	95
64	79
47	69
45	83
69	81
85	87
50	75

a. Draw a scatter diagram. Does there appear to be a relationship between the 2 variables? Is it direct or inverse?
b. Determine the regression equation. Estimate the number of meals to be served if there are 100 reservations.

c. There are 100 reservations for this Saturday night. Determine a 95% confidence interval for the number of meals to be served.
d. Using the material from Chapter 13, determine the coefficient of correlation. At the 0.05 significance level, can we conclude that there is a positive association between the number of reservations and the number of meals served?

27. The Bardi Trucking Company, located in Cleveland, Ohio, makes deliveries in the Great Lakes, southeastern, and northeastern regions of the United States. Ed Bardi, the owner, wants to study the relationship between the distance a shipment must travel and the time (in days) it takes the shipment to arrive at its destination. To investigate, Mr. Bardi, selected a random sample of 20 shipments made last month. The results are shown below. Shipping distance is the independent variable and shipping time the dependent variable.

Shipment	Distance (in miles)	Number of Days
1	656	5
2	853	14
3	646	6
4	783	11
5	610	8
6	841	10
7	785	9
8	639	9
9	762	10
10	762	9
11	862	7
12	679	5
13	835	13
14	607	3
15	665	8
16	647	7
17	685	10
18	720	8
19	652	6
20	828	10

a. Draw a scatter diagram. Based on these data, does it appear that there is a relationship between the distance a shipment must travel and the time it takes to arrive at its destination? Is there a positive or negative relationship between the variables?

b. Determine the regression equation. Discuss the meaning of the *a*- and *b*-values. How much longer does it take for a shipment to go 400 miles than 300 miles?

c. How long would you estimate that a shipment traveling 800 miles will take to arrive at its destination?

d. Develop a 95% confidence interval for all shipments traveling 800 miles.

e. Develop a 95% prediction interval for a particular shipment traveling 800 miles from Cleveland.

28. Listed at right are the mean high temperatures by month for Chicago and Los Angeles. Also listed are the mean amounts of rainfall (in inches) for each city by month.

Month	Chicago		Los Angeles	
	Temp	Precp	Temp	Precp
January	24.0	1.67	52.9	2.60
February	27.4	1.27	53.9	2.39
March	37.5	2.47	54.8	1.76
April	51.2	3.31	57.0	1.13
May	61.3	3.34	60.0	0.08
June	71.5	4.07	62.6	0.03
July	75.5	3.39	66.4	0.01
August	74.7	2.68	67.4	0.02
September	66.9	2.95	66.6	0.07
October	56.5	2.27	63.2	0.23
November	41.2	2.06	58.7	1.81
December	28.5	1.61	55.2	1.80

Let temperature be the independent variable and precipitation the dependent variable. Develop regression equations to estimate the monthly precipitation in Chicago and Los Angeles based on the temperature. Comment on the differences and the similarities in the two equations. Also, using the material in Chapter 13, is one of the equations a better predictor of precipitation than the other? Conduct appropriate tests of hypothesis.

DATA EXERCISES

29. Refer to the real estate data set, which reports information on homes sold in Alabama during 1996.

a. Determine the regression equation with the selling price as the dependent variable and the area of the home as the independent variable. Write a brief interpretation of the regression equation.

   (1) Estimate the selling price of a home with an area of 2500 square feet.

   (2) Develop a 95% confidence interval for the mean selling price of all homes with 2500 square feet.

   (3) Develop a 95% prediction interval for the mean selling price of a particular home with an area of 2500 square feet.

b. Determine the regression equation with the selling price as the dependent variable and the distance from the center of the city as the independent variable. Write a brief interpretation of the regression equation.

   (1) Estimate the selling price of a home located 20 miles from the center of the city.

   (2) Develop a 95% confidence interval for the mean selling price of all homes located 20 miles from the center of the city.

(3) Develop a 95% prediction interval for the mean selling price of a particular home located 20 miles from the center of the city.

30. Refer to the school data set, which includes information on the 94 school districts in northwestern Ohio.

a. Let the independent variable be the percentage of families on welfare in the school district and the number of students in the district be the dependent variable.

(1) Determine the regression equation. Comment on the values you found for *a* and *b*. Are they reasonable?

(2) Estimate the number of students in a school district if 10% of the families are on welfare.

(3) Develop a 95% confidence interval and a 95% prediction interval for all school districts that have 10% of the families on welfare.

b. Let the independent variable be the attendance rate and the percentage passing the proficiency examination be the dependent variable.

(1) Determine the regression equation. Comment on the values you found for *a* and *b*. Are they reasonable?

(2) Estimate the percentage of students passing the proficiency test if the attendance rate is 90 percent.

(3) Develop a 95% confidence interval and a 95% prediction interval for the percentage passing the proficiency test if the attendance rate is 90%.

CHAPTER ACHIEVEMENT TEST

The answers are at the back of the book.

MULTIPLE-CHOICE QUESTIONS

Select the response that best answers each of the questions.

1. A variable about which predictions or estimates are made is called the
a. dependent variable.
b. discrete variable.
c. independent variable.
d. correlation variable.

2. In the regression equation, the value that gives the amount by which *Y* changes for every unit in *X* is called the
a. coefficient of correlation.
b. coefficient of determination.
c. slope.
d. *Y*-intercept.

3. In the equation $Y' = a + bX$, the letter *a* stands for the

a. coefficient of correlation.
b. coefficient of determination.
c. slope of the regression line.
d. *Y*-intercept of the regression line.

4. What kind of relationship exists if *Y* decreases as *X* increases?
a. Inverse
b. Direct
c. Significant
d. No relationship
e. None of the above

Questions 5–7 are based on the following situation. Boys of varying ages were asked to perform chin-ups. The data were then used to develop the regression equation $Y' = -1.0 + 1.5X$.

Boy	Age (X)	Chin-Ups (Y)
Adam	16	22
Bob	8	12
Chuck	14	22
Denny	10	12
Felix	12	17

5. Which of the five data points comes closest to the regression line?
  a. Adam's
  b. Bob's
  c. Chuck's
  d. Denny's
  e. Felix's
6. The maximum difference between the actual and the predicted numbers of chin-ups is
  a. 0.
  b. 1.
  c. 2.

d. 3.
e. 4.
7. What is the sum of the squared errors?
  a. 0
  b. 3.16
  c. 6
  d. 10
  e. 100
8. The standard error of estimate is a measure of
  a. the variation around the mean of X.
  b. the variation around the regression line.
  c. the explained variation.
  d. All of the above
9. Which of the following does not affect the width of a confidence interval?
  a. The standard error of estimate
  b. The size of the sample
  c. The value of X for which the estimate is being made
  d. The Y-intercept

COMPUTATION PROBLEM

10. A commercial grower wants to study the relationship between the mean height of a new variety of dahlia and the number of days since emergence above ground for a sample of seven plants:

a. Draw a scatter diagram.
b. Determine the least-squares regression equation.
c. Determine the standard error of estimate.
d. Compute a 90% confidence interval for the mean height of all dahlias that have been above ground 25 days.

Number of Days Above Ground X	Height (in centimeters) Y
6	10
22	19
34	31
42	39
45	47
48	58
47	66

ANSWERS TO SELF-REVIEW PROBLEMS

14-1 a. Don has 5 years of education beyond high school; his weekly income is $450.

    b. $a = 250$

    c. $b = 100$. One way to compute: When $X = 3$, $Y = 550$, and when $X = 4$, $Y = 650$; then, $650 - 550 = 100$. It indicates that for each 1 additional year beyond high school, weekly income increases $100.

    d. $Y' = 250 + 100X$ (in dollars)

    e. $Y' = \$650$, found by $250 + 100(4)$

14-2 a. 104 found by

X	Deviation	Squared Deviation
3	8	64
4	−2	4
5	−6	36
		104

The sum of the squared deviations is 104.

    b. The units sold is the dependent variable.

14-3 a.

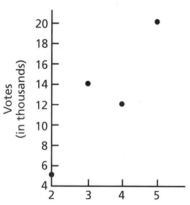

Campaign Expenditures
(in thousands of dollars)

b. $b = \dfrac{4(200) - (14)(51)}{4(54) - (14)^2} = \dfrac{86}{20} = 4.3$

$a = \dfrac{51 - 4.3(14)}{4}$

$= -2.3$

$Y' = -2.3 + 4.3X$ (in thousands of votes)

c. For each additional $1000, the candidate could expect 4300 votes.

14-4 $s_{Y \cdot X} = \sqrt{\dfrac{\Sigma Y^2 - a\Sigma Y - b\Sigma XY}{n - 2}}$

$= \sqrt{\dfrac{765 - (-2.3)(51) - 4.3(200)}{4 - 2}}$

$= \sqrt{\dfrac{765 + 117.3 - 860}{2}}$

$= 3.3$

14-5 a. $Y' = -2.3 + 4.3(3.8) = 14.0$

Then:

$14.0 \pm 2.920(3.3) \sqrt{\dfrac{1}{4} + \dfrac{(3.8 - 3.5)^2}{54 - \dfrac{(14)^2}{4}}}$

$= 9.0$ and 19.0

b. The probability is 0.90 (that is, the odds are 9 to 1) that a candidate who spends $3800 on his or her campaign will receive between 9000 and 19,000 votes.

c. $14.0 \pm 2.920(3.3) \sqrt{1 + \dfrac{1}{4} + \dfrac{(3.8 - 3.5)^2}{54 - \dfrac{(14)^2}{4}}}$

$= 14.0 \pm 10.85$

$= 3.15$ and 24.85

d. The probability is 0.90 that, if Sarah Norbett spends $3800 on her campaign, she will receive between 3150 and 24,850 votes.

e. The wider interval is the result of comparing individuals and comparing means.

CHAPTER 15

Multiple Regression and Correlation Analysis

OBJECTIVES

When you have completed this chapter, you will be able to

- describe the relationship between one dependent variable and two or more independent variables;
- interpret multiple regression and correlation output from a computer software package;
- conduct a test to determine if any of the net regression coefficients differ from zero;
- conduct a test to determine which, if any, of the regression coefficients are zero.

CHAPTER PROBLEM   Predicting Prejudice

Dr. Susan Welch is a noted sociologist doing research on racial prejudice. She has developed a test in which a racial prejudice "score" is given to respondents based on their answers to a series of questions. An equally important problem for her is to identify factors that are related to this score. Dr. Welch would like to be able to predict which respondents will have low prejudice scores and which will have high prejudice scores. Is age a factor? Marital status? How about political orientation; that is, are liberals more or less likely to exhibit racial prejudice than conservatives? What other factors should be considered?

INTRODUCTION

Our study of correlation and regression began in Chapter 13, where we showed that the relationship between 2 sets of interval-scaled measurements can be described by Pearson's product-moment coefficient of correlation. A coefficient of 0.91, for example, indicates that the relationship between the 2 variables is strong. A coefficient near 0, such as 0.17, indicates a very weak relationship. In Chapter 14 we developed an equation to express the relationship between two variables. If the relationship is linear, we can then use the equation $Y' = a + bX$ to predict the dependent variable, Y, based on the independent variable, X.

In this chapter, we will continue our study of correlation and regression. However, instead of using only one independent variable to predict the dependent variable, we will consider *two or more* independent variables. The purpose of using several independent variables is to obtain a more accurate prediction or to explain more of the variation in the dependent variable. When more than one independent variable is used, we refer to the procedure as multiple regression and correlation analysis.

ASSUMPTIONS ABOUT MULTIPLE REGRESSION AND CORRELATION

Multiple regression and correlation analysis is based on five assumptions:

1. The independent variables and the dependent variable have a linear (straight-line) relationship.
2. The dependent variable is continuous and at least of interval scale.
3. The variation around the regression line is the same for all values of X. This means that Y varies the same amount when X is a low value as when it is a high value. Statisticians call this assumption homoscedasticity.
4. Successive observations of the dependent variable are uncorrelated. Violation of this assumption is called autocorrelation. Autocorrelation often occurs when data are collected over a period of time.
5. The independent variables are not highly correlated. When the independent variables are correlated, this is called a multicollinearity.

Even when one or more of these assumptions are violated, which often happens in practice, multiple regression and correlation techniques seem to work well, and the results are still adequate. The regression equation and measures of correlation provide the researcher with predictions that are better than any that could otherwise be made.

THE MULTIPLE REGRESSION EQUATION

Recall from Chapter 14 (formula 14–1) that for *one* independent variable, the linear regression equation has the form $Y' = a + bX$. The multiple regression case extends this equation to include other independent variables. For two independent variables, the regression equation is

$$Y' = a + b_1X_1 + b_2X_2$$

15–1

where

X_1 is the first independent variable.
X_2 is the second independent variable.
a is the point of intercept with the Y-axis.
b_1 is the rate of change in Y for each unit change in X_1, with X_2 held constant. It is called a net regression coefficient.
b_2 is the rate of change in Y for each unit change in X_2, with X_1 held constant. It, too, is called a net regression coefficient.

For any number of independent variables (k), the general multiple regression equation is

$$Y' = a + b_1X_1 + b_2X_2 + b_3X_3 + \cdots + b_kX_k$$

15–2

Problem   The director of admissions at McLean University uses high-school grade point averages (X_1) and IQ scores (X_2) to predict first-quarter grade point averages (GPAs) of entering students. The director determined the multiple regression equation to be

$$Y' = 0.60 + 0.75X_1 + 0.001X_2$$

Predict the first-quarter GPA for a student whose high-school GPA is 3.0 and whose IQ is 100. Also predict the first-quarter GPA for an entering student with a high-school GPA of 4.0 and an IQ of 100. What do 0.60 and the regression coefficient 0.75 in the equation indicate?

Solution   The predicted college GPA for a student with a 3.0 GPA in high school and with an IQ of 100 is 2.95, computed as follows:

$$Y' = 0.60 + 0.75X_1 + 0.001X_2$$

$$= 0.60 + (0.75)(3.0) + (0.001)(100)$$

$$= 2.95$$

The predicted college GPA for a student with a 4.0 GPA in high school and with an IQ of 100 is 3.70, computed as follows:

$$Y' = 0.60 + (0.75)(4.0) + 0.001(100)$$

The 0.60 indicates that the regression equation crosses through the Y-axis at 0.60. This is also called the Y-intercept. The regression coefficient of

0.75 indicates that for each increase of 1 in the high-school grade point average, the first-quarter GPA at McLean will increase 0.75, *regardless of the student's IQ.*

Self-Review 15–1

Answers to the Self-Review problems are at the end of the chapter.

An economist is studying a sample of households to determine the weekly amount saved by each. Three independent variables seem to hold promise as predictors of the amount saved. These are: weekly income, weekly amount spent on food, and weekly amount spent on entertainment. The multiple regression equation was computed to be

$$Y' = 20.0 + 0.50X_1 - 1.20X_2 - 1.05X_3$$

where
Y' is the weekly amount saved.
X_1 is the weekly income of the household.
X_2 is the weekly amount spent on food.
X_3 is the weekly amount spent on entertainment.

If a household had an income of $300 per week and spent $70 on food and $50 on entertainment, how much would be saved per week? Interpret the regression equation.

Exercises

Answers to the even-numbered Exercises are at the back of the book.

1. Refer to the following regression equation:

$$Y' = 4.00 + 5.00X_1 - 2.00X_2$$

a. Compute the value of Y' if $X_1 = 10$ and $X_2 = 5$.
b. How much does Y' change for a unit change in X_1 if X_2 is held constant?

2. Refer to the following regression equation:

$$Y' = 55.0 - 3.20X_1 + 2.3X_2$$

a. Compute the value of Y' if $X_1 = 20$ and $X_2 = 30$.
b. How much does Y' change for a unit change in X_1 if X_2 is held constant?

3. A medical researcher is studying the systolic blood pressure of business executives. Two independent variables—income and age—are being used as predictors. The following multiple regression equation has been computed:

$$Y' = 130 + 0.32X_1 + 0.22X_2$$

where
Y' is the systolic pressure.
X_1 is the income (in thousands of dollars).
X_2 is the age.

a. What is the expected systolic pressure reading of a 60-year-old executive earning $80,000 a year?
b. Regardless of age, how much does the systolic pressure increase for each additional $10,000 in annual income?

4. A sample of women was polled to determine the degrees of satisfaction they felt in their marriages. All the women were between the ages of 35 and 50, and all had at least 1 child. An index of satisfaction was developed for the study. Three factors were used as predictor variables: the number of children (X_1), the number of years since the birth of the last child (X_2), and the mother's yearly earnings outside the home (X_3). The regression equation is:

$$Y' = 30 + 1.05X_1 + 6.5X_2 + 0.005X_3$$

a. Predict the index of marriage satisfaction for a woman with 3 children who earns $12,000 outside the home and had her last child 8 years ago.

b. Which would contribute more to marriage satisfaction: another year without a child or an additional $5000 in income?

MEASURING THE STRENGTH OF THE ASSOCIATION

If the multiple regression equation fits the data perfectly, there is no error in the predicted value of Y. Of course, this seldom occurs in practice, so as in simple linear regression we need measures that provide information on the strength of the association. In this section, we discuss two such measures—the coefficient of multiple determination and the multiple standard error of estimate.

Perhaps the easiest measure of association to interpret is the **coefficient of multiple determination**. It reports the proportion or percentage of the total variation in the dependent variable, Y, that is explained by the independent variables. It can range from 0 to 1.00 and is usually reported as R^2. An R^2 value of 0 indicates no association between the dependent variable and the independent variables. A value of 1.00 indicates that all the variation in the dependent variable is accounted for by the independent variables.

> **coefficient of multiple determination** The proportion (percentage) of the total variation in the dependent variable, Y, that is explained by the independent variables.

To explain further, suppose R^2 is 0.83 for a study of executives in the electronics industry. The dependent variable is salary and the independent variables are age and number of employees supervised. We conclude that 83% of the variation in salary is accounted for by the 2 independent variables. This is a strong association. On the other hand, an R^2 value of 0.23 is reported in a study where the dependent variable is the selling price of the home and the independent variables are the distance from the center of the city and the number of bedrooms in the home. This is a rather weak association.

A second measure of association, called the multiple standard error of estimate, is the natural extension to the standard error of estimate discussed in Chapter 14. It is based on the squared differences between Y and Y'. However, Y' is now based on *two or more* independent variables. It can be thought of as the square root of an average of the squared errors.

The multiple standard error of estimate is computed as follows:

$$S_{Y \cdot 12} = \sqrt{\frac{\Sigma(Y - Y')^2}{n - (k + 1)}}$$

15–3

where

$S_{Y \cdot 12}$ is the multiple standard error of estimate. The subscript $Y \cdot 12$ indicates that Y is the dependent variable. The 12 indicates that there are 2 independent variables, labeled X_1 and X_2.

n is the number of observations in the sample.

k is the number of independent variables.

What is the difference between the coefficient of multiple determination and the multiple standard error of estimate? The first, the coefficient of multiple determination, is a relative number between 0 and 1. On the other hand, the multiple standard error is an absolute number expressed in the same units as the dependent variable. So, if the dependent variable is in dollars, the standard error is also in dollars.

Problem An administrator at Valley General Hospital is studying patients' lengths of stay (in days) based on two variables: the patient's age (X_1) and gender (X_2). The administrator selects a random sample of nine patients who were in the hospital for at least one day during the last month. The results are as follows:

Age (in years)	Gender*	Stay (in days)
42	1	12
36	0	10
32	1	11
29	0	6
26	0	9
24	0	6
22	1	9
18	0	5
15	0	7

*A 1 indicates that the patient is female and a 0 that the patient is male.

a. Use a statistical software package to determine the regression equation.
b. Use this regression equation to estimate the length of stay for Ms. Kulla, a 42-year-old patient.
c. Identify and interpret the coefficient of multiple determination.
d. Identify and interpret the multiple standard error of estimate.

Solution MINITAB was used to determine the regression equation and measures of association. The dependent variable is length of stay. The patient's age and gender are the independent variables.

The MINITAB commands are:

Stat ▶ Regression ▶ Regression

In the dialog box enter "Stay" as the Response variable and "Gender" and "Stay" as the predictor variables, then click OK. The dialog box and results are shown on the next page.

a. The regression equation is

$$Y' = 3.30 + 0.157X_1 + 2.35X_2$$

The intercept with the Y axis is 3.30. This is the point at which both $X_1 = 0$ and $X_2 = 0$. An increase of 1 year in age in the same gender— say, from 40 to 41 years—means an increase in the length of stay of 0.157 day. A 41-year-old patient can be expected to stay 0.157 day longer in the hospital than a 40-year-old patient. Similarly, a female patient can be expected to stay an additional 2.35 days more than a male

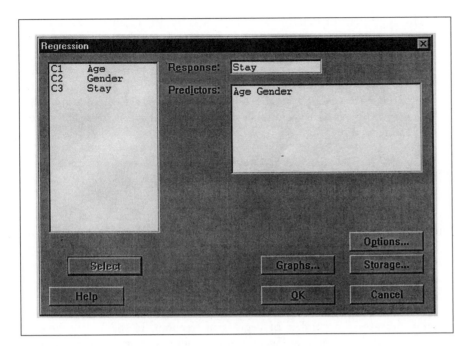

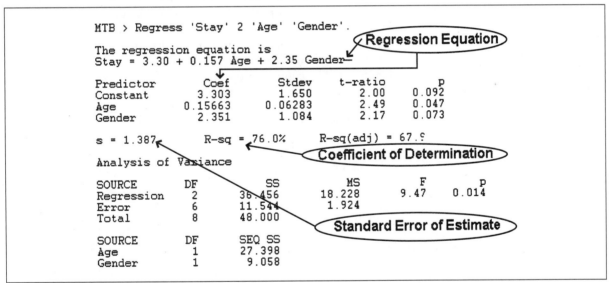

of the same age. How is this so? If the sampled patient is male, the variable X_2 is coded as 0, so 2.35 is multiplied by 0 and this term drops out of the equation. If the sampled patient is female, then 2.35 is multiplied by 1, so the length of stay is increased by 2.35 days. There are only 2 conditions—male and female—with this variable.

b. Ms. Kulla's stay is estimated to be 12.244 days, computed as follows:

$$Y' = 3.30 + 0.157(42) + 2.35(1)$$

$$= 12.244$$

c. The coefficient of determination is 0.760, which means that 76.0% of the variation in the length of a patient's stay can be explained by the variation in the patient's gender and age.

d. The multiple standard error of estimate is 1.387. This measure is interpreted much like the standard deviation. About 95% of the actual observations are within 2 standard errors of the regression equation. That is, we would expect to find most of the lengths of stay within 2.774 days of the regression equation, found by 2(1.387).

THE ANOVA TABLE

Included in many software programs is an ANOVA (analysis of variance) table similar to the one described in Chapter 12. In that chapter, the variation was divided into 2 components—that due to treatments and that due to other causes, including random variation. Here the total variation is also divided into 2 categories—that explained by the regression (the independent variables) and the error (unexplained variation). These two categories (regression and error) are identified in the Source column of the computer output. The DF column indicates the number of degrees of freedom. In the Valley General Hospital example above, there were 9 observations, so $n = 9$. The total number of degrees of freedom is $n - 1 = 9 - 1 = 8$. The degrees of freedom in the Regression row are equal to the number of independent variables. We use k to represent the number of independent variables, so $k = 2$. The degrees of freedom in the Error row are $n - (k + 1) = 9 - (2 + 1) = 6$.

The term SS (middle column of the ANOVA table) refers to the sum of the squares. These terms are computed as follows:

$$\text{Total sum of squares} = SS \text{ total} = \Sigma(Y - \overline{Y})^2 = 48$$

$$\text{Error sum of squares} = SSE = \Sigma(Y - Y')^2 = 11.544$$

$$\text{Regression sum of squares} = SSR = SS \text{ total} - SSE$$

$$= 48.000 - 11.544 = 36.456$$

The column headed MS, which refers to the mean square, is found by dividing SS by df. We obtain MSR by SSR/k and MSE by $SSE/[n - (k + 1)]$. Continuing, the F column is the ratio of the two mean squares—that is, MSR/MSE. The details of this calculation are discussed in the upcoming section on the global test of hypothesis. The P column refers to the p-value. As discussed in an earlier section, the p-value refers to the probability of obtaining a value of the test statistic this large or larger if the null hypothesis is true.

The general format of the computer-generated ANOVA table is

Analysis of Variance

Source	DF	SS	MS	F	P
Regression	k	SSR	MSR	MSR/MSE	
Error	$n - (k + 1)$	SSE	MSE		
Total	$n - 1$	SS total			

We calculate the coefficient of determination, R^2, using the numbers in the ANOVA table. Recall from Chapter 13 that the coefficient of determination is the percentage of the variation that is explained by the regression. In other words, the coefficient of multiple determination is the ratio of the regression sum of squares to the total sum of squares. It is the variation explained by the regression divided by the total variation:

$$R^2 = \frac{SSR}{SS \text{ total}}$$ **15–4**

In the Valley General Hospital example, R^2 is 0.760, found by formula 15–4:

$$R^2 = \frac{SSR}{SS \text{ total}} = \frac{36.456}{48.000} = 0.760$$

The ANOVA table can also be used to arrive at the multiple standard error of estimate. The formula is

$$s_{Y \cdot 12} = \sqrt{\frac{SSE}{n - (k + 1)}}$$ **15–5**

Using this formula, the multiple standard error of estimate is

$$s_{Y \cdot 12} = \sqrt{\frac{SSE}{n - (k + 1)}} = \sqrt{\frac{11.544}{9 - (2 + 1)}} = 1.387$$

This is also reported in the MINITAB output on page 469.

> **REAL STAT**
>
> Financial advisors use regression analysis to estimate future trends in the stock market based on historical data patterns. For example, according to the Hemline Theory, a rise in the hemlines of women's skirts will forecast an upturn in the Dow Jones Industrial Average. According to the Super Bowl Theory, if the winning team is from the American Conference, the Dow Jones Industrial Average will be down the following year. If the winning team is from the National Conference, the index will be up. Incidentally, the Super Bowl Index was correct 21 out of 23 years. Would you want to invest a large sum of your money on the basis of either of these 2 theories?

Exercises

5. Refer to the following ANOVA output:

SOURCE	DF	SS	MS	F
Regression	3	21.000	7.000	2.333
Error	15	45.000	3.000	
Total	18	66.000		

a. What is the sample size?
b. How many independent variables are there?
c. Compute the coefficient of multiple determination.
d. Compute the multiple standard error of estimate.

6. Refer to the following ANOVA output:

SOURCE	DF	SS	MS	F
Regression	5	60.000	12.000	1.714
Error	20	140.000	7.000	
Total	25	200.000		

a. What is the sample size?
b. How many independent variables are there?
c. Compute the coefficient of multiple determination.

d. Compute the multiple standard error of estimate.

A Computer Example

Applied research often deals with a large number of independent variables. Frequently, however, just a few of them will account for most of the variation in the dependent variable. How can the researcher determine which independent variable or set of independent variables is most useful in the predicting equation? The following Problem/Solution will illustrate how several independent variables can be considered. The variables will be tested both individually and collectively to see if they have a significant connection to the dependent variable.

Problem   Recall from the Chapter Problem that Dr. Welch, a noted sociologist, is studying factors related to racial prejudice. She has isolated three variables that she believes are related to racial prejudice.

TABLE 15–1	Degrees of Racial Prejudice, Mobility, Age, and Socioeconomic Status for a Sample of 21 People		
Prejudice Score (Prej)	**Mobility**	**Age**	**Status**
80	38	23	135
86	44	30	170
48	28	38	82
40	35	30	77
66	26	19	72
64	38	23	122
94	33	17	159
32	42	70	68
48	14	51	129
32	23	31	67
88	38	21	129
76	24	22	126
52	33	33	53
40	25	31	75
76	43	22	123
14	22	65	43
62	18	26	153
40	49	38	79
72	24	22	116
60	34	29	91
76	24	52	133

Mobility is the person's mobility, measured in number of months since his or her last change of address.

Age is the person's age in years.

Status is the person's "index of social and economic status." The base for this index is 100. Someone who is very poor and has practically no social status might have an index score of, say, 10. A person who has a high income and high social status might have a score of 192.

Prej is the degree of racial prejudice as measured by a test. This is the dependent variable.

Dr. Welch gave her test to a sample of 21 people. The sample data are shown in Table 15–1. Using these sample data, determine the multiple regression equation.

Solution   There are three independent variables under consideration: mobility, age, and status. To illustrate the relationships between the dependent variable, prejudice, and each of the three independent variables, the following scatter diagrams are drawn. It appears that mobility and status may both be positively related to prejudice and that age may be negatively related. The strongest relationship appears to be between prejudice and status.

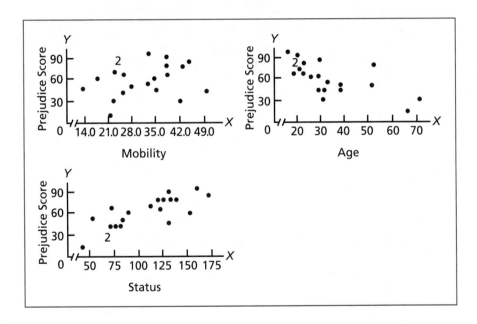

A correlation matrix is used to display all possible simple coefficients of correlation; that is, it shows the correlations between each independent variable and the dependent variable as well as the correlation among the independent variables. The commands to produce the MINITAB output are:

Stat ▶ Basic Statistics ▶ Correlation

Within the dialog box for variables, select all four of the variables:

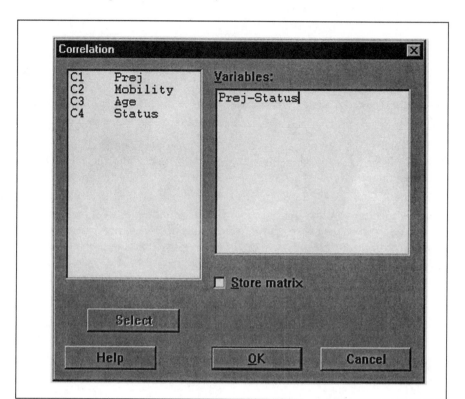

The MINITAB output is:

```
                Prej  Mobility       Age
Mobility       0.207
Age           -0.662    -0.127
Status         0.825     0.052    -0.423
```

The strength and direction of the relationships noted in the scatter diagrams are confirmed numerically in the correlation matrix. The correlation is strongest between prejudice and status, and that correlation is positive. The correlation matrix can also be used to look for multicollinearity, or correlation among the independent variables. Such correlation is undesirable because it can cause distortion in the multiple standard error of estimate and can lead to incorrect conclusions about which of the independent variables have significant net regression coefficients. Here the largest correlation among the independent variables—that between status and age—is -0.423, which is not large enough to cause a problem. How do we determine when the correlation is too large? One way is to conduct a test of hypothesis on the correlation coefficient as described in Chapter 13. A second method is to use the rule of thumb that the correlation must be stronger than -0.70 or $+0.70$ to be a problem.

The Skaff Appliance Company currently has 30 retail outlets across the United States. They retail name-brand electronic products, such as TVs, stereos, and VCRs. Skaff Appliance is considering opening several additional stores in other metropolitan areas. Paul Skaff, president, would like to study the relationship between his sales at the 30 existing locations and several factors regarding each existing store or its region. This information will be useful in selecting metropolitan areas in which to locate new stores. The following sample data were recorded:

Population (in thousands)	Percentage Unemployed	Yearly Advertising Expense (in thousands of dollars)	Monthly Sales (in thousands of dollars)
7.500	5.1	$59.0	$ 5.170
8.710	6.3	62.5	5.780
10.000	4.7	61.0	4.840
7.450	5.4	61.0	6.000
8.670	5.4	6.1	6.000
11.000	7.2	12.5	6.120
13.180	5.8	35.8	6.400
13.810	5.8	59.9	7.100
14.430	6.2	57.2	8.500
10.000	5.5	35.8	7.500
13.210	6.8	27.9	9.300
17.100	6.2	24.1	8.800
15.120	6.3	27.7	9.960
18.700	5.0	24.0	9.830
20.200	5.5	57.2	10.120
15.000	5.8	44.3	10.700
17.600	7.1	49.2	10.450
19.800	7.5	23.0	11.320
14.400	8.2	62.7	11.870
20.350	7.8	55.8	11.910
18.900	6.2	50.0	12.600
21.600	7.1	47.6	12.600
25.250	4.0	43.5	14.240
27.500	4.2	55.9	14.410
21.000	7.0	51.2	13.730
19.700	6.4	76.6	13.730
24.150	5.0	63.0	13.800
17.650	8.5	68.1	14.920
22.300	7.1	74.4	15.280
24.000	8.0	70.1	14.410

The correlation matrix from the statistical software package is given below.

	Sales	Popul	Unemp%
Popul	0.894		
Unemp%	0.312	0.075	
Advexp	0.385	0.260	0.128

a. What is the correlation between population (Popul) and the percentage of unemployment (Unemp%)?
b. Which of the three independent variables has the strongest correlation with the dependent variable?
c. Does there appear to be any problem with correlation among the independent variables? What is the condition of strong correlation among the independent variables called?

A GLOBAL TEST FOR THE MULTIPLE REGRESSION EQUATION

The overall ability of the set of independent variables to explain the variation of the dependent variable can be tested. To put it another way, can the hypothesis that all the regression coefficients are zero be rejected? A test of several regression coefficients at a time is referred to as a global test.

To relate the question to our Chapter Problem on racial prejudice the null hypothesis is that the net regression coefficients for mobility, age, and social status are all equal to zero; that is, none of these variables is useful in explaining the variations in prejudice. Open the following dialog box:

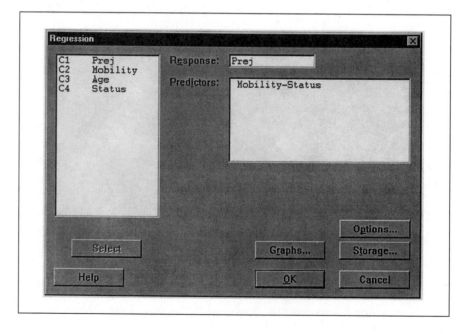

The MINITAB commands to open the dialog box are:

Stat ▶ Regression ▶ Regression

Within the dialog box, select "Prej" as the Response variable, select "Mobility," "Age," and "Status" as Predictors, then click OK.

The following computer output gives the regression equation, the ANOVA table, and other statistics for the data in Table 15–1 on page 472.

```
MTB > Regress 'Prej' 3  'Mobility'-'Status';
SUBC>   Constant.
```

Regression Analysis

```
The regression equation is
Prej = 27.1 + 0.290 Mobility - 0.530 Age + 0.388 Status

Predictor       Coef       StDev         T         P
Constant       27.13       13.58      2.00     0.062
Mobility      0.2897      0.2407      1.20     0.245
Age          -0.5297      0.1680     -3.15     0.006
Status       0.38763     0.06722      5.77     0.000

S = 10.01      R-Sq = 81.5%     R-Sq(adj) = 78.3%

Analysis of Variance

Source          DF          SS          MS         F         P
Regression       3       7508.6      2502.9     25.00     0.000
Error           17       1702.1       100.1
Total           20       9210.7
```

The net regression coefficients of 0.2897 for mobility, -0.5297 for age, and so on are estimates of the population values β_1, β_2, and β_3. We want to test whether these population values could be zero or not. The null and alternate hypotheses are

$$H_0: \beta_1 = \beta_2 = \beta_3 = 0$$

$$H_a: \text{Not all the } \beta\text{s equal 0.}$$

Not to reject the null hypothesis implies that the regression coefficients are all zero. This means that none of them is useful in explaining the variation in prejudice. Should this be the case, we would need to search for other independent variables.

To perform the test, we employ the F distribution, which, you will recall from Chapter 12, has the following major characteristics:

1. It is positively skewed.
2. It is based on two sets of degrees of freedom.
3. It cannot take on negative values.

The value of the test statistic is determined from the following formula:

$$F = \frac{SSR \,/\, k}{SSE \,/\, [n - (k + 1)]}$$

15–6

The degrees of freedom for the numerator of the F distribution are equal to k, the number of independent variables. In this example, there are 3 independent variables, so $k = 3$. The degrees of freedom in the denominator are equal to $n - (k + 1) = 21 - (3 + 1) = 17$, as reported in the Analysis of Variance section of the computer printout on page 477. The column headed DF reports the degrees of freedom. Using the 0.05 significance level and Appendix E, we find the critical value of F to be 3.20. The null hypothesis is rejected if the computed value of F exceeds 3.20. The value of F is computed as follows:

$$F = \frac{SSR \,/\, k}{SSE \,/\, [n - (k + 1)]} = \frac{7508.6 \,/\, 3}{1702.1 \,/\, [21 - (3 + 1)]} = 25.0$$

This information can also be found in the computer printout on page 477.

Since the computed F value of 25.00 exceeds the critical value of 3.20, the null hypothesis is rejected, and we conclude that at least 1 of the regression coefficients is not equal to zero. The p-value is 0.000, so there is little likelihood that H_0 is actually true. At least 1 of the independent variables is useful in explaining the variation in prejudice. The next problem is to find which of the variables are useful.

EVALUATING INDIVIDUAL REGRESSION COEFFICIENTS

If there are regression coefficients that could be equal to zero, we want to consider eliminating them from the multiple regression equation. To do so we will conduct individual tests of hypothesis for each regression coefficient.

For Mobility	For Age	For Status
$H_0: \beta_1 = 0$	$H_0: \beta_2 = 0$	$H_0: \beta_3 = 0$
$H_a: \beta_1 \neq 0$	$H_a: \beta_2 \neq 0$	$H_a: \beta_3 \neq 0$

The test statistic follows the t distribution with $n - (k + 1)$ degrees of freedom in each case. There were 21 people in the sample, so there are $n - (k + 1) = 21 - (3 + 1) = 17$ degrees of freedom in the test. Assuming the 0.05 significance level and using a two-tailed test, the critical values are -2.110 and 2.110. The decision rule is to reject H_0 if $t < -2.110$ or $t > 2.110$.

Refer again to the computer output on page 477. The column headed Coef, which refers to the coefficient, reports the multiple regression equation. In the next column, StDev, the standard deviation of the regression coefficient is given. How is this value interpreted? We selected a sample of 21 people and obtained the information on the dependent variable and each of the three independent variables for each person. If we were to select another random sample of 21 people and compute the regression equation on that sample, we would probably obtain similar, but slightly different, regression coefficients. If we were to repeat this process many times, we could obtain a sampling distribution for each of the regression coefficients. The value reported in the StDev column reflects the variability in these coefficients.

The sampling distribution of Coef/StDev follows the t distribution with $n - (k + 1)$ degrees of freedom. Hence, by testing whether the corresponding net regression coefficient is significantly different from zero, we are able to determine whether an independent variable should be in the regression equation.

From the output on page 477, we see that the t-values for age and status exceed the critical values and that the t-value for mobility does not. Therefore, the variables age and status are included in the regression and mobility is dropped.

We can now run the MINITAB regression procedure using only the independent variables that have significant regression coefficients. The commands are the same as on page 476, except "mobility" is not selected as a predictor variable. The dialog box is shown below. The results are shown on the following page. The new equation, containing only 2 independent variables, explains 79.9% of the variation in prejudice. Using 3 independent variables, we explained 81.5% of the variation in prejudice, so by deleting 1 independent variable, mobility, we reduced the explained variation by 1.6 percentage points. This is a worthwhile because the new regression equation, with only 2 independent variables, is easier to understand and interpret. In addition, in the revised regression equation, both the independent variables have significant net regression coefficients. The details of the individual tests of hypothesis are not shown, but note that the p-values in the output are both less than 0.05, which indicates they are significant at the 0.05 level.

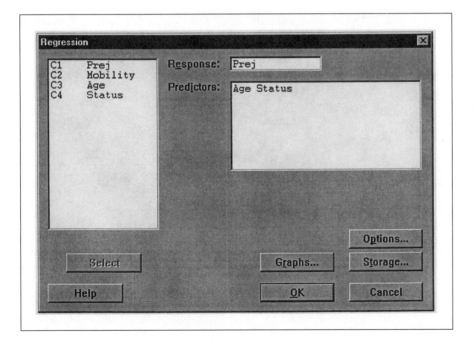

```
MTB > Regress 'Prej' 2 'Age' 'Status';
SUBC>    Constant.
```

Regression Analysis

```
The regression equation is
Prej = 37.0 - 0.553 Age + 0.387 Status

Predictor        Coef        StDev          T         P
Constant        36.96        10.98       3.37     0.003
Age            -0.5532       0.1690      -3.27     0.004
Status         0.38746      0.06805       5.69     0.000

S = 10.13      R-Sq = 79.9%      R-Sq(adj) = 77.7%

Analysis of Variance

Source        DF          SS           MS         F         P
Regression     2       7363.5       3681.8     35.88     0.000
Error         18       1847.1        102.6
Total         20       9210.7
```

Caution should be exercised when deleting independent variables. It is usually best to delete one variable and rerun the analysis to see if other variables can be deleted. Do not drop two variables at a time. Why? If two independent variables are strongly correlated, but neither is statistically significant when tested individually, removing one will sometimes make the other significant. How do you decide which variable to drop first? The usual practice is to drop the one with the largest p-value (or the smallest absolute t-value) first.

Self-Review 15–3

Refer to the Skaff Appliance example (Self-Review 15–2). The following multiple regression output was obtained.

```
Sales = -3.41 + 0.502 Popul + 0.663 Unemp% + 0.0244 Advexp

Predictor        Coef        StDev          T         P
Constant        -3.406       1.411       -2.41     0.023
Popul          0.50196      0.04253      11.80     0.000
Unemp%         0.6633       0.1982        3.35     0.002
Advexp         0.02442      0.01287       1.90     0.069

s = 1.229      R-Sq = 87.7%      R-Sq (adj) = 86.3%

Analysis of Variance

Source        DF          SS           MS         F         P
Regression     3       279.921      93.307     61.82     0.000
Error         26        39.245       1.509
Total         29       319.166
```

a. What percentage of the variation in sales is explained by the 3 independent variables?
b. In a region with a population of 25,000 and 5.0% unemployment, what sales would you estimate for a store with an advertising expense of $30,000? (Hint: Remember that some of the data are in thousands.)
c. Conduct the global test of hypothesis. Can we conclude that any of the regression coefficients are not equal to zero? Use the 0.05 significance level.
d. Conduct a test of hypothesis on each of the regression coefficients. Which can we conclude are not equal to zero? Use the 0.05 significance level.

Exercises

7. Refer to the following information:

Predictor	Coef	StDev
Constant	20.00	10.00
X_1	−1.00	0.25
X_2	12.00	8.00
X_3	−15.00	5.00

Analysis of Variance

Source	DF	SS	MS	F
Regression	3	7500.0	?	?
Error	18	?	?	
Total	21	10000.0		

a. Complete the ANOVA table.
b. Conduct a global test of hypothesis. Can you conclude that any of the regression coefficients are not equal to zero? Use the 0.05 significance level.
c. Conduct a test of hypothesis on each of the regression coefficients to determine which are not equal to zero. Use the 0.05 significance level.

8. Refer to the following information:

Predictor	Coef	StDev	T
Constant	−150.00	90.00	?
X_1	2000.00	500.00	?
X_2	−25.00	30.00	?
X_3	5.00	5.00	?
X_4	−300.00	100.00	?
X_5	0.60	0.15	?

Analysis of Variance

Source	DF	SS	MS	F
Regression	5	1500.0	?	?
Error	15	?	?	
Total	20	2000.0		

a. Complete the ANOVA table.
b. Conduct a global test of hypothesis. Can you conclude that any of the regression coefficients are not equal to zero? Use the 0.05 significance level.
c. Conduct a test of hypothesis on each of the regression coefficients to determine which are not equal to zero. Use the 0.05 significance level.

CHAPTER OUTLINE

I. Multiple regression and correlation analysis is based on the following assumptions:
 A. There is a linear relationship between the dependent variable and the independent variables.
 B. The dependent variable is continuous and of interval scale.
 C. The variation around the regression line is the same for all values of X.
  1. This means that Y varies the same amount when X is either large or small.
  2. When the variation is the same for all X-values, this is called homoscedasticity.
 D. Successive observations of the dependent variable are not correlated.
  1. Violation of this assumption is called autocorrelation.
  2. Autocorrelation frequently occurs when data are collected in order over time.
 E. The independent variables are not highly correlated.
  1. When independent variables are correlated, this is called multicollinearity.
  2. A correlation between 2 independent variables stronger than -0.70 or 0.70 can cause difficulty.

II. The general form of the multiple regression equation is

$$Y' = a + b_1X_1 + b_2X_2 + \cdots + b_kX_k \qquad \textbf{15–2}$$

where Y' is the estimated value, a is the Y-intercept, the bs are the sample regression coefficients, and the Xs represent the values of the various independent variables.
 A. There can be an unlimited number of independent variables.

 B. A computer is probably needed to determine the value of a and the various b values.
III. There are two measures of the effectiveness of the regression equation.
 A. The coefficient of multiple determination reports the fraction of the variation in the dependent variable that is explained by the independent variables.
  1. It can range from 0 to 1.
  2. It does not reveal the direction of the relationship.
 B. The multiple standard error of estimate is similar to the standard deviation.
  1. It is measured in the same units as the dependent variable.
  2. It uses deviations from the multiple regression instead of from the mean.
IV. A correlation matrix reports all possible simple coefficients of correlation among the variables.
 V. A global test determines if any of the net regression coefficients are different from zero.
 A. The null hypothesis is that all the net regression coefficients are equal to zero.
 B. The alternate hypothesis is that at least one of the net regression coefficients is different from zero.
 C. The test statistic follows the F distribution with k (the number of independent variables) and $n - (k + 1)$ degrees of freedom, where n is the total sample size.
VI. The test for individual variables determines which of the independent variables have net regression coefficients different from zero.
 A. The test statistic follows the t distribution with $n - (k + 1)$ degrees of freedom.
 B. The variables that could have zero as the net regression coefficient are usually deleted from the analysis.

Exercises

9. A real estate agent is studying the selling prices of homes in a certain district of the city. The following equation has been developed:

$$Y' = 24.2 + 9.80X_1 + 3.50X_2$$

where

Y' is the selling price in thousands of dollars.
X_1 is the number of bedrooms.
X_2 is the number of bathrooms.

a. What do you estimate the selling price to be for a three-bedroom, two-bath home in the area?
b. Does the addition of a bedroom add more to the selling price than the addition of a bathroom?

10. From a study of the number of points scored by teams in the National Football League, the following multiple regression equation was developed:

$$Y' = 6.0 + 0.06X_1 + 0.05X_2 - 2.75X_3$$

where

Y' is the number of points scored.
X_1 is the number of yards gained running the ball.
X_2 is the number of yards gained passing the ball.
X_3 is the number of turnovers (the number of fumbles plus the number of passes intercepted).

a. If the Buffalo Bills gained 150 yards running the ball, gained 310 yards passing, and had 3 turnovers, how many points would you expect them to score?
b. How many points are lost for each turnover?

11. In the course of a study of the yearly amount spent on food, a sociologist found that annual income and the number of persons in the family explained 86.3% of the variation in the yearly amount spent. The regression equation was computed to be

$$Y' = 500 + 0.023X_1 + 252X_2$$

where

X_1 is the annual family income.
X_2 is the number of family members.

a. How much does the food expenditure increase with each additional family member?
b. If a family's income increases by $1000, how much would you expect the expenditure on food to increase?

c. Predict the amount to be spent on food by a family of 5 with an annual income of $20,000.

12. An insurance company is analyzing the profiles of its individual policyholders as they relate to the amount of life insurance coverage carried. Three independent variables are being considered: the policyholder's annual income (X_1), the number of children under 21 (X_2), and age (X_3).

$$Y' = 2.43 + 0.32X_1 + 8.89X_2 + 0.14X_3$$

where Y' and X_1 are measured in thousands of dollars.

a. How much life insurance would a 40-year-old woman with an annual income of $50,000 and 3 children be expected to carry?
b. How much does the amount of coverage increase with each child?
c. If 2 policyholders were both 40 years old and each had 5 children, but 1 made $40,000 per year and the other $50,000, how much difference would you expect there to be in their respective insurance coverages?

13. A market research firm is studying the relationship between monthly income for a household (Y) and 3 factors: the mortgage payment for the household, the value of the family car, and the age of the husband. A random sample of 19 observations revealed

Monthly Income	Mortgage Payment	Car Value	Husband's Age
$3552	$656	$ 7800	36
2520	568	5100	37
3384	416	10,500	30
4020	504	9500	34
3168	736	6260	39
2676	584	4380	34
1956	432	3760	31
3684	544	7350	41
3540	784	6580	38
4152	896	7900	37
3816	504	9450	40
4020	608	12,600	35
3920	648	10,630	33
2544	568	5340	32
2736	416	4690	36

Monthly Income	Mortgage Payment	Car Value	Husband's Age
2556	536	1200	39
3120	440	9600	42
3720	736	8800	36
3192	608	11,400	33

a. The following correlation matrix was developed. What is the correlation between age and car? Does the fact that the correlation is negative surprise you? Explain.

	Income	Mortgage	Car
Mortgage	0.461		
Car	0.746	0.098	
Age	0.198	0.167	−0.119

b. Which of the set of independent variables has the strongest correlation with the dependent variable?

c. The following output was obtained. Compute the coefficient of multiple determination and the multiple standard error of estimate.

The regression equation is
Income = −424 + 1.69 Mortgage + 0.157 Car + 42.5 Age

Predictor	Coef	StDev	T	P
Constant	−424.3	928.8	−0.46	0.654
Mortgage	1.6863	0.6227	2.71	0.016
Car	0.15711	0.02741	5.73	0.000
Age	42.51	24.32	1.75	0.101

Analysis of Variance

Source	DF	SS	MS	F	P
Regression	3	5430941	1810314	15.66	0.000
Error	15	1733741	115583		
Total	18	7164682			

d. Referring to the computer output in part c, conduct a global test of hypothesis to determine if any of the net regression coefficients are not equal to zero. Use the 0.05 significance level.

e. Referring to the computer output in part c, conduct an individual test of hypothesis on each independent variable to determine which of the net regression coefficients are not equal to zero. Use the 0.05 significance level.

14. The planning director for Dhont Automatic Door Company has been asked by the president of the company to prepare an analysis of the product sales as they relate to advertising expense and bonuses paid to the sales force. The planning director selected a sample of 15 sales territories and determined, for each, the sales in that region, the advertising expense, and bonuses paid to the sales force. The results were as follows:

Sales (in thousands of dollars)	Advertising Expense (in thousands of dollars)	Bonuses Paid (in thousands of dollars)
$158.75	$16.2	$10
186.25	22.5	11
132.50	17.1	7
203.75	21.6	14
127.50	13.5	7
225.00	23.4	17
201.25	22.5	14
160.00	14.4	12
173.75	15.3	12
180.00	20.7	12
198.75	19.8	14
172.50	13.5	15
155.00	12.6	8
130.30	13.5	9
160.80	11.6	14

a. The following correlation matrix was developed. What is the correlation between bonuses and advertising expense? Would you expect this association to be positive or negative?

	Sales	Advexp
Advexp	0.780	
Bonus	0.849	0.447

b. Refer to the computer output in part a. Which of the independent variables has the strongest correlation with the dependent variable, sales?

c. The following output was obtained. Compute the coefficient of multiple determination and the multiple standard error of estimate.

The regression equation is
Sales = 41.2 + 3.49 Advexp + 5.95 Bonus

Predictor	Coef	StDev	T	P
Constant	41.17	11.29	3.65	0.003
Advexp	3.4925	0.6327	5.52	0.000
Bonus	5.9480	0.8614	6.90	0.000

Analysis of Variance

Source	DF	SS	MS	F	P
Regression	2	10740.8	5370.4	70.18	0.000
Error	12	918.3	76.5		
Total	14	11659.1			

d. Referring to the computer output in part c, conduct a global test of hypothesis to determine if any of the net regression coefficients are different from zero. Use the 0.05 significance level.

e. Referring to the computer output in part c, conduct an individual test of hypothesis on each independent variable to determine which of the net regression coefficients are not equal to zero. Use the 0.05 significance level.

15. Below are some data from the 1995 major league baseball season. Listed for each of the 28 teams is the number of games won (out of 144; recall that this season started late due to a players' strike), the team batting average, the number of home runs hit, the number of bases stolen, and the team ERA (earned run average, which is a measure of pitching effectiveness; a lower value indicates better pitching).

Team	Wins	Batting	Homers	Stolen	ERA
Atlanta	90	.250	168	73	3.44
Chicago Cubs	73	.265	158	105	4.13
Cincinnati	84	.270	161	190	4.03
Colorado	76	.282	200	125	4.97
Florida	66	.262	144	131	4.27
Houston	75	.275	109	176	4.06
Los Angeles	77	.264	140	127	3.66
Montreal	66	.259	118	120	4.11
New York Mets	68	.267	125	58	3.88
Philadelphia	69	.262	94	72	4.21
Pittsburgh	57	.259	125	84	4.70
St. Louis	62	.247	107	79	4.09
San Diego	70	.272	116	124	4.13
San Francisco	67	.253	152	138	4.86
Cleveland	100	.291	207	132	3.83
Chicago White Sox	68	.280	146	110	4.85
Boston	86	.280	175	99	4.39
Minnesota	56	.279	120	105	5.76
California	78	.277	186	57	4.52
New York Yankees	79	.276	122	50	4.56
Seattle	79	.276	182	110	4.50
Milwaukee	65	.266	128	105	4.82
Texas	74	.265	138	90	4.66
Oakland	67	.264	169	112	4.93
Baltimore	71	.262	173	92	4.31
Kansas City	70	.260	119	120	4.49
Toronto	56	.260	140	75	4.88
Detroit	60	.247	159	73	5.49

a. Using a statistical software package, develop a correlation matrix reporting the relationships between the dependent variable and each of the independent variables. Also show the correlations between all of the independent variables. Which independent variable has the strongest correlation with the dependent variable (the number of games won)? Are there any potential problems with the correlations among the independent variables? If so, which ones?

b. Determine the regression equation, and write it out. Does it surprise you that the coefficient for the independent variable ERA is negative? Explain.

c. Estimate the number of wins for a team that had a team batting average of .280, hit 150 home runs, stole 100 bases, and had an ERA of 4.00.

d. Find the coefficient of determination on your computer output. Comment on the value.

e. Conduct a global test of hypothesis on the set of independent variables. Use the 0.05 significance level. Is it reasonable to conclude that some of the independent variables have net regression coefficients different from zero?

f. Conduct a test of hypothesis on each regression coefficient. Would you consider deleting any of the independent variables? If so, which ones? Use the 0.05 significance level.

16. The *Times-Observer* is a daily newspaper in Metro City. Like many city newspapers, the *Times-Observer* is suffering through difficult financial times. The circulation manager is interested in studying other papers in similar cities in the United States and Canada. She is particularly interested in what variables relate to the number of subscribers to a paper. She is able to find sample information for 25 newspapers. The following notation is used:

Sub = Number of subscribers (in thousands)
Popul = Metropolitan population (in thousands)
Adv = Advertising budget of the paper (in hundreds of dollars)
Income = Median family income in the metropolitan area (in thousands of dollars)

Paper	Sub	Popul	Adv	Income
1	37.95	588.9	13.2	35.1
2	37.66	585.3	13.2	34.7
3	37.55	566.3	19.8	34.8

Paper	Sub	Popul	Adv	Income
4	38.78	642.9	17.6	35.1
5	37.67	624.2	17.6	34.6
6	38.23	603.9	15.4	34.8
7	36.90	571.9	11.0	34.7
8	38.28	584.3	28.6	35.3
9	38.95	605.0	28.6	35.1
10	39.27	676.3	17.6	35.6
11	38.3	587.4	17.6	34.9
12	38.84	576.4	22.0	35.4
13	38.14	570.8	17.6	35.0
14	38.39	586.5	15.4	35.5
15	37.29	544.0	11.0	34.9
16	39.15	611.1	24.2	35.0
17	38.29	643.3	17.6	35.3
18	38.09	635.6	19.8	34.8
19	37.83	598.9	15.4	35.1
20	39.37	657.0	22.0	35.3
21	37.81	595.2	15.4	35.1
22	37.42	520.0	19.8	35.1
23	38.83	629.6	22.0	35.3
24	38.33	680.0	24.2	34.7
25	40.24	651.2	33.0	35.8

a. Develop a correlation matrix. Do you see any problems with multicollinearity?

b. Determine the regression equation.

c. Conduct a global test of hypothesis to determine if any of the net regression coefficients are different from zero.

d. Conduct a test for the individual regression coefficients. Would you consider deleting any of the independent variables?

e. If the answer to part d was to delete independent variables, delete them one at a time until all the remaining variables are significant. Write out this equation.

17. Dr. Jamie Heins is an industrial psychologist. Currently, she is investigating the relationship between social interaction on the job and job morale. The Suppa Test, designed in the early 1960s by Dr. Victor Suppa, is the standard for measuring employee morale. When Dr. Heins gave this test to a sample of 27 clerical workers, the results were a mean score of 1860 and a standard deviation of 714. The scores ranged from 1220 up to 2330. To define social interaction each sampled employee maintained careful records of the number of social contacts each day for a week. Dr. Heins then determined the mean number of contacts per day per employee. The mean number of contacts was 25.70 per day, with a standard deviation of 7.85. The number of contacts per day ranged from 14 to 31.

Dr. Heins has a computer system with a statistical software package. She found the regression equation to be $Y' = -354.2 + 86.16X$; the dependent variable is the estimated employee morale on the Suppa scale, and the independent variable is the number of social contacts. In addition, she was able to determine that the number of social contacts explained 89.6% of the variation in morale. How would she interpret the equation? Can she conclude that there is a positive association between morale and the number of social contacts?

Some recent research in the area of employee morale suggests that, if employees have a clear understanding of their job requirements, their morale will be higher. To investigate this matter Dr. Heins compared the job description of each of the 27 employees in her sample with the duties they currently perform. She found that there was an agreement between duties and description in 48% of the cases.

She then developed a multiple regression equation with morale as the dependent variable and the number of social contacts and the agreement between duties and job description as the independent variables. The independent variable for agreement between duties and job description is an indicator variable, where 1 was used if there was agreement and 0 if there was not agreement. Using her statistical software system, Dr. Heins found the following regression equation: $Y' = -388.0 + 83.94X_1 + 188.83X_2$. The standard deviations of the two regression coefficients were

5.552 for the variable contacts and 85.58 for the indicator variable agreement. These 2 independent variables explained 91.4% of the variation in morale. She observed that the independent variable agreement increased the amount of explained variation by 1.8 percentage points. Is this additional explained variation worth the trouble?

Overall is it reasonable to conclude that the morale of an employee is affected by the amount of social interaction and the agreement between job description and job duties?

18. Select a dependent variable that you think can be predicted by several independent variables. One suggestion might involve predicting sales based on advertising expense, number of sales personnel, and amount spent on research and development. Another might involve predicting the salary of executives based on the number of employees supervised, their age, and their number of years of experience. Exercise 15 gave information on major league baseball, but football fans might estimate the number of points scored in college or professional games based on the number of yards gained running the ball, the number of yards gained passing the ball, the number of fumbles, and the number of passes intercepted. For basketball, you might try predicting points scored based on the percentage of field goals made, the number of foul shots attempted, and the identity of the home team. Car enthusiasts might estimate a car's mileage rating based on the engine size, weight, number of doors, and whether or not it was manufactured in the United States.

a. Collect the data and enter the information into the computer.

b. Analyze the computer printout, and write a summary of your findings. Include in your discussion an interpretation of such measures as the coefficient of determination and the multiple regression equation. Also conduct global and individual tests of hypothesis.

DATA EXERCISES

19. Refer to the real estate data set, which reports information on homes sold in Alabama during 1996. Develop a multiple regression equation using selling price as the dependent variable and the area of the home in square feet, the number of bedrooms, the number of bathrooms, whether or not there is a pool, whether or not there is an attached garage, and the distance from the center of the city as the independent variables.

a. Write out the regression equation. Interpret the regression coefficients. For example, how much does an extra bedroom add to the value of the home? How much is an attached garage worth to the value of the home? How much is a pool worth? Does it surprise you that the regression coefficient for distance is negative? Why?

b. What is the coefficient of determination? Interpret.

c. Conduct a global test of hypothesis to determine if any of the independent variables have regression coefficients different from zero. Use $\alpha = 0.05$.

d. Conduct a test of hypothesis for each independent variable to determine which of the regression coefficients are different from zero. Would you consider deleting any of the independent variables? If so, which variables would you delete? Use $\alpha = 0.05$.

e. Rerun the regression equation deleting the variables from part d. Comment on the change in the regression equation. Was there much change in the coefficient of determination?

20. Refer to the schools data set which includes information on the 94 school districts in northwestern Ohio. Let the dependent variable be the percentage passing the state proficiency test. The independent variables are the percentage on welfare in the district, the instructional cost per student in the district and the percentage in attendance in the district.

a. Using these three independent variables, what is the coefficient of determination? Is a reasonable amount of the variation explained by the three variables? Write out the regression equation, and interpret each of the regression coefficients.

b. Conduct a global test of hypothesis. Interpret your conclusion.

c. Conduct an individual test for each of the independent variables. Would you consider deleting any of the independent variables? If so, which ones?

d. Rerun the analysis without the variable or variables you recommended dropping in the part c. Comment on your new equation. Does it appear that all the variables are significant? Did your R^2 value change from the previous regression model?

CHAPTER ACHIEVEMENT TEST

The answers are at the back of the book.

MULTIPLE-CHOICE QUESTIONS

Select the response that best answers each of the questions.

1. Which of the following are both measures of association?

a. The regression equation and the standard error of estimate
b. The coefficient of determination and the standard error of estimate
c. The regression equation and the coefficient of determination
d. The regression coefficient and the standard error of estimate

2. The variation around the regression line must be the same for all values of X. This requirement is called
   a. stepwise.
   b. autocorrelation.
   c. homoscedasticity.
   d. correlation.
3. If successive observations of the dependent variable are themselves correlated, then this is called
   a. autocorrelation.
   b. homoscedasticity.
   c. regression.
   d. stepwise.
4. Multiple regression analysis assumes there
   a. is a linear relationship between the variables.
   b. is no correlation in the population.
   c. are never any negative signs in the regression coefficients.
   d. is at least the nominal scale of measurement.
5. For the multiple regression technique to be employed, the dependent variable must be at least of which scale of measurement?
   a. Interval
   b. Ratio
   c. Ordinal
   d. Nominal

Questions 6–11 are based on the following information. A study has been undertaken to predict annual income in thousands of dollars based on the number of years on the job, age, and the number of years of education beyond eighth grade. This equation has been developed: $Y' = 10 + 0.2X_1 + 0.1X_2 + 2.5X_3$. Variable 1 is the number of years on the job, variable 2 is age, and variable 3 is the number of years of education beyond eighth grade.

6. The equation is called a
   a. coefficient of regression equation.
   b. simple regression equation.
   c. multiple regression equation.
   d. dependent variable equation.
7. How many dependent variables are there?
   a. 0
   b. 1
   c. 2
   d. 3
8. How many independent variables are there?
   a. 0
   b. 1
   c. 2
   d. 3
   e. None of the above
9. What are the numbers 0.2, 0.1, and 2.5 called?
   a. Partial correlation coefficients
   b. Net regression coefficients
   c. Coefficients of determination
   d. Coefficients of standard errors
   e. None of the above
10. An employee has been on the job 20 years, is 50 years old, and has 8 years of education beyond eighth grade. What is the employee's predicted annual salary?
    a. \$12,800
    b. \$39,000
    c. \$88,000
    d. \$16,000
    e. None of the above
11. For each additional year of education beyond eighth grade, annual income increases
    a. \$2500.
    b. \$1000.
    c. \$1280.
    d. \$39,000.
    e. None of the above

COMPUTATION PROBLEMS

Refer to the following output:

Predictor	Coef	StDev	T
Constant	5.00	2.00	2.50
X_1	2.00	1.00	2.00
X_2	0.50	0.15	3.33
X_3	0.30	0.10	3.00
X_4	1.00	0.25	4.00

$s = 2.00$ $R-Sq = 66.7\%$

Analysis of Variance

Source	DF	SS	MS	F
Regression	4	200.0	50.0	12.50
Error	25	100.0	4.0	
Total	29	300.0		

12. Write the multiple regression equation.
13. Conduct a global test of hypothesis. Can you conclude that any of the net regression coefficients are not zero?
14. Test the net regression coefficients individually. Are there any that are not equal to zero? Should any of the variables be dropped?

. . . .

ANSWERS TO SELF-REVIEW PROBLEMS

15-1 $33.50 per week
$Y' = 20.0 + 0.50X_1 - 1.20X_2 - 1.05X_3$
$= 20.0 + 0.50(300) - 1.20(70)$
$- 1.05(50)$
$= 33.5$ (in dollars)
Households with a weekly income of $300 that spend $70 on food and $50 on entertainment are expected to save $33.50 per week. Note that, as the amount earned increases, so does the amount to be saved. Similarly, as the amounts spent on food and entertainment increase, the amount saved decreases.

15-2 a. 0.075
b. Population
c. No. The largest correlation is 0.260, between advertising expense and population. This is less than the rule-of-thumb limit of 0.70. The condition is called multicollinearity.

15-3 a. $R^2 = \dfrac{279.921}{319.166} = 0.877$, or 87.7%
b. 13.187, computed as follows:
$Y' = -3.41 + 0.502(25) + 0.663(5.0) +$
$0.0244(30)$
c. $H_0: \beta_1 = \beta_2 = \beta_3 = 0$
H_a: At least one coefficient is not zero.
H_0 is rejected if $F > 2.95$ (approximately).
$F = 93.307/1.509 = 61.82$
H_0 is rejected. At least one of the regression coefficients does not equal zero.
d. $H_0: \beta_1 = 0; H_a: \beta_1 \neq 0$
$H_0: \beta_2 = 0; H_a: \beta_2 \neq 0$
$H_0: \beta_3 = 0; H_a: \beta_3 \neq 0$
In each case, H_0 is rejected if $t < -2.056$ or $t > 2.056$.
H_0 is rejected for population and unemployment. Advertising expense is dropped from the analysis.

Unit Review

In the last three chapters, we introduced the fundamental concepts of correlation and regression analysis. We examined various measures used to describe the degree of relationship between a dependent variable and one or more independent variables. We also developed a mathematical equation that allows us to predict a dependent variable based on one or more independent variables.

KEY CONCEPTS

1. **Correlation analysis** allows us to describe the strength of the linear relationship between two or more variables and to determine what proportion of the total variation is explained by the independent variable(s).

   a. **Pearson's product-moment coefficient of correlation** (r). This measure reports the strength of the linear relationship between variables. Its use assumes the data are of an interval scale. For two variables, r can have any value from -1.00 to $+1.00$, inclusive. The *strength* of the relationship is not dependent on the *direction* of the relationship. For example, correlation coefficients of -0.09 and $+0.09$ are equal in strength—both are very weak.

   b. **The coefficient of (multiple) determination** reports the proportion of the total variation that is explained by the independent variable (or variables). For 2 variables, it is the proportion of the total variation in 1 variable explained by the other variable. Likewise, for more than 2 independent variables, it is the proportion of the total variation in the dependent variable explained by the independent variables. It varies from 0 to $+1.00$. A coefficient of 0.80, for example, indicates that 80% of the total variation in the dependent variable is explained by the independent variable or variables.

   c. **Spearman's rank-order correlation coefficient** reports the relationship between 2 ordinal-level variables—that is, ranked data. Spearman's coefficient can assume any value between -1.00 and $+1.00$.

2. **Regression analysis** is important because, by using a mathematical equation, we can estimate the value of one variable based on another variable or variables.

   a. **Regression equation.** For one dependent and one independent variable, the equation has the form $Y' = a + bX$. For one dependent and three independent variables, it is $Y' = a + b_1X_1 + b_2X_2 + b_3X_3$.

   b. **The standard error of estimate** is a measure that allows us to assess the accuracy of the regression equations. It is reported in the same units as the dependent variable.

   c. The **regression assumptions** are: For each value of X, the Y observations are normally distributed around the regression line; the variation around the regression line is the same for all values of X; and the deviations from the regression line are not related.

   d. There are two types of interval estimates. In a **confidence interval** the value of Y is based on all values of X, whereas in a **prediction interval** we use a particular value of X.

3. In **multiple regression and correlation analysis** we use a single dependent variable and several independent variables.

   a. A **correlation matrix** reports all possible simple coefficients of correlation among all the variables. It is useful for finding which independent variables are strongly related to the dependent variable and which of the independent variables are correlated with each other.

   b. The **global test** is useful for determining if any

491

of the net regression coefficients are different from zero. The test statistic follows the F distribution.

c. In the **individual test,** each of the variables are examined to determine which of the net regression coefficients are different from zero. The test statistic follows the *t* distribution. If the net regression coefficient is not shown to be different from zero, it is dropped from the regression equation.

d. The assumptions for multiple regressions are: There is a linear relationship between the dependent variable and the independent variables, the dependent variable is continuous and at least interval scale, successive observations of the dependent variables are not correlated, and the independent variables are not highly correlated.

KEY TERMS

Correlation analysis
Scatter diagram
Dependent variable
Independent variable
Pearson's product-moment
  correlation coefficient
Homoscedasticity
Regression coefficient
Autocorrelation
Multicollinearity

Coefficient of determination
Spearman's rank-order
  correlation coefficient
Regression equation
Slope
Y-intercept
Multiple coefficient of
  correlation
Coefficient of multiple
  determination

Least-squares principle
Standard error of estimate
Confidence interval
Prediction interval
Multiple regression and
  correlation analysis
Multiple standard error of
  estimate
Correlation matrix
Global test

KEY SYMBOLS

r The coefficient of correlation for two variables.

r^2 The coefficient of determination.

R^2 The coefficient of multiple determination.

Y' The dependent variable of a regression equation.

X One of the independent variables.

$s_{Y \cdot X}$ The standard error of estimate.

r_s Spearman's rank-order correlation coefficient.

a The Y-intercept—that is, the value of Y when $X = 0$.

b The slope of the line. The variable b is the amount Y changes for a unit change in X.

 **CASE STUDY** • Delivering Medical Kits

Terry and Associates is a specialized medical testing center in Denver, Colorado. One of the major sources of their revenue is a kit used to test for elevated amounts of lead in the blood. Workers in auto body shops, those in the lawn care industry, and commercial house painters are possibly exposed to large amounts of lead and thus must be randomly tested. Because it is quite expensive to conduct the test, the kits are sold and delivered on demand to a variety of locations throughout the Denver area.

Kathleen Terry, the owner, is concerned about appropriate costing of each delivery. To investigate the problem, Ms. Terry gathered information on a sample of 50 recent deliveries. Factors thought to be related to the cost of delivering the kit are

Prep—the time between when the customized order is phoned to the company and when it is ready for delivery.
Delivery—the actual travel time from Terry to the customer.
Mileage—the distance (in miles) from Terry to the customer.

Cost	Prep	Delivery	Mileage
32.60	10	51	20
23.37	11	33	12
31.49	6	47	19
19.31	9	18	8
28.35	8	88	17
22.63	9	20	11
22.63	9	39	11
21.53	10	23	10
21.16	13	20	8
21.53	10	32	10
28.17	5	35	16
20.42	7	23	9
21.53	9	21	10
27.55	7	37	16
23.37	9	25	12
17.10	15	15	6
27.06	13	34	15
15.99	8	13	4
17.96	12	12	4
25.22	6	41	14
24.29	3	28	13
22.76	4	26	10
28.17	9	54	16
19.68	7	18	8
25.15	6	50	13

Cost	Prep	Delivery	Mileage
20.36	9	19	7
21.16	3	19	8
25.95	10	45	14
18.76	12	12	5
18.76	8	16	5
24.29	7	35	13
19.56	2	12	6
22.63	8	30	11
21.16	5	13	8
21.16	11	20	8
19.68	5	19	8
18.76	5	14	7
17.96	5	11	4
23.37	10	25	12
25.22	6	32	14
27.06	8	44	16
21.96	9	28	9
22.63	8	31	11
19.68	7	19	8
22.76	8	28	10
21.96	13	18	9
25.95	10	32	14
26.14	8	44	15
24.29	8	34	13
24.35	3	33	12

a. Develop and test a regression equation that describes the relationship among the cost of the delivery and the other factors. Ms. Terry, of course, wants the simplest statistically accurate equation available.
b. Estimate the delivery cost for a kit that takes 10 minutes for prep, must travel 30 minutes, and covers a distance of 14 miles.
c. Describe each of the components of your estimation equation—that is, slope (or slopes) and intercept values. Explain to Ms. Terry, in the simplest possible terms, how the price of the delivery is calculated.

CHAPTER 16

Analysis of Nominal-Level Data: The Chi-Square Distribution

OBJECTIVES

When you have completed this chapter, you will be able to

- explain the characteristics of the chi-square distribution;
- test a hypothesis regarding the difference between an observed set of frequencies and a corresponding expected set of frequencies;
- test whether two criteria of classification are related.

CHAPTER PROBLEM What Melts in Your Mouth, Not in Your Hands?

Have you ever wondered about the distribution of colors in a bag of M&M chocolate peanut candies? After checking with the manufacturer, the *Lansing State Journal* reported that in a bag, there are 30% brown, 30% yellow, 10% blue, 10% red, 10% green, and 10% orange candies. You purchase a bag of M&M chocolate peanut candies at the nearby Kroger Food and Drug and find 17 brown, 20 yellow, 13 blue, 7 red, 6 green, and 9 orange candies, for a total of 72 candies. Does the bag you purchased agree with the distribution suggested by the newspaper?

INTRODUCTION

The hypothesis tests we examined in Chapters 10–12 dealt with the interval and ratio levels of measurement for problems in which the population was assumed to be normal. This chapter begins our study of statistical tests for *nominal* and *ordinal* scales of measurement or for cases in which no assumptions can be made about the shape of the population. Recall from Chapter 1 that the nominal level of data is the most "primitive," or the "lowest," type of measurement. Nominal-level information, such as male or female, can be classified only into categories. The ordinal level of measurement assumes one category is ranked higher than the next one. To illustrate, each staff therapist in a clinic might be rated as being superior, good, fair, or poor. A rating of "superior" is higher than a "good" rating, a "good" rating is higher than a "fair" rating, and so on.

Tests involving nominal or ordinal levels of measurement are called non-parametric or distribution-free tests. In this chapter, we will examine two tests that employ the chi-square distribution as the test statistic. In the next chapter, we will present several other nonparametric tests using the ordinal level of measurement.

THE CHI-SQUARE (χ^2) DISTRIBUTION

The chi-square distribution is appropriate for both nominal-level and ordinal-level data. It is designated χ^2, and, because it involves squared observations, *it is never negative*. Like the t and F distributions discussed earlier, there are many chi-square distributions, each with a different shape that depends on the number of degrees of freedom. Figure 16–1 shows the shape of various chi-square distributions for selected degrees of freedom (df). Note that, as the number of degrees of freedom increases, the distribution approaches a symmetric distribution, and the peak moves to the right.

The number of degrees of freedom is determined by the number of categories minus 1—that is, $k - 1$—and not by the size of the sample. For example, if a sample of 150 undergraduate students is classified as freshmen, sophomores, juniors, or seniors, there are $k - 1 = 4 - 1 = 3$ degrees of freedom.

There are three major properties of a chi-square distribution:

1. Chi-square is nonnegative; that is, it is either zero or positive.
2. A chi-square distribution is not symmetrical. Its skewness is positive. However, as the number of degrees of freedom increases, chi-square approaches a symmetric distribution.
3. There is a family of chi-square distributions. There is a particular distribution for each degree of freedom.

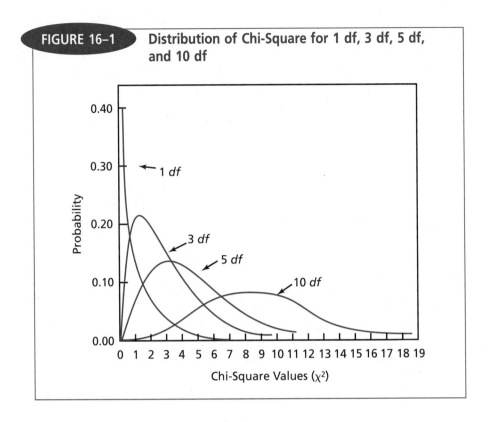

FIGURE 16–1   Distribution of Chi-Square for 1 df, 3 df, 5 df, and 10 df

. . . .

THE GOODNESS-OF-FIT TEST: EQUAL EXPECTED FREQUENCIES

We use the chi-square distribution as a test statistic to determine how well a set of observations "fits" a theoretical, or expected, set of observations. To put it another way, the objective is to find out how well an *observed* set of frequencies, abbreviated f_0, fits an *expected* set of frequencies, f_e. This test is called the goodness-of-fit test.

The general nature of the goodness-of-fit test can be shown by a specific application. The following Problem/Solution illustrates the procedure when the expected frequencies, f_e, are all the same.

Problem   A traffic engineer wished to study whether drivers have a preference for certain tollbooths at a bridge during a nonrush hour. Are drivers equally likely to approach all tollbooths, or do they tend to prefer certain ones? For the study, the engineer selected eastbound traffic on the Walt Whitman Bridge, which spans the Delaware River between Philadelphia and New Jersey. The number of automobiles passing through each tollbooth lane was counted during randomly selected 5-minute intervals. Nonrush-hour periods were sampled when there were 5 tollbooths operating. The sample information is as follows. Can we conclude that there are differences in the numbers of cars selecting each of the lanes? Use the 0.05 significance level.

Tollbooth Lane	Number of Cars Observed
1	171
2	224
3	211
4	180
5	214
Total	1000

Solution   We will use the five-step hypothesis-testing procedure.

Step 1. State the null and alternate hypotheses. The null hypothesis, H_0, is that there is no preference among the five lanes. The alternate, H_a, is that there is a preference among the five tollbooths. We could also report the null and alternate hypotheses in a symbolic form similar to that used in Chapter 10 in the section regarding tests of proportions. In that chapter, we let π indicate the proportion of the population having the trait of interest. So in this case, we let π_1 represent the proportion of drivers who select the first tollbooth, π_2 the proportion who select the second tollbooth, and so on.

$$H_0 : \pi_1 = \pi_2 = \pi_3 = \pi_4 = \pi_5$$

$$H_a : \text{All } \pi_i \text{ are not equal.}$$

Step 2. Select a level of significance. The 0.05 level has been chosen. Recall that this is the probability of making a Type I error. It is the probability of incorrectly rejecting a true null hypothesis.

Step 3. Choose an appropriate test statistic. In this case, the level of measurement is the nominal scale because the numbering of the tollbooths is a means of identification. The test statistic is the chi-square distribution.

$$\chi^2 = \sum \frac{(f_o - f_e)^2}{f_e}$$

16–1

where

f_o is the observed frequency in a category (a tollbooth lane in this problem).
f_e is the expected frequency in a particular category.

Step 4. Formulate a decision rule. As in other hypothesis-testing situations, we look at the sampling distribution of the test statistic (chi-square in this case) in order to arrive at a *critical value*. Recall that the critical value is the number that separates the region in which H_0 is not rejected from the region where it is rejected. The shape of the chi-square distribution depends on the number of degrees of freedom. In a goodness-of-fit test, there are $k - 1$ degrees of freedom, where k represents the number of categories. In this case, there are 5 tollbooths, so there are $k - 1 = 5 - 1 = 4$ degrees of freedom. The critical value of chi-square is found in Appendix F, a portion of which is shown in Table 16–1. To locate the critical value for this problem begin by locating the degrees of freedom in the left margin. Then move across the column headings to the appropriate significance level. The various column headings, such as 0.10, 0.05, and 0.01, represent the area, or probability, to the right of a particular value of χ^2. In this example, go down the left-hand column to 4 degrees of freedom and then move across to the column headed 0.05, the level of significance selected for this example. The critical value is 9.488. The decision rule is to reject the null hypothesis if the computed value of χ^2 is greater than 9.488.

TABLE 16–1   Critical Values of the Chi-Square Distribution

Degrees of Freedom df	Possible Values of χ^2 Right-Tail Area			
	0.10	0.05	0.02	0.01
1	2.706	3.841	5.412	6.635
2	4.605	5.991	7.824	9.210
3	6.251	7.815	9.837	11.345
4	7.779	9.488	11.668	13.277
5	9.236	11.070	13.388	15.086

The decision rule is shown graphically in the diagram at the top of the next page. Note that all the rejection region is in the right tail of the distribution. A large computed value of χ^2 (more than 9.488 in this problem) occurs when there is a substantial difference between the observed and the expected frequencies. If that happens, we reject the null hypothesis that there is no difference between the observed and the expected distributions. If the differences between f_o and f_e are small, then the computed value of χ^2 is also small (less than 9.488 in this problem), indicating that the difference between the two distributions *occurred by chance*.

Step 5. Compute the value of χ^2, and make a decision. Recall that the traffic engineer observed a total of 1000 automobiles going through the tollbooths. He found that 171 selected tollbooth 1, 224 selected tollbooth 2, and so on. These are the *observed* frequencies. How do we find the expected frequencies? Recall in Step 1 that the null hypothesis indicated there was no preference among the 5 tollbooths. If the null hypothesis is true, we would expect $\frac{1}{5}$

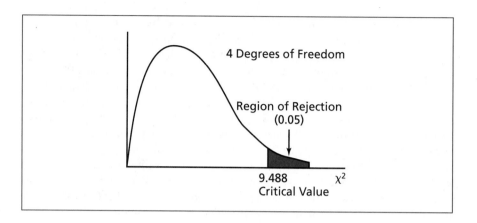

of the 1000 automobiles—that is, 200—to select tollbooth 1, another 200 to select tollbooth 2, and so on. Thus, in this problem, each of the expected frequencies (f_e) is equal to 200.

The observed and the expected frequencies and the calculations needed to determine the value of χ^2 are in Table 16–2. The steps are

1. Subtract the expected frequency from the observed frequency—that is, $f_o - f_e$.
2. Square the difference: $(f_o - f_e)^2$.
3. Divide each squared difference by the corresponding expected frequency:

$$\frac{(f_o - f_e)^2}{f_e}$$

4. Sum these quantities for each tollbooth to find the value of χ^2:

$$\sum \frac{(f_o - f_e)^2}{f_e}$$

From Table 16–2, the computed value of χ^2 is 10.670. So the decision is to reject the null hypothesis. We conclude that there is not an equal utilization of the tollbooths.

TABLE 16–2		Calculations for Chi-Square			
Tollbooth	f_o	f_e	$f_o - f_e$	$(f_o - f_e)^2$	$(f_o - f_e)^2/f_e$
1	171	200	−29	841	4.205
2	224	200	24	576	2.880
3	211	200	11	121	0.605
4	180	200	−20	400	2.000
5	214	200	14	196	0.980
Total	1000	1000	0		10.670

In earlier chapters on hypothesis testing, we often reported p-values. Recall that a p-value indicates the probability of finding a value of the test statistic as large or larger when the null hypothesis is true. Recall also that there are 4 degrees of freedom in this problem. From Appendix F, or the portion of Appendix F shown as Table 16–1, we find the row associated with 4 degrees of freedom. We then move across the row until we find the values that bracket the computed value of χ^2. If the significance level had been set at 0.02, the critical value of χ^2 would have been 11.668. Our computed value falls between the critical values for the 0.05 and 0.02 significance levels. Hence, we report that the p-value is less than 0.05, but greater than 0.02, as illustrated below.

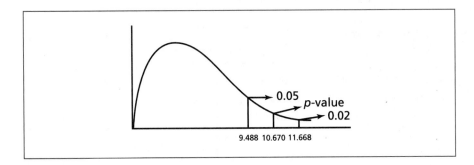

Answers to the Self-Review problems are at the end of the chapter.

Self-Review 16–1

A toothpaste manufacturer hopes to market 1 or more toothpastes with these unusual flavors: cranberry, root beer, lime, vanilla, and orange. The manufacturer wonders if consumers have a preference among the flavors. Small tubes of each flavor are given to 200 consumers, and each consumer is asked to state his or her preference. The 0.01 level of significance is to be used.

a. What are the null and alternate hypotheses?
b. What is the decision rule?

The preferences of the 200 in the sample are

Flavor	Number Preferring Flavor
Cranberry	32
Root beer	30
Lime	28
Vanilla	58
Orange	52
Total	200

c. Calculate the value of chi-square.
d. What is your decision with respect to the null hypothesis?

Exercises

Answers to the even-numbered Exercises are at the back of the book.

1. Find the critical value of χ^2 under the following conditions:
   a. the 0.05 significance level with 10 degrees of freedom.
   b. the 0.01 significance level with 3 degrees of freedom.
   c. the 0.02 significance level with 15 degrees of freedom.
   d. the 0.10 significance level with 30 degrees of freedom.
2. List the characteristics of the χ^2 distribution.
3. In a study, there were 5 categories and 225 sample observations. At the 0.01 significance level, what is the critical value of χ^2?
4. Determine the p-value under each of the following conditions:
   a. There were 6 degrees of freedom, and the computed value of χ^2 was 13.786.
   b. There were 10 degrees of freedom, and the computed value of χ^2 was 26.365.
   c. There were 3 degrees of freedom, and the computed value of χ^2 was 3.276.
   d. There were 12 degrees of freedom, and the computed value of χ^2 was 19.587.

Use the five-step hypothesis-testing procedure for all of the exercises.

5. A city has 3 television stations, each with its own evening news program from 6:00 P.M. to 6:30 P.M. The Acklin Survey Group is hired to determine if the viewing audience has a preference for any station. A random sample of 150 viewers revealed that 53 watched the evening news on WNAE-TV, 64 on WMWM-TV, and 33 on WRRN-TV. At the 0.05 level of significance, is there sufficient evidence to show that the 3 stations do not have equal shares of the evening news audience?
6. It is suspected that a particular 6-sided die is "loaded," making it not a "true" die. As an experiment, this die is rolled 120 times. The results are

Face	1	2	3	4	5	6	Total
Observed Frequency	15	29	14	17	28	17	120

Using the five-step procedure and the 0.05 significance level, test the null hypothesis that there is no difference between the distribution of observed frequencies and the distribution of expected frequencies.

7. A sociologist is studying retired people living in Florida. A sample of 150 who had lived in the Northeast revealed that they were from the following metropolitan regions:

Region	Observed Frequency
North Jersey–New York	60
South Jersey–Philadelphia	40
Buffalo	25
Pittsburgh	25

At the 0.05 significance level, can we conclude that equal numbers of retirees were from the 4 regions?

8. In the Michigan lottery "Daily Game," officials select numbers by having plastic balls with digits on them blown at random, mechanically, through a plastic tube. It is argued that each of the 10 digits, 0 through 9, has the same chance of occurrence. To test for potential bias, researchers recorded the number of times each number was selected during 200 trials.
   a. State the null and alternate hypotheses.
   b. Determine the expected frequencies for each digit.
   c. For a test at the 0.01 level of significance, what is the critical value?
   d. Here are the results:

Digit	Frequency
0	10
1	23
2	15
3	24
4	21
5	23
6	19

Digit	Frequency
7	18
8	25
9	22

Compute the chi-square test statistic.

e. Make a decision about the potential bias. Interpret your results.

9. In a study of work patterns, an organizational behavior specialist collected the data below on absenteeism by day of the week.

Day	Frequency
Monday	124
Tuesday	74
Wednesday	104
Thursday	98
Friday	120

a. If 520 absences were recorded over a 5-day workweek, what is the null hypothesis, the alternate hypothesis, and what is the expected number of absences each day?

b. What is the critical value for a test at the 0.05 level of significance?

c. Conduct a test using chi-square, and arrive at a decision.

THE GOODNESS-OF-FIT TEST: UNEQUAL EXPECTED FREQUENCIES

The expected frequencies in the preceding Problem/Solution were all the same; that is, we expected the same number of drivers to select tollbooth 1, tollbooth 2, and so on. What if the expected frequencies are not equal? The χ^2 distribution can still be applied, but the calculation of the expected frequencies is changed. The following Problem/Solution will serve as an illustration.

Problem   Recall from the Chapter Problem that the *Lansing State Journal* reported that in a bag of M&M chocolate peanut candies, there are 30% brown, 30% yellow, 10% blue, 10% red, 10% green, and 10% orange candies. You purchase a bag of M&M chocolate peanut candies at the nearby Kroger Food and Drug and find 17 brown, 20 yellow, 13 blue, 7 red, 6 green, and 9 orange candies, for a total of 72 candies. At the 0.01 significance level, does the bag you purchased agree with the distribution suggested by the newspaper?

Solution   The usual hypothesis-testing procedure is used:

H_0: The distribution is the same as that given by the newspaper.

H_a: The distribution is not the same as that given by the newspaper.

If the null hypothesis were true, we would expect 30% of the 72 candies or 21.6 to be brown. Likewise, we would expect 7.2 of the candies to be orange, found by 72(0.10). The observed and expected frequencies are:

Color	f_o	f_e	f_e, Found by
Brown	17	21.6	0.30(72)
Yellow	20	21.6	0.30(72)
Blue	13	7.2	0.10(72)
Red	7	7.2	0.10(72)
Green	6	7.2	0.10(72)
Orange	9	7.2	0.10(72)
Total	72	72.0	

The computed value of chi-square, using formula 16–1, is 6.426.

$$\chi^2 = \Sigma \frac{(f_o - f_e)^2}{f_e}$$

$$= \frac{(17 - 21.6)^2}{21.6} + \cdots + \frac{(9 - 7.2)^2}{7.2} = 6.426$$

Because there are 6 colors, or categories, there are 5 degrees of freedom. The critical value at the 0.01 significance level is 15.086. The null hypothesis is rejected if the computed value of chi-square exceeds 15.086. Because the value of chi-square is less than the critical value, the null hypothesis is not rejected. We conclude that the distribution of candies in the bag you purchased at the local Kroger Food and Drug could have come from the population described in the newspaper. The p-value is larger than 0.10.

The versatility of the chi-square distribution is further demonstrated in the following Problem/Solution.

Problem   A national study revealed that within 5 years of their release from prison, 20% of criminals had not been arrested again, 38% had been arrested once, and so on. Table 16–3 shows the complete distribution.

TABLE 16–3	Numbers of Arrests and Percentages of the Total
Number of Arrests After Release from Prison	**Percentage of Total**
0	20.0
1	38.0
2	18.0
3	13.5
4 or more	10.5
Total	100.0

A social agency in a large city has developed a guidance program for former prisoners who settle there. Anxious to compare the local results with the national figures in Table 16–3, the director of the social agency selected at random former prisoners who were in the guidance program. The distributions of the frequencies observed and the national experience are shown in Table 16–4.

TABLE 16–4	Comparison of the Local and National Distributions	
Number of Arrests After Release from Prison	Local Experience (number)	National Experience (percentage of total)
0	58	20.0
1	62	38.0
2	28	18.0
3	16	13.5
4 or more	36	10.5
Total	200	100.0

How would the director of the social agency use chi-square to compare the local experience with the national experience? Use the 0.01 level of significance.

Solution   The *number* in each category resulting from local experience cannot be directly compared with the *percentage* from the national study. The national percentages can, however, be converted to expected frequencies (f_e). Logically, if there is no difference between the local experience and the national experience, 20% of the 200 sampled, or 40, would never be arrested again after being released from prison. Likewise, 38% of the 200, or 76, would be arrested once, and so on. For a complete set of observed and expected frequencies, see Table 16–5.

TABLE 16–5	Comparison of the Local and National Frequencies	
Number of Arrests After Release from Prison	Local Experience (number)	Expected Frequency
0	58	40
1	62	76
2	28	36
3	16	27
4 or more	36	21
Total	200	200

In the 1860s, an Austrian monk named Gregor Mendel (1822–1884) suggested the existence of *genes*. He believed that these genes were passed on from generation to generation. According to 1 of Mendel's theories, the cross-fertilization of 2 pure strains of pea plants—1 producing only round yellow seeds and 1 only wrinkled green seeds—would produce a first generation of hybrids having only round yellow seeds. Yet a mating of these hybrids with each other would yield plants with round yellow, round green, wrinkled yellow, and wrinkled green seeds in proportions of 9:3:3:1; that is, 1 pea plant out of 16 would yield wrinkled green seeds.

Mendel conducted several experiments, one yielding the following results:

Pea Type	Observed Frequency
Round yellow	315
Round green	108
Wrinkled yellow	101
Wrinkled green	32
Total	556

If we establish the null hypothesis that the proportions in the population are as proposed by Mendel and compute the value of χ^2, we do not reject the null hypothesis; that is, we conclude that the proportions suggested by Mendel are reasonable. The computed value of χ^2 is 0.47.

continued

This very small value of χ^2 led R. A. Fisher (1890–1962), a noted statistician, to wonder about the data. In other words, Fisher wondered if the results were too good to be true. Fisher looked over some of Mendel's other experiments and found the observed and expected results to be similarly close. Fisher pooled many of Mendel's results and found it extremely unlikely that such close matches between observed and expected results could be achieved. Fisher believed that Mendel's gardener, who was very eager to please the boss, was responsible for fudging the results.

The null hypothesis (H_0) is that there is no difference between the local experience and the national experience. That is, any differences between the observed and the expected frequencies are due to chance (sampling). The alternate hypothesis (H_a) is that there is a difference between the local experience and the national experience. There are 5 categories in Table 16–5, so there are $k - 1 = 5 - 1 = 4$ degrees of freedom. The critical value of χ^2 from Appendix F is 13.277. The decision rule is shown graphically below.

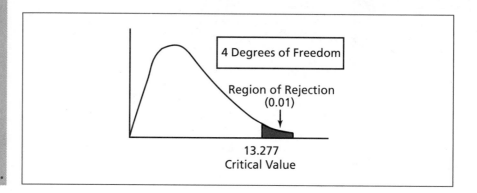

The computed value of chi-square is 27.653 (see Table 16–6).

TABLE 16–6   Computation of Chi-Square

Number of Arrests After Release from Prison	Local Experience f_o	National Experience f_e	$f_o - f_e$	$(f_o - f_e)^2$	$\dfrac{(f_o - f_e)^2}{f_e}$
0	58	40	+18	324	324/40 = 8.100
1	62	76	−14	196	196/76 = 2.579
2	28	36	−8	64	64/36 = 1.778
3	16	27	−11	121	121/27 = 4.482
4 or more	36	21	+15	225	225/21 = 10.714
Total	200	200	0		27.653

Must be 0 χ^2

The computed value of 27.653 is in the region beyond the critical value of 13.277. At the 0.01 level of significance, therefore, the null hypothesis is rejected, and the alternate hypothesis is accepted. The conclusion is that there is a difference between the local and the national experiences. The director now has evidence that his program results in significantly fewer arrests. The p-value is less than 0.01.

National figures revealed that 42% of vacationers who travel outside the United States go to Europe, 20% to the Far East, 16% to South and Central America, 6% to the Middle East, 12% to the South Pacific, and 4% to all other destinations. A local travel agency wondered if its customers differ significantly from this breakdown with respect to their travel destinations. A sample of the files of 200 of its customers revealed the following:

Destination	Number of Vacationers
Europe	80
Far East	44
South and Central America	34
Middle East	16
South Pacific	20
All others	6

a. What level of measurement is involved?
b. State both the null and the alternate hypotheses.
c. Show the decision rule graphically. Use the 0.05 level.
d. Compute χ^2 and arrive at a decision.

Exercises

10. The occupations of parents are thought to influence the occupation choices of their children. The percentage distribution of career employment by categories of professional, technical, and service for a population of parents is shown here:

Occupation	Percentage
Professional	26
Technical	64
Service	10

A sample of 360 children revealed the following about their occupations:

Occupation	Number
Professional	117
Technical	206
Service	37

Use the 0.01 level of significance to test the hypothesis that there has been no change in the distribution of occupations.

11. The law guarantees a defendant the right to be tried before a jury of his or her peers. However, many students of American law complain that the wealthy and powerful are easily excused from jury duty. As part of a study of jury compositions, the following data were developed for Logan County. The population is 80% white, 15% black, and 5% Hispanic. Three hundred persons who recently served on juries were found to fall into those categories as follows:

White	Black	Hispanic
233	53	14

a. State the null and alternate hypotheses.
b. Find the expected frequency for each group.
c. Determine the critical value for a test at the 0.01 level of significance.

d. What is the value of your test statistic?

e. What conclusion do you draw?

12. According to figures previously released by the Census Bureau, health insurance coverage for American citizens was distributed in the following percentages:

Insurance	Percentage of U.S. Citizens
Private	65
Medicare/Medicaid	19
Uninsured	16

A random sample of 235 U.S. citizens conducted several years later revealed the following distribution (in number of U.S. citizens):

Insurance	Number of U.S. Citizens
Private	137
Medicare/Medicaid	53
Uninsured	45

Use the 0.05 level of significance to test the hypothesis that there has been no change in the distribution.

13. Aldini Medical Services studied a sample of 3459 abortions. The numbers of previous abortions reported by these patients were

Number of Abortions	Number of Patients
0	2441
1	784
2	184
3 or more	50

Are these data consistent with national data that indicate that 67% have never had a previous abortion, 27% have had 1 previous abortion, 5% have had 2 previous abortions, and 1% have had 3 or more previous abortions? Use the 0.05 level of significance.

14. Professor Graham teaches at least 1 section of sophomore-level statistics each quarter. He selected 50 students from his previous class and found 30 sophomores, 10 freshmen, and 10 juniors. Can Professor Graham conclude that there are twice as many sophomores as freshmen and juniors? Use the 0.05 significance level.

CONTINGENCY-TABLE ANALYSIS

The goodness-of-fit test discussed in the previous section was concerned with only a *single* trait, such as the color of an M&M candy. What testing procedure is followed if we are interested in the relationship between two characteristics, such as a person's adjustment to retirement and whether or not that person moved to a retirement community? By classifying adjustment as excellent, good, fair, or poor, the research data may be tallied into a table:

	Adjustment to Retirement			
Status	Excellent	Good	Fair	Poor
Moved to retirement community	𝚝𝚑𝚕 𝚝𝚑𝚕 𝚝𝚑𝚕 𝚝𝚑𝚕 𝚝𝚑𝚕 ///	𝚝𝚑𝚕 𝚝𝚑𝚕 𝚝𝚑𝚕 //	𝚝𝚑𝚕 𝚝𝚑𝚕 𝚝𝚑𝚕 𝚝𝚑𝚕	𝚝𝚑𝚕 𝚝𝚑𝚕 𝚝𝚑𝚕
Did not move to retirement community	𝚝𝚑𝚕 𝚝𝚑𝚕 𝚝𝚑𝚕 ///	𝚝𝚑𝚕 𝚝𝚑𝚕 𝚝𝚑𝚕 //	𝚝𝚑𝚕 𝚝𝚑𝚕 𝚝𝚑𝚕 /	𝚝𝚑𝚕 𝚝𝚑𝚕 𝚝𝚑𝚕 𝚝𝚑𝚕 𝚝𝚑𝚕 //

This table is called a **contingency table.** It is also popularly referred to as cross-tabulated data.

We can also use the chi-square statistic to determine whether two traits—home status and adjustment to retirement—are related. The null hypothesis is that the two traits are independent (not related). This is consistent with our earlier use of the null hypothesis to represent the case in which there was no change. We always took as the H_0 that $\mu_1 = \mu_2$ or that μ was equal to some particular value. Like the goodness-of-fit test, the contingency-table analysis uses observed frequencies (f_o) and expected frequencies (f_e). The observed frequencies are recorded in the preceding contingency table in the form of tallies. The expected frequencies must be computed.

> **contingency table**
> Frequency data resulting from the simultaneous classification of *more than one* variable or trait of the observed item.

Problem   A metropolitan law enforcement agency classifies crimes committed within its jurisdiction as either "violent" or "nonviolent." An investigation has been ordered to find out whether the type of crime (violent or nonviolent) depends on the age of the person who committed it. A sample of 100 crimes was selected at random from the police files. The results are reported in Table 16–7.

TABLE 16–7	Classification of Crimes by Types of Crimes and by Ages of Persons			
	Age			
Type of Crime	Under 25	25 up to 50	50 and Over	**Total**
Violent	15	30	10	55
Nonviolent	5	30	10	45
Total	20	60	20	100

Does it appear that there is any relationship between the age of a criminal and the nature of the crime? Use the 0.05 level of significance.

Solution   As usual, the initial step is to state the null hypothesis, H_0, and the alternate hypothesis, H_a:

H_0: There is no relationship between the type of crime committed and the age of the criminal.

H_a: There is a relationship between the type of crime committed and the age of the criminal.

The observed frequencies are shown in Table 16–7. To determine the expected frequencies note first the number of crimes being studied (100). In the marginal totals on the right side of the contingency table, observe that 55 of the 100 crimes, or 55%, were violent. If the null hypothesis is true (that there is no relationship between type of crime and age), one would expect that 11 of the 20 criminals under age 25 (55%) committed a violent crime, found by 0.55×20. Likewise, if the null hypothesis is true, one would expect 33 of the criminals in the 25-up-to-50 age bracket, or 55%, to have committed a violent crime. The expected frequency in the 50-and-Over age group is the same as in the Under-25 age group.

The same logic can be followed to find the expected frequencies for the nonviolent crimes. Again, from the right-hand marginal totals, 45/100, or 45%, were nonviolent crimes. Then

$$45/100 \times 20 = 9 \text{ expected frequencies for the Under-25 age group}$$

$$45/100 \times 60 = 27 \text{ expected frequencies in the 25-up-to-50 age group}$$

By now, no doubt you will have noticed that under the assumption of independence, an expected frequency, f_e, can be computed as follows:

$$\text{Expected frequency for a cell} = \frac{(\text{Row total})(\text{Column total})}{\text{Grand total}} \quad \boxed{16\text{--}2}$$

For the nonviolent, 25-up-to-50 age cell—just computed to be 27 expected frequencies—the formula would be

$$f_e = \frac{(\text{Row total})(\text{Column total})}{\text{Grand total}} = \frac{(45)(60)}{100}$$

= 27, the same answer as the one computed earlier

The observed frequencies and the corresponding expected frequencies are shown in the form of a contingency table (see Table 16–8).

| TABLE 16–8 | Observed Frequencies and Expected Frequencies |

	Age							
	Under 25		25 up to 50		50 and Over		**Total**	
Type of Crime	f_o	f_e	f_o	f_e	f_o	f_e	f_o	f_e
Violent	15	11	30	33	10	11	55	55
Nonviolent	5	9	30	27 ←	10	9	45	45
Total	20	20	60	60	20	20	100	100

Must be equal

$$\frac{(45)(60)}{100}$$

Must be equal

We formulate a decision rule by first determining the number of degrees of freedom. For a contingency table, we find the number of degrees of freedom by multiplying the number of rows minus 1 by the number of columns minus 1. This contingency table has 2 rows and 3 columns. To find the degrees of freedom we compute as follows: (Rows − 1)(Columns − 1) = (2 − 1)

$(3 - 1) = 2$ *df*. The critical value can now be located in Appendix F. Recall that the 0.05 level of significance had been selected. Go down the left margin in Appendix F to 2 df and read the critical value under the 0.05 column. It is 5.991. Thus, the decision rule is to reject the null hypothesis if the computed value of χ^2 is greater than 5.991.

As before, chi-square is computed by formula 16–1:

$$\chi^2 = \Sigma \frac{(f_o - f_e)^2}{f_e}$$

Inserting the observed frequencies, f_o, and the expected frequencies, f_e, from Table 16–8 in the formula for χ^2, we obtain

$$\chi^2 = \frac{(15 - 11)^2}{11} + \frac{(5 - 9)^2}{9} + \frac{(30 - 33)^2}{33} + \frac{(30 - 27)^2}{27}$$

$$+ \frac{(10 - 11)^2}{11} + \frac{(10 - 9)^2}{9}$$

$$= 1.455 + 1.778 + 0.273 + 0.333 + 0.091 + 0.111$$

$$= 4.041$$

The computed chi-square value of 4.041 is less than the critical value of 5.991. Therefore, we fail to reject the null hypothesis. The law enforcement agency cannot conclude that there is any relationship between the age of the criminal and whether the crime was a violent or nonviolent occurrence. To put it another way, the age of the criminal is independent of the occurrence of violence.

CAUTION.   In the formula for χ^2, note that the expected frequencies, f_e, are in the denominator. If the expected frequency for any cell is quite small, the corresponding $(f_o - f_e)^2/f_e$ value for that cell may be disproportionately large. In turn, this one large cell value might result in a computed χ^2 greater than the critical value, causing the null hypothesis to be rejected. If the expected frequency for that one cell were larger, the null hypothesis would probably not be rejected. To avoid this problem a useful rule of thumb is to look for an expected frequency of at least five in each cell. Whenever the number is smaller than that, you may want to consider combining several adjacent cells. The obvious alternative, of course, would be to increase the size of the sample.

We will now use MINITAB for the same Problem/Solution. To begin we enter the data using the Read command and identify these columns using Name.

Data			
	C1	C2	C3
	Under 25	25 to 50	Over 50
1	15	30	10
2	5	30	10

From the MINITAB toolbar, we select

Stat ▶ Tables ▶ Chisquare Test

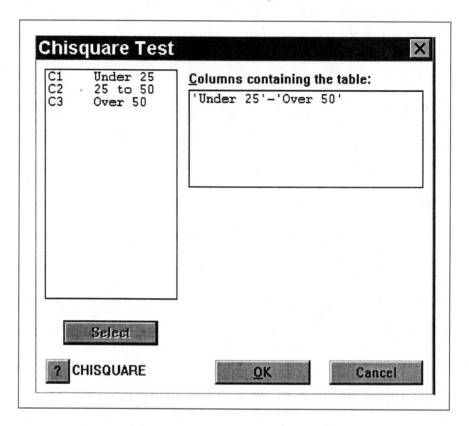

Finally, we highlight all three columns, click on **Select,** and then click on **OK.** Here is the MINITAB output:

```
     Expected counts are printed below observed counts

           Under 25  25 to 50   Over 50    Total
       1         15        30        10        55
              11.00     33.00     11.00

       2          5        30        10        45
               9.00     27.00      9.00

Total            20        60        20       100

Chi-Sq =  1.455 +  0.273 +  0.091 +
          1.778 +  0.333 +  0.111 =   4.040
DF = 2, P-Value = 0.133
```

From the MINITAB output, the computed value of χ^2 is the same as that computed earlier. In addition to computing the value of χ^2, MINITAB outputs the p-value. Recall that a p-value is the likelihood that we can find a value of the test statistic as large or larger than that computed when the null hypothesis is true. In this case, the p-value is the likelihood of finding a value of χ^2 larger than 4.040 with 2 degrees of freedom. That likelihood is 0.133, according to MINITAB. To put it another way, we could reject the null hypothesis that there is no relationship between the type of crime committed and the age of the criminal had we selected a significance level of 0.133 or more.

- -

Self-Review 16–3

In a city with a maximum-security prison, the residents have been polled to determine if a relationship exists between marital status and a resident's stand on capital punishment.

a. What are the null and alternate hypotheses? A random sample of 200 residents was asked for their opinions and for their marital statuses. The results were cross-classified into the following table:

Stand on Capital Punishment	Marital Status		Total
	Married	Not Married	
Favor	100	20	120
Oppose	50	30	80
Total	150	50	200

b. What is this table called?
c. How many cells does the table contain?
d. How many degrees of freedom are there?
e. For the 0.01 level of significance, what is the critical value of chi-square?
f. What is the computed value of chi-square?
g. What is your decision regarding the null hypothesis?

REAL STAT

A study of 1000 Americans over the age of 24 showed that 28% never married. Of those who never married, 22% completed college. Twenty-three percent of those studied have married and completed college. Can we conclude from the information given that being married is related to completing college? The study indicated that the 2 variables were related, that the computed value of chi-square was 11.525, and that the p-value was less than 0.001.

- -

Exercises

15. In a particular study, we have a sample of 200 observations classified into 3 rows and 5 columns. How many degrees of freedom are there in the study?

16. In a study of undergraduate enrollment at a major university, a sample of 500 students was classified by gender and class rank (freshman, sophomore, junior, and senior). How many degrees of freedom are there in the study?

17. Five hundred persons were divided into two groups to be sampled—one group consisting of

people with religious affiliations and the other of people without religious affiliations. Each person was given a test to determine his or her degree of racial prejudice. The question to be explored is whether religious affiliation and racial prejudice are related.

a. State the null and alternate hypotheses. The results of the survey were cross-classified into the following contingency table:

Religious Status	Degree of Racial Prejudice		
	Highly Prejudiced	Somewhat Prejudiced	Not Prejudiced
Church-affiliated	70	160	170
Not church-affiliated	20	50	30

b. Using the 0.05 level, determine the critical value of chi-square.

c. Compute χ^2 and arrive at a decision.

18. A random sample of 780 registered voters was asked to evaluate President Clinton's handling of the abortion issue. They were cross-classified by party affiliation into the following table:

	Democrat	Republican	Independent
Approved	200	125	70
Disapproved	130	230	25

Does the sample support the hypothesis that one's party affiliation is independent of one's opinion of President Clinton's handling of the abortion issue? Use the 0.05 level of significance.

19. The table below shows the distribution of hair colors for a sample of 422 people. You wish to see if hair color and gender are related.

Hair Color	Number of Men	Number of Women
Black	56	32
Blonde	37	66
Brown	84	90
Red	19	38
Total	196	226

Use the 0.01 level of significance to test whether these distributions are significantly different. Explain your findings.

20. A random sample of voter registrations for several adjacent precincts led to the following data:

	Precinct 16	Precinct 23	Precinct 27
Democrat	300	420	270
Independent	25	17	10
Republican	150	300	200

At the 0.01 significance level, can we conclude that precinct and party affiliation are related?

21. A sample of people was asked by the Gallup organization to rate the honesty and ethical standards of various groups. Some of the results are shown below.

	High	Average	Low
Medical doctors	53	35	10
Lawyers	24	43	27
Business executives	18	55	20
Members of Congress	14	43	38

Use the 0.10 significance level to determine if the selected professions are related to honesty and ethical standards.

CHAPTER OUTLINE

I. The major characteristics of the χ^2 distribution are as follows:

A. The value of χ^2 is never negative.

B. The χ^2 distribution is positively skewed.

C. There is a family of χ^2 distributions.

1. Each time the degrees of freedom change, a new distribution is formed.

2. As the degrees of freedom increase, the distribution approaches a normal distribution.

II. A goodness-of-fit test will show if an observed set of frequencies could have come from a hypothesized distribution.

A. The degrees of freedom are $k - 1$, where k is the number of categories.

B. The formula for computing the value of chi-square is

$$\chi^2 = \Sigma \frac{(f_o - f_e)^2}{f_e} \qquad \boxed{\text{16–1}}$$

III. A contingency table is used to test whether two traits or characteristics are related.

A. Each observation is classified according to the two traits.

B. The expected frequencies are computed as follows:

$$f_e = \frac{(\text{Row total})(\text{Column total})}{(\text{Grand total})} \qquad \boxed{\text{16–2}}$$

C. The degrees of freedom are computed as follows:

$$df = (\text{Rows} - 1)(\text{Columns} - 1)$$

D. Formula 16–1 is used to find the value of χ^2.

Exercises

22. Vehicles heading west on Front Street may turn right, turn left, or go straight ahead at Elm Street. The city traffic engineer believes that half of the vehicles will continue straight through the intersection. Of the remaining half, equal proportions will turn right and left. Two hundred vehicles were observed with the following results:

	Straight	Right Turn	Left Turn
Frequency	112	48	40

Use the 0.10 significance level to investigate the claim by the city engineer.

23. There are 4 entrances to the OCF Building in downtown Philadelphia. The building maintenance supervisor would like to know if the entrances are equally utilized. As an experiment, 400 people are observed as they enter the building. The numbers using each entrance are reported below. At the 0.01 significance level, can we conclude that there is a difference in the use of the 4 entrances?

Entrance	Number Entering
Main Street	140
Broad Street	120
Cherry Street	90
Walnut Street	50
Total	400

24. The publisher of a sports magazine plans to offer new subscribers 1 of 3 free gifts—a sweatshirt with their favorite team logo, a coffee cup with their favorite team logo, or a pair of earrings with their favorite team logo. In a sample of 500 new subscribers, the numbers selecting each gift are reported below. At the 0.05 significance level, is there a preference for the gifts, or should the publisher conclude the gifts are equally well liked?

Gift	Frequency
Sweatshirt	183
Coffee cup	175
Earrings	142

25. The owner of a national mail-order catalog would like to compare catalog sales with the geographical distribution of the population. According to the U.S. Bureau of the Census, 21% of the population live in the Northeast, 24% in the Midwest, 35% in the South, and 20% in the West. Listed below is a breakdown of a sample of 400 orders randomly selected from those shipped last month.

Region	Frequency
Northeast	68
Midwest	104
South	155
West	73
Total	400

At the 0.01 significance level, is the distribution of the destinations of the orders shipped reflective of that of the population?

26. A group of 385 mental patients has been classified according to parental social class, with the following results:

Social Class	Frequency
Upper	18
Upper-middle	31
Middle	46
Lower-middle	126
Lower	164

Test at the 0.05 level of significance to verify that the data are consistent with the assumption that all social classes are equally likely to be represented.

27. Refer to Exercise 26. Test the hypothesis that 30% of the patients are lower class, 30% are lower-middle class, 20% are middle class, 10% are upper-middle class, and 10% are upper class. Use the 0.05 significance level.

28. Each human being has blood of type A, B, O, or AB. The proportions of the general population having each type follow:

Type	Proportion
A	0.39
B	0.11
O	0.46
AB	0.04
	1.00

A sample of 700 people revealed that 250 have type A, 70 have type B, 340 have type O, and 40 have type AB. At the 0.05 significance level, can we conclude that this group of people differs from the general population?

29. As part of a study of air traffic at the Express Airport, a record has been kept of the numbers of aircraft arrivals during half-hour periods. The table that follows shows the numbers of half-hour periods in which there were 0, 1, 2, 3, or 4 or more arrivals. Test the hypothesis that the numbers of such arrivals follow the expected frequencies shown. Use the 0.05 significance level.

Number of Arrivals	Observed Frequency	Expected Frequency
0	19	14
1	23	27
2	28	27
3	18	18
4 or more	12	14
Total	100	100

30. The U.S. Department of Education reports that the educational levels of adults in the United States conform to the distribution that follows. The Phoenix, Arizona Chamber of Commerce wishes to compare a sample of 500 residents against this norm. What conclusion should be drawn from the results of this analysis? Use the 0.05 significance level.

Number of Years of Education	Expected Frequency	Observed Frequency
1 up to 9	100	77
9 up to 12	150	198
12 up to 15	200	178
15 and over	50	47

31. Students claim not to like morning classes. As a test, a college statistics department offered sections of a basic statistics course at various times during the day. No limit was set on class size. The following table shows the numbers of students who selected each class. At the 0.01 level of significance, is there sufficient evidence to show that students have a time preference?

Time	Number of Students
Early A.M.	40
Late A.M.	62
Early P.M.	60
Late P.M.	35
Early evening	58
Late evening	45
Total	300

32. A group of executives was classified according to total income and age. Test the hypothesis, at the 0.01 level, that age is not related to level of income.

	Income		
Age	Less than $100,000	$100,000 up to $400,000	$400,000 or More
Under 40	6	9	5
40 up to 55	18	19	8
55 or older	11	12	17

33. A manufacturer of women's apparel wants to know if a woman's age is a factor in determining whether she will buy a particular garment. Accordingly, the firm surveyed three age groups and asked each woman to rate the garment as excellent, average, or poor. The results follow. Test the hypothesis, at the 0.05 level, that rating is not related to age group.

	Age Group			
Rating	15 up to 25	25 up to 40	40 up to 55	Total
Excellent	40	47	46	133
Average	51	74	57	182
Poor	29	19	37	85
Total	120	140	140	400

34. The following results were reported in Japan's parliamentary elections for the lower house:

	1990	1996
Liberal Democratic Party	250	300
Socialist Party	112	85
Clean Government Party	58	56
Democratic Socialist Party	38	26
Communist Party	26	26
New Liberal Club	8	6
Independents	19	13

Use the 0.01 level of significance to test if there has been a significant change in the distribution.

35. A study was conducted of 400 families, each having exactly 3 children, to determine whether or not boy and girl births were equally likely. The results were as given. What is the value of χ^2? Are the results significantly different from those expected at the 0.05 significance level?

Number of Boys	0	1	2	3
Number of Cases	65	140	148	47

(Hint: Use the binomial distribution to find the expected frequencies.)

36. Five different brands of canned tuna were examined for quality. A total of 24 cans of each brand was checked. The numbers of acceptable and unacceptable cans are as follows:

	Brands				
	A	B	C	D	E
Acceptable	3	10	5	3	9
Unacceptable	21	14	19	21	15

At the 0.05 significance level, can we conclude that the 5 brands are of the same quality?

37. A manufacturer claims that a certain drug is effective in treating arthritis. In an experiment, 176 people were given the drug, and 176 people were given sugar pills. The results are shown below. At the 0.05 significance level, is there a difference in the effectiveness of the 2 drugs?

	Better	No Effect	Worse	Total
Drug	88	50	38	176
Sugar	60	65	51	176
Total	148	115	89	352

DATA EXERCISES

38. Refer to the real estate data, which reports information on homes sold in Alabama during 1996.
  a. Develop a contingency table that shows whether or not a home has a pool and the township in which the house is located. Is there an association between the variables of pool and township? Use the 0.05 significance level.
  b. Develop a contingency table that shows whether on not a home has an attached garage and the township in which the home is located. Is there an association between the variables of attached garage and township? Use the 0.05 significance level.
39. Refer to the schools data set in Appendix, which has information on 94 school districts in northwest Ohio.

  a. Group the districts based on the percentage of students on welfare: "low" (less than 5%), "moderate" (5 up to 10%), and "high" (10% or more). Conduct a test of hypothesis to determine if the 3 categories are equally represented in the sample. Use the 0.05 significance level.
  b. Identify each district as "small" (less than 1000 students), medium (1000 up to 3000 students), and "large" (more than 3000 students). Create a table that shows the districts cross-classified by both the percentage of students on welfare and the size of the district. At the 0.05 significance level, can we conclude that the two characteristics are related?

CHAPTER ACHIEVEMENT TEST

The answers are at the back of the book.

MULTIPLE-CHOICE QUESTIONS

Select the response that best answers each of the questions.

1. For a study to determine if region of birth (4 categories) is related to age (6 categories), the number of degrees of freedom in the test statistic would be
  a. 10.
  b. 24.
  c. 15.
  d. 2.

2. If the chi-square test is to be applied to data cross-classified in a contingency table,
 a. the expected frequencies cannot be determined.
 b. the observed frequencies must total 100.
 c. there must be at least 10 expected frequencies in each cell.
 d. None of the above

3. The chi-square distribution
 a. is positively skewed.
 b. never assumes a negative value.
 c. changes when the number of degrees of freedom changes.
 d. All of the above

4. A sample of 100 students consists of 60 females and 40 males. The purpose of the study is to determine if the gender of the student is related to passing or failing the first test. Out of the 100 students in the sample, 70 students passed. The expected number of males passing is
 a. 70.
 b. 40.
 c. 28.
 d. 12.
 e. None of the above

Questions 5–10 are based on the following problem. A supervisor wants to determine if unexcused absences in her department are distributed evenly throughout the workweek. A random sample of personnel files revealed the following:

Day	Frequency
Monday	20
Tuesday	16
Wednesday	17
Thursday	21
Friday	26
	100

5. How many degrees of freedom are there?
 a. 4
 b. 99
 c. 5
 d. 25
 e. None of the above

6. If the null hypothesis is true, what is the expected number of absences on Friday?
 a. 25
 b. 20
 c. 26
 d. 16
 e. None of the above

7. At the 0.01 significance level, the critical value of chi-square is
 a. 13.277.
 b. 15.086.
 c. 9.488.
 d. 11.070.
 e. None of the above

8. The computed value of χ^2 is
 a. 9.
 b. 3.1.
 c. 62.
 d. None of the above

9. The correct decision is to
 a. reject H_0.
 b. not reject H_0.
 c. There is not enough information available to make a determination.

10. The correct statistical conclusion is that
 a. the unexcused absences are evenly distributed throughout the workweek.
 b. the unexcused absences are not evenly distributed throughout the workweek.
 c. Neither is a correct conclusion.

COMPUTATION PROBLEMS

11. Ninety congresspersons were surveyed to determine if there is a relationship between party affiliation and position on a proposed Social Security tax increase. At the 0.05 significance level, test the null hypothesis that there is no relationship. The survey results were

	Favor	Oppose	Undecided
Democrats	22	13	28
Republicans	8	7	12

12. An analysis of last year's automotive sales revealed that for every 1 full-sized automobile purchased, 2 medium-sized, 3 compact, and 4 sub-

compact cars were purchased. A sample of purchases made during the last month shows car sales of 38 full-sized, 62 medium-sized, 41 compact, and 59 subcompact cars. Test at the 0.01 significance level to determine whether a change has occurred in the types of automobile purchased.

. . . .

ANSWERS TO SELF-REVIEW PROBLEMS

16-1 a. H_0: There is no preference.
H_a: There is a preference.
b. Reject H_0 if $\chi^2 > 13.277$.

c.

$f_o - f_e$	$(f_o - f_e)^2$	$\frac{(f_o - f_e)^2}{f_e}$
32–40	$(-8)^2$	64/40 = 1.6
30–40	$(-10)^2$	100/40 = 2.5
28–40	$(-12)^2$	144/40 = 3.6
58–40	$(18)^2$	324/40 = 8.1
52–40	$(12)^2$	144/40 = 3.6
		$\chi^2 = 19.4$

d. Reject H_0 because 19.4 is in the region beyond 13.277; accept H_a. There is a difference with respect to taste preference.

16-2 a. Nominal
b. H_0: There is no difference between local and national vacation destinations.
H_a: There is a difference between local and national destinations.

c.

Region of Rejection

11.070

d.

f_o	f_e	f_o-f_e	$(f_o$-$f_e)^2$	$\frac{(f_o - f_e)^2}{f_e}$
80	84	−4	16	16/84 = 0.190
44	40	4	16	16/40 = 0.400
34	32	2	4	4/32 = 0.125
16	12	4	16	16/12 = 1.333
20	24	−4	16	16/24 = 0.667
6	8	−2	4	4/8 = 0.500
				$\chi^2 = 3.215$

Do not reject the null hypothesis because 3.215 is less than 11.070.

16-3 a. H_0: There is no relationship between a person's stand on capital punishment and his or her marital status.
H_a: There is a relationship between a person's stand on capital punishment and his or her marital status.
b. A contingency table
c. 4, found by 2 × 2
d. 1, found by $(2 - 1)(2 - 1)$
e. 6.635, from Appendix F
f. $\chi^2 = \frac{(100 - 90)^2}{90} + \frac{(20 - 30)^2}{30} + \frac{(50 - 60)^2}{60} + \frac{(30 - 20)^2}{20}$
$= 11.11$
g. Reject the null hypothesis. Conclusion: At the 0.01 level, there is a relationship between a person's stand on capital punishment and his or her marital status.

CHAPTER 17

Nonparametric Methods: Analysis of Ranked Data

OBJECTIVES

When you have completed this chapter, you will be able to

- perform the sign test and describe its applications;
- calculate the Wilcoxon signed-rank test and describe its applications;
- describe the Wilcoxon rank-sum test;
- compute the Kruskal-Wallis analysis of variance by ranks test.

CHAPTER PROBLEM Mirror, Mirror on the Wall . . .

The Shelbyville Career Training Center specializes in helping recently widowed homemakers reenter the job market. Of particular concern is the degree of self-confidence exhibited. A sample of 13 widows was selected at random. Before the start of the training program, each of the widows was rated by a staff of psychiatrists as having no self-confidence, some self-confidence, or great self-confidence. After the training program, the same staff rated each widow's degree of self-confidence on the same scale. Do the data in Table 17–1 demonstrate an increase in self-confidence?

INTRODUCTION

In Chapter 16, we began examining nonparametric hypothesis tests, of which the chi-square goodness-of-fit test is one example. We noted that such tests are especially useful if no assumptions—such as normality—can be made about the shape of the population.

All four nonparametric tests in *this* chapter—namely, the sign test, the Wilcoxon signed-rank test, the Wilcoxon rank-sum test, and the Kruskal-Wallis analysis of variance by ranks test—assume the data are at least of the ordinal level of measurement, which is to say that the observations can be ordered (ranked) from lowest to highest. For example, if families were to be classified according to socioeconomic status (lower, lower-middle, middle, upper-middle, or upper class), the level of measurement would be ordinal scale.

THE SIGN TEST

The sign test is one of the more widely used nonparametric tests. As the name implies, it is based on the signs of the differences in paired observations. The pairs frequently represent "before" and "after" measurements taken on the same individual or item. Before the experiment, a person is given a rating, asked for an opinion, or asked to perform a specific task, the results of which are recorded. Then the experiment or treatment takes place, after which another observation is obtained and any changes are recorded. An increase may be given a plus sign and a decrease a minus sign. The underlying assumption is that, if the null hypothesis is true, there will be an equal chance for plus and minus signs. If the treatment had no effect, the distribution of positive differences would be binomial, with $\pi = 0.50$. If the treatment had *some* impact, the number of positive differences would be either significantly larger or significantly smaller, and the null hypothesis would be rejected.

The test statistic is the number of plus signs. When the null hypothesis is true, the sampling distribution follows the binomial distribution, with $\pi = 0.50$. Recall from Chapter 6 that the principal features of the binomial are that (1) the probability of a success remains constant from trial to trial, (2) successive trials are independent, and (3) we count the number of successes in a given number of trials. These traits are generally consistent with the sign test. The following Problem/Solution illustrates the details of a one-tailed sign test.

Problem   Recall from the Chapter Problem that the Shelbyville Career Training Center specializes in programs designed to increase the levels of self-confidence among recently widowed homemakers. The sample information is reported in Table 17–1. At the 0.10 significance level, can we conclude that self-confidence has increased?

Solution   A plus sign indicates that the special training program was a *success;* that is, the widow had more confidence after the program. Note, for example, that widow Angler had no confidence in herself before the training program, but showed some confidence after the program. Therefore, the sign of the difference for her is plus. A minus sign indicates a *failure;* that is, the widow was less self-confident about reentering the job market after the training program than before it. If a subject felt *no change* in self-confidence, she

TABLE 17–1	Self-Confidence of Widows Before and After a Training Program		
		Self-Confidence		**Sign of Difference**
Widow	**Before**	**After**	
Angler	None	Some	+
Farad	Some	Great	+
Flame	None	Great	+
Goodman	Some	Great	+
Herringbone	Some	None	−
Isenhath	None	Some	+
Kent	Some	Great	+
Kline	None	Great	+
Niger	None	Some	+
Sander	None	Some	+
Sikkenga	Great	None	−
Slayer	None	Great	+
~~Starlings~~	~~None~~	~~None~~	~~0~~

Excluded from further analysis ←

was dropped from the study and thus omitted from further analysis. Widow Starlings had no confidence in her chances in the job market either before or after the program. Thus, she was excluded, and the sample was reduced from 13 to 12 observations.

It does seem logical that, if the special training program were *not* effective in developing self-confidence among the widows, half of them would show increased self-confidence and the other half reduced self-confidence; that is, there would be an equal number of pluses and minuses, and the *probability* of a success in the experiment would be $\pi = 1/2 = 0.50$. Therefore, the null hypothesis to be tested is $\pi \leq 0.50$. The null and alternate hypotheses are

H_0: There has been no increase in a widow's degree of self-confidence as a result of the training program.

H_a: A widow's confidence is increased.

We can also write the hypotheses as

$$H_0: \pi \leq 0.50$$

$$H_a: \pi > 0.50$$

The 0.10 level of significance is chosen. The appropriate test statistic is the number of + signs, which follows the binomial distribution, with π, the probability of a success, equal to 0.50 and n, the size of the sample, equal to 12.

The alternate hypothesis, stated earlier, is $\pi > 0.50$. Recall from previous hypothesis-testing problems that, if a direction is predicted (such as that π is greater than 0.50), a *one-tailed* test is used. And because only the tails of the sampling distribution are used in determining the critical region, only the right tail of the binomial distribution will be needed in this problem (because it is stated that π is *greater* than 0.50).

We locate the region of rejection by accumulating the probabilities starting with 12 successes, then 11 successes, and so on until *we come as close as possible to the level of significance of 0.10 without exceeding it.* In this case, the probability of 12 successes is 0.000, the probability of 11 successes is 0.003, and so on, as shown in the middle column of Table 17–2. (The probabilities are found in Appendix A for a π of 0.50 and an n of 12.)

TABLE 17–2	Cumulative Probabilities for 12, 11, 10, 9, and 8 Successes		
Number of Successes	Probability of Success		Cumulative Probability
⋮	⋮		⋮
8	0.121		0.194 = 0.073 + 0.121
9	0.054		0.073 = 0.019 + 0.054
10	0.016	add up	0.019 = 0.003 + 0.016
11	0.003		0.003 = 0.000 + 0.003
12	0.000		0.000

The probabilities of success are cumulated starting with 12 successes. The cumulative probability for 10 or more successes, for example, is found by 0.000 + 0.003 + 0.016. Refer to the cumulative probabilities in the right column of Table 17–2. A decision rule can now be formulated. The cumulative probability of 0.073 is as close as possible to the significance level of 0.10 without exceeding it. Thus, if 9 or more pluses (increases in self-confidence among the widows) appear in the sample of 12 usable pairs of ratings, the null hypothesis that $\pi \leq 0.50$ is rejected. The rejection region is shown in Figure 17–1. The probabilities on the Y-axis are from Appendix A.

A count of the pluses and minuses in Table 17–1 shows that there are 10 pluses (successes) and 2 minuses (failures). The 10 successes are in the region of rejection. The null hypothesis that $\pi \leq 0.50$ is rejected; the alternate hypothesis that $\pi > 0.50$ is accepted. This means the special training program was a success. Out of 12 widows, 10 had more self-confidence after the program than before it. In rejecting the null hypothesis, we conclude that it is highly unlikely that such a large number of successes (10 successes out of 12 in the sample) *could be due to chance.* It is reasonable to assume therefore that this large number of successes is due to the effects of the special training program.

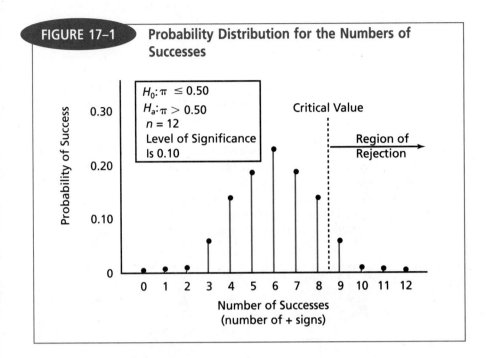

FIGURE 17–1   **Probability Distribution for the Numbers of Successes**

$H_0: \pi \leq 0.50$
$H_a: \pi > 0.50$
$n = 12$
Level of Significance Is 0.10

Critical Value

Region of Rejection

Probability of Success

0.30

0.20

0.10

0

0 1 2 3 4 5 6 7 8 9 10 11 12

Number of Successes
(number of + signs)

· ·

Self-Review 17–1

Answers to the Self-Review problems are at the end of the chapter.

A physical-fitness instructor believes that his planned exercise program is effective in weight reduction. To investigate this claim, a random sample of seven persons enrolled in his program was selected. All were weighed before they started the program and again after completing it. These were the results:

Student	Before	After
Dave	192	185
Rick	198	185
Jim	170	162
Kelly	130	118
Stacey	113	118
Rob	165	159
Andy	168	164

The sign test is to be applied to determine if there is evidence to demonstrate that the weight-reduction program is effective. The 0.10 significance level is used. (Note: In solving the problem, we consider a *reduction* in a person's weight a "success" and designate it by a plus sign. An increase in a person's weight is considered a "failure" and is designated by a minus sign.)

a. State the null and alternate hypotheses.

b. State the decision rule.

c. Should the null hypothesis be rejected? Explain.

d. Interpret your findings.

The Problem/Solution that follows illustrates the use of the sign test in formulating a decision rule for a *two-tailed* hypothesis test.

Problem   Twenty police officers are paired into 10 teams that include 1 black and 1 white officer each. The "partners" are then observed and rated as to their degrees of mutual trust. A high score indicates a great deal of trust. After this observation period, the 10 pairs are subjected to a "wilderness experience" and then retested on their degrees of mutual trust. Do the data in Table 17–3 suggest any significant changes attributable to the experience?

TABLE 17–3	Degrees of Trust Before and After a Wilderness Experience		
Pair	**Rating Before Experience**	**Rating After Experience**	**Sign of Difference**
Braid-Ahern	68	32	−
Chadwick-Gibbon	55	55	0
Falloff-Konrath	49	55	+
Unman-Bernstein	40	75	+
Chamberlain-Sultan	20	50	+
Lively-Julia	18	25	+
Lucite-Austin	30	23	−
Mangos-Chasm	70	49	−
Moharter-Klieg	52	62	+
O'Neill-Burner	50	41	−

Solution   Use the five-step hypothesis-testing procedure.

Step 1. The null hypothesis is that there is no difference in the ratings before and after the wilderness experience. This will be tested against the alternate hypothesis that there is a difference. Note that this is a two-sided alternative. We are looking for *any* significant change.

Again, it seems logical that, if the wilderness experience had *not* affected mutual trust among the police officers, half of the pairs would show increased trust and the other half less trust; that is, there would be an equal proportion of pluses and minuses. Thus, if the wilderness experience has no impact, the probability of a success is $\pi = 0.50$. The null and alternate hypotheses are

$$H_0: \pi = 0.50$$

$$H_a: \pi \neq 0.50$$

Step 2. For this problem, the researcher selected a 20% level of significance.

Step 3. The decision rule is based on the binomial distribution. Note that, because the rating for partners Chadwick and Gibbon was neither an increase nor a decrease, the pair is removed from further calculations. Thus, nine partner pairs remain. As in the Chapter Problem, we construct the sampling distribution applicable for a true null hypothesis by referring to Appendix A, with $n = 9$ and $\pi = 0.50$. These probabilities are shown in Table 17–4.

TABLE 17–4	Binomial Probability Distribution ($n = 9$, $\pi = 0.50$)	
Number of Successes	Probability of Success	Cumulative Probability
0	.002	.002
1	.018	.020
2	.070	.090
		·········· Critical value
3	.164	.254
4	.246	.500
5	.246	.500
6	.164	.254
		·········· Critical value
7	.070	.090
8	.018	.020
9	.002	.002

(For successes 0, 1, 2: "add down"; for successes 7, 8, 9: "add up")

Step 4. The test statistic is the number of plus signs. When the null hypothesis is true, it follows the binomial distribution. In this illustration, the number of plus signs is 5 (from Table 17–3).

Step 5. Because this is a two-tailed test with a 20% level of significance, each tail of the sampling distribution should contain a probability of 10%. We can readily see the probability of 0.090 that there will be 2 or fewer plus signs and, by symmetry, the probability of 0.090 that there will be 7 or more plus signs. Therefore, the decision rule for a significance level of approximately 20% is to reject the null hypothesis if there are 2 or fewer or 7 or more successes (plus signs). Otherwise, we do not reject the null hypothesis. This decision rule is illustrated in Figure 17–2. The probabilities are based on Table 17–4.

Referring to Table 17–3, note again that there are five plus signs. This number of plus signs does not fall in the rejection region, so H_0 is not rejected. We conclude therefore that the wilderness experience did *not* change the degree of mutual trust between the partners. Essentially, we are saying that five successes could reasonably be attributed to chance.

REAL STAT

There is a simple, but time-consuming, way to convince a group of people that you are a genius. Start with a list of, say, 512, people. Contact half of them, and tell them it is going to rain next Thursday. Tell the other half it is not going to rain. After next Thursday, there will be 256 people for whom you correctly predicted the weather. Now contact half of this smaller group, and tell them it is going to rain on the following Monday; tell the other half it is not going to rain on the following Monday. When Monday arrives, there will be 128 people for whom you have now made 2 correct predictions. Repeat this process, contacting only those for whom you provided "correct" predictions. After 5 predictions, there will be 16 people for whom you have been right 5 times in a row. From their point of view, you will have done something truly unusual!

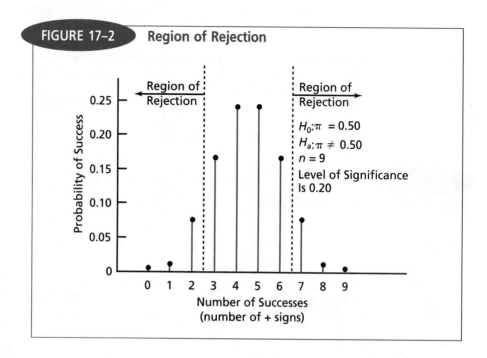

FIGURE 17–2 Region of Rejection

- -

Self-Review 17–2 Two groups, Calorie Counters and Fitness Fanatics, take different approaches to weight loss. Our interest is in determining if there is a difference in the amount of weight loss experienced by the members of the 2 groups. A sample of 15 members from each of the 2 groups is selected and matched on the basis of age, current weight, and several other factors. The pounds of weight lost by each person is recorded. Based on this information, shown below, would you conclude that the 2 weight-loss methods differ in their success rates? Use the 0.05 significance level.

a. What is the null hypothesis? The alternate hypothesis?
b. What is the sample size?
c. What is the decision rule?
d. What is your decision?

Pair	Calorie Counters	Fitness Fanatics	Sign*
A	12	4	+
B	8	17	−
C	26	5	+
D	19	6	+
E	7	12	−
F	2	24	−
G	15	18	−

H	23	4	+
I	5	15	−
J	8	21	−
K	3	25	−
L	18	14	+
M	4	15	−
N	17	8	+
O	5	21	−

*+ if Calorie Counters loses more weight, − in all other cases.

. . . .

Exercises

Answers to the even-numbered Exercises are at the back of the book.

1. A random sample of 8 college students and their parents of the same sex were administered a questionnaire that yields a score on their attitudes toward premarital sex. Of the parents, 7 were classified as more conservative than their offspring and 1 as less conservative. Use the sign test to determine if there is sufficient evidence at the 10% level of significance to claim that the attitudes of the 2 groups are significantly different.

2. The salaries paid 6 women and 6 men in similar positions are paired as shown. Use the sign test to see if there is a significant difference in salaries between the males and females at the 5% level of significance.

Pair	Males	Females
1	$33,963	$28,673
2	37,271	28,697
3	26,130	18,180
4	33,016	38,265
5	60,718	57,653
6	17,957	13,214

3. A course was offered to help students improve their reading comprehension. A sample of 12 students was tested before and after this course, and each student's comprehension percentage was recorded:

Student	Before	After
Tom	62	65
Alice	70	68
Greg	73	70
Alan	59	71
Cindy	67	68
Jackie	79	73
Scott	46	52
Janet	80	76
Chris	59	64
Dan	67	72
Sara	69	66
Norm	75	80

Use the sign test and the 0.05 significance level to determine if students' comprehension increased after they took the course.

4. The numbers of emergency patients at 2 emergency care facilities in the same city are to be compared. The numbers of cases at each facility are obtained for the first 15 days of November. At the 0.05 significance level, can it be concluded that St. Vincent handles more emergency patients than St. Rose?

Date	St. Rose	St. Vincent
1	15	21
2	18	18
3	19	22

Date	St. Rose	St. Vincent
4	16	20
5	19	21
6	24	19
7	16	20
8	19	15
9	18	30
10	16	32
11	12	17
12	18	21
13	23	25
14	17	20
15	32	22

Date	East Coast	West Coast
1	4	15
2	8	6
3	3	7
4	0	7
5	2	5
6	6	1
7	8	9
8	5	26
9	3	6
10	1	6
11	5	8
12	3	6
13	5	9
14	9	0

5. A new type of radar is installed at air traffic control towers on the West Coast, and there is concern that controllers may not be as accurate as they were on the East Coast. The number of "near misses" is recorded each day during the first two weeks of October. They are

Use the sign test and the 0.05 level of significance to test if there were significantly more "near misses" on the West Coast than on the East Coast.

THE SIGN TEST WITH LARGE SAMPLES

As discussed in Chapter 7, the binomial probability distribution closely approximates the normal probability distribution when both $n\pi$ and $n(1 - \pi)$ exceed 5. The mean and standard deviation of the binomial distribution (see formulas 6–4 and 6–5) are computed as follows:

Mean: $\mu = n\pi$

Standard deviation: $\sigma = \sqrt{n\pi(1 - \pi)}$

The test statistic for the number of plus signs or minus signs (X) is given by the formula

$$z = \frac{X - n\pi}{\sqrt{n\pi(1 - \pi)}}$$

17–1

where

X is the number of plus signs or minus signs.
π is the probability of a plus sign, which is 0.50 if the null hypothesis is true.
n is the number of paired observations.
z is the test statistic.

Because $\pi = 0.50$, these formulas for the mean and standard deviation reduce to $\mu = 0.50n$ and $\sigma = 0.50\sqrt{n}$. To meet the requirement that $n\pi$ is at least 5, n must be at least 10. Hence, the large sample methods can be employed when the sample size is at least 10. Another consideration remains— the continuity correction factor (described in Chapter 7). Recall that a continuity correction factor of 0.50 is used when a continuous distribution approximates a discrete distribution. In this case, the binomial is approximated by the normal. The test statistic simplifies to

$$z = \frac{(X \pm 0.50) - 0.50n}{0.50\sqrt{n}} \qquad \boxed{17\text{-}2}$$

where X is the number of plus signs or minus signs, whichever is being tested or counted. $X + 0.50$ is used if X is less than $n/2$, and $X - 0.50$ is used if X is more than $n/2$. The following illustrative example will help to clarify these points further.

Problem   Recall the Chapter Problem about the self-confidence level of widows before and after a training program (see Table 17–1). Did the training program increase the widows' self-confidence? As before, use the 0.10 significance level.

Solution   We can simplify the calculations if we use the normal approximation to the binomial. Since $n\pi = 12(0.5) = 6.0$ and $n(1 - \pi) = 12(0.5) = 6.0$, the qualifications for using the normal approximation to the binomial are met. The null and alternate hypotheses are as before:

$$H_0 : \pi \leq 0.50$$

$$H_a : \pi > 0.50$$

We determine the decision rule by using the normal probability distribution table (Appendix C), with 0.10 in the right tail of the curve. The decision rule is to reject H_0 if z is greater than 1.28; otherwise, we do not reject H_0. The decision rule is pictured in the following graph.

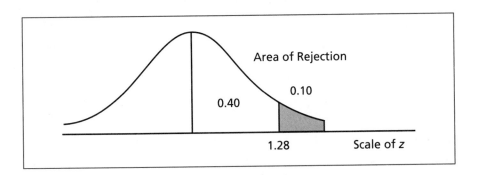

Reviewing the components in this problem, we find

$X = 10$ the number of widows exhibiting increased self-confidence after the training program.

$n = 12$ the size of the sample.

Substituting these values into formula 17–2,

$$z = \frac{(X \pm 0.50) - 0.50n}{0.50\sqrt{n}}$$

we get

$$z = \frac{(10 - 0.50) - 0.50(12)}{0.50\sqrt{12}}$$

$$= 2.02$$

Note that 0.50 is subtracted from 10 because 10, the number of "successes," exceeds half the sample size. The steps are shown in the following diagram:

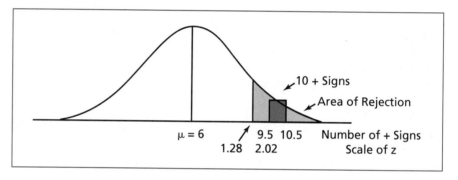

The computed value of 2.02 is in the rejection region beyond 1.28. Therefore, we reject H_0 and accept H_a. Again we conclude that the training program does enhance the widows' self-confidence. In this case, the p-value is 0.0217, found by $0.5000 - 0.4783$.

Self-Review 17–3

Refer to Self-Review 17–2, in which 15 members from each of 2 health clubs were matched on the basis of similar characteristics. The percentage of weight loss for each person was determined. Six plus signs were noted. Use the sign test, the normal approximation to the binomial, and the 0.05 significance level to determine if there is a difference in the effectiveness of the 2 weight-control programs.

Exercises

6. A gerontologist examined 20 elderly couples and rated both the husband and the wife on a scale designed to measure various degrees of opti-mism. On this scale, 0 represents absolutely no optimism and 100 extreme optimism. The fol-lowing data show the paired ratings. Use the sign test to determine if there is a significant difference between husbands and wives with re-spect to optimism. Use the 0.05 significance level.

Husband	70	85	73	75	65	50	80	71	80	51	72	76	79	65	59	72	84	90	56	57
Wife	65	41	45	80	84	50	71	52	42	78	62	38	80	65	54	67	87	90	38	43
Difference	5	44	28	−5	−19	0	9	19	38	−27	10	38	−1	0	5	5	−3	0	18	14

7. Refer to Exercise 1, but assume the sample size to be 28, with 20 of the parents classified as more conservative than their offspring and 8 as less conservative. Use the 0.05 significance level, and apply the sign test. Can the normal approximation be used? If so, solve the problem using the normal approximation.

8. Refer to Exercise 3. Use the normal approximation to the sign test to solve the problem.
9. Refer to Exercise 4. Use the normal approximation to the sign test to solve the problem.
10. Refer to Exercise 5. Use the normal approximation to the sign test to solve the problem.

THE WILCOXON SIGNED-RANK TEST

The sign test, discussed in the previous section, considers only the sign of the difference between two paired observations. This appears to waste information because it ignores the value of the difference between the two paired observations. The **Wilcoxon signed-rank test,** developed by Frank Wilcoxon in 1945, considers not only the sign of the difference, but also the magnitude (amount) of the difference between two paired observations. Thus, it is considered more efficient than the sign test. The Wilcoxon signed-rank test is applied only to paired interval-level data. It is a replacement for the paired *t* test discussed in Chapter 11. If you suspect that the distribution of differences between the paired observations is not normal, then you should avoid the paired *t* test and use the Wilcoxon signed-rank test in its place.

To apply the Wilcoxon signed-rank test follow these steps:

1. Compute the difference between each pair of observations. Whenever the difference between the paired values is zero, discard that pair of observations.
2. Rank these differences from lowest to highest, without regard to their signs. That is, rank the smallest absolute difference a 1, the next largest absolute difference a 2, and so on. If ties occur, average the ranks involved, and award each tied observation the average value.
3. Put the ranks with positive differences in one column and those with negative differences in another.
4. Determine the sums of the positive (R^+) and negative (R^-) ranks.
5. Compare the smaller of the two sums (R^+ or R^-) with the critical values found in Appendix G. Based on this comparison, make your decision regarding the null hypothesis.

Problem   Consider again the earlier Problem/Solution dealing with the wilderness experience of paired police officers. The before-and-after ratings are repeated in Table 17–5. The numerical differences between the paired ratings,

TABLE 17–5	Degrees of Trust Before and After a Wilderness Experience						
	Rating						
Pair	Before Experience	After Experience	Difference	Absolute Value	Rank	R^+	R^-
Braid-Ahern	68	32	−36	36	9		9
Chadwick-Gibbon	55	55	0				
Falloff-Konrath	49	55	+6	6	1	1	
Unman-Bernstein	40	75	+35	35	8	8	
Chamberlain-Sultan	20	50	+30	30	7	7	
Lively-Julia	18	25	+7	7	2.5	2.5	
Lucite-Austin	30	23	−7	7	2.5		2.5
Mangos-Chasm	70	49	−21	21	6		6
Moharter-Klieg	52	62	+10	10	5	5	
O'Neill-Burner	50	41	−9	9	4		4
						23.5	21.5

the ranks, and the ranks' signs have been added. Use the Wilcoxon signed-rank test to determine if there is a difference in the degrees of trust before and after the police officers' wilderness experience.

Solution   The null and alternate hypotheses are

H_0: The trust ratings are the same after the wilderness experience as before it.

H_a: The trust ratings are not the same after the wilderness experience as before it.

If the null hypothesis is true, then it is to be expected that the total of the positive ranks is about equal to the total of the negative ranks. A large difference in the two sums is an indication that the null hypothesis is false. The Wilcoxon test serves as a basis for determining how large a difference between the positive and negative sums can reasonably be attributed to chance.

For a two-tailed alternative hypothesis, the smaller of R^+ and R^- is compared with the critical value. Appendix G, a portion of which is repeated at the top of the following page, is used to formulate the decision rule. Suppose the 0.05 significance level is to be used; to locate the critical value find the column headed by 0.05. Next, move down that column to the row corresponding to the sample size. In this case, find the two-tailed probability 0.05 and move down to $n = 9$. (Recall that we started out with 10 pairs, but because 1 pair—Chadwick and Gibbon—showed no difference, that pair was omitted, leaving a sample of only 9.) The critical value from Appendix G is 5. The decision rule is to reject H_0 if the smaller of R^+ and R^- is less than or equal to 5.

One-Tailed Value	0.10	0.05	0.025	0.01	0.005
Two-Tailed Value	0.20	0.10	0.05	0.02	0.01
n					
5	2	0	—	—	—
6	3	2	0	—	—
7	5	3	2	0	—
8	8	5	3	1	0
9	10	8	5	3	1
10	14	10	8	5	3
11	17	13	10	7	5
12	21	17	13	9	7

Referring to Table 17–5, the sum of R^+ is 23.5, and the sum of R^- is 21.5. The smaller of the two (21.5) is greater than 5; hence, the null hypothesis cannot be rejected. Thus, we conclude that the trust ratings are the same after the wilderness experience as before it.

The following Problem/Solution illustrates a case where a one-tailed test is appropriate.

Problem   A study is to be conducted of the differences in ages between husbands and wives at their first marriages. The question is: Are husbands older than their wives?

Solution   Use the five-step hypothesis-testing procedure.

Step 1. The null hypothesis is that there is no significant difference between the two ages. It is suspected that in first marriages, the husbands are older than their wives; this is the *alternate hypothesis.* Because we are interested in determining if the husbands are *older* than their wives (only one direction), a one-tailed test is indicated.

Step 2. A 10% level of significance has been chosen.

Step 3. The Wilcoxon signed-rank test is an appropriate test to determine if there is a significant difference between two paired observations.

Step 4. Eight married couples are randomly selected for inclusion in the study. We formulate the decision rule by referring to Appendix G. Go down the n column until you locate the number in the sample (8). Next, find the critical one-tail value in the column headed 0.10; it is 8. Hence, the decision rule is as follows: If the sum of the negative ranks is 8 or less, reject the null hypothesis, and accept the alternate hypothesis. Otherwise, do not reject H_0.

Step 5. The ages of both husbands and wives are listed in Table 17–6. The age differences are shown in column 4 and are *ranked* in column 5, where we

TABLE 17–6	Comparison of Husbands' and Wives' Ages					
(1) **Couple**	**(2)** **Husband's Age**	**(3)** **Wife's Age**	**(4)** **Difference**	**(5)** **Rank**	**(6)** **R^+**	**(7)** **R^-**
A	24	21	3	6	6	
B	25	23	2	4	4	
C	27	19	8	8	8	
D	22	24	−2	4		4
E	20	18	2	4	4	
F	30	25	5	7	7	
G	18	17	1	1.5	1.5	
H	21	22	−1	1.5		1.5
					30.5	5.5

disregarded the sign of the difference. There is a tie for first position: Couple G had a difference of 1, couple H, − 1 (see column 4). When the differences were ranked, we resolved this tie by giving each of the two couples a rank of 1.5, which we found by adding ranks 1 and 2 and then dividing by 2. There is also a 3-way tie for third place (couples B, D, and E). We resolved this tie by adding ranks 3, 4, and 5 and dividing by 3. Each of these 3 couples was assigned a rank of 4.

Note that the sum of the negative ranks is 5.5, less than the critical value of 8. Based on the decision rule in Step 4, the null hypothesis is rejected. It is reasoned that such a large difference between the sum of the positive signed ranks (30.5) and the sum of the negative signed ranks (5.5) *could not be due to chance alone.* The alternate hypothesis—that in first marriages, husbands are older than their wives—is accepted. We conclude that in first marriages, a preponderance of husbands are older than their wives.

Self-Review 17–4

The burglary rates for 1995 and last year for a sample of 10 major cities are shown in the following tabulation:

1995	Last Year
10.1	20.4
10.6	22.1
8.2	10.2
4.9	9.8
11.5	13.7
17.3	24.7
15.4	12.4
11.1	12.7
8.6	13.3
10.0	18.4

Is there sufficient evidence, at the 0.05 level of significance, to show that the rates have increased between 1995 and last year?

a. State the null and alternative hypotheses.
b. State the decision rule.
c. State your conclusion and interpret.

· ·

Exercises

11. Ten sets of identical twins were separated at the age of 6. One twin in each set was educated at a private school, while the other twin attended a public school. At the end of 8 years, each of the twins' academic achievement was measured. Is there sufficient evidence, at the 0.05 level of significance, to claim that private schools produce a higher level of educational achievement?

Twin Pair	Public School Education	Private School Education
A	12	19
B	14	13
C	8	6
D	11	24
E	14	12
F	12	16
G	13	10
H	15	18
I	18	24
J	17	22

a. State the null and alternate hypotheses.
b. Using the sign test, state the decision rule.
c. What conclusion would you reach using the sign test?
d. If the Wilcoxon signed-rank test were used, what would the decision rule be?

e. What conclusion would be reached from the Wilcoxon signed-rank test?

12. Apply the Wilcoxon signed-rank test to the data on weight reduction presented in Self-Review 17–2. Use the 0.05 significance level.

13. A random sample of seven young professionals and their parents was asked the size of their homes (expressed in square feet of usable space).

City	Young Professional	Parent
Atlanta	1725	1275
Baltimore	1310	1130
Chicago	1647	1476
Dallas/Ft. Worth	1672	1762
New York	1520	1250
San Francisco	1886	1688
Washington, D.C.	1672	1762

Use the Wilcoxon signed-rank test to determine if there is statistical evidence at the .10 significance level to claim that "yuppies" live in larger homes than do their parents.

14. Refer to Exercise 4. Use the signed-rank test to determine if St. Vincent Hospital handles more emergency patients than St. Rose.

 ·

THE WILCOXON RANK-SUM TEST

One test specifically designed to determine whether two independent samples come from the same population is the **Wilcoxon rank-sum test.** It is an alternative to the two-sample t test described in Chapter 11, which actually compares only two means. Recall that the t test required that the two populations be normally distributed and have equal variances. These conditions need not be met in order for the Wilcoxon ranksum test to apply.

The test is based on the sum of the ranks. The data from the two independent samples are ranked as if they were part of a single sample. If the null hypothesis is true, then the ranks will be nearly evenly distributed between the two samples; that is, the low, medium, and high ranks should be about equally divided between the two samples. If the alternate hypothesis is true, one of the samples will have more of the lower ranks and thus a smaller total. The other sample will have the higher ranks and a larger total. If both samples are eight or larger, then the normal distribution may be used as the test statistic. The formula is

$$z = \frac{W - \dfrac{n_1(n_1 + n_2 + 1)}{2}}{\sqrt{\dfrac{n_1 n_2(n_1 + n_2 + 1)}{12}}} \qquad \boxed{\textbf{17-3}}$$

where

n_1 is the number of observations in the first sample.
n_2 is the number of observations in the second sample.
W is the sum of the ranks in the first population.
z is the standard normal deviate.

Problem   The Phoenix Health Maintenance Organization (PHMO) wants to compare the lengths of stay for men and women who have joint replacements. Many physicians claim the hospital stay after the surgery is longer for men than women. To investigate, a PHMO staff analyst selects a random sample of 17 patients who had joint replacements in the last month. The patients are classified by gender and length of stay. The data are reported in Table 17–7. At the 0.05 significance level, can we say men are confined longer?

Solution   If we could assume the population distributions of the lengths of confinement for men and women were normally distributed and that the variances of these two populations were the same, then the *t* test for independent samples discussed in Chapter 11 on page 344 would be appropriate. However, the distribution of the lengths of stay for men is positively skewed. There are two males whose hospital stay is quite long, so the assumption of a normal population—and, for that matter, equal variances—is not believable. Hence, the nonparametric Wilcoxon rank-sum test is appropriate.

Our strategy is to replace the actual data with their ranks. If the lengths of stay are the same for men and women, then we expect the sum of the *ranks* to be about the same. If the lengths of stay are not the same, then we expect the sum of the ranks for the men to be more than that for the women.

The Problem suggests the men stay longer, so we use a one-tailed test. We state the null and alternate hypotheses is terms of the probability distribution.

TABLE 17–7	Length of Hospital Confinement (in days) for a Sample of Men and Women

Men	Women
5	4
7	7
2	3
10	1
3	9
12	2
19	10
14	14
24	

To put it another way, we assume the distributions of the lengths of stay for the men and the women have the same shape, but not the same location on the horizontal axis. See the graph below.

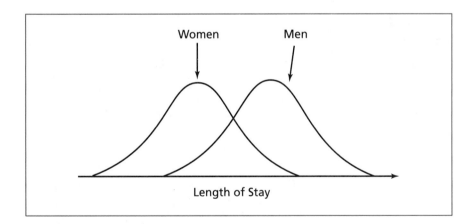

The null and alternate hypotheses are

H_0: The distributions of lengths of stay are the same for men and women.

H_a: The distribution of lengths of stay is larger for men than for women.

The test statistic is the standard normal distribution z. At the 0.05 significance level, the decision rule is to reject the null hypothesis if z is larger than 1.645.

The alternate hypothesis is that the lengths of confinement are longer for men, which means the distribution for men is to the right of that of women. The lengths of stay for both men and women are ranked as if they were a single group. We begin by observing that the fourth woman had the shortest stay (1 day). She is thus given the rank of 1. The third man and the sixth woman each had a stay of 2 days, so they are tied. As usual, we resolve the tie by averaging the ranks. These patients are each ranked 2.5. Other ties are resolved in the same manner. The man whose confinement is 24 days receives the highest rank (17). The final rankings are given in Table 17–8. The sum of the ranks is 94.5 for the 9 men and 58.5 for the 8 women. We represent the sum of the men as W. There are 9 men and 8 women in the study, so $n_1 = 9$ and $n_2 = 8$. Computing z from formula 17–3,

$$z = \frac{W - \dfrac{n_1(n_1 + n_2 + 1)}{2}}{\sqrt{\dfrac{n_1 n_2(n_1 + n_2 + 1)}{12}}} = \frac{94.5 - \dfrac{9(9 + 8 + 1)}{2}}{\sqrt{\dfrac{9(8)(9 + 8 + 1)}{12}}} = 1.30$$

Because 1.30 is less than the critical value of 1.65, the null hypothesis is not rejected. This evidence does not prove there is a difference in the distributions of the lengths of stay. To put it another way, we have not shown there is a difference in the distributions of the lengths of the hospital stays for men and women having joint replacements. The p-value is the probability of finding a z-value greater than 1.30. From Appendix C, it is 0.0968, found by $0.5000 - 0.4032$.

TABLE 17–8	Ranked Lengths of Hospital Confinement		
Men		**Women**	
Length	Rank	Length	Rank
5	7.0	4	6.0
7	8.5	7	8.5
2	2.5	3	4.5
10	11.5	1	1.0
3	4.5	9	10.0
12	13.0	2	2.5
19	16.0	10	11.5
14	14.5	14	14.5
24	17.0		
	94.5		58.5

When using the Wilcoxon rank-sum test, you may name either population as number 1. However, once you make a choice, W is the sum of ranks for that population. In the previous example, suppose we called the population of women number 1. Then $n_1 = 8$, $n_2 = 9$, and we may change the wording in our statement of the null and alternate hypotheses.

H_0: The distributions of the lengths of stay are the same for men and women.

H_a: The distribution of the lengths of stay is smaller for women than for men.

The decision rule is to reject the null hypothesis if the computed z is less than -1.65. The computed z (with the opposite sign), the decision regarding the null hypothesis, and the p-value are all the same as before.

$$z = \frac{W - \dfrac{n_1(n_1 + n_2 + 1)}{2}}{\sqrt{\dfrac{n_1 n_2(n_1 + n_2 + 1)}{12}}} = \frac{58.5 - \dfrac{8(8 + 9 + 1)}{2}}{\sqrt{\dfrac{8(9)(8 + 9 + 1)}{12}}} = -1.30$$

Self-Review 17–5

The research director for a golf ball manufacturer wants to know if there is any difference in the distance two of the company's golf balls travel. Eight of their Dino brand and eight of their Maxi brand were hit by an automatic driving machine. The results (distances in yards traveled) were as follows:

Dino: 252, 263, 279, 273, 271, 265, 257, 280
Maxi: 262, 242, 256, 260, 258, 243, 239, 265

Using the 0.05 level and the Wilcoxon rank-sum test,

a. state the null and alternate hypotheses.
b. state the decision rule.
c. compute z.
d. state and interpret your decision.

Exercises

15. A study dealing with the social mobility of women gathered data on 2 groups of women, 1 group composed of women who had never married and 1 group consisting of married women. Each group was rated on an upward-mobility scale from 0 to 100, with a higher score denoting greater mobility. At the 0.10 level of significance, is there a difference between these 2 groups with respect to social mobility?

Never married: 42, 49, 53, 54, 55, 56, 58, 60, 65, 68, 70, 75
Married: 30, 35, 42, 45, 55, 60, 60, 70

16. Samples of 8 socially active students and 10 socially nonactive students were randomly selected to determine if there was a difference between the 2 groups in a measure of social prestige. Each student was evaluated on a 20-point scale on which a low score denotes lack of prestige. The findings are recorded below. Test the null hypothesis of no difference between the 2 populations using a 0.10 level of significance.

Active students: 11, 13, 14, 15, 8, 10, 5, 6
Nonactive students: 8, 9, 10, 16, 17, 1, 7, 2, 5, 4

17. A supervisor in the Hamtramck, Michigan, auto plant examined the quality of the work of several male and female employees, with the following results (higher scores represent better quality):

Males: 68, 72, 78, 83, 84, 85, 86, 90, 96, 99
Females: 71, 74, 79, 80, 82, 85, 86, 88

Is there any difference between these two groups? Use the 0.10 level of significance.

18. The time until a light bulb failed for each of 2 brands is obtained (in hours). At the 0.05 significance level, can we conclude there is a difference in the 2 bulbs?

Longlast: 372, 283, 712, 849, 623, 382, 427, 821

Everlast: 645, 682, 913, 742, 691, 689, 842, 751

19. *Consumer Reports* magazine wants to compare the fuel efficiency of 2 similar makes of golf carts. Each is driven on a random basis over a 17-day period, and the mileages are recorded:

Day	Make	Miles per Gallon
1	Go-Go	24.1
2	Go-Go	25.0
3	Putt-Putt	24.8
4	Putt-Putt	24.3
5	Go-Go	24.2
6	Putt-Putt	25.3
7	Putt-Putt	24.2
8	Putt-Putt	23.6
9	Go-Go	24.5
10	Go-Go	24.4
11	Go-Go	24.5
12	Putt-Putt	23.2
13	Putt-Putt	24.0
14	Putt-Putt	23.8
15	Go-Go	23.8
16	Go-Go	23.0
17	Go-Go	22.9

Using the 0.05 level of significance and the Wilcoxon rank-sum test, can we conclude that there is a difference in the two makes?

20. Heights (in inches) of children of comparable ages from developed and underdeveloped countries are shown below. Is there a significant difference in the 2 distributions? Use the rank-sum test with a significance level of 0.05.

Developed: 30.2, 34.6, 37.8, 40.8, 43.4, 45.9, 39.2, 40.4, 40.0

Undeveloped: 29.4, 33.8, 37.5, 40.7, 43.4, 31.2, 32.1, 33.4

ANALYSIS OF VARIANCE BY RANKS: THE KRUSKAL–WALLIS TEST

As noted at the beginning of Chapter 12, if we want to apply the ANOVA method to test if the means of several populations are equal, it is assumed that (1) the populations are normally distributed, (2) the standard deviations of the populations are equal, and (3) the populations are independent. If these assumptions cannot be met—or if the level of measurement is of ordinal scale (ranked)—then we can use an analysis of variance technique developed in 1952 by W. H. Kruskal and W. A. Wallis. It is referred to as the **Kruskal-Wallis one-way analysis of variance by ranks.**

The procedure for the Kruskal-Wallis test is to substitute the *rankings* of the items in each sample for the actual values. Also, instead of working with the means of each treatment—as in ANOVA—we use the sum of the ranks for each treatment. The test statistic, designated as *H,* is

$$H = \frac{12}{n(n + 1)}\left(\frac{S_1^2}{n_1} + \frac{S_2^2}{n_2} + \cdots + \frac{S_k^2}{n_k}\right) - 3(n + 1)$$

17-4

where

S_1 is the sum of the ranks for the sample designated 1, S_2 is the sum of the ranks for the sample designated 2, and so on.

n_1 is the number in the sample designated 1, n_2 is the number in the sample designated 2, and so on.

n is the combined number of observations for all samples.

k is the number of populations.

In practice, this distribution is approximated by the chi-square distribution with $k - 1$ degrees of freedom, where k is the number of populations. This approximation is accurate whenever each sample consists of at least 5 observations. Should any one of the samples include fewer than 5 observations, a special table of critical values for the analysis of variance by ranks (Kruskal-Wallis test) will have to be consulted. Many advanced texts contain such tables; for our purposes, an analysis of samples greater than or equal to 5 will suffice.

Problem   Five persons from each of 3 different ethnic groups were randomly selected and their perceptions of their own social mobility recorded. The groups are identified as Black, Asian, and Hispanic. Each person's response was recorded on a scale on which 100 indicates extreme mobility and 0 indicates no mobility. The actual responses by group were

Black: 73, 51, 55, 64, 71
Asian: 63, 74, 53, 92, 84
Hispanic: 81, 93, 84, 91, 84

TABLE 17–9   Rankings of Perceived Mobility

Black		Asian		Hispanic	
Score	Rank	Score	Rank	Score	Rank
73	7	63	4	81	9
51	1	74	8	93	15
55	3	53	2	84	11
64	5	92	14	91	13
71	6	84	11	84	11
	$S_1 = 22$		$S_2 = 39$		$S_3 = 59$

Are there differences in the perceived mobility scores among the 3 groups? Test at the 0.05 significance level.

Solution   We begin by stating the hypotheses:

H_0: The score distributions are the same for all groups.

H_a: At least one distribution is different.

Then we rank the responses without regard to ethnic background and determine the sum of each of the three ranks (see Table 17–9). Note that there was a tie for the 11th rank—3 responses of 84. To break the tie and still have the entire sample ordered from 1 through 15, we assigned the averaged rank $[(10 + 11 + 12) \div 3 = 11]$ to each of the 3 tied values.

The sums of the ranks for the samples are $S_1 = 22$, $S_2 = 39$, and $S_3 = 59$. These values are inserted into formula $17 - 4$, and the value of H is computed:

$$
\begin{aligned}
H &= \frac{12}{n(n + 1)}\left(\frac{S_1^2}{n_1} + \frac{S_2^2}{n_2} + \frac{S_3^2}{n_3}\right) - 3(n + 1) \\
&= \frac{12}{15(15 + 1)}\left(\frac{(22)^2}{5} + \frac{(39)^2}{5} + \frac{(59)^2}{5}\right) - 3(15 + 1) \\
&= 54.86 - 48 \\
&= 6.86
\end{aligned}
$$

The critical value of H is located in Appendix F. Recall that there are $k - 1$ degrees of freedom, where k is the number of populations. There are 3 populations (Black, Asian, and Hispanic), so $k - 1 = 3 - 1 = 2$ degrees of freedom. The critical value for 2 degrees of freedom at the 0.05 significance level is 5.991. The decision rule is reject H_0 if $H > 5.991$. $H = 6.86$, so H_0 is rejected. We conclude that at least 1 population perceives its social mobility in a manner significantly different from that of the others.

The Kruskal-Wallis test is available in the MINITAB software package. The 15 mobility scores from the previous problem are entered in the first column and labeled "Score." The second column, labeled "Group," identifies the minority group. See the screen at the top of the next page.

The Blacks are identified as Level 1, the Asians as Level 2, and the Hispanics as Level 3. The mouse commands are

Stat ▶ Nonparametric ▶ Kruskal-Wallis

Within the dialog box (shown at the bottom of the next page), select Score as the response variable and Group as the factor; then click **OK**.

The output appears at the top of page 546.

Notice that the computed value of H is 6.86, the same as computed earlier. MINITAB also reports the p-value, which is 0.032. The probability of finding a value of H this large or larger when the null hypothesis is true is 0.032.

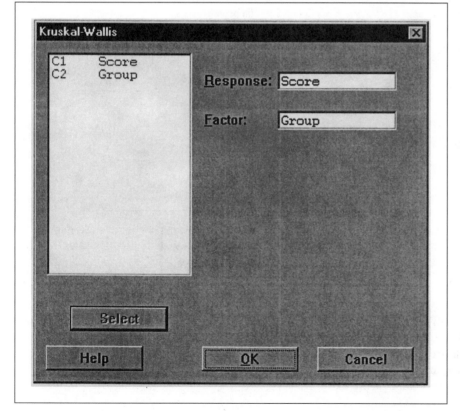

↓	Score	Group
1	73	1
2	51	1
3	55	1
4	64	1
5	71	1
6	63	2
7	74	2
8	53	2
9	92	2
10	84	2
11	81	3
12	93	3
13	84	3
14	91	3
15	84	3

Kruskal-Wallis

C1 Score
C2 Group

Response: Score

Factor: Group

Select

Help   OK   Cancel

REAL STAT

A quality control inspector at MJP Electronics was told to select 3 items out of a box of 30, test them, and accept the box if all 3 were satisfactory. He grabbed 3 items off the top, found them to be acceptable, and sent the box on. Much to his dismay, the person at the first workstation found that the box contained 25 defective items. How could the quality control inspector and the employee at the first workstation reach such different conclusions? Is it really possible that 25 out of 30 items in the box were defective? If so, how could this happen? Perhaps someone knew he was going to start at the top so they hid the defects on the bottom!

```
MTB > Kruskal-Wallis 'Score' 'Group'.

LEVEL      NOBS      MEDIAN   AVE. RANK   Z VALUE
  1          5       64.00       4.4       -2.20
  2          5       74.00       7.8       -0.12
  3          5       84.00      11.8        2.33
OVERALL     15                   8.0

H = 6.86  d.f. = 2  p = 0.033
H = 6.91  d.f. = 2  p = 0.032 (adjusted for ties)
```

Self-Review 17–6

In a study of the effect of packaging on sales of fresh produce, 24 supermarkets in Cambridge were each assigned randomly to 1 of 3 groups. The first group of stores sells its produce in bulk; customers select their own produce and place it in paper bags. The second group of stores places produce in mesh bags, visible to the customer. The third group's produce is prepackaged in kraft bags, through which the customer cannot see it. The amount of produce (in pounds) sold in 1 week by each group is shown in the table that follows:

Produce Sold Each Week (in pounds)		
In Kraft Bags	In Mesh Bags	In Paper Bags
576	464	272
640	480	356
752	720	224
784	756	335
368	208	279
596	512	304
608	288	192
448	496	
	672	

Use the Kruskal-Wallis test and the 0.05 significance level to determine if the different packaging methods result in the same weekly produce sales.

Exercises

21. The Orange County Sheriff's Department is planning to purchase a large number of patrol cars. Four different models are being considered. Samples of all 4 makes were driven 10,000 miles each, and separate records were kept of the expenditures for repairs. At the 0.05 significance level, are there differences in the amounts spent on repairs for the 4 models?

Model A	Model B	Model C	Model D
$220	$120	$180	$168
290	250	215	247
256	173	231	241
273	235	200	210
265	179	190	276
	249	258	193
	236		273
	147		

22. An insurance company is preparing a new advertising campaign aimed at professionals and wants to know the amount of term insurance carried by lawyers, physicians, veterinarians, and college professors. The amount of term insurance (in thousands of dollars) carried by a sample taken from each of these 4 categories is shown in the following table. Are there differences in the amounts of insurance carried by the 4 groups? Use the 0.01 significance level.

	Amount of Insurance Carried (in thousands of dollars)		
By Lawyers	By Physicians	By Veterinarians	By College Professors
$200	$120	$ 20	$ 30
120	190	105	25
75	200	90	120
65	195	70	190
140	130	200	45
135		125	
		175	

23. In an attempt to justify budget requests for more officers in some precincts, a police commissioner collected the following data on the number of crimes committed weekly in several randomly selected weeks and precincts:

Precinct	Number of Crimes
Central	73, 68, 64, 65, 80
North	42, 39, 51, 38, 39, 40
South	49, 60, 52, 51, 54

At the 0.01 significance level, are there differences in the numbers of crimes in the 3 precincts?

24. A drug company tests 2 drugs by administering them, along with a placebo, to 3 different treatment groups. The data below are reaction times (in minutes).

Drug	Time
Drug A	21, 25, 24, 28, 32
Drug B	19, 22, 21, 25, 23
Placebo	28, 28, 26, 35, 30

Are there differences among the treatment groups at the 0.10 level of significance?

25. To test 4 types of baseball bats, a professional player with the Florida Marlins used each for 1 week in rotation during 20 weeks of play. He recorded his batting average each week with each bat:

Bat	Average				
A	.251	.262	.237	.228	.242
B	.232	.255	.282	.248	.256
C	.230	.225	.209	.216	.236
D	.259	.222	.277	.250	.238

At the 0.01 significance level, are there differences among the bats?

CHAPTER OUTLINE

I. Each of the hypothesis tests in this chapter uses ranked- or ordinal-level data and does not assume that the populations from which the samples are drawn follow any particular distribution.

II. The sign test uses the signs of the differences in a set of paired observations.

A. Since there are two equally likely outcomes, the binomial distribution determines the critical values.

B. For large samples, the normal approximation to the binomial with the correction for continuity is used.

$$z = \frac{(X \pm 0.50) - 0.50n}{0.50\sqrt{n}} \quad \boxed{17\text{-}2}$$

III. The Wilcoxon signed-rank test considers not only the sign, but also the magnitude of the differences in a set of paired observations.

A. The absolute values of the differences are ranked from smallest to highest.

B. The ranks in each of the two signed groups are totaled.

C. The smaller of the two sums is compared with a special table of critical values to make the decision.

IV. The Wilcoxon rank-sum test determines if two independent samples came from the same or equal populations.

A. The data from the two samples are ranked as if they were part of one sample.

B. The sum of the ranks from one of the populations is then inserted in the following formula, which follows the standard normal distribution:

$$z = \frac{W - \dfrac{n_1(n_1 + n_2 + 1)}{2}}{\sqrt{\dfrac{n_1 n_2 (n_1 + n_2 + 1)}{12}}} \quad \boxed{17\text{-}3}$$

V. The Kruskal-Wallis test does an analysis of variance using ranks to test if two or more samples are from identical populations.

A. The data from the samples are ranked as if they were part of a single sample.

B. Sums of the ranks by sample are used in the following formula, which follows the chi-square distribution:

$$H = \frac{12}{n(n + 1)}\left(\frac{S_1^2}{n_1} + \frac{S_2^2}{n_2} + \cdots + \frac{S_k^2}{n_k}\right) - 3(n + 1) \quad \boxed{17\text{-}4}$$

Exercises

26. The work-methods department of an electronics manufacturer has suggested that soft background music might increase productivity. A stereo system has been set up, and the productivity of a random sample of 8 workers is checked against their productivity before music was piped in. Can we conclude that the workers are more productive with the background music on? Use the 0.05 significance level and the sign test.

Worker	With Music	Without Music
Smith	34	32
Clark	34	37
Craig	41	28
Berry	34	33
Roberts	31	28
Nardelli	36	33
Palmer	29	27
Tang	39	38

27. A study to determine the effect of nursing intervention on the behavior modification of patients over a 2-month period produced the following test results. Use a sign test to determine whether nursing intervention produced a significant change in behavior, as indicated by an increased score. Use the 0.05 level of significance.

Subject	Initial Score	Final Score
1	31	32
2	40	39
3	41	42
4	118	134
5	57	73
6	12	22
7	120	132
8	44	60
9	47	29
10	53	72

28. Refer to Exercise 27. Examine the effect of nursing intervention using the normal approximation to the sign test. Use the 0.05 significance level.

29. Apply the Wilcoxon signed-rank test to the data in Exercise 27 regarding the effect of nursing intervention. Use the 0.05 significance level.

30. In an attempt to judge the effectiveness of an antismoking campaign, researchers exposed 12 smokers to a series of presentations. The numbers of cigarettes the subjects smoked the week before the presentations and the week after the presentations were obtained as follows:

	Subject											
	1	2	3	4	5	6	7	8	9	10	11	12
Before	23	23	21	24	21	27	22	19	19	23	26	27
After	23	12	21	15	19	15	17	21	18	14	18	19

Use the sign test to determine whether the antismoking campaign was effective at the 0.05 level of significance.

31. Refer to Exercise 30. Examine the antismoking campaign using the normal approximation to the sign test. Use the 0.05 significance level.

32. Apply the Wilcoxon signed-rank test to the data in Exercise 30 regarding the antismoking campaign. Use the 0.05 significance level.

33. Two major league baseball fans were comparing the relative strengths of the American and the National Leagues. One of the comparisons was the number of home runs hit by the home run leader in each league. Listed below are the results for the period from 1980 through 1996. At the 0.05 significance level, use the sign test to determine if there is a difference between the 2 leagues.

| Year | National League | | | American League | |
	Player	HR		Player	HR
1980	Mike Schmidt	48		Reggie Jackson, Ben Oglivie	41
1981	Mike Schmidt	31		4 players tied	22
1982	Dave Kingman	37		Gorman Thomas	39
1983	Mike Schmidt	40		Jim Rice	39
1984	Mike Schmidt, Dale Murphy	36		Tony Armas	43
1985	Dale Murphy	37		Darrell Evans	40
1986	Mike Schmidt	37		Jesse Barfield	40
1987	Andre Dawson	49		Mark McGwire	49
1988	Darryl Strawberry	39		Jose Canseco	42
1989	Kevin Mitchell	47		Fred McGriff	36
1990	Ryne Sandberg	40		Cecil Fielder	51
1991	Howard Johnson	38		Cecil Fielder, Jose Canseco	44
1992	Fred McGriff	35		Juan Gonzalez	43
1993	Barry Bonds	46		Juan Gonzalez	46
1994	Matt Williams	43		Ken Griffey, Jr.	40
1995	Dante Bichette	40		Albert Belle	50
1996	Andres Gallarraga	47		Mark McGwire	52

34. The numbers of traffic citations given by 2 police officers are to be compared. The following is a 25-day sample of citations. At the 0.01 level of significance, can it be concluded that Officer A gives more citations than Officer B? Use the sign test.

Date	A	B	Date	A	B	Date	A	B
5–1	10	7	5–11	8	1	5–23	7	9
5–2	11	5	5–12	7	2	5–24	9	3
5–3	7	4	5–15	11	8	5–25	12	8
5–4	8	8	5–16	12	10	5–26	10	7
5–5	8	9	5–17	11	7	5–29	10	8
5–8	7	3	5–18	7	2	5–30	9	7
5–9	4	2	5–19	5	6	5–31	8	4
5–10	5	8	5–22	6	5	6–1	5	2
						6–2	6	7

35. A study conducted by the National Weather Service on the change in temperature following a thunderstorm found that after 40 storms, the temperature increased on 5 occasions, stayed the same on 10 occasions, and decreased on 25 occasions. Is this sufficient evidence, at the 0.05 significance level, to conclude that there is a significant decrease in temperature after a thunderstorm?

36. Refer to Exercise 26. Use the Wilcoxon signed-rank test to determine if workers are more productive with background music.

37. Refer to Exercise 33. Use the Wilcoxon signed-rank test to determine if there is a difference in the numbers of home runs hit by the leaders in the American and National Leagues.

38. Refer to Exercise 34. Use the Wilcoxon signed-rank test to determine if Officer A gives more citations.

39. Seventeen randomly selected married men with incomes greater than $100,000 are asked whether they consider their marriages to be satisfactory. Each man is also asked to report the length of his marriage in years. The results of the study follow. At the 0.01 significance level, use the Wilcoxon rank-sum test to determine if there is a difference in years of marriage between the satisfied and the dissatisfied groups.

Length of Marriage	
Satisfied	Not Satisfied
15	8
3	5
9	9
12	12
23	10
29	13
14	4
6	7
16	

40. Two judges normally preside over the family court where all divorce hearings are held. Ms. Kerger, a young lawyer new to the area, represents the husband in an upcoming case. She gathers the following data on the monthly alimony payments (in dollars per month) recently awarded by the 2 judges. At the 0.01 significance level, use the Wilcoxon rank-sum test to determine if there is a difference in the alimony settlements awarded by the 2 judges. The samples for each group have been arranged from low to high.

Judge Cain	Judge Stevens
$ 80	$120
160	140
220	192
520	204
580	252
640	440
840	500
920	560
1200	
1360	
2000	

41. The owner of G-Mart, a large retailer, is concerned about the effectiveness of the store's advertising. G-Mart advertises 3 different special articles in 3 different media and observes sales over the next few weeks. The data are from 5 or 6 randomly selected weeks and are not necessar-

ily the same weekly sales for all media. Here are the results. Are there significant differences among the media at the 0.01 level? (Sales are in thousands of dollars.)

Radio	TV	Newspaper
$14	$46	$39
21	49	48
36	59	51
12	52	50
18	50	49
	51	

42. A study has been made of the murder rates (per 1000 population) for a sample of medium-sized cities in the Northeast, South, and Far West. Use the 0.05 significance level to determine if there are differences in the murder rates of different geographical areas. The sample for each region has been arranged from low to high.

Northeast	South	Far West
2.3	1.6	1.7
4.5	1.9	3.7
6.7	3.0	3.8
7.8	6.5	4.3
9.5	7.2	5.9
12.7	11.6	6.2
13.1	13.1	7.9
	14.5	8.4
	15.1	

43. You want to use a sign test to find out whether or not the experience of playing softball on a team made up of both boys and girls is successful in reducing sexism among boys. You test the boys for sexism both before league play begins and at the end of the season. The scores are

Boy	Score	
	Before Season	After Season
Carter	22	20
Jones	30	26

	Score	
Boy	Before Season	After Season
Ford	28	28
Yamamoto	38	30
Cosell	12	11
West	40	41
Archer	37	32
Oreon	32	29
Giles	50	40
Raggi	42	36
Nice	36	29
Cork	33	21
Mueller	36	21

Use the 0.05 significance level to test whether the softball experience has been effective in reducing sexism among boys.

44. An oil company advertises that customers will obtain higher gas mileage using "Super" lead-free gasoline than using their regular lead-free. To investigate this claim, a sample of 14 cars of various makes and models was selected. The 14 cars were driven the same distances over the same roads with both the Super lead-free and the regular lead-free gasolines. The results (in miles per gallon) follow:

Car	Super	Regular
1	21.5	20.9
2	32.5	32.9
3	29.7	30.9
4	35.4	35.7
5	26.2	26.0
6	40.4	41.7
7	25.9	25.3
8	25.1	25.0
9	24.0	23.8
10	18.9	18.5
11	23.4	23.2
12	30.2	28.5
13	26.8	26.2
14	27.6	27.0

At the 0.01 significance level, can it be concluded that the lead-free mileage is greater? Use the sign test.

45. Refer to Exercise 43. Use the Wilcoxon signed-rank test to investigate whether the softball experience reduced sexism among boys. Use the 0.05 significance level.

46. Refer to Exercise 44. Use the Wilcoxon signed-rank test to investigate whether Super lead-free gasoline yields higher mileage. Use the 0.01 significance level.

47. The amounts of money spent for lunches by working men and women are being compared. The amounts spent by a sample of 10 men and 12 women on a particular day are tabulated; the results are shown in the table that follows. Can it be concluded that men spend more than women? Use the 0.01 significance level.

Men	Women
$3.73	$3.85
4.40	4.05
4.72	4.10
3.95	4.15
5.10	3.90
3.95	3.65
4.00	3.55
4.35	3.20
4.45	3.05
3.85	3.55
	2.90
	4.62

48. A study of the amount of overpayment to welfare recipients is made in 3 regions. Are there differences in the amounts of overpayment in the 3 regions? Use the 0.05 level of significance.

West	Northwest	Midwest
$20	$24	$33
18	27	32
26	26	40
14	29	23
25	30	28

West	Northwest	Midwest
19	31	35
22	21	38
17	28	30

DATA EXERCISES

49. Refer to the real estate data set, which reports information on homes sold in Alabama during 1996.
  a. Consider only the homes with attached garages. Use a nonparametric test to determine if there is a difference in the distributions of the selling prices of the homes with fireplaces and those without. Use the 0.05 significance level.
  b. Consider only the homes without fireplaces. Use a nonparametric test to determine if there is a difference in the distributions of the selling prices of the homes with garages and those without. Use the 0.05 significance level.

50. Refer to the schools data set, which has information on 94 school districts.
  a. Group the data by size: "large" (more than 3000 students), "medium" (1000 to 3000), and "small" (fewer than 1000 students). Then use an appropriate nonparametric test to determine if there are differences in the mean amounts spent on instruction. Use the 0.05 significance level. Write a brief report summarizing your findings.
  b. Use the same groupings and significance level to test if there are differences in salaries by size of district. Write a brief report summarizing your findings.

CHAPTER ACHIEVEMENT TEST

The answers are at the back of the book.

MULTIPLE-CHOICE QUESTIONS

Select the response that best answers each of the questions.

1. Which of the following is *not* a nonparametric test?
  a. Sign test
  b. t test
  c. Wilcoxon signed-rank test
  d. Wilcoxon rank-sum test
  e. All are nonparametric tests.

2. The sign test should be employed instead of the paired t test when
  a. the sample is greater than 30.
  b. the population from which the sample is drawn is known to be nonnormal.
  c. both $n\pi$ and $n(1 - \pi)$ are less than 5.
  d. All of the above
  e. None of the above

3. When the sign test is employed, the null hypothesis is that the probability of a positive difference

in a paired observation is 0.50, and the sampling distribution for testing follows a binomial distribution.
a. True
b. False

4. In comparing the sign test and the Wilcoxon signed-rank test, we find that the Wilcoxon signed-rank test
   a. considers the magnitudes of the differences between paired observations.
   b. considers only the signs of the differences.
   c. is actually the same as the sign test.
   d. assumes a normal distribution.
   e. must always have 30 or more observations.

5. In comparing the sign test and the Wilcoxon signed-rank test to the Wilcoxon ranked-sum test, we find that the Wilcoxon ranked-sum test assumes
   a. a normal distribution.
   b. a positive difference.
   c. dependent samples.
   d. independent samples.

6. If we are using the Wilcoxon signed-rank test on a set of data with a sample size of 15 and we want a two-tailed test with the 0.01 significance level, then the null hypothesis will be rejected if
   a. R^+ is greater than 19.
   b. R^- is greater than 19.
   c. the smaller of R^+ and R^- is less than or equal to 15.

d. the smaller of R^+ and R^- is less than or equal to 25.

7. In the Wilcoxon signed-rank test, if the absolute differences between paired observations are equal (tied) and greater than zero, the suggested procedure is to
   a. discard the tied observations.
   b. use the smaller value of the corresponding ranks.
   c. use the larger value of the corresponding ranks.
   d. assign the average value of the corresponding ranks.
   e. None of the above

8. In the Wilcoxon rank-sum test, the differences between pairs of scores are ranked.
   a. True
   b. False

9. A principal advantage of the nonparametric tests is that the underlying assumptions are often less restrictive.
   a. True
   b. False

10. The region of rejection for the Wilcoxon rank-sum test can be in
    a. the upper tail only.
    b. the lower tail only.
    c. the lower tail or the upper tail or both tails.
    d. None of the above

COMPUTATION PROBLEMS

11. A medical researcher wishes to determine if the lung capacities of nonsmokers are greater than those of persons who smoke. A random sample of 15 matched pairs has been selected and tested. For each pair, the lung capacity of the smoker has been subtracted from that of the nonsmoking partner. A total of 13 positive differences was obtained (2 were negative). Is this sufficient evidence to show that the lung capacities of nonsmokers are greater? Use a 0.01 level of significance and the sign test.

12. Several married couples were asked to estimate the ages at which the husband achieved emotional maturity. Use the Wilcoxon signed-rank

test at the 0.01 level of significance to test for a difference between the perceptions of husbands and wives.

Couple	Husband	Wife
A	43	31
B	28	32
C	36	34
D	39	39
E	43	35
F	28	22
G	31	28

Couple	Husband	Wife
H	23	24
I	29	34
J	39	39
K	26	19
L	34	29
M	35	35

13. A study compared the achievement scores of a group of randomly selected inner-city sixth-graders to the scores of similar groups of students from suburban and rural areas within the same school district. Use the 0.05 significance level and the Kruskal-Wallis test to determine if differences exist among the 3 groups. Note that the data have been ordered to assist you in your calculations.

Inner-city
students: 95, 100, 100, 106, 108, 115, 120, 125, 130
Suburban
students: 108, 114, 118, 122, 124, 126, 130, 136, 142, 150
Rural
students: 102, 109, 112, 120, 127, 129, 140

ANSWERS TO SELF-REVIEW PROBLEMS

17-1 a. H_0: There is no change as a result of the planned exercise program.
H_a: The program is successful in reducing weight.
  b. Refer to Appendix A for a π of 0.50 and an n of 7. The probability of 7 successes is 0.008; of 6 successes, 0.055. Adding gives 0.063, which is as close as possible to 0.10 without exceeding it. Therefore, reject the null hypothesis if there are 6 or 7 pluses.
  c. Yes; there are 6 pluses.
  d. The planned exercise program is successful in reducing a person's weight.

17-2 a. H_0: $\pi = 0.50$
H_a: $\pi \neq 0.50$
  b. $n = 15$
  c. The probability of 3 or fewer plus signs is 0.017. The probability of 12 or more plus signs is also 0.017; hence, the probability of 3 or fewer or 12 or more plus signs is 0.034. Do not reject H_0 if the number of $+$ signs is between 4 and 11.
  d. Since there are 6 plus signs, H_0 cannot be rejected.

17-3 Since $n\pi = 15(0.5) = 7.5$ and $n(1 - \pi) = 15(0.5) = 7.5$, the normal approximation may be used.

H_0: $\pi = 0.50$
H_a: $\pi \neq 0.50$

The null hypothesis is rejected if z is less than -1.96 or greater than 1.96. Six plus signs were noted.

$$z = \frac{6 + 0.5 - 0.5(15)}{0.5\sqrt{15}} = -0.52$$

H_0 is not rejected. There is no difference in the 2 groups. Results are the same as those we obtained earlier using the sign test.

17-4 a. H_0: The burglary rates are the same for the two years.
H_a: There were more burglaries last year.
  b. Reject the null hypothesis if R^- is 10 or less.
  c. $R^- = 0$; hence, we reject the null hypothesis. There has been a significant increase in the burglary rates during the period.

17-5 a. H_0: There is no difference in the distances traveled by Dino and by Maxi.
H_a: There is a difference in the distances traveled by Dino and by Maxi.
  b. Do not reject H_0 if the computed z is between 1.96 and -1.96 (from Appendix C); otherwise, reject H_0 and accept H_a.
  c. $n_1 = 8$, the number of observations in the first sample

Dino		Maxi	
Distance	Rank	Distance	Rank
252	4	262	9
263	10	242	2
279	15	256	5
273	14	260	8
271	13	258	7
265	11.5	243	3
257	6	239	1
280	16	265	11.5
Total	$\overline{89.5}$		$\overline{46.5}$

$$W = 89.5$$

$$z = \frac{89.5 - \dfrac{8(8 + 8 + 1)}{2}}{\sqrt{\dfrac{(8)(8)(8 + 8 + 1)}{12}}}$$

$$= \frac{21.5}{9.52} = 2.26$$

d. Reject H_0; accept H_a. There is a difference in the distances traveled by the two golf balls.

17-6 H_0: The sales of produce are the same.
H_a: The sales of produce are not the same.

Rankings of Produce Sold Weekly		
Kraft Rank	Mesh Rank	Paper Rank
16	12	4
19	13	9
22	21	3
24	23	8
10	2	5
17	15	7
18	6	1
11	14	
	20	
$\overline{137}$	$\overline{126}$	$\overline{37}$

$$H = \frac{12}{24(24 + 1)}\left[\frac{(137)^2}{8} + \frac{(126)^2}{9} + \frac{(37)^2}{7}\right] - 3(24 + 1)$$

$$= 86.11 - 75 = 11.11$$

Since the computed value of H is greater than 5.991, H_0 is rejected and H_a is accepted. The conclusion is that the sales of produce are not the same under the various packaging conditions.

CHAPTER 18

Quality Charting

OBJECTIVES

When you have completed this chapter, you will be able to

- discuss the role of quality control in production and service operations;
- define the terms *chance cause, assignable cause, in control, and out of control;*
- construct and interpret a Pareto chart;
- construct and interpret a fishbone diagram;
- construct two charts for variables—namely, a mean chart and a range chart;
- construct two charts for attributes—namely, a percentage defective chart and a chart for the number of defects per unit.

CHAPTER PROBLEM

Statistical Software, Inc., offers a toll-free number where from 7 A.M. until 11 P.M. customers call about problems involving the use of their products. It is impossible to have every call answered immediately by a technical representative, but it is important that customers do not wait too long for a person to come on the line. How could Statistical Software visually portray this information to understand the variation in the call waiting process?

INTRODUCTION

Throughout this text you have seen many applications of hypothesis testing. In Chapter 10 we described methods of testing an hypothesis about a population mean when the sample was large. In Chapter 11 we discussed testing an hypothesis about a population mean when the sample was small. In Chapter 13 we described tests of hypothesis about the correlation coefficient, and so on. In this chapter we present another, somewhat different application of hypothesis testing, called statistical process control or SPC. We also refer to this as statistical quality control.

What is statistical process control? It is the collection of strategies, techniques, and actions taken by an organization to assure themselves that they are producing a quality product or providing a quality service. It begins at the product planning or design stage when the attributes of the product or service are specified. It continues through the production stage. Each attribute throughout the process is a contributor to the overall product quality. To determine if there is quality control, measurable attributes and specifications must be developed against which the actual attributes of the product or service can be compared.

A BRIEF HISTORY OF QUALITY CONTROL

Prior to the 1900s, American industry was largely characterized by small shops making relatively simple products, such as candles or furniture. In these small shops, the individual worker was generally a craftsman who was completely responsible for the "quality" of the work. The worker could ensure the quality through the personal selection of the materials, skillful manufacturing, and selective fitting and adjustment.

In the early 1900s, factories sprang up, where people with limited training were formed into large assembly lines. Products became much more complex. The individual worker no longer had complete control over the quality of the product. A semiprofessional staff, usually called the inspection department, became responsible for the quality of the product. The quality responsibility was usually fulfilled by a 100% inspection of all the important characteristics. If there were any discrepancies noted, these problems were handled by the manufacturing department supervisor. In essence, quality was attained by "inspecting the quality into the product."

During the 1920s, Dr. Walter A. Shewhart, of Bell Telephone Laboratories, developed the concepts of statistical quality control. He introduced the concept of "controlling" the quality of a product as it was being manufactured, rather than inspecting the quality into the product after it was manufactured. For the purpose of controlling quality, Shewhart developed charting techniques for controlling in-process manufacturing operations. In addition, he introduced the concept of statistical sample inspection to estimate the quality of the product as it was being manufactured, replacing the old method of inspecting each part after completing the production operation.

Statistical quality control really came into its own during World War II. The need for mass-produced war-related items, such as bomb sights, accurate radar, and other electronic equipment, at the lowest possible cost hastened the use of statistical sampling and quality control charts. Since World War II, these statistical techniques have been refined and sharpened. The use of computers in the last decade has also widened their use.

World War II virtually destroyed the Japanese production capability. Rather than retool their old production methods, the Japanese enlisted the aid of the late Dr. W. Edwards Deming, of the U.S. Department of Agriculture, to help them develop an overall plan. In a series of seminars with Japanese planners, he stressed a philosophy that is known today as Deming's 14 points. He emphasized that quality originates from improving the process, not from inspecting. Also, he stressed that the quality is determined by the customers. The manufacturer must be able, via market research, to anticipate the needs of customers. Upper management has the responsibility for long-term improvement. Another of his points, and one that the Japanese strongly endorsed, is that every member of the company must contribute to the long-term improvement. To achieve this improvement, ongoing education and training are necessary.

Deming had some ideas that did not mesh with management philosophies in the United States. Two areas where Deming's ideas differed from conventional American management philosophy were production quotas and merit ratings. He claimed these two practices, which are both common in the United States, are not productive and should be eliminated. He also pointed out that American managers are mostly interested in good news. Good news, however, does not provide an opportunity for improvement. On the other hand, bad news opens the door for new products and allows for company improvement.

Interest in quality has accelerated dramatically in the United States in the past 10 years. We need only turn on the TV and watch the commercials sponsored by GM, Ford, and Chrysler to verify the emphasis on quality control on the assembly line. It is now one of the "in" topics in all facets of business. Quoting from just a few recently published articles: "Quality is entering a new dimension"; "Quality: The way to export success"; "Quality starts with design"; "Futuristic agile factories must be based on work performed with quality foremost"; and "Lumbering corporate hierarchies must be streamlined to focus on teamwork, quality, and speed." To emphasize the need for better quality, V. Daniel Hunt, president of Technology Research Corporation, said in his book *Quality in America* that in the United States, 20 to 25% of the costs of production are currently spent finding and correcting mistakes. And, he added, the additional cost incurred in repairing or replacing faulty products in the field drives the total cost of poor quality to nearly 30%. In Japan, he indicates, this cost is about 3%!

In recent years, companies have been motivated to improve quality by the challenge of being recognized for their quality achievements. The Malcolm Baldridge National Quality Award, established in 1988, is awarded annually to U.S. companies that demonstrate excellence in quality achievement and management. The award categories include manufacturing, service, and small business, and past winners are Motorola, Xerox, IBM, Federal Express, and Cadillac. One of the most recent winners is a very small firm in Minneapolis, Minnesota, that employs only 100 workers.

What is quality? There is no commonly agreed upon definition of quality. We quote a few diverse definitions: From Westinghouse, "Total quality is performance leadership in meeting the customer requirements by doing the right things right the first time." From AT&T, "Quality is meeting the customer expectations." "Quality is achieving or reaching the highest standard as against being satisfied with the sloppy or fraudulent," said historian Barbara W. Tuchman.

CAUSES OF VARIATION

No 2 parts are *exactly* the same. There is some variation. The weight of each McDonald's Quarter Pounder is not exactly 0.25 pound. Some will weigh more than 0.25 pound, others less. The standard time for the Toledo Area Regional Transit Authority (TARTA) run from downtown Toledo to Perrysburg, Ohio, is 25 minutes. However, each run does not take *exactly* 25 minutes. Some runs take longer. Other times the TARTA driver must wait in Perrysburg before returning to Toledo. In some cases, there is a reason for the bus being late—an accident on the expressway or a snowstorm, for example. In other cases, the driver may not "hit" the green lights, or the traffic may be unusually heavy or slow for no apparent reason.

There are two general causes of variation in a process—chance and assignable variations. Chance variations are usually large in number and random in nature. They cannot be eliminated unless there is a major change in the equipment or material used in the process. Internal machine friction, slight variations in material or process conditions (such as the temperature of the mold being used to make glass bottles), atmospheric conditions (such as temperature, humidity, and the dust content of the air), and vibrations transmitted to a machine from a passing forklift are a few examples of sources of chance variation.

Assignable variations are usually nonrandom in nature and can be eliminated or reduced. If the hole drilled in a piece of steel is too large due to a dull drill, the drill may be sharpened or a new drill inserted. An operator who continually sets up the machine incorrectly can be replaced or retrained. If the roll of steel to be used in the process does not have the correct tensile strength, it can be rejected.

There are several reasons why we should be concerned with variation.

1. It will change the shape, dispersion, and central tendency of the distribution of the product characteristic being measured.
2. Assignable variation is usually correctable, whereas chance variation usually cannot be corrected or stabilized economically.

DIAGNOSTIC CHARTS

A variety of diagnostic techniques is available to investigate quality problems. Two of the more prominent of these techniques are Pareto charts and fishbone diagrams.

Pareto analysis is a technique for tallying the numbers and types of defects that happen within a product or service. The chart is named after 19th-century Italian scientist Vilfredo Pareto. He noted that most of the "activity" in a process is caused by relatively few of the "factors." His concept, often called the 80–20 rule, is that 80% of the activity is caused by 20% of the factors. By concentrating on 20% of the factors, managers can attack 80% of the problem. Suppose the activity a restaurant is investigating is "customer complaints" and the factor most often cited is "discourteous service." Then the manager knows that he needs to discuss this with the waitstaff.

Pareto Charts

To develop a Pareto chart, we begin by tallying the types of defects. Next, we rank the types of defects from largest to smallest. Finally, we produce a vertical bar chart, with the heights of the bars corresponding to the numbers of each type of defect. The following example illustrates these ideas.

Problem   The city manager of a large western city is concerned with water usage, particularly in single-family homes. She would like to develop a plan to reduce the water usage in her community. To investigate, she selects a sample of 100 homes and determines the mean daily water usage for various purposes. These sample results are as follows.

Reasons for Water Usage	Gallons per Day
Laundering	24.9
Lawn watering	143.7
Personal bathing	106.7
Cooking	5.1
Swimming pool usage	28.3
Dish washing	12.3
Car washing	10.4
Drinking	7.9

What is the area of greatest usage? Where should she concentrate her efforts to reduce the water usage?

Solution   A Pareto chart is useful for identifying the major areas of water usage and focusing on those areas where the greatest reduction can be achieved. The first step is to convert each of the activities to a percentage and then to order them from largest to smallest. The total water usage per day is 339.3 gallons, found by totaling the gallons used in the 8 activities. The activity with the largest use is lawn watering. It accounts for 143.7 gallons of water per day, or 42.4% of the amount of water used. The next largest category is personal bathing, which accounts for 31.4% of the water used. These 2 activities account for 73.8% of the water usage. This information is summarized in the Pareto chart on page 563, developed using the MINITAB software package. The scale of the vertical axis on the left side corresponds to the count, or frequency, of each activity, and the percentage the activity is of the whole is scaled on the right side. From this chart, the city manager can conclude that the largest reduction in water use can be achieved by reducing the water used in watering and bathing. We entered the data on the screen below.

	C1-T	C2
↓	Useage	Gallons
1	Laundering	24.9
2	Lawn watering	143.7
3	Personal bathing	106.7
4	Cooking	5.1
5	Swimming pool	28.3
6	Dish washing	12.3
7	Car washing	10.4
8	Drinking	7.9
9		

The commands are:

Stat ▶ Quality Tools ▶ Pareto Chart

Within the dialog box below, we selected the Charts defects table; the labels are in column 1, also called "Useage"; the frequencies are in column 2, also called "Gallons." We left the default option to combine defects after the first 95% and added the title. Finally, we click OK.

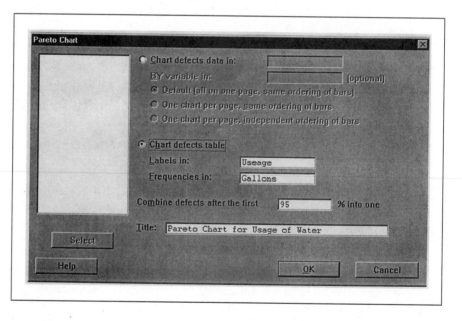

Fishbone Diagram

The second type of diagnostic chart is a **cause-and-effect diagram,** or a **fishbone diagram.** It is called a cause-and-effect diagram to emphasize the relationship between an effect and a set of possible causes that produce that particular effect. This diagram is useful to help organize ideas and to identify relationships. By identifying these relationships, we can isolate the factors that cause variability in our process. The name "fishbone" comes from the manner in which the various causes and effects are organized on the diagram. The effect is usually a particular problem, or perhaps a goal, and it is shown on

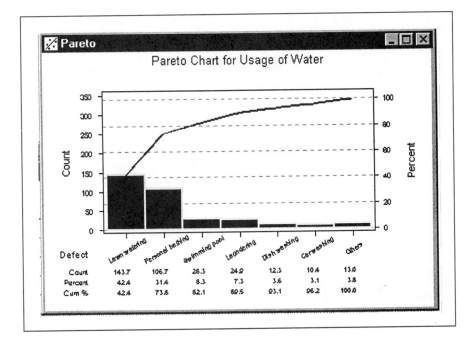

the right-hand side of the diagram. The major causes are listed on the left-hand side of the diagram.

The usual approach to a fishbone diagram is to consider four problem areas: methods, materials, equipment, and personnel. The problem, or the effect, is the head of the fish. See Figure 18–1. Under each of the possi-

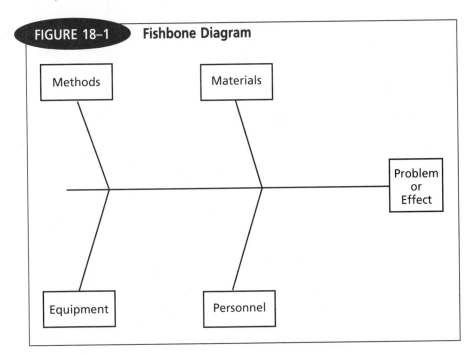

FIGURE 18–1   **Fishbone Diagram**

ble causes are subcauses that are identified and investigated. The sub-causes are factors that may or may not be producing the particular effect. Information is gathered about the problem and used to fill in the fish-bone diagram. Each of the subcauses is investigated, and those that are not important are eliminated until the real cause of the problem is iden-tified.

Figure 18–2, adapted from *Operations Management* by Vonderembse and White, illustrates the details of a fishbone diagram. Suppose a family restau-rant, such as those found along interstate highways, has recently been expe-riencing complaints from customers that the food being served is cold. Notice that each of the subcauses is listed as an assumption. Each of these subcauses must be investigated to find the real problem regarding the cold food. In a fishbone diagram, there is no weighting of the subcauses. It is a total which encourages brainstorming for ideas.

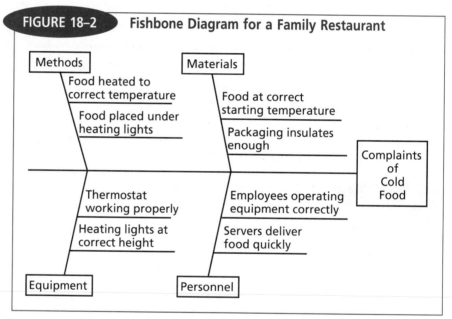

FIGURE 18–2   **Fishbone Diagram for a Family Restaurant**

Source: Adapted from Vonderembse MA, and White GP. *Operations Management,* 3rd ed. (South-Western College Publishing, 1996) p. 489.

Self-Review 18–1

Answers to the Self-Review problems are at the end of the chapter.

Patients at Rouse Home have been complaining recently about the conditions there. The administrator would like to use a Pareto chart to investigate. When a patient or patient's relative has a complaint, he or she is asked to complete a complaint form. Below is a summary of the complaint forms received during the last 12 months.

Complaint	Number
Nothing to do	45
Poor care by staff	71
Medication error	2
Dirty conditions	63
Poor-quality food	84
Lack of respect by staff	38

Develop a Pareto chart. What complaints would you suggest the administrator work on to achieve the most significant improvement?

Exercises

Answers to the even-numbered Exercises are at the back of the book.

1. Tom Sharkey is the owner of Sharkey Chevy. At the start of the year, Tom instituted a customer opinion program to find ways to improve service. One week after the service is performed Tom's administrative assistant calls the customer to find out if the service was performed satisfactorily and how the service might be improved. Below is a summary of the complaints for the first six months.

Complaint	Frequency
Problem not corrected	38
Error on invoice	8
Unfriendly atmosphere	12
Price too high	23
Wait too long for service	10

Develop a Pareto chart. What complaints would you suggest that Tom work on to improve the quality of service?

2. Out of 500 diesel engines tested, the manufacturer found that 9 had leaky radiators, 15 had faulty cylinders, 4 had ignition problems, 52 had oil leaks, and 30 had cracked blocks. Draw a Pareto chart to identify the key problem in the production process.

PURPOSES AND TYPES OF QUALITY CONTROL CHARTS

Control charts identify when assignable causes of variation or changes have entered the process. For example, the Wheeling Company makes vinyl-coated aluminum replacement windows for older homes. The vinyl coating must have a thickness between certain limits. If the coating becomes too thick, it will cause the window to jam. On the other hand, if the coating becomes too thin, the window will not seal properly. The mechanism that determines how much coating is to be put on each window becomes worn and begins making the coating too thick. Thus, a change has occurred in the process. Control charts are useful for detecting the change in process conditions. It is important to know when changes have entered the process so that the cause may be identified and corrected before a large number of unacceptable items are produced.

Control charts may be compared to the scoreboard in a baseball game. By looking at the scoreboard, the fans, coaches, and players can tell which team is winning the game. However, the scoreboard can do nothing to win or lose the game. Control charts serve a similar function. These charts indicate to the workers, group leaders, quality control engineers, production supervisors, and management whether the production of the part or service is "in control" or "out of control." If the production is out of control, the control chart will not fix the situation; it is just a piece of paper with figures and dots on it. Instead, the person responsible will adjust the machine manufacturing the part or do what is necessary to return production to the in control condition.

There are two types of control charts. A **variable control chart** portrays measurements, such as the amount of cola in a 2-liter bottle or the time it takes a nurse at Mt. Carmel Hospital to respond to a patient's call. A variable control chart requires the interval or the ratio scale of measurement. An **attribute control chart** classifies a product or service as either acceptable or unacceptable. It is based on the nominal scale of measurement. Patients in a hospital are asked to rate the meals served as acceptable or unacceptable; bank loans are either repaid, or they are defaulted.

Control Charts for Variables

To develop control charts for variables we rely on the sampling theory discussed in connection with the central limit theorem in Chapter 9. Suppose a sample of five pieces is selected each hour from the production process and the mean of each sample computed. The sample means are $\overline{X}_1$, $\overline{X}_2$, $\overline{X}_3$, and so on. The mean of these sample means is denoted as $\overline{\overline{X}}$. We use k to indicate the number of sample means. The overall or grand mean is computed as follows:

$$\overline{\overline{X}} = \frac{\sum \text{ of the means of the subgroups}}{\text{Number of sample means}} = \frac{\sum \overline{X}}{k}$$

18–1

The standard error of the distribution of the individual sample means is designated by $s_{\overline{x}}$. It is computed as follows:

$$s_{\overline{x}} = \frac{s}{\sqrt{n}}$$

18–2

These relationships allow limits to be set up around the sample means in order to show how much variation can be expected for a given sample size. These expected limits are called the **upper control limit** (UCL) and the **lower control limit** (LCL). An example will illustrate how control limits are determined and used.

Problem   Recall the Chapter Problem, where Statistical Software, Inc., offers a number customers can call about problems involving the use of their products. Customers become upset when they hear the message "Your call is important to us; the next available representative will be with you shortly" too many times. To understand its process, Statistical Software decides to develop a control chart portraying the total time from when a call is received until the representative answers the caller's question. Yesterday, for the 16 hours of operation, 5 calls were sampled each hour. This information is reported in Table 18–1. Based on this information, develop a control chart for the mean duration of the call. Does there appear to be a trend in the calling times? Is there any period where it appears that customers wait longer than others?

TABLE 18–1		Duration of 16 Samples of 5 Help Sessions (in minutes)					
Time		1	2	3	4	5	Mean	Range
A.M.	7	8	9	15	4	11	9.4	11
	8	7	10	7	6	8	7.6	4
	9	11	12	10	9	10	10.4	3
	10	12	8	6	9	12	9.4	6
	11	11	10	6	14	11	10.4	8
	12	7	7	10	4	11	7.8	7
P.M.	1	10	7	4	10	10	8.2	6
	2	8	11	11	7	7	8.8	4
	3	8	11	8	14	12	10.6	6
	4	12	9	12	17	11	12.2	8
	5	7	7	9	17	13	10.6	10
	6	9	9	4	4	11	7.4	7
	7	10	12	12	12	12	11.6	2
	8	8	11	9	6	8	8.4	5
	9	10	13	9	4	9	9.0	9
	10	9	11	8	5	11	8.8	6
Total							150.6	102

Solution   A mean chart has two limits, an upper control limit (UCL) and a lower control limit (LCL). These upper and lower control limits are computed as follows:

$$\text{UCL} = \overline{\overline{X}} + 3\,\frac{s}{\sqrt{n}} \quad \text{and} \quad \text{LCL} = \overline{\overline{X}} - 3\frac{s}{\sqrt{n}} \qquad \boxed{\textbf{18–3}}$$

where s is an estimate of the standard deviation of the population, σ. Notice that in the calculation of the upper and lower control limits, the number 3 appears. It represents the 99.74% confidence limits. The limits are often called the 3-sigma limits. However, other levels of confidence (such as 90 and 95%) can be used. Rather than calculate the standard deviation from each sample as a measure of variation, it is easier to use the range. For small samples, there is a constant relationship between the range and the standard deviation, so we can use the following to determine the 99.74% control limits:

$$\text{UCL} = \overline{\overline{X}} + A_2\overline{R} \quad \text{and} \quad \text{LCL} = \overline{\overline{X}} - A_2\overline{R} \qquad \boxed{18\text{-}4}$$

where

A_2 is a constant used in the computation of the upper and lower control limits. It is based on the average range, $\overline{R}$. The factors for various sample sizes can be found in Appendix I. (Note: n in this table refers to the number in the sample.) A portion of Appendix I is shown below. To locate the A_2 factor for this problem, find the sample size for n in the left margin. It is 5 in this example. Then move horizontally to the A_2 column, and read the factor. It is 0.577.

n	A_2	d_2	D_3	D_4
2	1.880	1.128	0	3.267
3	1.023	1.693	0	2.575
4	0.729	2.059	0	2.282
5	0.577	2.326	0	2.115
6	0.483	2.534	0	2.004

$\overline{\overline{X}}$ is the mean of the sample means, found by $\Sigma\overline{X}/k$, where k is the number of samples selected. In this problem, a sample of 5 observations is taken each hour for 16 hours, so $k = 16$.

$\overline{R}$ is the mean of the ranges of the sample. It is found by $\Sigma R/k$. Remember that the range is the difference between the largest and the smallest values in each sample and describes the amount of variability occurring in each sample.

The centerline ($\overline{\overline{X}}$) for the mean chart, the mean of the sample means ($\overline{X}$), is 9.4125 minutes, found by 150.60/16. The mean of the ranges ($\overline{R}$) is 6.375 minutes, found by 102/16. See Table 18–1. Thus, the upper control limit of the X bar chart is

$$\text{UCL} = \overline{\overline{X}} + A_2\overline{R} = 9.4125 + 0.577(6.375) = 13.0909$$

The lower control limit of the X bar chart is

$$\text{LCL} = \overline{\overline{X}} - A_2\overline{R} = 9.4125 - 0.577(6.375) = 5.7341$$

$\overline{\overline{X}}$, UCL, LCL, and the sample means are portrayed in Figure 18–3. Note that the mean, $\overline{\overline{X}}$, is 9.4125 minutes; the upper control limit is located at 13.0909 minutes; and the lower control limit is located at 5.7341 minutes. There is some variation in the duration of the calls, but all sample means are within the control limits. So based on 16 samples of 5 calls, we conclude that 99.74% of the time the mean length of a sample of 5 calls will be between 5.7341 and 13.0909 minutes.

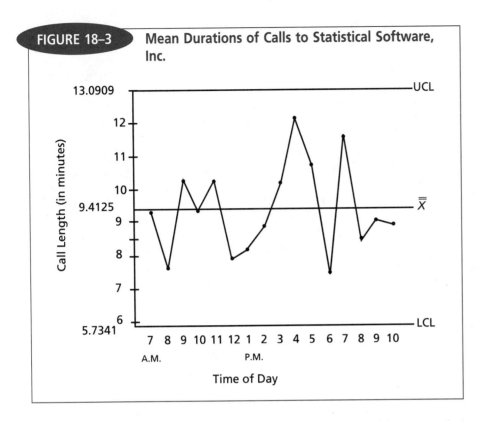

FIGURE 18–3   **Mean Durations of Calls to Statistical Software, Inc.**

Because the statistical theory is based on the normality of large samples, control charts should be based on a stable process—that is, a fairly large sample, taken over a long period of time. One rule of thumb is to design the chart after at least 25 samples have been selected.

In addition to the central tendency in a sample, we can also monitor the amount of variation from sample to sample.

RANGE CHART.   A **range chart** shows the variation in the sample ranges. If the points representing the ranges fall between the upper and the lower control limits, it is concluded that the operation is in control as far as variability is concerned. According to chance, about 997 times out of 1000 the range of the samples will fall within the limits. If the range should fall above the limits, we conclude that an assignable cause affected the operation and that an adjustment to the process is needed. Why are we not as concerned about the lower limit of the range? For small samples, the lower limit is often zero. Actually, for any sample of six or less, the lower control limit is zero. If the range is zero, then all the parts are the same, and there is no problem with the variability of the operation.

The upper and lower control limits of the range chart are determined from the following equations:

$$\text{UCL} = D_4\overline{R} \quad \text{and} \quad \text{LCL} = D_3\overline{R}$$

18–5

The values for D_3 and D_4 are found in Appendix I or in the table on page 568.

Problem The lengths of time customers of Statistical Software, Inc., waited from the time their calls were answered until technical representatives answered their questions or solved their problems are recorded in Table 18–1. Develop a control chart for the range. Does it appear that there is any time when there is too much variation in the operation?

Solution The first step is to find the mean range. The range for the 5 calls sampled in the 7 A.M. hour is 11 minutes. The longest call selected from that hour was 15 minutes, and the shortest was 4 minutes; the difference in the lengths is 11 minutes. In the 8 A.M. hour, the range is 4 minutes. The total of the 16 ranges is 102 minutes, so the average range is 6.375 minutes, computed as follows: $\overline{R} = 102/16$. Referring to Appendix I or the partial table on page xxx, D_3 and D_4 are 0 and 2.115, respectively. The lower and upper control limits are 0 and 13.4831.

$$\text{UCL} = D_4\overline{R} = 2.115(6.375) = 13.4831$$

$$\text{LCL} = D_3\overline{R} = 0(6.375) = 0$$

The range chart with the 16 sample ranges plotted is shown in Figure 18–4. In reviewing, we see that all the ranges are well within the control limits. Hence, we conclude that the variation in the time needed to service the customers' calls is within normal limits; that is, it is in control.

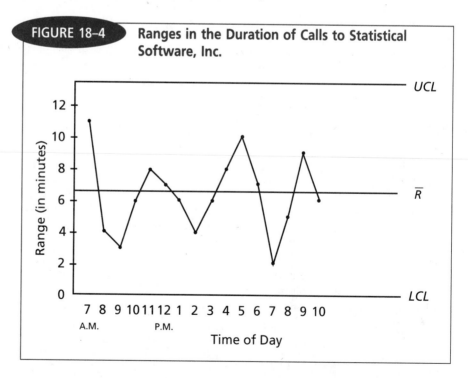

FIGURE 18–4 **Ranges in the Duration of Calls to Statistical Software, Inc.**

MINITAB will draw control charts for the mean and the range. The output from the Statistical Software Problem/Solution is on page 572. There are some minor differences in the limits, which are due to rounding. To begin we enter the following data.

↓	C1	C2
	Time	
1	8	
2	9	
3	15	
4	4	
5	11	
6	7	
7	10	
8	7	
9	6	
10	8	
11	11	
12	12	
13	10	
14	9	
15	10	

Give the following commands:

Stat ▶ Quality Control ▶ Control Charts ▶ Xbar–R

In the dialog box below, select Single colum and enter "Time," for the Subgroup size enter "5," and then click OK.

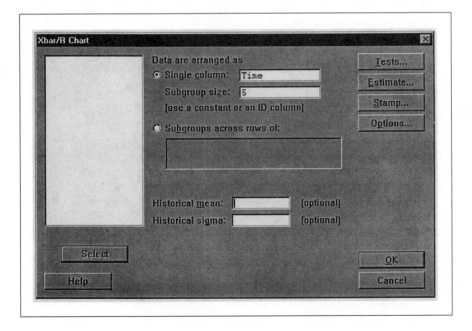

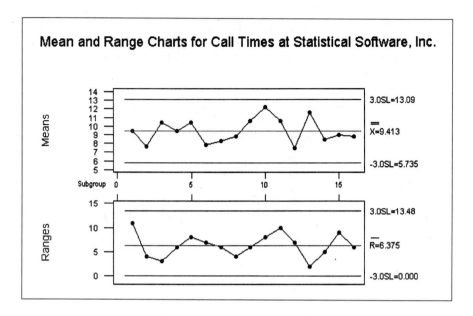

Mean and Range Charts for Call Times at Statistical Software, Inc.

SOME IN-CONTROL AND OUT-OF-CONTROL SITUATIONS. Following are three illustrations of in-control and out-of-control production processes.

1. The mean chart and the range chart below together indicate that the process is in control. Note that the sample means and sample ranges are clustered close to the centerlines. Some are above and some below the centerlines, indicating that the process is quite stable. That is, there is no visible tendency for the means and ranges to move toward the out-of-control areas.

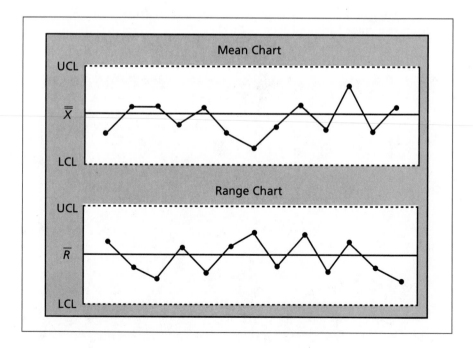

2. The sample means are in control, but the ranges of the last two samples are out of control. This indicates there is considerable variation from piece to piece. Some pieces are large; others are small. An adjustment in the process is necessary.

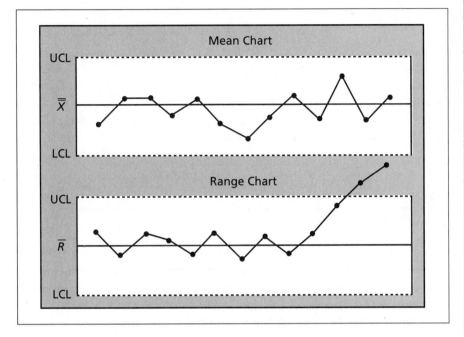

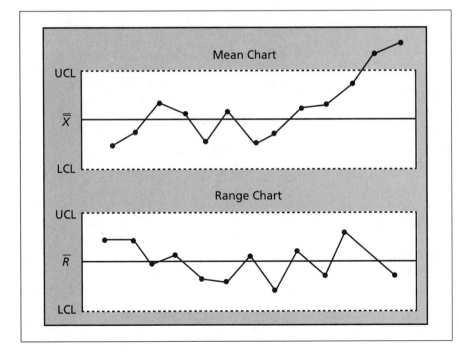

3. The arithmetic mean was in control for the first samples, but starting with the eighth sample, there has been an upward trend in the mean. The means of the last two samples are out of control. An adjustment in the process is indicated.

The top chart on page 573 for the mean is an example in which the control chart offers some additional information. Note the direction of the last five observations of the mean. They are all above $\overline{\overline{X}}$ and increasing, and, in fact, the last two observations are out of control. The fact that the sample means were increasing for six consecutive observations is an indication that the process is out of control.

Self-Review 18–2

Every hour the quality control inspector measures the outside diameters of four parts. The results of the measurements are shown below.

	Sample Piece			
Time	1	2	3	4
9 A.M.	1	4	5	2
10 A.M.	2	3	2	1
11 A.M.	1	7	3	5

a. Compute the mean outside diameter and the mean range, and determine the control limits for the mean and the range.
b. Are the measurements within the control limits? Interpret the chart.

Exercises

3. Describe the difference between assignable variation and chance variation.
4. Describe the difference between an attribute control chart and a variable control chart.
5. Samples of size $n = 4$ are selected from a production line.
   a. What is the value of the A_2 factor used to determine the upper and lower control limits for the mean?
   b. What are the values of the D_3 and D_4 factors used to determine the upper and lower control limits for the range?
6. Samples of size $n = 5$ are selected from a manufacturing process. The mean of the sample ranges is 0.50. What is the estimate of the standard deviation of the population?
7. A new industrial oven has just been installed at the Piatt Bakery. In order to develop experience regarding the oven temperature, an inspector reads the temperature at 4 different places inside the oven each half hour. The first reading, taken at 8:00 A.M., was 1040°F. (Only the last 2 digits are given in the following table to make the computations easier.)

	Reading			
Time	1	2	3	4
8:00 A.M.	40	50	55	39
8:30 A.M.	44	42	38	38
9:00 A.M.	41	45	47	43
9:30 A.M.	39	39	41	41
10:00 A.M.	37	42	46	41
10:30 A.M.	39	40	39	40

a. Based on this initial experience, determine the control limits for the mean temperature. Determine the grand mean. Plot the experience on a chart.

b. Interpret the chart. Does there seem to be a time when the temperature is out of control?

8. Refer to Exercise 7.

  a. Based on this initial experience, determine the control limits for the range. Plot the experience on a chart.

  b. Does there seem to be a time when there is too much variation in the temperature?

Attribute Control Charts

Sometimes the data we collect come simply from counting, rather than measuring; that is, we observe the presence or absence of some attribute. For example, the screw top on a bottle of shampoo either fits onto the bottle and does not leak (an in-control condition), or it does not seal and a leak results (an out-of-control condition); a bank makes a loan to a customer, and the loan is either repaid or not repaid. In other cases, we are interested in the number of defects per unit. An airline might count the number of late arrivals *per day* at Gatwick Airport in London. In this section, we discuss two types of attribute charts: the *p* (percentage defective) and the *c* bar (number of defectives).

PERCENTAGE DEFECTIVE CHART.   If the item recorded is the fraction of unacceptable parts made in a batch of larger parts, the appropriate control chart is the percentage defective chart. This chart is based on the binomial distribution, discussed in Chapter 6, and proportions, discussed in Chapter 10. The centerline is drawn at $\bar{p}$, the mean percentage defective. The $\bar{p}$ replaces the $\bar{X}$ of the variable control chart. The mean proportion defective is computed as follows:

$$\bar{p} = \frac{\sum \text{ of the numbers defective}}{\text{Number of samples}} \qquad \boxed{18\text{--}6}$$

The variation in the sample proportion is described by the standard error of a proportion. It is computed as follows:

$$s_p = \sqrt{\frac{\bar{p}(1 - \bar{p})}{n}} \qquad \boxed{18\text{--}7}$$

Hence, the upper control limit (UCL) and the lower control limit (LCL) are computed as the mean percentage defective plus or minus three times the standard error of the percentages (proportions). The formula for the control limits is

$$\text{UCL, LCL} = \bar{p} \pm 3\sqrt{\frac{\bar{p}(1 - \bar{p})}{n}} \qquad \boxed{18\text{--}8}$$

An example will show the details of the calculations and the conclusions.

Problem   The credit department at Global National Bank is responsible for entering each transaction charged to the customer's monthly statement. Of course, accuracy is critical, and errors will make the customer very unhappy! To guard against errors each data entry clerk keys a sample of 1500 transactions from his or her batch of work a second time, and a computer program checks to see that the numbers match. The program also prints out a report of the numbers and sizes of any discrepancies. Seven people were working last hour, and here are their results:

Inspector	Number Inspected	Number Mismatched
Mullins	1500	4
Rider	1500	6
Gankowski	1500	6
Smith	1500	2
Reed	1500	15
White	1500	4
Reading	1500	4

Construct the percentage defective chart for this process. What are the upper and lower control limits? Interpret the results. Does it appear that any of the data entry clerks are out of control?

Solution   The first step is to determine the mean proportion defective, $\bar{p}$, using formula 18–6. It is 0.00390, found by 41/1050.

Inspector	Number Inspected	Number Mismatched	Proportion Defective
Mullins	150	4	.0267
Rider	150	6	.0400
Gankowski	150	6	.0400
Smith	150	2	.0133
Reed	150	15	.1000
White	150	4	.0267
Reading	150	4	.0267
Total	1050	41	

The upper and lower control limits are computed using formula 18–8.

$$\text{LCL, UCL} = \bar{p} \pm 3\sqrt{\frac{\bar{p}(1 - \bar{p})}{n}}$$

$$= 0.00390 \pm 3\sqrt{\frac{0.00390(1 - 0.00390)}{1500}} = 0.00390 \pm 0.00475$$

From the above calculations, the upper control limit is 0.0865, found by 0.0390 + 0.0475. The lower control limit is 0. Why? The lower limit by formula is determined by $0.0039 - 0.00475$, which is equal to -0.0084. A negative proportion defective is not possible; the smallest value possible is 0, so we set the control limit at 0. Thus, any data entry clerk whose proportion defective is between 0 and 0.00864 is in control. Clerk 5 (Reed) is therefore out of control. Her proportion defective is 0.01, which is outside the upper control limit. Perhaps she should receive additional training or be transferred to another position. This information is summarized in Figure 18–5.

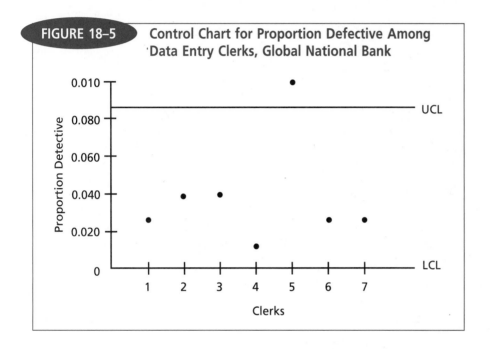

FIGURE 18–5   **Control Chart for Proportion Defective Among Data Entry Clerks, Global National Bank**

c BAR CHART.   The c bar chart plots the number of defects or failures per unit. It is based on the Poisson distribution, discussed in Chapter 6. The number of bags mishandled on a flight by Southwest Airlines might be monitored by a c bar chart. The "unit" under consideration is the flight. On some flights, there may not be any bags mishandled. On others, there may be only one; on others, two; and so on. The Internal Revenue Service might count and develop a control chart for the number of errors in arithmetic per tax return. Most returns will not have any errors, some returns will have a single error, others will have two, and so on. We let $\bar{c}$ (read as "c bar") be the mean number of defects per unit. So $\bar{c}$ is the mean number of bags mishandled by Southwest Airlines per flight or the mean number of arithmetic errors per tax return. Recall from Chapter 6 that the standard deviation of a Poisson distribution is the square root of the mean. Thus, we can determine the 3-sigma, or 99.74%, limits on a c bar chart by

$$\text{UCL, LCL} = \bar{c} \pm 3\sqrt{\bar{c}} \qquad \boxed{18\text{–}9}$$

Problem   The publisher of the *Oak Harbor Daily Telegraph* is concerned about the number of misspelled words in the daily newspaper. In an effort to control the problem and promote the need for correct spelling a control chart is to be instituted. The numbers of misspelled words found in the final editions of the paper for the last 10 days are 5, 6, 3, 0, 4, 5, 1, 2, 7, and 4. A paper is not published on Saturday or Sunday. Determine the appropriate control limits, and interpret the chart. Were there any days during the period that the numbers of misspelled words were out of control?

Solution   The sum of the numbers of misspelled words over the 10-day period is 37, so the mean number of defects, $\bar{c}$, is 3.7. The square root of this number is 1.924, so the upper control limit is

$$\text{UCL} = \bar{c} + 3\sqrt{\bar{c}} = 3.7 + 3\sqrt{3.7} = 3.7 + 5.77 = 9.47$$

The computed lower control limit would be $3.7 - 3(1.924) = -2.07$. However, the number of defects cannot be less than 0, so we use 0 as the lower limit. The lower control limit is 0, and the upper control limit is 9.47. When we compare each of the data points to the value of 9.47, we see they are all less than the upper control limit. Thus, the number of misspelled words is in control. Of course, newspapers are going to strive to eliminate all misspelled words, but control charting techniques offer a means of tracking daily results and determining if there has been a change. For example, if a new proofreader was hired, his or her work could be compared with that of others. These results are summarized as follows:

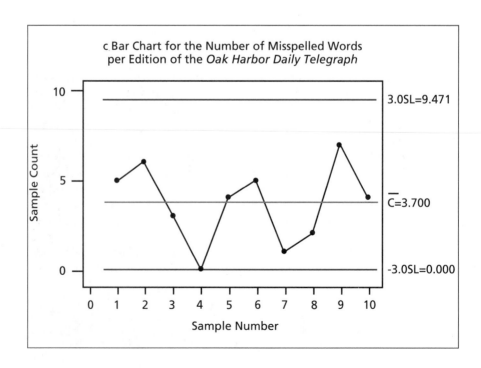

c Bar Chart for the Number of Misspelled Words per Edition of the *Oak Harbor Daily Telegraph*

A battery manufacturer randomly samples 20 batteries at the end of each shift in order to test for defectives. The numbers of defectives found over the last 12 shifts are 2, 1, 0, 2, 1, 1, 7, 1, 1, 2, 6, and 1. Construct a control chart for this process, and comment on whether the process is in control.

Self-Review 18–3

Exercises

9. A bicycle manufacturer randomly selects 10 frames each day in order to test for defectives. The numbers of defectives found over the last 14 days are 3, 2, 1, 3, 2, 2, 8, 2, 0, 3, 5, 2, 0, and 4. Construct a control chart for this process, and comment on whether the process is in control.

10. Scott Paper tests its toilet paper by subjecting 15 rolls to a wet stress test to see whether and how often the paper tears during the test. Following are the numbers of tears in each sampled roll of paper: 2, 3, 1, 2, 2, 1, 3, 2, 2, 1, 2, 2, 1, 0, and 0. Construct a control chart for this process, and comment on whether the process is in control.

11. Sam's Supermarkets test their checkout clerks by randomly examining the printout receipts for keying errors. Following are the numbers of errors on each receipt for July 23, last Wednesday: 0, 1, 1, 0, 0, 1, 1, 0, 1, 1, and 0. Construct a control chart for this process, and comment on whether the process is in control.

12. Dave Christi runs a car-wash chain scattered throughout Chicago. He is concerned that some local managers are giving away free washes to their friends. So he decides to collect data on the number of "voided" sales receipts. Of course, some of them are legitimate voids. Would the following data indicate a reasonable number of voids at his facilities: 3, 8, 3, 4, 6, 5, 0, 1, 2, and 4? Construct a control chart for this process, and comment on whether the process is in control.

CHAPTER OUTLINE

I. The objective of statistical quality control is to control the quality of the product or service as it is being developed.

II. A Pareto chart is a technique for tallying the numbers and types of defects that happen within a product or service.
  A. This chart was named after Italian scientist Vilfredo Pareto.
  B. The concept of the chart is that 80% of the activity is caused by 20% of the factors.

III. A fishbone diagram emphasizes the relationships between an effect and possible causes of that effect.
  A. It is also called a cause-and-effect diagram.
  B. The usual approach is to consider four problem areas: methods, materials, equipment, and personnel.

IV. The purpose of a control chart is to portray graphically the quality of a product or service.
  A. There are two types of control charts.
    1. A variable control chart is the result of a measurement.
    2. An attribute control chart shows whether the product or service is acceptable or not acceptable.
  B. There are two sources of variation in the quality of a product or service.
    1. Chance variation is random in nature and cannot be controlled or eliminated.
    2. Assignable variation is not due to random causes and can be eliminated.
  C. Three control charts were considered in this chapter.

1. A mean chart shows the mean of a variable, and a range chart shows the range of the variable.
   a. The upper and lower control limits are set at plus or minus three standard errors from the sample mean.
   b. The formulas for the upper and lower control limits for the mean are

$$\text{UCL} = \overline{\overline{X}} + A_2\overline{R} \quad \text{and} \quad \text{LCL} = \overline{\overline{X}} - A_2\overline{R} \qquad \boxed{\text{18-4}}$$

   c. The formulas for the upper and lower control limits for the range are

$$\text{UCL} = D_4\overline{R} \quad \text{and} \quad \text{LCL} = D_3\overline{R} \qquad \boxed{\text{18-5}}$$

2. A percentage defective chart is an attribute chart that shows the proportion of the product or service that does not conform to the standard.
   a. The mean percentage defective is computed as follows:

$$\overline{p} = \frac{\Sigma \text{ of the numbers defective}}{\text{Number of samples}} \qquad \boxed{\text{18-6}}$$

   b. The control limits for the proportion defective are determined from the equation

$$\text{UCL, LCL} = \overline{p} \pm 3\sqrt{\frac{\overline{p}(1 - \overline{p})}{n}} \qquad \boxed{\text{18-8}}$$

3. A c bar chart refers to the number of defects per unit.
   a. It is based on the Poisson distribution.
   b. The mean number of defects per unit is $\overline{c}$.
   c. The control limits are determined from the following equation:

$$\text{UCL, LCL} = \overline{c} \pm 3\sqrt{\overline{c}} \qquad \boxed{\text{18-9}}$$

Exercises

13. The production supervisor at Westburg Electric, Inc., noted an increase in the number of electric motors rejected at the time of the final inspection. From the last 200 motors rejected, 40% of the defects were due to poor wiring, 30% contained a short in the coil, 25% involved a defective plug, and 5% involved other defects. Develop a Pareto chart to show the major problem areas.

14. A manufacturer of athletic shoes conducted a study on its newly developed jogging shoe. Listed below are the types and frequencies of the nonconformities and failures found. Develop a Pareto chart to show the major problem areas.

Type of Nonconformity	Frequency
Sole separation	34
Heel separation	98
Sole penetration	62
Lace breakage	14
Eyelet failure	10
Other	16

15. The McBurger Restaurant dispenses its soft drinks with an automatic machine that operates based on the weight of the soft drink. When the process is in control, the machine fills each cup such that the grand mean is 10.0 ounces and the mean range is 0.25 ounce for samples of 5.
    a. Determine the upper and lower control limits for the process for both the mean and the range.
    b. The manager of the I-280 store tested 5 soft drinks served last hour and found that the mean was 10.16 ounces and the range was 0.35 ounce. Is the process in control? Should other action be taken?

16. A new machine has just been installed to cut and rough-shape large slugs. The slugs are then transferred to a precision grinder. One of the critical measurements is the outside diameter. The quality control inspector randomly selected 5 slugs each hour, measured the outside diameter, and recorded the results. The measurements for the period from 8:00 A.M. to 10:30 A.M. follow.

Time	Outside Diameter (in millimeters)				
	1	2	3	4	5
8:00 A.M.	87.1	87.3	87.9	87.0	87.0
8:30 A.M.	86.9	88.5	87.6	87.5	87.4
9:00 A.M.	87.5	88.4	86.9	87.6	88.2
9:30 A.M.	86.0	88.0	87.2	87.6	87.1
10:00 A.M.	87.1	87.1	87.1	87.1	87.1
10:30 A.M.	88.0	86.2	87.4	87.3	87.8

a. Determine the control limits for the mean and the range.

b. Plot the control limits for the mean outside diameter and the range.

c. Are there any points on the mean chart or the range chart that are out of control? Comment on the charts.

17. The Long Last Tire Company, as part of its inspection process, tests its tires for tread wear under simulated road conditions. A total of 20 samples of 3 tires each was selected from different shifts over the last month of operation. The tread wear is reported below (in hundredths of an inch).

Sample	Tread Wear		
1	44	41	19
2	39	31	21
3	38	16	25
4	20	33	26
5	34	33	36
6	28	23	39
7	40	15	34
8	36	36	34
9	32	29	30
10	29	38	34
11	11	33	34
12	51	34	39
13	30	16	30
14	22	21	35
15	11	28	38
16	49	25	36
17	20	31	33
18	26	18	36
19	26	47	26
20	34	29	32

a. Determine the control limits for the mean and the range.

b. Plot the control limits for the mean tread wear and the range.

c. Are there any points on the mean chart or the range chart that are out of control? Comment on the charts.

18. The Charter National Bank has a staff of loan officers located in its branch offices throughout the southwestern United States. The vice-president in charge of the loan officers would like some information on the typical amounts of the loans and on the range in the amounts of the loans. A staff analyst selected a sample of ten loan officers and from each officer selected a sample of five loans he or she made last month. The data are reported below. Develop control charts for the mean and the range of each loan officer. Do any of the officers appear to be out of control? Comment on your findings.

Officer	Loan Amount (in thousands of dollars)				
	1	2	3	4	5
Weinraub	59	74	53	48	65
Visser	42	51	70	47	67
Moore	52	42	53	87	85
Brunner	36	70	62	44	79
Wolf	34	59	39	78	61
Bowyer	66	80	54	68	52
Kuhlman	74	43	45	65	49
Ludwig	75	53	68	50	31
Longnecker	42	65	70	41	52
Simonetti	43	38	10	19	47

19. The producer of a candy bar, called the Mickey Mantle Bar, reports on the package that the calorie content is 420 per 2-ounce bar. A sample of 10 bars from each of the last 5 days is sent for a chemical analysis of the calorie content. The results are shown below. Does it appear that there are any days where the calorie count is out of control? Develop an appropriate control chart, and analyze your findings.

	Calorie Count				
Sample	1	2	3	4	5
1	426	406	418	431	432
2	421	422	415	412	411
3	425	420	406	409	414
4	424	419	402	400	417
5	421	408	423	410	421
6	427	417	408	418	422
7	422	417	426	435	426
8	419	417	412	415	417
9	417	432	417	416	422
10	420	422	421	415	422

20. The Early Morning Delivery Service guarantees delivery of small packages by 10:30 A.M. Of course, some of the packages are not delivered by 10:30 A.M. For a sample of 200 packages delivered on each of the last 15 working days, the following numbers of packages were delivered after the deadline: 9, 14, 2, 13, 9, 5, 9, 3, 4, 3, 4, 3, 3, 8, and 4.
   a. Determine the mean proportion of packages delivered after 10:30 A.M.
   b. Determine the control limits for the proportion of packages delivered after 10:30 A.M. Were any of the sampled days out of control?
   c. If 10 packages out of 200 in the sample were delivered after 10:30 A.M. today, is this sample within the control limits?
21. An automatic machine produces 5.0-millimeter bolts at a high rate of speed. A quality control program has been started to control the number of defectives. The quality control inspector selects 50 bolts at random and determines how many are defective. The numbers of defectives in the first 10 samples are 3, 5, 0, 4, 1, 2, 6, 5, 7, and 7.
   a. Design a percentage defective chart. Insert the mean percentage defective, UCL, and LCL.
   b. Plot the percentage defective for the first ten samples on the chart.
   c. Interpret the chart.
22. The Inter State Moving and Storage Company is setting up a control chart to monitor the proportion of residential moves that result in written complaints due to late delivery, lost items, or damaged items. Samples of 50 moves are selected for each of the last 12 months. The numbers of written complaints in each sample are 8, 7, 4, 8, 2, 7, 11, 6, 7, 6, 8, and 12.
   a. Design a percentage defective chart. Insert the mean percentage defective, UCL, and LCL.
   b. Plot the proportion of written complaints in the last 12 months.
   c. Interpret the chart. Does it appear that the number of complaints is out of control for any of the months?
23. Eric's Cookie House sells chocolate chip cookies in shopping malls. Of concern is the number of chocolate chips in each cookie. Eric, the owner and president, would like to establish a control chart for the number of chocolate chips per cookie. He selects a sample of 15 cookies from today's production and counts the number of chocolate chips in each. The results are as follows: 6, 8, 20, 12, 20, 19, 11, 23, 12, 14, 15, 16, 12, 13, and 12.
   a. Determine the centerline and the control limits.
   b. Develop a control chart, and plot the number of chocolate chips per cookie.
   c. Interpret the chart. Does it appear that the number of chocolate chips is out of control in any of the cookies sampled?
24. The numbers of "near misses" recorded for the last 20 months at the Lima International Airport are 3, 2, 3, 2, 2, 3, 5, 1, 2, 2, 4, 4, 2, 6, 3, 5, 2, 5, 1, and 3. Develop an appropriate control chart. Determine the mean number of misses per month and the limits on the number of misses per month. Are there any months when the number of near misses is out of control?

25. The following numbers of robberies were reported during the last 10 days to the Robbery Division of the Metro City Police: 10, 8, 8, 7, 8, 5, 8, 5, 4, and 7. Develop an appropriate control chart. Determine the mean number of robberies reported per day, and determine the control limits. Are there any days when the number of robberies reported is out of control?

CHAPTER ACHIEVEMENT TEST

The answers are at the end of the book.

MULTIPLE-CHOICE QUESTIONS

Select the response that best answers each of the questions.

1. Statistical quality control was developed
   a. in the Dark Ages.
   b. around the time of the Civil War.
   c. in the 1920s.
   d. after the 1980s.
2. The underlying theme of Deming's philosophy is that
   a. production and quota systems should be used.
   b. cooperation and teamwork are important.
   c. good news means good ideas.
   d. all companies should adopt a merit rating system.
3. Chance causes are
   a. random in nature.
   b. due to some specific cause, such as a worn tool.
   c. usually eliminated by the use of control charts.
   d. highlighted by the use of hypothesis testing.
4. A variable control chart is
   a. used to estimate the probability of acceptance.
   b. used to estimate the standard deviation.
   c. based on a measurement or a reading.
   d. based on whether the product is acceptable or unacceptable.
5. Which of the following is an example of a variable control chart?
   a. c bar chart
   b. Mean and range charts
   c. Percentage defective chart
   d. All of the above

6. The upper and lower control limits are usually how far from the mean?
   a. 99.74%
   b. Plus or minus 2 standard deviations
   c. A_2 times the mean
   d. Plus or minus 3 standard deviations
7. Which type of control chart is used to graphically portray the number of defects per unit?
   a. c bar chart
   b. Mean chart
   c. Percentage defective chart
   d. Range chart
8. When we find a point that is out of control for the range, this means that there
   a. are too many defectives per unit.
   b. is too much variation in the process.
   c. has been a change in the average number of defects.
   d. None of the above
9. How do we designate the mean of the sample means?
   a. $\overline{\overline{X}}$
   b. A_2
   c. p
   d. $\overline{R}$
10. Which of the following charts reports the proportion of defective units?
   a. c bar chart
   b. Mean chart
   c. Percentage defective chart
   d. Range chart

COMPUTATIONAL PROBLEMS

11. A new assembly operation has just been started. A small group of assembly-line operators inserts parts, solders, and performs other tasks to produce a radio chassis. The numbers of defects per chassis for the first 10 produced are 0, 1, 0, 2, 3, 0, 1, 1, 2, and 4. Develop an appropriate control chart. Interpret the chart. Are there any points that seem to be out of control?

12. The inspection department at Bolger's Coffee Company is charged with the responsibility to monitor the weights of its 1-pound jars of instant coffee. The chief inspector selected a sample of 5 jars each hour for the first 6 hours of production today. The information is reported below.

Time	Sample				
	1	2	3	4	5
8:00 A.M.	1.00	1.01	1.00	0.99	1.00
9:00 A.M.	1.01	1.00	1.01	1.02	1.00
10:00 A.M.	1.00	1.02	0.99	1.00	1.00
11:00 A.M.	1.00	1.00	0.99	1.01	1.02
12:00 P.M.	1.00	1.00	1.01	1.03	1.00
1:00 P.M.	1.01	1.00	0.98	0.97	1.00

a. Determine the control limits for the mean weight per jar.

b. Determine the control limits for the range.

c. Interpret the chart. Does it appear that the process is out of control?

13. Ms. Holler teaches fifth grade at Lacy Elementary School. She would like to keep a control chart for the percentage of students absent each day. For the first 10 days of school, the following numbers of students were absent each day: 5, 4, 3, 4, 5, 3, 7, 2, 3, and 4. There are 40 children in the class. Develop a control chart for the proportion of students absent each day. Were there any days when the number of absences seemed to be out of control?

ANSWERS TO SELF-REVIEW PROBLEMS

18-1 About 72% of the complaints involve poor food, poor care, and dirty conditions. These are the factors the administrator should address.

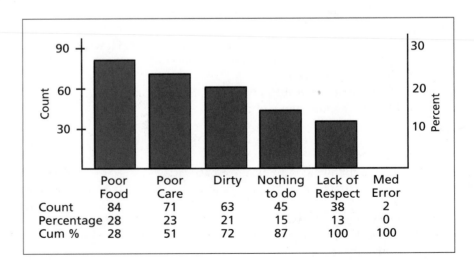

	Poor Food	Poor Care	Dirty	Nothing to do	Lack of Respect	Med Error
Count	84	71	63	45	38	2
Percentage	28	23	21	15	13	0
Cum %	28	51	72	87	100	100

18-2 a. $\overline{\overline{X}} = \dfrac{9}{3} = 3.00$, $\overline{R} = \dfrac{12}{3} = 4.00$

UCL = 3.00 + 0.729(4) = 5.916
LCL = 3.00 − 0.729(4) = 0.084

For the range chart:

UCL = $D_4 \overline{R}$ = 2.282 (4.0) = 9.128
LCL = $D_3 \overline{R}$ = 0

b. The measurements are in control.

18-3 $\overline{c} = \dfrac{25}{12} = 2.083$

UCL = $2.083 + 3\sqrt{2.083}$ = 2.083 + 4.330
    = 6.413
LCL = $2.083 - 3\sqrt{2.083}$ = 2.083 − 4.330
    = 0

The shift with 7 defects is out of control.

Unit Review

Chapters 16 and 17 presented some of the basic concepts of hypothesis testing that are appropriate when the scale of measurement is nominal or ordinal. We described several tests. In Chapter 16, we discussed the chi-square goodness-of-fit test and the chi-square test for independence. Both of these use the chi-square distribution as the test statistic and require only the nominal scale of measurement.

In Chapter 17, we began by describing the sign test. It is generally used in association with small dependent samples where we cannot assume the distribution of the differences follows the normal probability distribution. In this chapter, we also described three tests that require ordinal (ranked) data. The Wilcoxon signed-rank test requires dependent samples, and the Wilcoxon rank-sum test requires independent samples. The Kruskal-Wallis test is an alternative to the one-way analysis of variance where there is concern regarding the normality assumption.

Chapter 18 presented a description of control charts. We began by describing Pareto charts, which are used to tally the numbers and types of defects, and fishbone diagrams, which are used to emphasize the relationship between possible problem causes and a particular effect. We also presented variable control charts for the mean and the range and attribute charts for the proportion defective and number of defects per unit.

KEY CONCEPTS

1. The **nominal level of measurement** is considered the "lowest" level of measurement. Data on this level can be classified only into categories; there is no particular order for the categories, and the categories are mutually exclusive. Recall that "mutually exclusive" means a respondent can be placed in one and only one category.

2. The **ordinal level of measurement** is the next highest level of measurement. For ordinal-scaled data, one category is rated higher than the previous one. The categories are also mutually exclusive.

3. Hypothesis tests applied to nominal-level and ordinal-level data are often referred to as **nonparametric tests**, or **distribution-free tests**. These tests do not make any assumptions about the distribution of the population from which the sample or samples are selected; that is, to apply these nonparametric tests we need not have a normally distributed population.

4. The **chi-square goodness-of-fit test** requires only nominal-scaled data. It is concerned with a single trait—the religious affiliation of the respondent, for example. The purpose of the test is to find out how well an observed set of data compares to an expected set of data.

5. **Contingency tables,** an application of the chi-square statistic, are concerned with the simultaneous classification of two traits; that is, are the two traits related or not? The chi-square statistic compares observed frequencies with expected frequencies.

6. The **chi-square distribution** has the following major characteristics:
   a. Its value is nonnegative.
   b. There is a different chi-square distribution for each number of degrees of freedom.
   c. It is positively skewed, but approaches a symmetrical distribution as the number of degrees of freedom increases.

7. The **sign test** is used to investigate changes in paired or related observations. The sign of the difference may be either positive or negative, and the binomial distribution is used as the test statistic.

8. The **Wilcoxon signed-rank test** is an extension of the sign test. It considers not only the signs of

587

the differences in related observations, but also the magnitudes of the differences.

9. The **Wilcoxon rank-sum test** concerns random samples taken from two independent populations. It is an alternative to the t test, but the assumption of a normal population is not required. If both samples are at least of eight items, we use a test statistic which follows the standard normal distribution.

10. The **Kruskal-Wallis test** concerns random samples obtained from more than two independent samples. It is a nonparametric alternative to the ANOVA test, but the normality assumption is not required. The chi-square distribution is used as the test statistic.

11. A **Pareto chart** is a quantitative technique used to tally the numbers and types of defects that occur within a product or service. The concept of the chart is that about 80% of the problems are caused by about 20% of the factors.

12. A **fishbone diagram** is a diagnostic chart used to emphasize the relationship between an effect and a set of possible causes that produce that particular effect. The diagram is particularly useful to organize ideas and identify relationships.

13. There are two types of variable control charts. In the **control chart for the mean,** periodic samples of a particular size are taken from the process, and the mean is computed. The mean is compared to the control limits to determine if the variation is random or can be attributed to a change in the process. The **control chart for the range,** also based on periodic samples from the process, is used to determine if the variation can be attributed to randomness or if there has been a change in the process that produces too much or too little variation.

14. There are two attribute charts. In the **percentage defective chart,** a sampled item is classified as acceptable or unacceptable. The proportion of defective items in the sample is then compared to the control limits to determine if the proportion defective in the process has changed. In the **c bar chart,** we study the number of defects per unit. Control limits are established based on the number of defects per unit. The number of defects per unit is then compared to the limits to determine if there has been a change in the process.

KEY TERMS

Nonparametric tests	Sign test	Attribute
Distribution-free tests	Wilcoxon signed-rank test	Variable
Chi-square distribution	Wilcoxon rank-sum test	Pareto chart
Observed frequency	Kruskal-Wallis one-way analysis	Fishbone diagram
Expected frequency	of variance by ranks	Mean chart
Goodness-of-fit test	Chance cause	Range chart
Contingency table	Assignable cause	p chart
		c bar chart

KEY SYMBOLS

f_e	Frequency expected.		R^-	Sum of the negative signed differences.
f_0	Frequency observed.		W	Test statistic for the Wilcoxon-rank sum test.
k	Number of categories.			
χ^2	Chi-square.		H	Test statistic for the Kruskal-Wallis test.
R^+	Sum of the positive signed differences.		$\overline{\overline{X}}$	Mean of the sample means.

$\overline{R}$	Mean of the sample ranges.	
UCL	Upper control limit.	
LCL	Lower control limit.	
A_2	Constant used to determine control limits for means.	

D_3, D_4	Constants used to determine control limits for ranges.
$\overline{p}$	Mean percentage defective.
$\overline{c}$	Mean number of defects per unit.

CASE STUDY · Driving Drunk

Professor John Thomas has taught health education at the college level for many years. Most of his research in recent years has focused on the problem of drunk driving. Recently, with some funds obtained as part of a summer research grant, he has been able to purchase the Rupple Driving Simulator. This device allows a subject to take a "road test" and provides a score indicating the number of driving errors committed during the test drive. Higher scores indicate more driving errors. Driving errors include not coming to a complete stop at a stop sign, not using turn signals, not exercising caution on wet or snowy pavement, and so on. During the road test, problems appear at random, and not all problems appear in each road test. These are major advantages to the Rupple Driving Simulator because subjects do not gain any advantage by taking the test several times.

With the new driving simulator, Professor Thomas would like to study in detail the problem of drunk driving. He begins by selecting a random sample of 25 drivers. He then asks each of the selected individuals to take the test drive on the Rupple Driving Simulator. The numbers of errors for each driver are recorded below. Next, he has each of the individuals in the group drink 3 16-ounce cans of beer in a 60-minute period and return to the Rupple Driving Simulator for another test drive. The numbers of driving errors after drinking the beer are also shown below. Professor Thomas' research question: Does alcohol impair the driver's ability and therefore increase the number of driving errors?

Professor Thomas believes that the distribution of scores on the test drive does not follow a normal distribution and therefore that a nonparametric test should be used. Because the observations are paired, he decides to use both the sign test and the Wil-

coxon signed-rank test. Compare the results using these two procedures. Which statistical test would you suggest? What conclusion would you reach regarding the effects of drunk driving? Write a brief report summarizing your findings.

Subject	Without Alcohol	With Alcohol
1	75	89
2	78	83
3	89	80
4	100	90
5	85	84
6	70	68
7	64	84
8	79	104
9	83	81
10	82	88
11	83	93
12	84	92
13	80	103
14	72	106
15	83	89
16	99	89
17	75	77
18	58	78
19	93	108
20	69	69
21	86	84
22	97	86
23	65	92
24	96	97
25	85	94

A   Binomial Probability Distribution

n = 1
Probability

X	0.05	0.10	0.20	0.30	0.40	0.50	0.60	0.70	0.80	0.90	0.95
0	0.950	0.900	0.800	0.700	0.600	0.500	0.400	0.300	0.200	0.100	0.050
1	0.050	0.100	0.200	0.300	0.400	0.500	0.600	0.700	0.800	0.900	0.950

n = 2
Probability

X	0.05	0.10	0.20	0.30	0.40	0.50	0.60	0.70	0.80	0.90	0.95
0	0.903	0.810	0.640	0.490	0.360	0.250	0.160	0.090	0.040	0.010	0.003
1	0.095	0.180	0.320	0.420	0.480	0.500	0.480	0.420	0.320	0.180	0.095
2	0.003	0.010	0.040	0.090	0.160	0.250	0.360	0.490	0.640	0.810	0.903

n = 3
Probability

X	0.05	0.10	0.20	0.30	0.40	0.50	0.60	0.70	0.80	0.90	0.95
0	0.857	0.729	0.512	0.343	0.216	0.125	0.064	0.027	0.008	0.001	0.000
1	0.135	0.243	0.384	0.441	0.432	0.375	0.288	0.189	0.096	0.027	0.007
2	0.007	0.027	0.096	0.189	0.288	0.375	0.432	0.441	0.384	0.243	0.135
3	0.000	0.001	0.008	0.027	0.064	0.125	0.216	0.343	0.512	0.729	0.857

n = 4
Probability

X	0.05	0.10	0.20	0.30	0.40	0.50	0.60	0.70	0.80	0.90	0.95
0	0.815	0.656	0.410	0.240	0.130	0.063	0.026	0.008	0.002	0.000	0.000
1	0.171	0.292	0.410	0.412	0.346	0.250	0.154	0.076	0.026	0.004	0.000
2	0.014	0.049	0.154	0.265	0.346	0.375	0.346	0.265	0.154	0.049	0.014
3	0.000	0.004	0.026	0.076	0.154	0.250	0.346	0.412	0.410	0.292	0.171
4	0.000	0.000	0.002	0.008	0.026	0.063	0.130	0.240	0.410	0.656	0.815

n = 5
Probability

X	0.05	0.10	0.20	0.30	0.40	0.50	0.60	0.70	0.80	0.90	0.95
0	0.774	0.590	0.328	0.168	0.078	0.031	0.010	0.002	0.000	0.000	0.000
1	0.204	0.328	0.410	0.360	0.259	0.156	0.077	0.028	0.006	0.000	0.000
2	0.021	0.073	0.205	0.309	0.346	0.313	0.230	0.132	0.051	0.008	0.001
3	0.001	0.008	0.051	0.132	0.230	0.313	0.346	0.309	0.205	0.073	0.021
4	0.000	0.000	0.006	0.028	0.077	0.156	0.259	0.360	0.410	0.328	0.204
5	0.000	0.000	0.000	0.002	0.010	0.031	0.078	0.168	0.328	0.590	0.774

$n = 6$
Probability

X	0.05	0.10	0.20	0.30	0.40	0.50	0.60	0.70	0.80	0.90	0.95
0	0.735	0.531	0.262	0.118	0.047	0.016	0.004	0.001	0.000	0.000	0.000
1	0.232	0.354	0.393	0.303	0.187	0.094	0.037	0.010	0.002	0.000	0.000
2	0.031	0.098	0.246	0.324	0.311	0.234	0.138	0.060	0.015	0.001	0.000
3	0.002	0.015	0.082	0.185	0.276	0.313	0.276	0.185	0.082	0.015	0.002
4	0.000	0.001	0.015	0.060	0.138	0.234	0.311	0.324	0.246	0.098	0.031
5	0.000	0.000	0.002	0.010	0.037	0.094	0.187	0.303	0.393	0.354	0.232
6	0.000	0.000	0.000	0.001	0.004	0.016	0.047	0.118	0.262	0.531	0.735

$n = 7$
Probability

X	0.05	0.10	0.20	0.30	0.40	0.50	0.60	0.70	0.80	0.90	0.95
0	0.698	0.478	0.210	0.082	0.028	0.008	0.002	0.000	0.000	0.000	0.000
1	0.257	0.372	0.367	0.247	0.131	0.055	0.017	0.004	0.000	0.000	0.000
2	0.041	0.124	0.275	0.318	0.261	0.164	0.077	0.025	0.004	0.000	0.000
3	0.004	0.023	0.115	0.227	0.290	0.273	0.194	0.097	0.029	0.003	0.000
4	0.000	0.003	0.029	0.097	0.194	0.273	0.290	0.227	0.115	0.023	0.004
5	0.000	0.000	0.004	0.025	0.077	0.164	0.261	0.318	0.275	0.124	0.041
6	0.000	0.000	0.000	0.004	0.017	0.055	0.131	0.247	0.367	0.372	0.257
7	0.000	0.000	0.000	0.000	0.002	0.008	0.028	0.082	0.210	0.478	0.698

$n = 8$
Probability

X	0.05	0.10	0.20	0.30	0.40	0.50	0.60	0.70	0.80	0.90	0.95
0	0.663	0.430	0.168	0.058	0.017	0.004	0.001	0.000	0.000	0.000	0.000
1	0.279	0.383	0.336	0.198	0.090	0.031	0.008	0.001	0.000	0.000	0.000
2	0.051	0.149	0.294	0.296	0.209	0.109	0.041	0.010	0.001	0.000	0.000
3	0.005	0.033	0.147	0.254	0.279	0.219	0.124	0.047	0.009	0.000	0.000
4	0.000	0.005	0.046	0.136	0.232	0.273	0.232	0.136	0.046	0.005	0.000
5	0.000	0.000	0.009	0.047	0.124	0.219	0.279	0.254	0.147	0.033	0.005
6	0.000	0.000	0.001	0.010	0.041	0.109	0.209	0.296	0.294	0.149	0.051
7	0.000	0.000	0.000	0.001	0.008	0.031	0.090	0.198	0.336	0.383	0.279
8	0.000	0.000	0.000	0.000	0.001	0.004	0.017	0.058	0.168	0.430	0.663

n = 9
Probability

X	0.05	0.10	0.20	0.30	0.40	0.50	0.60	0.70	0.80	0.90	0.95
0	0.630	0.387	0.134	0.040	0.010	0.002	0.000	0.000	0.000	0.000	0.000
1	0.299	0.387	0.302	0.156	0.060	0.018	0.004	0.000	0.000	0.000	0.000
2	0.063	0.172	0.302	0.267	0.161	0.070	0.021	0.004	0.000	0.000	0.000
3	0.008	0.045	0.176	0.267	0.251	0.164	0.074	0.021	0.003	0.000	0.000
4	0.001	0.007	0.066	0.172	0.251	0.246	0.167	0.074	0.017	0.001	0.000
5	0.000	0.001	0.017	0.074	0.167	0.246	0.251	0.172	0.066	0.007	0.001
6	0.000	0.000	0.003	0.021	0.074	0.164	0.251	0.267	0.176	0.045	0.008
7	0.000	0.000	0.000	0.004	0.021	0.070	0.161	0.267	0.302	0.172	0.063
8	0.000	0.000	0.000	0.000	0.004	0.018	0.060	0.156	0.302	0.387	0.299
9	0.000	0.000	0.000	0.000	0.000	0.002	0.010	0.040	0.134	0.387	0.630

n = 10
Probability

X	0.05	0.10	0.20	0.30	0.40	0.50	0.60	0.70	0.80	0.90	0.95
0	0.599	0.349	0.107	0.028	0.006	0.001	0.000	0.000	0.000	0.000	0.000
1	0.315	0.387	0.268	0.121	0.040	0.010	0.002	0.000	0.000	0.000	0.000
2	0.075	0.194	0.302	0.233	0.121	0.044	0.011	0.001	0.000	0.000	0.000
3	0.010	0.057	0.201	0.267	0.215	0.117	0.042	0.009	0.001	0.000	0.000
4	0.001	0.011	0.088	0.200	0.251	0.205	0.111	0.037	0.006	0.000	0.000
5	0.000	0.001	0.026	0.103	0.201	0.246	0.201	0.103	0.026	0.001	0.000
6	0.000	0.000	0.006	0.037	0.111	0.205	0.251	0.200	0.088	0.011	0.001
7	0.000	0.000	0.001	0.009	0.042	0.117	0.215	0.267	0.201	0.057	0.010
8	0.000	0.000	0.000	0.001	0.011	0.044	0.121	0.233	0.302	0.194	0.075
9	0.000	0.000	0.000	0.000	0.002	0.010	0.040	0.121	0.268	0.387	0.315
10	0.000	0.000	0.000	0.000	0.000	0.001	0.006	0.028	0.107	0.349	0.599

n = 11
Probability

X	0.05	0.10	0.20	0.30	0.40	0.50	0.60	0.70	0.80	0.90	0.95
0	0.569	0.314	0.086	0.020	0.004	0.000	0.000	0.000	0.000	0.000	0.000
1	0.329	0.384	0.236	0.093	0.027	0.005	0.001	0.000	0.000	0.000	
2	0.087	0.213	0.295	0.200	0.089	0.027	0.005	0.001	0.000	0.000	0.000
3	0.014	0.071	0.221	0.257	0.177	0.081	0.023	0.004	0.000	0.000	0.000
4	0.001	0.016	0.111	0.220	0.236	0.161	0.070	0.017	0.002	0.000	0.000
5	0.000	0.002	0.039	0.132	0.221	0.226	0.147	0.057	0.010	0.000	0.000
6	0.000	0.000	0.010	0.057	0.147	0.226	0.221	0.132	0.039	0.002	0.000
7	0.000	0.000	0.002	0.017	0.070	0.161	0.236	0.220	0.111	0.016	0.001
8	0.000	0.000	0.000	0.004	0.023	0.081	0.177	0.257	0.221	0.071	0.014
9	0.000	0.000	0.000	0.001	0.005	0.027	0.089	0.200	0.295	0.213	0.087
10	0.000	0.000	0.000	0.000	0.001	0.005	0.027	0.093	0.236	0.384	0.329
11	0.000	0.000	0.000	0.000	0.000	0.000	0.004	0.020	0.086	0.314	0.569

$n = 12$
Probability

X	0.05	0.10	0.20	0.30	0.40	0.50	0.60	0.70	0.80	0.90	0.95
0	0.540	0.282	0.069	0.014	0.002	0.000	0.000	0.000	0.000	0.000	0.000
1	0.341	0.377	0.206	0.071	0.017	0.003	0.000	0.000	0.000	0.000	0.000
2	0.099	0.230	0.283	0.168	0.064	0.016	0.002	0.000	0.000	0.000	0.000
3	0.017	0.085	0.236	0.240	0.142	0.054	0.012	0.001	0.000	0.000	0.000
4	0.002	0.021	0.133	0.231	0.213	0.121	0.042	0.008	0.001	0.000	0.000
5	0.000	0.004	0.053	0.158	0.227	0.193	0.101	0.029	0.003	0.000	0.000
6	0.000	0.000	0.016	0.079	0.177	0.226	0.177	0.079	0.016	0.000	0.000
7	0.000	0.000	0.003	0.029	0.101	0.193	0.227	0.158	0.053	0.004	0.000
8	0.000	0.000	0.001	0.008	0.042	0.121	0.213	0.231	0.133	0.021	0.002
9	0.000	0.000	0.000	0.001	0.012	0.054	0.142	0.240	0.236	0.085	0.017
10	0.000	0.000	0.000	0.000	0.002	0.016	0.064	0.168	0.283	0.230	0.099
11	0.000	0.000	0.000	0.000	0.000	0.003	0.017	0.071	0.206	0.377	0.341
12	0.000	0.000	0.000	0.000	0.000	0.000	0.002	0.014	0.069	0.282	0.540

$n = 13$
Probability

X	0.05	0.10	0.20	0.30	0.40	0.50	0.60	0.70	0.80	0.90	0.95
0	0.513	0.254	0.055	0.010	0.001	0.000	0.000	0.000	0.000	0.000	0.000
1	0.351	0.367	0.179	0.054	0.011	0.002	0.000	0.000	0.000	0.000	0.000
2	0.111	0.245	0.268	0.139	0.045	0.010	0.001	0.000	0.000	0.000	0.000
3	0.021	0.100	0.246	0.218	0.111	0.035	0.006	0.001	0.000	0.000	0.000
4	0.003	0.028	0.154	0.234	0.184	0.087	0.024	0.003	0.000	0.000	0.000
5	0.000	0.006	0.069	0.180	0.221	0.157	0.066	0.014	0.001	0.000	0.000
6	0.000	0.001	0.023	0.103	0.197	0.209	0.131	0.044	0.006	0.000	0.000
7	0.000	0.000	0.006	0.044	0.131	0.209	0.197	0.103	0.023	0.001	0.000
8	0.000	0.000	0.001	0.014	0.066	0.157	0.221	0.180	0.069	0.006	0.000
9	0.000	0.000	0.000	0.003	0.024	0.087	0.184	0.234	0.154	0.028	0.003
10	0.000	0.000	0.000	0.001	0.006	0.035	0.111	0.218	0.246	0.100	0.021
11	0.000	0.000	0.000	0.000	0.001	0.010	0.045	0.139	0.268	0.245	0.111
12	0.000	0.000	0.000	0.000	0.000	0.002	0.011	0.054	0.179	0.367	0.351
13	0.000	0.000	0.000	0.000	0.000	0.000	0.001	0.010	0.055	0.254	0.513

n = 14
Probability

X	0.05	0.10	0.20	0.30	0.40	0.50	0.60	0.70	0.80	0.90	0.95
0	0.488	0.229	0.044	0.007	0.001	0.000	0.000	0.000	0.000	0.000	0.000
1	0.359	0.356	0.154	0.041	0.007	0.001	0.000	0.000	0.000	0.000	0.000
2	0.123	0.257	0.250	0.113	0.032	0.006	0.001	0.000	0.000	0.000	0.000
3	0.026	0.114	0.250	0.194	0.085	0.022	0.003	0.000	0.000	0.000	0.000
4	0.004	0.035	0.172	0.229	0.155	0.061	0.014	0.001	0.000	0.000	0.000
5	0.000	0.008	0.086	0.196	0.207	0.122	0.041	0.007	0.000	0.000	0.000
6	0.000	0.001	0.032	0.126	0.207	0.183	0.092	0.023	0.002	0.000	0.000
7	0.000	0.000	0.009	0.062	0.157	0.209	0.157	0.062	0.009	0.000	0.000
8	0.000	0.000	0.002	0.023	0.092	0.183	0.207	0.126	0.032	0.001	0.000
9	0.000	0.000	0.000	0.007	0.041	0.122	0.207	0.196	0.086	0.008	0.000
10	0.000	0.000	0.000	0.001	0.014	0.061	0.155	0.229	0.172	0.035	0.004
11	0.000	0.000	0.000	0.000	0.003	0.022	0.085	0.194	0.250	0.114	0.026
12	0.000	0.000	0.000	0.000	0.001	0.006	0.032	0.113	0.250	0.257	0.123
13	0.000	0.000	0.000	0.000	0.000	0.001	0.007	0.041	0.154	0.356	0.359
14	0.000	0.000	0.000	0.000	0.000	0.000	0.001	0.007	0.044	0.229	0.488

n = 15
Probability

X	0.05	0.10	0.20	0.30	0.40	0.50	0.60	0.70	0.80	0.90	0.95
0	0.463	0.206	0.035	0.005	0.000	0.000	0.000	0.000	0.000	0.000	0.000
1	0.366	0.343	0.132	0.031	0.005	0.000	0.000	0.000	0.000	0.000	0.000
2	0.135	0.267	0.231	0.092	0.022	0.003	0.000	0.000	0.000	0.000	0.000
3	0.031	0.129	0.250	0.170	0.063	0.014	0.002	0.000	0.000	0.000	0.000
4	0.005	0.043	0.188	0.219	0.127	0.042	0.007	0.001	0.000	0.000	0.000
5	0.001	0.010	0.103	0.206	0.186	0.092	0.024	0.003	0.000	0.000	0.000
6	0.000	0.002	0.043	0.147	0.207	0.153	0.061	0.012	0.001	0.000	0.000
7	0.000	0.000	0.014	0.081	0.177	0.196	0.118	0.035	0.003	0.000	0.000
8	0.000	0.000	0.003	0.035	0.118	0.196	0.177	0.081	0.014	0.000	0.000
9	0.000	0.000	0.001	0.012	0.061	0.153	0.207	0.147	0.043	0.002	0.000
10	0.000	0.000	0.000	0.003	0.024	0.092	0.186	0.206	0.103	0.010	0.001
11	0.000	0.000	0.000	0.001	0.007	0.042	0.127	0.219	0.188	0.043	0.005
12	0.000	0.000	0.000	0.000	0.002	0.014	0.063	0.170	0.250	0.129	0.031
13	0.000	0.000	0.000	0.000	0.000	0.003	0.022	0.092	0.231	0.267	0.135
14	0.000	0.000	0.000	0.000	0.000	0.000	0.005	0.031	0.132	0.343	0.366
15	0.000	0.000	0.000	0.000	0.000	0.000	0.000	0.005	0.035	0.206	0.463

$n = 16$
Probability

X	0.05	0.10	0.20	0.30	0.40	0.50	0.60	0.70	0.80	0.90	0.95
0	0.440	0.185	0.028	0.003	0.000	0.000	0.000	0.000	0.000	0.000	0.000
1	0.371	0.329	0.113	0.023	0.003	0.000	0.000	0.000	0.000	0.000	0.000
2	0.146	0.275	0.211	0.073	0.015	0.002	0.000	0.000	0.000	0.000	0.000
3	0.036	0.142	0.246	0.146	0.047	0.009	0.001	0.000	0.000	0.000	0.000
4	0.006	0.051	0.200	0.204	0.101	0.028	0.004	0.000	0.000	0.000	0.000
5	0.001	0.014	0.120	0.210	0.162	0.067	0.014	0.001	0.000	0.000	0.000
6	0.000	0.003	0.055	0.165	0.198	0.122	0.039	0.006	0.000	0.000	0.000
7	0.000	0.000	0.020	0.101	0.189	0.175	0.084	0.019	0.001	0.000	0.000
8	0.000	0.000	0.006	0.049	0.142	0.196	0.142	0.049	0.006	0.000	0.000
9	0.000	0.000	0.001	0.019	0.084	0.175	0.189	0.101	0.020	0.000	0.000
10	0.000	0.000	0.000	0.006	0.039	0.122	0.198	0.165	0.055	0.003	0.000
11	0.000	0.000	0.000	0.001	0.014	0.067	0.162	0.210	0.120	0.014	0.001
12	0.000	0.000	0.000	0.000	0.004	0.028	0.101	0.204	0.200	0.051	0.006
13	0.000	0.000	0.000	0.000	0.001	0.009	0.047	0.146	0.246	0.142	0.036
14	0.000	0.000	0.000	0.000	0.000	0.002	0.015	0.073	0.211	0.275	0.146
15	0.000	0.000	0.000	0.000	0.000	0.000	0.003	0.023	0.113	0.329	0.371
16	0.000	0.000	0.000	0.000	0.000	0.000	0.000	0.003	0.028	0.185	0.440

$n = 17$
Probability

X	0.05	0.10	0.20	0.30	0.40	0.50	0.60	0.70	0.80	0.90	0.95
0	0.418	0.167	0.023	0.002	0.000	0.000	0.000	0.000	0.000	0.000	0.000
1	0.374	0.315	0.096	0.017	0.002	0.000	0.000	0.000	0.000	0.000	0.000
2	0.158	0.280	0.191	0.058	0.010	0.001	0.000	0.000	0.000	0.000	0.000
3	0.041	0.156	0.239	0.125	0.034	0.005	0.000	0.000	0.000	0.000	0.000
4	0.008	0.060	0.209	0.187	0.080	0.018	0.002	0.000	0.000	0.000	0.000
5	0.001	0.017	0.136	0.208	0.138	0.047	0.008	0.001	0.000	0.000	0.000
6	0.000	0.004	0.068	0.178	0.184	0.094	0.024	0.003	0.000	0.000	0.000
7	0.000	0.001	0.027	0.120	0.193	0.148	0.057	0.009	0.000	0.000	0.000
8	0.000	0.000	0.008	0.064	0.161	0.185	0.107	0.028	0.002	0.000	0.000
9	0.000	0.000	0.002	0.028	0.107	0.185	0.161	0.064	0.008	0.000	0.000
10	0.000	0.000	0.000	0.009	0.057	0.148	0.193	0.120	0.027	0.001	0.000
11	0.000	0.000	0.000	0.003	0.024	0.094	0.184	0.178	0.068	0.004	0.000
12	0.000	0.000	0.000	0.001	0.008	0.047	0.138	0.208	0.136	0.017	0.001
13	0.000	0.000	0.000	0.000	0.002	0.018	0.080	0.187	0.209	0.060	0.008
14	0.000	0.000	0.000	0.000	0.000	0.005	0.034	0.125	0.239	0.156	0.041
15	0.000	0.000	0.000	0.000	0.000	0.001	0.010	0.058	0.191	0.280	0.158
16	0.000	0.000	0.000	0.000	0.000	0.000	0.002	0.017	0.096	0.315	0.374
17	0.000	0.000	0.000	0.000	0.000	0.000	0.000	0.002	0.023	0.167	0.418

$n = 18$
Probability

X	0.05	0.10	0.20	0.30	0.40	0.50	0.60	0.70	0.80	0.90	0.95
0	0.397	0.150	0.018	0.002	0.000	0.000	0.000	0.000	0.000	0.000	0.000
1	0.376	0.300	0.081	0.013	0.001	0.000	0.000	0.000	0.000	0.000	0.000
2	0.168	0.284	0.172	0.046	0.007	0.001	0.000	0.000	0.000	0.000	0.000
3	0.047	0.168	0.230	0.105	0.025	0.003	0.000	0.000	0.000	0.000	0.000
4	0.009	0.070	0.215	0.168	0.061	0.012	0.001	0.000	0.000	0.000	0.000
5	0.001	0.022	0.151	0.202	0.115	0.033	0.004	0.000	0.000	0.000	0.000
6	0.000	0.005	0.082	0.187	0.166	0.071	0.015	0.001	0.000	0.000	0.000
7	0.000	0.001	0.035	0.138	0.189	0.121	0.037	0.005	0.000	0.000	0.000
8	0.000	0.000	0.012	0.081	0.173	0.167	0.077	0.015	0.001	0.000	0.000
9	0.000	0.000	0.003	0.039	0.128	0.185	0.128	0.039	0.003	0.000	0.000
10	0.000	0.000	0.001	0.015	0.077	0.167	0.173	0.081	0.012	0.000	0.000
11	0.000	0.000	0.000	0.005	0.037	0.121	0.189	0.138	0.035	0.001	0.000
12	0.000	0.000	0.000	0.001	0.015	0.071	0.166	0.187	0.082	0.005	0.000
13	0.000	0.000	0.000	0.000	0.004	0.033	0.115	0.202	0.151	0.022	0.001
14	0.000	0.000	0.000	0.000	0.001	0.012	0.061	0.168	0.215	0.070	0.009
15	0.000	0.000	0.000	0.000	0.000	0.003	0.025	0.105	0.230	0.168	0.047
16	0.000	0.000	0.000	0.000	0.000	0.001	0.007	0.046	0.172	0.284	0.168
17	0.000	0.000	0.000	0.000	0.000	0.000	0.001	0.013	0.081	0.300	0.376
18	0.000	0.000	0.000	0.000	0.000	0.000	0.000	0.002	0.018	0.150	0.397

$n = 19$
Probability

X	0.05	0.10	0.20	0.30	0.40	0.50	0.60	0.70	0.80	0.90	0.95
0	0.377	0.135	0.014	0.001	0.000	0.000	0.000	0.000	0.000	0.000	0.000
1	0.377	0.285	0.068	0.009	0.001	0.000	0.000	0.000	0.000	0.000	0.000
2	0.179	0.285	0.154	0.036	0.005	0.000	0.000	0.000	0.000	0.000	0.000
3	0.053	0.180	0.218	0.087	0.017	0.002	0.000	0.000	0.000	0.000	0.000
4	0.011	0.080	0.218	0.149	0.047	0.007	0.001	0.000	0.000	0.000	0.000
5	0.002	0.027	0.164	0.192	0.093	0.022	0.002	0.000	0.000	0.000	0.000
6	0.000	0.007	0.095	0.192	0.145	0.052	0.008	0.001	0.000	0.000	0.000
7	0.000	0.001	0.044	0.153	0.180	0.096	0.024	0.002	0.000	0.000	0.000
8	0.000	0.000	0.017	0.098	0.180	0.144	0.053	0.008	0.000	0.000	0.000
9	0.000	0.000	0.005	0.051	0.146	0.176	0.098	0.022	0.001	0.000	0.000
10	0.000	0.000	0.001	0.022	0.098	0.176	0.146	0.051	0.005	0.000	0.000
11	0.000	0.000	0.000	0.008	0.053	0.144	0.180	0.098	0.017	0.000	0.000
12	0.000	0.000	0.000	0.002	0.024	0.096	0.180	0.153	0.044	0.001	0.000
13	0.000	0.000	0.000	0.001	0.008	0.052	0.145	0.192	0.095	0.007	0.000
14	0.000	0.000	0.000	0.000	0.002	0.022	0.093	0.192	0.164	0.027	0.002
15	0.000	0.000	0.000	0.000	0.001	0.007	0.047	0.149	0.218	0.080	0.011
16	0.000	0.000	0.000	0.000	0.000	0.002	0.017	0.087	0.218	0.180	0.053
17	0.000	0.000	0.000	0.000	0.000	0.000	0.005	0.036	0.154	0.285	0.179
18	0.000	0.000	0.000	0.000	0.000	0.000	0.001	0.009	0.068	0.285	0.377
19	0.000	0.000	0.000	0.000	0.000	0.000	0.000	0.001	0.014	0.135	0.377

$n = 20$
Probability

X	0.05	0.10	0.20	0.30	0.40	0.50	0.60	0.70	0.80	0.90	0.95
0	0.358	0.122	0.012	0.001	0.000	0.000	0.000	0.000	0.000	0.000	0.000
1	0.377	0.270	0.058	0.007	0.000	0.000	0.000	0.000	0.000	0.000	0.000
2	0.189	0.285	0.137	0.028	0.003	0.000	0.000	0.000	0.000	0.000	0.000
3	0.060	0.190	0.205	0.072	0.012	0.001	0.000	0.000	0.000	0.000	0.000
4	0.013	0.090	0.218	0.130	0.035	0.005	0.000	0.000	0.000	0.000	0.000
5	0.002	0.032	0.175	0.179	0.075	0.015	0.001	0.000	0.000	0.000	0.000
6	0.000	0.009	0.109	0.192	0.124	0.037	0.005	0.000	0.000	0.000	0.000
7	0.000	0.002	0.055	0.164	0.166	0.074	0.015	0.001	0.000	0.000	0.000
8	0.000	0.000	0.022	0.114	0.180	0.120	0.035	0.004	0.000	0.000	0.000
9	0.000	0.000	0.007	0.065	0.160	0.160	0.071	0.012	0.000	0.000	0.000
10	0.000	0.000	0.002	0.031	0.117	0.176	0.117	0.031	0.002	0.000	0.000
11	0.000	0.000	0.000	0.012	0.071	0.160	0.160	0.065	0.007	0.000	0.000
12	0.000	0.000	0.000	0.004	0.035	0.120	0.180	0.114	0.022	0.000	0.000
13	0.000	0.000	0.000	0.001	0.015	0.074	0.166	0.164	0.055	0.002	0.000
14	0.000	0.000	0.000	0.000	0.005	0.037	0.124	0.192	0.109	0.009	0.000
15	0.000	0.000	0.000	0.000	0.001	0.015	0.075	0.179	0.175	0.032	0.002
16	0.000	0.000	0.000	0.000	0.000	0.005	0.035	0.130	0.218	0.090	0.013
17	0.000	0.000	0.000	0.000	0.000	0.001	0.012	0.072	0.205	0.190	0.060
18	0.000	0.000	0.000	0.000	0.000	0.000	0.003	0.028	0.137	0.285	0.189
19	0.000	0.000	0.000	0.000	0.000	0.000	0.000	0.007	0.058	0.270	0.377
20	0.000	0.000	0.000	0.000	0.000	0.000	0.000	0.001	0.012	0.122	0.358

$n = 25$
Probability

X	0.05	0.10	0.20	0.30	0.40	0.50	0.60	0.70	0.80	0.90	0.95
0	0.277	0.072	0.004	0.000	0.000	0.000	0.000	0.000	0.000	0.000	0.000
1	0.365	0.199	0.024	0.001	0.000	0.000	0.000	0.000	0.000	0.000	0.000
2	0.231	0.266	0.071	0.007	0.000	0.000	0.000	0.000	0.000	0.000	0.000
3	0.093	0.226	0.136	0.024	0.002	0.000	0.000	0.000	0.000	0.000	0.000
4	0.027	0.138	0.187	0.057	0.007	0.000	0.000	0.000	0.000	0.000	0.000
5	0.006	0.065	0.196	0.103	0.020	0.002	0.000	0.000	0.000	0.000	0.000
6	0.001	0.024	0.163	0.147	0.044	0.005	0.000	0.000	0.000	0.000	0.000
7	0.000	0.007	0.111	0.171	0.080	0.014	0.001	0.000	0.000	0.000	0.000
8	0.000	0.002	0.062	0.165	0.120	0.032	0.003	0.000	0.000	0.000	0.000
9	0.000	0.000	0.029	0.134	0.151	0.061	0.009	0.000	0.000	0.000	0.000
10	0.000	0.000	0.012	0.092	0.161	0.097	0.021	0.001	0.000	0.000	0.000
11	0.000	0.000	0.004	0.054	0.147	0.133	0.043	0.004	0.000	0.000	0.000
12	0.000	0.000	0.001	0.027	0.114	0.155	0.076	0.011	0.000	0.000	0.000
13	0.000	0.000	0.000	0.011	0.076	0.155	0.114	0.027	0.001	0.000	0.000
14	0.000	0.000	0.000	0.004	0.043	0.133	0.147	0.054	0.004	0.000	0.000
15	0.000	0.000	0.000	0.001	0.021	0.097	0.161	0.092	0.012	0.000	0.000
16	0.000	0.000	0.000	0.000	0.009	0.061	0.151	0.134	0.029	0.000	0.000
17	0.000	0.000	0.000	0.000	0.003	0.032	0.120	0.165	0.062	0.002	0.000
18	0.000	0.000	0.000	0.000	0.001	0.014	0.080	0.171	0.111	0.007	0.000
19	0.000	0.000	0.000	0.000	0.000	0.005	0.044	0.147	0.163	0.024	0.001
20	0.000	0.000	0.000	0.000	0.000	0.002	0.020	0.103	0.196	0.065	0.006
21	0.000	0.000	0.000	0.000	0.000	0.000	0.007	0.057	0.187	0.138	0.027
22	0.000	0.000	0.000	0.000	0.000	0.000	0.002	0.024	0.136	0.226	0.093
23	0.000	0.000	0.000	0.000	0.000	0.000	0.000	0.007	0.071	0.266	0.231
24	0.000	0.000	0.000	0.000	0.000	0.000	0.000	0.001	0.024	0.199	0.365
25	0.000	0.000	0.000	0.000	0.000	0.000	0.000	0.000	0.004	0.072	0.277

B   Poisson Distribution: Probability of Exactly X Occurrences

X	μ 0.1	0.2	0.3	0.4	0.5	0.6	0.7	0.8	0.9
0	0.9048	0.8187	0.7408	0.6703	0.6065	0.5488	0.4966	0.4493	0.4066
1	0.0905	0.1637	0.2222	0.2681	0.3033	0.3293	0.3476	0.3595	0.3659
2	0.0045	0.0164	0.0333	0.0536	0.0758	0.0988	0.1217	0.1438	0.1647
3	0.0002	0.0011	0.0033	0.0072	0.0126	0.0198	0.0284	0.0383	0.0494
4	0.0000	0.0001	0.0003	0.0007	0.0016	0.0030	0.0050	0.0077	0.0111
5	0.0000	0.0000	0.0000	0.0001	0.0002	0.0004	0.0007	0.0012	0.0020
6	0.0000	0.0000	0.0000	0.0000	0.0000	0.0000	0.0001	0.0002	0.0003
7	0.0000	0.0000	0.0000	0.0000	0.0000	0.0000	0.0000	0.0000	0.0000

X	μ 1.0	2.0	3.0	4.0	5.0	6.0	7.0	8.0	9.0
0	0.3679	0.1353	0.0498	0.0183	0.0067	0.0025	0.0009	0.0003	0.0001
1	0.3679	0.2707	0.1494	0.0733	0.0337	0.0149	0.0064	0.0027	0.0011
2	0.1839	0.2707	0.2240	0.1465	0.0842	0.0446	0.0223	0.0107	0.0050
3	0.0613	0.1804	0.2240	0.1954	0.1404	0.0892	0.0521	0.0286	0.0150
4	0.0153	0.0902	0.1680	0.1954	0.1755	0.1339	0.0912	0.0573	0.0337
5	0.0031	0.0361	0.1008	0.1563	0.1755	0.1606	0.1277	0.0916	0.0607
6	0.0005	0.0120	0.0504	0.1042	0.1462	0.1606	0.1490	0.1221	0.0911
7	0.0001	0.0034	0.0216	0.0595	0.1044	0.1377	0.1490	0.1396	0.1171
8	0.0000	0.0009	0.0081	0.0298	0.0653	0.1033	0.1304	0.1396	0.1318
9	0.0000	0.0002	0.0027	0.0132	0.0363	0.0688	0.1014	0.1241	0.1318
10	0.0000	0.0000	0.0008	0.0053	0.0181	0.0413	0.0710	0.0993	0.1186
11	0.0000	0.0000	0.0002	0.0019	0.0082	0.0225	0.0452	0.0722	0.0970
12	0.0000	0.0000	0.0001	0.0006	0.0034	0.0113	0.0263	0.0481	0.0728
13	0.0000	0.0000	0.0000	0.0002	0.0013	0.0052	0.0142	0.0296	0.0504
14	0.0000	0.0000	0.0000	0.0001	0.0005	0.0022	0.0071	0.0169	0.0324
15	0.0000	0.0000	0.0000	0.0000	0.0002	0.0009	0.0033	0.0090	0.0194
16	0.0000	0.0000	0.0000	0.0000	0.0000	0.0003	0.0014	0.0045	0.0109
17	0.0000	0.0000	0.0000	0.0000	0.0000	0.0001	0.0006	0.0021	0.0058
18	0.0000	0.0000	0.0000	0.0000	0.0000	0.0000	0.0002	0.0009	0.0029
19	0.0000	0.0000	0.0000	0.0000	0.0000	0.0000	0.0001	0.0004	0.0014
20	0.0000	0.0000	0.0000	0.0000	0.0000	0.0000	0.0000	0.0002	0.0006
21	0.0000	0.0000	0.0000	0.0000	0.0000	0.0000	0.0000	0.0001	0.0003
22	0.0000	0.0000	0.0000	0.0000	0.0000	0.0000	0.0000	0.0000	0.0001

C Areas Under the Normal Curve

Example:
If $z = 1.96$, then
.4750 of the area is
between 0 and 1.96

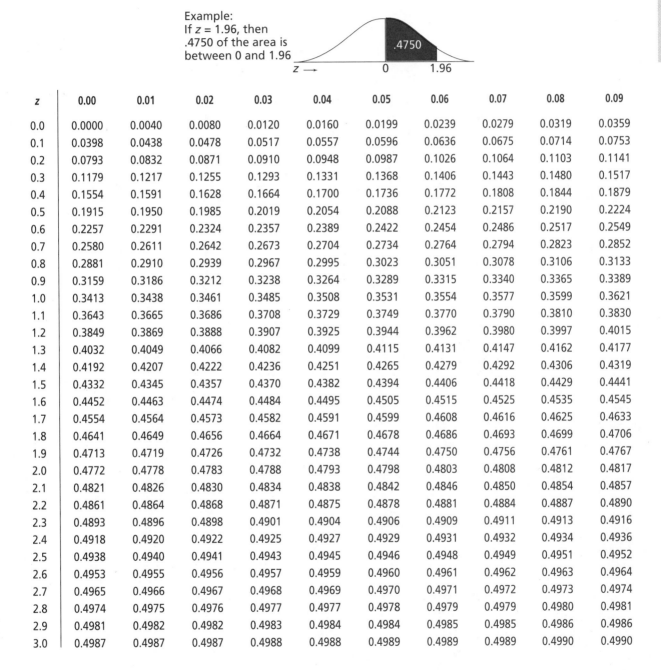

.4750

$z \longrightarrow$ 0   1.96

z	0.00	0.01	0.02	0.03	0.04	0.05	0.06	0.07	0.08	0.09
0.0	0.0000	0.0040	0.0080	0.0120	0.0160	0.0199	0.0239	0.0279	0.0319	0.0359
0.1	0.0398	0.0438	0.0478	0.0517	0.0557	0.0596	0.0636	0.0675	0.0714	0.0753
0.2	0.0793	0.0832	0.0871	0.0910	0.0948	0.0987	0.1026	0.1064	0.1103	0.1141
0.3	0.1179	0.1217	0.1255	0.1293	0.1331	0.1368	0.1406	0.1443	0.1480	0.1517
0.4	0.1554	0.1591	0.1628	0.1664	0.1700	0.1736	0.1772	0.1808	0.1844	0.1879
0.5	0.1915	0.1950	0.1985	0.2019	0.2054	0.2088	0.2123	0.2157	0.2190	0.2224
0.6	0.2257	0.2291	0.2324	0.2357	0.2389	0.2422	0.2454	0.2486	0.2517	0.2549
0.7	0.2580	0.2611	0.2642	0.2673	0.2704	0.2734	0.2764	0.2794	0.2823	0.2852
0.8	0.2881	0.2910	0.2939	0.2967	0.2995	0.3023	0.3051	0.3078	0.3106	0.3133
0.9	0.3159	0.3186	0.3212	0.3238	0.3264	0.3289	0.3315	0.3340	0.3365	0.3389
1.0	0.3413	0.3438	0.3461	0.3485	0.3508	0.3531	0.3554	0.3577	0.3599	0.3621
1.1	0.3643	0.3665	0.3686	0.3708	0.3729	0.3749	0.3770	0.3790	0.3810	0.3830
1.2	0.3849	0.3869	0.3888	0.3907	0.3925	0.3944	0.3962	0.3980	0.3997	0.4015
1.3	0.4032	0.4049	0.4066	0.4082	0.4099	0.4115	0.4131	0.4147	0.4162	0.4177
1.4	0.4192	0.4207	0.4222	0.4236	0.4251	0.4265	0.4279	0.4292	0.4306	0.4319
1.5	0.4332	0.4345	0.4357	0.4370	0.4382	0.4394	0.4406	0.4418	0.4429	0.4441
1.6	0.4452	0.4463	0.4474	0.4484	0.4495	0.4505	0.4515	0.4525	0.4535	0.4545
1.7	0.4554	0.4564	0.4573	0.4582	0.4591	0.4599	0.4608	0.4616	0.4625	0.4633
1.8	0.4641	0.4649	0.4656	0.4664	0.4671	0.4678	0.4686	0.4693	0.4699	0.4706
1.9	0.4713	0.4719	0.4726	0.4732	0.4738	0.4744	0.4750	0.4756	0.4761	0.4767
2.0	0.4772	0.4778	0.4783	0.4788	0.4793	0.4798	0.4803	0.4808	0.4812	0.4817
2.1	0.4821	0.4826	0.4830	0.4834	0.4838	0.4842	0.4846	0.4850	0.4854	0.4857
2.2	0.4861	0.4864	0.4868	0.4871	0.4875	0.4878	0.4881	0.4884	0.4887	0.4890
2.3	0.4893	0.4896	0.4898	0.4901	0.4904	0.4906	0.4909	0.4911	0.4913	0.4916
2.4	0.4918	0.4920	0.4922	0.4925	0.4927	0.4929	0.4931	0.4932	0.4934	0.4936
2.5	0.4938	0.4940	0.4941	0.4943	0.4945	0.4946	0.4948	0.4949	0.4951	0.4952
2.6	0.4953	0.4955	0.4956	0.4957	0.4959	0.4960	0.4961	0.4962	0.4963	0.4964
2.7	0.4965	0.4966	0.4967	0.4968	0.4969	0.4970	0.4971	0.4972	0.4973	0.4974
2.8	0.4974	0.4975	0.4976	0.4977	0.4977	0.4978	0.4979	0.4979	0.4980	0.4981
2.9	0.4981	0.4982	0.4982	0.4983	0.4984	0.4984	0.4985	0.4985	0.4986	0.4986
3.0	0.4987	0.4987	0.4987	0.4988	0.4988	0.4989	0.4989	0.4989	0.4990	0.4990

APPENDIX TABLES

D   Student *t* Distribution

df	Level of Significance for One-Tailed Test					
	.10	.05	.025	.01	.005	.0005
	Level of Significance for Two-Tailed Test					
	.20	.10	.05	.02	.01	.001
1	3.078	6.314	12.706	31.821	63.657	636.619
2	1.886	2.920	4.303	6.965	9.925	31.599
3	1.638	2.353	3.182	4.541	5.841	12.924
4	1.533	2.132	2.776	3.747	4.604	8.610
5	1.476	2.015	2.571	3.365	4.032	6.869
6	1.440	1.943	2.447	3.143	3.707	5.959
7	1.415	1.895	2.365	2.998	3.499	5.408
8	1.397	1.860	2.306	2.896	3.355	5.041
9	1.383	1.833	2.262	2.821	3.250	4.781
10	1.372	1.812	2.228	2.764	3.169	4.587
11	1.363	1.796	2.201	2.718	3.106	4.437
12	1.356	1.782	2.179	2.681	3.055	4.318
13	1.350	1.771	2.160	2.650	3.012	4.221
14	1.345	1.761	2.145	2.624	2.977	4.140
15	1.341	1.753	2.131	2.602	2.947	4.073
16	1.337	1.746	2.120	2.583	2.921	4.015
17	1.333	1.740	2.110	2.567	2.898	3.965
18	1.330	1.734	2.101	2.552	2.878	3.922
19	1.328	1.729	2.093	2.539	2.861	3.883
20	1.325	1.725	2.086	2.528	2.845	3.850
21	1.323	1.721	2.080	2.518	2.831	3.819
22	1.321	1.717	2.074	2.508	2.819	3.792
23	1.319	1.714	2.069	2.500	2.807	3.768
24	1.318	1.711	2.064	2.492	2.797	3.745
25	1.316	1.708	2.060	2.485	2.787	3.725
26	1.315	1.706	2.056	2.479	2.779	3.707
27	1.314	1.703	2.052	2.473	2.771	3.690
28	1.313	1.701	2.048	2.467	2.763	3.674
29	1.311	1.699	2.045	2.462	2.756	3.659
30	1.310	1.697	2.042	2.457	2.750	3.646
40	1.303	1.684	2.021	2.423	2.704	3.551
60	1.296	1.671	2.000	2.390	2.660	3.460
120	1.289	1.658	1.980	2.358	2.617	3.373
∞	1.282	1.645	1.960	2.326	2.576	3.291

E   The F Distribution at a 5 Percent Level of Significance

Degrees of Freedom for the Numerator

	1	2	3	4	5	6	7	8	9	10	12	15	20	24	30	40
1	161	200	216	225	230	234	237	239	241	242	244	246	248	249	250	251
2	18.5	19.0	19.2	19.2	19.3	19.3	19.4	19.4	19.4	19.4	19.4	19.4	19.4	19.5	19.5	19.5
3	10.1	9.55	9.28	9.12	9.01	8.94	8.89	8.85	8.81	8.79	8.74	8.70	8.66	8.64	8.62	8.59
4	7.71	6.94	6.59	6.39	6.26	6.16	6.09	6.04	6.00	5.96	5.91	5.86	5.80	5.77	5.75	5.72
5	6.61	5.79	5.41	5.19	5.05	4.95	4.88	4.82	4.77	4.74	4.68	4.62	4.56	4.53	4.50	4.46
6	5.99	5.14	4.76	4.53	4.39	4.28	4.21	4.15	4.10	4.06	4.00	3.94	3.87	3.84	3.81	3.77
7	5.59	4.74	4.35	4.12	3.97	3.87	3.79	3.73	3.68	3.64	3.57	3.51	3.44	3.41	3.38	3.34
8	5.32	4.46	4.07	3.84	3.69	3.58	3.50	3.44	3.39	3.35	3.28	3.22	3.15	3.12	3.08	3.04
9	5.12	4.26	3.86	3.63	3.48	3.37	3.29	3.23	3.18	3.14	3.07	3.01	2.94	2.90	2.86	2.83
10	4.96	4.10	3.71	3.48	3.33	3.22	3.14	3.07	3.02	2.98	2.91	2.85	2.77	2.74	2.70	2.66
11	4.84	3.98	3.59	3.36	3.20	3.09	3.01	2.95	2.90	2.85	2.79	2.72	2.65	2.61	2.57	2.53
12	4.75	3.89	3.49	3.26	3.11	3.00	2.91	2.85	2.80	2.75	2.69	2.62	2.54	2.51	2.47	2.43
13	4.67	3.81	3.41	3.18	3.03	2.92	2.83	2.77	2.71	2.67	2.60	2.53	2.46	2.42	2.38	2.34
14	4.60	3.74	3.34	3.11	2.96	2.85	2.76	2.70	2.65	2.60	2.53	2.46	2.39	2.35	2.31	2.27
15	4.54	3.68	3.29	3.06	2.90	2.79	2.71	2.64	2.59	2.54	2.48	2.40	2.33	2.29	2.25	2.20
16	4.49	3.63	3.24	3.01	2.85	2.74	2.66	2.59	2.54	2.49	2.42	2.35	2.28	2.24	2.19	2.15
17	4.45	3.59	3.20	2.96	2.81	2.70	2.61	2.55	2.49	2.45	2.38	2.31	2.23	2.19	2.15	2.10
18	4.41	3.55	3.16	2.93	2.77	2.66	2.58	2.51	2.46	2.41	2.34	2.27	2.19	2.15	2.11	2.06
19	4.38	3.52	3.13	2.90	2.74	2.63	2.54	2.48	2.42	2.38	2.31	2.23	2.16	2.11	2.07	2.03
20	4.35	3.49	3.10	2.87	2.71	2.60	2.51	2.45	2.39	2.35	2.28	2.20	2.12	2.08	2.04	1.99
21	4.32	3.47	3.07	2.84	2.68	2.57	2.49	2.42	2.37	2.32	2.25	2.18	2.10	2.05	2.01	1.96
22	4.30	3.44	3.05	2.82	2.66	2.55	2.46	2.40	2.34	2.30	2.23	2.15	2.07	2.03	1.98	1.94
23	4.28	3.42	3.03	2.80	2.64	2.53	2.44	2.37	2.32	2.27	2.20	2.13	2.05	2.01	1.96	1.91
24	4.26	3.40	3.01	2.78	2.62	2.51	2.42	2.36	2.30	2.25	2.18	2.11	2.03	1.98	1.94	1.89
25	4.24	3.39	2.99	2.76	2.60	2.49	2.40	2.34	2.28	2.24	2.16	2.09	2.01	1.96	1.92	1.87
30	4.17	3.32	2.92	2.69	2.53	2.42	2.33	2.27	2.21	2.16	2.09	2.01	1.93	1.89	1.84	1.79
40	4.08	3.23	2.84	2.61	2.45	2.34	2.25	2.18	2.12	2.08	2.00	1.92	1.84	1.79	1.74	1.69
60	4.00	3.15	2.76	2.53	2.37	2.25	2.17	2.10	2.04	1.99	1.92	1.84	1.75	1.70	1.65	1.59
120	3.92	3.07	2.68	2.45	2.29	2.18	2.09	2.02	1.96	1.91	1.83	1.75	1.66	1.61	1.55	1.50
∞	3.84	3.00	2.60	2.37	2.21	2.10	2.01	1.94	1.88	1.83	1.75	1.67	1.57	1.52	1.46	1.39

Degrees of Freedom for the Denominator

The *F* Distribution at a 1 Percent Level of Significance

Degrees of Freedom for the Numerator

	1	2	3	4	5	6	7	8	9	10	12	15	20	24	30	40
1	4052	5000	5403	5625	5764	5859	5928	5981	6022	6056	6106	6157	6209	6235	6261	6287
2	98.5	99.0	99.2	99.2	99.3	99.3	99.4	99.4	99.4	99.4	99.4	99.4	99.4	99.5	99.5	99.5
3	34.1	30.8	29.5	28.7	28.2	27.9	27.7	27.5	27.3	27.2	27.1	26.9	26.7	26.6	26.5	26.4
4	21.2	18.0	16.7	16.0	15.5	15.2	15.0	14.8	14.7	14.5	14.4	14.2	14.0	13.9	13.8	13.7
5	16.3	13.3	12.1	11.4	11.0	10.7	10.5	10.3	10.2	10.1	9.89	9.72	9.55	9.47	9.38	9.29
6	13.7	10.9	9.78	9.15	8.75	8.47	8.26	8.10	7.98	7.87	7.72	7.56	7.40	7.31	7.23	7.14
7	12.2	9.55	8.45	7.85	7.46	7.19	6.99	6.84	6.72	6.62	6.47	6.31	6.16	6.07	5.99	5.91
8	11.3	8.65	7.59	7.01	6.63	6.37	6.18	6.03	5.91	5.81	5.67	5.52	5.36	5.28	5.20	5.12
9	10.6	8.02	6.99	6.42	6.06	5.80	5.61	5.47	5.35	5.26	5.11	4.96	4.81	4.73	4.65	4.57
10	10.0	7.56	6.55	5.99	5.64	5.39	5.20	5.06	4.94	4.85	4.71	4.56	4.41	4.33	4.25	4.17
11	9.65	7.21	6.22	5.67	5.32	5.07	4.89	4.74	4.63	4.54	4.40	4.25	4.10	4.02	3.94	3.86
12	9.33	6.93	5.95	5.41	5.06	4.82	4.64	4.50	4.39	4.30	4.16	4.01	3.86	3.78	3.70	3.62
13	9.07	6.70	5.74	5.21	4.86	4.62	4.44	4.30	4.19	4.10	3.96	3.82	3.66	3.59	3.51	3.43
14	8.86	6.51	5.56	5.04	4.69	4.46	4.28	4.14	4.03	3.94	3.80	3.66	3.51	3.43	3.35	3.27
15	8.68	6.36	5.42	4.89	4.56	4.32	4.14	4.00	3.89	3.80	3.67	3.52	3.37	3.29	3.21	3.13
16	8.53	6.23	5.29	4.77	4.44	4.20	4.03	3.89	3.78	3.69	3.55	3.41	3.26	3.18	3.10	3.02
17	8.40	6.11	5.18	4.67	4.34	4.10	3.93	3.79	3.68	3.59	3.46	3.31	3.16	3.08	3.00	2.92
18	8.29	6.01	5.09	4.58	4.25	4.01	3.84	3.71	3.60	3.51	3.37	3.23	3.08	3.00	2.92	2.84
19	8.18	5.93	5.01	4.50	4.17	3.94	3.77	3.63	3.52	3.43	3.30	3.15	3.00	2.92	2.84	2.76
20	8.10	5.85	4.94	4.43	4.10	3.87	3.70	3.56	3.46	3.37	3.23	3.09	2.94	2.86	2.78	2.69
21	8.02	5.78	4.87	4.37	4.04	3.81	3.64	3.51	3.40	3.31	3.17	3.03	2.88	2.80	2.72	2.64
22	7.95	5.72	4.82	4.31	3.99	3.76	3.59	3.45	3.35	3.26	3.12	2.98	2.83	2.75	2.67	2.58
23	7.88	5.66	4.76	4.26	3.94	3.71	3.54	3.41	3.30	3.21	3.07	2.93	2.78	2.70	2.62	2.54
24	7.82	5.61	4.72	4.22	3.90	3.67	3.50	3.36	3.26	3.17	3.03	2.89	2.74	2.66	2.58	2.49
25	7.77	5.57	4.68	4.18	3.85	3.63	3.46	3.32	3.22	3.13	2.99	2.85	2.70	2.62	2.54	2.45
30	7.56	5.39	4.51	4.02	3.70	3.47	3.30	3.17	3.07	2.98	2.84	2.70	2.55	2.47	2.39	2.30
40	7.31	5.18	4.31	3.83	3.51	3.29	3.12	2.99	2.89	2.80	2.66	2.52	2.37	2.29	2.20	2.11
60	7.08	4.98	4.13	3.65	3.34	3.12	2.95	2.82	2.72	2.63	2.50	2.35	2.20	2.12	2.03	1.94
120	6.85	4.79	3.95	3.48	3.17	2.96	2.79	2.66	2.56	2.47	2.34	2.19	2.03	1.95	1.86	1.76
∞	6.63	4.61	3.78	3.32	3.02	2.80	2.64	2.51	2.41	2.32	2.18	2.04	1.88	1.79	1.70	1.59

Degrees of Freedom for the Denominator

F   Critical Values of Chi-Square

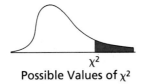

Possible Values of χ^2

Degrees of Freedom	Right-Tail Area			
	.10	.05	.02	.01
df				
1	2.706	3.841	5.412	6.635
2	4.605	5.991	7.824	9.210
3	6.251	7.815	9.837	11.345
4	7.779	9.488	11.668	13.277
5	9.236	11.070	13.388	15.086
6	10.645	12.592	15.033	16.812
7	12.017	14.067	16.622	18.475
8	13.362	15.507	18.168	20.090
9	14.684	16.919	19.679	21.666
10	15.987	18.307	21.161	23.209
11	17.275	19.675	22.618	24.725
12	18.549	21.026	24.054	26.217
13	19.812	22.362	25.472	27.688
14	21.064	23.685	26.873	29.141
15	22.307	24.996	28.259	30.578
16	23.542	26.296	29.633	32.000
17	24.769	27.587	30.995	33.409
18	25.989	28.869	32.346	34.805
19	27.204	30.144	33.687	36.191
20	28.412	31.410	35.020	37.566
21	29.615	32.671	36.343	38.932
22	30.813	33.924	37.659	40.289
23	32.007	35.172	38.968	41.638
24	33.196	36.415	40.270	42.980
25	34.382	37.652	41.566	44.314
26	35.563	38.885	42.856	45.642
27	36.741	40.113	44.140	46.963
28	37.916	41.337	45.419	48.278
29	39.087	42.557	46.693	49.588
30	40.256	43.773	47.962	50.892

G Wilcoxon *T* Values

n	2α α	.15 .075	.10 .050	.05 .025	.04 .020	.03 .015	.02 .010	.01 .005
4		0						
5		1	0					
6		2	2	0	0			
7		4	3	2	1	0	0	
8		7	5	3	3	2	1	0
9		9	8	5	5	4	3	1
10		12	10	8	7	6	5	3
11		16	13	10	9	8	7	5
12		19	17	13	12	11	9	7
13		24	21	17	16	14	12	9
14		28	25	21	19	18	15	12
15		33	30	25	23	21	19	15
16		39	35	29	28	26	23	19
17		45	41	34	33	30	27	23
18		51	47	40	38	35	32	27
19		58	53	46	43	41	37	32
20		65	60	52	50	47	43	37
21		73	67	58	56	53	49	42
22		81	75	65	63	59	55	48
23		89	83	73	70	66	62	54
24		98	91	81	78	74	69	61
25		108	100	89	86	82	76	68
26		118	110	98	94	90	84	75
27		128	119	107	103	99	92	83
28		138	130	116	112	108	101	91
29		150	140	126	122	117	110	100
30		161	151	137	132	127	120	109
31		173	163	147	143	137	130	118
32		186	175	159	154	148	140	128
33		199	187	170	165	159	151	138
34		212	200	182	177	171	162	148
35		226	213	195	189	182	173	159
40		302	286	264	257	249	238	220
50		487	466	434	425	413	397	373
60		718	690	648	636	620	600	567
70		995	960	907	891	872	846	805
80		1318	1276	1211	1192	1168	1136	1086
90		1688	1638	1560	1537	1509	1471	1410
100		2105	2045	1955	1928	1894	1850	1779

Source: Abridged from Robert L. McCormack, "Extended Tables of the Wilcoxon Matched-Pair Signed Rank Statistic," *Journal of the American Statistical Association,* September 1965, pp. 866–67.

H Table of Random Numbers

02711	08182	75997	79866	58095	83319	80295	79741	74599	84379
94873	90935	31684	63952	09865	14491	99518	93394	34691	14985
54921	78680	06635	98689	17306	25170	65928	87709	30533	89736
77640	97636	37397	93379	56454	59818	45827	74164	71666	46977
61545	00835	93251	87203	36759	49197	85967	01704	19634	21898
17147	19519	22497	16857	42426	84822	92598	49186	88247	39967
13748	04742	92460	85801	53444	65626	58710	55406	17173	69776
87455	14813	50373	28037	91182	32786	65261	11173	34376	36408
08999	57409	91185	10200	61411	23392	47797	56377	71635	08601
78804	81333	53809	32471	46034	36306	22498	19239	85428	55721
82173	26921	28472	98958	07960	66124	89731	95069	18625	92405
97594	25168	89178	68109	05043	17407	48201	83917	11413	72920
73881	67176	93504	42636	38233	16154	96451	57925	29667	30859
46071	22912	90326	42453	88108	72064	58601	32357	90610	32921
44492	19686	12495	93135	95185	77799	52441	88272	22024	80631
31864	72170	37722	55794	14636	05148	54505	50113	21119	25228
51574	90692	43339	65689	76539	27909	05467	21727	51141	72949
35350	76132	92925	92124	92634	35681	43690	89136	35599	84138
46943	36502	01172	46045	46991	33804	80006	35542	61056	75666
22665	87226	33304	57975	03985	21566	65796	72915	81466	89205
39437	97957	11838	10433	21564	51570	73558	27495	34533	57808
77082	47784	40098	97962	89845	28392	78187	06112	08169	11261
24544	25649	43370	28007	06779	72402	62632	53956	24709	06978
27503	15558	37738	24849	70722	71859	83736	06016	94397	12529
24590	24545	06435	52758	45685	90151	46516	49644	92686	84870
48155	86226	40359	28723	15364	69125	12609	57171	86857	31702
20226	53752	90648	24362	83314	00014	19207	69413	97016	86290
70178	73444	38790	53626	93780	18629	68766	24371	74639	30782
10169	41465	51935	05711	09799	79077	88159	33437	68519	03040
81084	03701	28598	70013	63794	53169	97054	60303	23259	96196
69202	20777	21727	81511	51887	16175	53746	46516	70339	62727
80561	95787	89426	93325	86412	57479	54194	52153	19197	81877
08199	26703	95128	48599	09333	12584	24374	31232	61782	44032
98883	28220	39358	53720	80161	83371	15181	11131	12219	55920
84568	69286	76054	21615	80883	36797	82845	39139	90900	18172
04269	35173	95745	53893	86022	77722	52498	84193	22448	22571
10538	13124	36099	13140	37706	44562	57179	44693	67877	01549
77843	24955	25900	63843	95029	93859	93634	20205	66294	41218
12034	94636	49455	76362	83532	31062	69903	91186	65768	55949
10524	72829	47641	93315	80875	28090	97728	52560	34937	79548
68935	76632	46984	61772	92786	22651	07086	89754	44143	97687
89450	65665	29190	43709	11172	34481	95977	47535	25658	73898
90696	20451	24211	97310	60446	73530	62865	96574	13829	72226
49006	32047	93086	00112	20470	17136	28255	86328	07293	38809
74591	87025	52368	59416	34417	70557	86746	55809	53628	12000
06315	17012	77103	00968	07235	10728	42189	33292	51487	64443
62386	09184	62092	46617	99419	64230	95034	85481	07857	42510
86848	82122	04028	36959	87827	12813	08627	80699	13345	51695
65643	69480	46598	04501	40403	91408	32343	48130	49303	90689
11084	46534	78957	77353	39578	77868	22970	84349	09184	70603

I Factors for Control Charts

Number of Items in Sample, n	Chart for Averages Factors for Control Limits A_2	Chart for Ranges		
		Factors for Central Line d_2	Factors for Control Limits	
			D_3	D_4
2	1.880	1.128	0	3.267
3	1.023	1.693	0	2.575
4	.729	2.059	0	2.282
5	.577	2.326	0	2.115
6	.483	2.534	0	2.004
7	.419	2.704	.076	1.924
8	.373	2.847	.136	1.864
9	.337	2.970	.184	1.816
10	.308	3.078	.223	1.777
11	.285	3.173	.256	1.744
12	.266	3.258	.284	1.716
13	.249	3.336	.308	1.692
14	.235	3.407	.329	1.671
15	.223	3.472	.348	1.652

Source: Adapted from American Society for Testing and Materials, *Manual on Quality Control of Materials,* 1951, Table B2, p. 115. For a more detailed table and explanation, see Acheson J. Duncan, *Quality Control and Industrial Statistics,* 3d ed. (Homewood, Ill.: Richard D. Irwin, 1974), Table M, p. 927.

DATA SET

1 Sales of Homes in Alabama in 1996

X_1 = Selling price in $000
X_2 = Number of bedrooms
X_3 = Size of home on square feet
X_4 = Pool (1 = yes, 0 = no)
X_5 = Distance from the center of the city
X_6 = Township
X_7 = Garage attached (1 = yes, 0 = no)
X_8 = Number of bathrooms

X_1	X_2	X_3	X_4	X_5	X_6	X_7	X_8
194.9	4	2349	0	17	5	1	2.0
135.1	4	2102	1	19	4	0	2.0
179.3	3	2271	1	12	3	0	2.0
158.2	2	2188	1	16	2	0	2.5
103.6	2	2148	1	28	1	0	1.5
181.8	2	2117	0	12	1	1	2.0
242.4	6	2484	1	15	3	1	2.0
201.3	2	2130	1	9	2	1	2.5
163.8	3	2254	0	18	1	0	1.5
197.5	4	2385	1	13	4	1	2.0
216.6	4	2108	1	14	3	1	2.0
154.8	2	1715	1	8	4	1	1.5
200.6	6	2495	1	7	4	1	2.0
182.3	4	2073	1	18	3	1	2.0
144.0	2	2283	1	11	3	0	2.0
208.4	3	2119	1	16	2	1	2.0
127.9	4	2189	0	16	3	0	2.0
153.7	5	2316	0	21	4	0	2.5
147.3	3	2220	0	10	4	1	2.0
155.0	6	1901	0	15	4	1	2.0
186.9	4	2624	1	8	4	1	2.0
142.9	4	1938	0	14	2	1	2.5
155.0	5	2101	1	20	5	0	1.5
255.8	8	2644	1	9	4	1	2.0
241.7	6	2141	1	11	5	1	3.0
128.2	2	2198	0	21	5	1	1.5
138.5	2	1912	1	26	4	0	2.0
190.5	2	2117	1	9	4	1	2.0
172.6	3	2162	1	14	3	1	1.5
133.6	2	2041	1	11	5	0	2.0

X_1	X_2	X_3	X_4	X_5	X_6	X_7	X_8
173.3	2	1712	1	19	3	1	2.0
153.4	2	1974	1	11	5	1	2.0
183.5	5	2438	1	16	2	1	2.0
123.1	3	2019	0	16	2	1	2.0
131.2	2	1919	1	10	5	1	2.0
135.3	4	2023	0	14	4	0	2.5
160.0	4	2310	1	19	2	0	2.0
231.2	6	2639	1	7	5	1	2.5
148.0	3	2069	1	19	3	1	2.0
202.4	5	2182	1	16	2	1	3.0
152.6	3	2090	0	9	3	0	1.5
172.0	3	1928	0	16	1	1	1.5
146.9	4	2056	0	19	1	1	1.5
151.9	3	2012	0	20	4	0	2.0
130.1	4	2262	0	24	4	1	2.0
228.0	3	2431	0	21	2	1	3.0
199.4	5	2217	1	8	5	1	3.0
166.5	3	2157	1	17	1	1	2.5
127.1	3	2014	0	16	4	0	2.0
160.6	3	2221	1	15	1	1	2.0
142.7	6	2236	0	14	1	0	2.0
175.1	5	2189	1	20	3	1	2.0
127.7	3	2218	1	23	3	0	2.0
186.2	3	1937	1	12	2	1	2.0
182.2	6	2296	1	7	3	1	3.0
109.2	6	1749	0	12	1	0	2.0
130.4	4	2230	1	15	1	1	2.0
169.2	3	2263	1	17	5	1	1.5
123.3	3	1593	0	19	3	0	2.5
140.3	4	2221	1	24	1	1	2.0
231.2	7	2403	1	13	3	1	3.0
214.7	6	2036	1	21	3	1	3.0
199.9	5	2170	0	11	4	1	2.5
114.3	2	2007	1	13	2	0	2.0
164.5	2	2054	1	9	5	1	2.0
155.3	5	2247	0	13	2	1	2.0
141.4	3	2190	0	18	3	1	2.0
188.4	4	2495	0	15	3	1	2.0
153.7	3	2080	0	10	2	0	2.0
155.3	4	2210	0	19	2	1	2.0
217.8	2	2133	1	13	2	1	2.5
130.6	2	2037	0	17	3	0	2.0
218.0	7	2448	1	8	4	1	2.0
165.9	3	1900	0	6	1	1	2.0
92.6	2	1871	1	18	4	0	1.5

2 School Districts in Northwest Ohio

X_1 = Number of students
X_2 = Mean family income in the district
X_3 = Mean property value in district
X_4 = Percent of families receiving welfare
X_5 = Mean salary of classroom teachers
X_6 = Amount spent per pupil
X_7 = Mean daily attendance
X_8 = Percent passing 12th grade proficiency examination

School District	X_1	X_2	X_3	X_4	X_5	X_6	X_7	X_8
Bluffton	1132	24487	62678	1.8	31221	2130	95.7	85
Shawnee	2472	29777	130910	2.6	34860	2570	94.7	73
Spencerville	1026	23161	51645	5.4	30155	2262	95.5	68
Delphos	1104	21792	88453	6.2	32273	2506	96.5	65
Elida	3204	24446	65550	8.9	32876	2250	94.1	62
Lima	5963	18394	44138	33.8	33142	2657	92.3	40
Northeastern	1194	26428	88789	1.7	30919	2431	96.1	72
Ayersville	921	28228	82707	3.9	32850	2693	95.6	68
Defiance	3046	23812	56333	11.2	34750	2438	94.2	63
Hicksville	990	22448	56411	7.0	34224	2351	95.7	59
Central	1216	24189	53923	4.6	34430	2496	94.8	56
Berlin-Milan	1593	25223	76878	4.0	32166	2564	96.1	77
Perkins	2038	25586	117545	3.3	39352	2861	95.8	74
Huron	1494	27135	105588	4.6	33433	2968	95.4	74
Margaretta	1560	23849	74601	2.8	37084	2464	95.5	66
Sandusky	4426	19529	72425	25.2	36042	2766	93.0	37
Kelleys Island	20	19854	802081	11.4	27144	11226	95.0	100
Pettisville	503	25079	53948	3.9	31159	2834	96.1	78
Wauseon	1864	23408	60896	4.9	32499	2252	95.3	75
Evergreen	1238	23826	69432	5.2	32353	2250	95.0	72
Archbold	1401	26706	107547	4.1	35982	2837	96.2	69
Pike-Delta-York	1559	23396	48638	6.1	31310	2309	94.8	66
Gorham-Fayette	487	22405	57221	6.9	33166	2492	94.6	51
Swanton	1725	24596	69320	6.3	33690	2615	94.9	50
Arlington	685	26175	55478	2.7	31821	2205	96.5	84
Vanlue	346	24709	49606	4.1	28411	2420	96.2	83
Liberty-Benton	954	28718	77503	1.5	30330	2063	96.7	78
Van Buren	840	28964	151992	1.8	33447	2584	96.4	75
Cory-Rawson	794	23904	69242	2.4	31241	2416	96.4	73

School District	X_1	X_2	X_3	X_4	X_5	X_6	X_7	X_8
Arcadia	597	24305	78102	2.5	30738	2752	95.9	64
McComb	805	23754	69347	5.7	28986	2321	95.9	61
Findlay	5758	24269	92648	7.6	35879	2860	94.9	60
Ada	855	23029	52655	5.3	28479	2380	95.9	69
Kenton	2228	20418	61155	13.8	30907	2512	93.7	54
Liberty Center	1009	24723	57685	4.1	30904	2431	95.8	82
Patrick Henry	1176	23061	63134	5.2	31895	2552	95.9	75
Napoleon	2331	25304	84245	6.6	32773	2422	94.6	73
Holgate	605	23962	49709	6.0	31324	2454	94.9	71
Monroeville	686	22942	63103	4.3	30838	2474	95.8	64
Bellevue	2276	24025	66912	6.9	32164	2374	95.1	55
Willard	2300	23304	58832	14.3	35042	2347	94.6	53
Norwalk	2650	21551	72266	12.2	37145	2384	94.0	50
Ottawa Hills	933	45723	122356	0.2	43256	4150	95.7	95
Anthony Wayne	3178	29215	88004	3.1	35617	2844	95.3	75
Sylvania	7822	32114	101503	3.8	39684	2943	95.5	72
Maumee	3009	27604	117921	3.9	41634	3933	95.0	69
Oregon	3594	24525	123599	8.7	35848	2941	94.7	52
Washington	7154	23507	102485	11.8	39155	2997	93.7	51
Springfield	3575	26048	98346	12.2	34437	2774	93.8	49
Toledo	36790	21079	62668	42.8	36190	2611	90.7	28
Benton-Carroll-Salem	2063	23899	237206	4.7	42734	3444	95.6	81
Danbury	635	21325	182360	3.2	34971	3158	95.3	65
Genoa	1584	25321	53120	4.8	34661	2845	95.9	65
Port Clinton	2238	20941	129961	10.5	39542	2926	94.5	50
Put-in-Bay	70	19266	426419	1.4	30242	7824	94.5	33
Paulding	1993	22677	46163	8.6	32928	2560	94.5	59
Ottoville	610	24128	46582	0.2	26125	2588	99.8	86
Columbus Grove	866	23562	55568	8.4	30476	2174	96.2	84
Kalida	775	24456	44267	0.7	28962	2274	96.8	81
Continental	792	23625	37277	10.0	28945	2225	95.5	79
Ottawa-Glandorf	1749	25363	64288	6.5	31185	2154	96.5	77
Pandora-Gilboa	632	23806	55446	1.2	27693	2078	96.5	67
Leipsic	748	20941	62648	16.1	30282	2811	93.9	47
Gibsonburg	983	23312	46098	17.8	31244	2242	94.4	71
Lakota	1332	22678	55933	10.5	30765	2306	95.2	66
Fremont	5156	22327	74874	16.2	37759	2616	94.7	57
Woodmore	1141	26460	90484	10.9	32296	2227	96.5	56
Clyde-Green Springs	2368	23854	55724	7.0	33998	2383	94.6	47
Bettsville	347	22103	37269	6.1	27466	2394	95.8	83
Seneca East	1183	22656	50895	4.8	29940	2435	95.9	80
Old Fort	540	23208	50712	19.3	28195	2743	96.0	78
Hopewell-Loudon	870	22103	72201	4.7	30644	2564	94.6	74

School District	X_1	X_2	X_3	X_4	X_5	X_6	X_7	X_8
New Riegel	459	23314	41376	2.7	29099	2501	97.0	73
Tiffin	3632	21246	65291	9.0	35513	2506	94.7	67
Fostoria	2742	20809	62268	20.4	34241	2455	92.7	34
Van Wert	2504	22728	67932	7.1	33885	2511	95.3	61
Edon-Northwest	744	23035	38462	2.7	30833	1916	95.1	73
Millcreek-West Unity	788	21302	40239	3.2	31582	2382	95.9	71
Bryan	2266	22607	81152	4.2	32643	2706	95.2	69
North Central	757	21871	59396	2.9	31978	2349	95.6	68
Montpelier	1172	20787	44383	8.2	33243	2654	94.8	67
Edgerton	767	22429	54040	4.1	28975	2470	95.9	61
Stryker	579	24084	65532	3.4	33855	2617	95.7	47
Perrysburg	3839	32773	97888	2.6	40320	3011	96.1	98
Elmwood	1237	22179	47644	7.5	30434	2643	95.2	68
Bowling Green	3534	21307	84682	7.5	37983	2849	94.7	68
Otsego	1643	24614	57601	5.0	36065	2539	95.4	64
Northwood	1091	25905	77077	7.9	35536	2979	94.9	61
Eastwood	1739	25043	67929	4.7	35742	2499	95.9	60
Lake	1665	23559	95859	6.2	38046	2820	95.2	55
Rossford	2087	25360	123725	7.2	39476	3258	94.9	54
North Baltimore	839	22075	58383	10.6	29579	2331	94.5	50
Upper Sandusky	1801	21063	68348	3.9	32778	2267	95.6	62
Carey	915	21658	51497	6.0	30968	2513	95.5	59

Answers to Chapter Achievement Tests

CHAPTER ONE

1. False, ordinal
2. True
3. False, nominal is the lowest level of measurement
4. True
5. False, inferential statistics
6. False, a sample
7. True
8. True

CHAPTER TWO

1. b   **2.** b   **3.** a   **4.** b   **5.** c   **6.** a   **7.** a   **8.** c

9.

Time	Patients
1 up to 6	3
6 up to 11	6
11 up to 16	5
16 up to 21	6
21 up to 26	5
	25

10.

Tread Depth (1/32")	Frequency	Percent Frequency	Less Than Cumulative Frequency
0 up to 4	4	7	4
4 up to 8	15	29	19
8 up to 12	25	48	44
12 up to 16	5	10	49
16 up to 20	3	6	52
	52	100	

a.

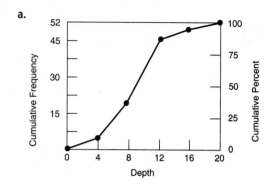

b. 19 out of 52, or 37%, have less than 8/32″ Tread Depth.

c. 50 percent of the tires have less than 9/32″ tread.

11. _____

Expense	Amount	Percent of Total
Housing	$ 400	38%
Utilities	140	13
Medical	25	2
Food	190	18
Transportation	150	14
Clothing	50	5
Savings	50	5
Miscellaneous	50	5
	$1,055	

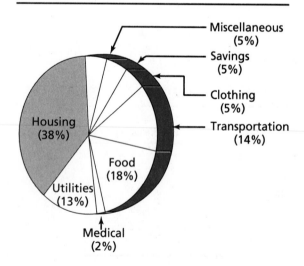

CHAPTER THREE

1. a   **2.** c   **3.** c   **4.** b   **5.** d   **6.** d   **7.** d   **8.** b   **9.** d   **10.** a

11. a. $\dfrac{\$1860}{4} = \$465$ **b.** $432.50   **12.** $Md = \$76.10$

Pounds	f	X	fX	CF
0 up to 0.5	3	0.25	0.75	3
0.5 up to 1.0	8	0.75	6.00	11
1.0 up to 1.5	20	1.25	25.00	31
1.5 up to 2.0	10	1.75	17.50	41
2.0 up to 2.5	6	2.25	13.50	47
2.5 up to 3.0	3	2.75	8.25	50
	50		71.00	

13. a. $\overline{X} = \dfrac{71.00}{50.00} = 1.42$

14. Median $= 1.00 + \dfrac{(50/2) - 11}{20}(0.5) = 1.35$

15. Mode $= 1.25$

CHAPTER FOUR

1. c   **2.** c   **3.** c   **4.** d   **5.** b   **6.** d   **7.** b   **8.** b   **9.** b   **10.** a

11. a. range $= 6 - 2 = 4$

b. $\overline{X} = 20/5 = 4$; $MD = \dfrac{2 + 2 + 1 + 0 + 1}{5} = 1.2$

c. Sample variance: $s^2 = \dfrac{90 - \dfrac{(20)^2}{5}}{5 - 1} = 2.5$

d. $s = \sqrt{2.5} = 1.58$

12. a. $80 - 20 = 60$

b. $Q_1 = 40 + \dfrac{30 - 23}{20}(10) = 43.5$

$Q_3 = 50 + \dfrac{90 - 43}{50}(10) = 59.4$

$QD = \dfrac{59.4 - 43.5}{2} = 7.95$

c. $s^2 = \dfrac{341,800 - \dfrac{(6220)^2}{120}}{120 - 1} = 163.0$

$s = \sqrt{163.0} = 12.77$

13. a. $CV = \dfrac{9}{87}(100) = 10.3$ in the Sanford district versus $\dfrac{12.77}{51.83}(100) = 24.9$ in the Jefferson district.

b. $sk = \dfrac{3(\$87,000 - \$84,000)}{\$9,000} = 1.00$ in the Sanford district

$sk = \dfrac{3(\$51,833 - \$53,400)}{\$12,800} = -0.37$ in the Jefferson district

A moderate positive skew in Sanford versus a slight negative skew in Jefferson.

CHAPTER FIVE

1. c   **2.** c   **3.** d   **4.** c   **5.** a   **6.** b   **7.** b   **8.** b   **9.** b   **10.** c   **11.** b   **12.** a   **13.** c
14. d   **15.** b
16. $\dfrac{325}{500} = 0.65$

17. a.

	Men	Women	Totals
In State	5	18	23
Out of State	5	2	7
Totals	10	20	30

b. $\dfrac{5}{30} = 0.167$

c. $\dfrac{5}{7} = 0.714$

d. $\dfrac{12}{30} = 0.4$

18. $\dfrac{6}{10} \cdot \dfrac{5}{9} = \dfrac{1}{3} = 0.333$

19. $(0.35)(0.25) = 0.0875$

20. 0.60

21. $5/8 = 0.625$

22. a. 0.33   **b.** $0.67 + 0.23 - 0.12 = 0.78$

23. $_{30}P_3 = 30 \cdot 29 \cdot 28 = 24{,}360$

24. $_{15}C_3 = \dfrac{15!}{12!\ 3!} = \dfrac{15 \cdot 14 \cdot 13}{3 \cdot 2} = 455$

25. $\dfrac{1}{_{30}C_5} = \dfrac{5!\ 25!}{30!} = \dfrac{5 \cdot 4 \cdot 3 \cdot 2}{30 \cdot 29 \cdot 28 \cdot 27 \cdot 26} = 0.000007$

CHAPTER SIX

1. a   **2.** b   **3.** b   **4.** c   **5.** a   **6.** c   **7.** b   **8.** a   **9.** b   **10.** c
11. $\pi = 0.9, n = 8$

a. $P(X = 6) = \dfrac{8!}{6!\ 2!}(0.9)^6(0.1)^2 = 0.1488$

b. $P(X \geq 6) = 0.149 + 0.383 + 0.430 = 0.962$

c. $P(3) + P(4) + P(5) = 0.000 + 0.005 + 0.033 = 0.038$

12. $\mu = 150(\frac{1}{50}) = 3$

   a. $P(0) = \dfrac{3^0 e^{-3}}{0!} = 0.0498$

   b. $P(2) = \dfrac{3^2 e^{-3}}{2!} = 0.2240$

   c. $P(X < 4) = P(0) + P(1) + P(2) + P(3)$

      $P(1) = \dfrac{3^1 e^{-3}}{1!} = 0.1494 \qquad P(2) = \dfrac{3^2 e^{-3}}{2} = 0.2240$

      $P(3) = \dfrac{3^3 e^{-3}}{3!} = 0.2240$

      $P(X < 4) = 0.0498 + 0.1494 + 0.2240 + 0.2240 = 0.6472$

   d. $1 - P(0) = 0.9502$

13.

X	$P(X)$	$X \cdot P(X)$	$(X - \mu)^2$	$(X - \mu)^2 P(X)$
0	0.05	0.00	5.5225	0.2761
1	0.15	0.15	1.8225	0.2734
2	0.30	0.60	0.1225	0.0368
3	0.40	1.20	0.4225	0.1690
4	0.10	0.40	2.7225	0.2723
		2.35		1.0275

$\mu = 2.35 \qquad \sigma = \sqrt{1.0275} = 1.0137$

CHAPTER SEVEN

1. a   **2.** d   **3.** b   **4.** d   **5.** c   **6.** b   **7.** a   **8.** a   **9.** b   **10.** d

11. $\mu = 3.00, \sigma = 1.00$

                                             Answer

   a. $z = (4.5 - 3.00)/1.00 = 1.50$ 0.4332

   b. $z = (4.00 - 3.00)/1.00 = 1.00$ $0.5000 - 0.3413 = 0.1587$

   c. $z = (2.00 - 3.00)/1.00 = -1.00$

      $z = (3.50 - 3.00)/1.00 = 0.50$ $0.3413 + 0.1915 = 0.5328$

   d. $z = (1.00 - 3.00)/1.00 = -2.00$ $0.5000 - 0.4772 = 0.0228$

   e. $1.28 = (x - 3.00)/1.00$

      $x = 3.00 + 1.28(1.00) = 4.28$

12. $\mu = n\pi = 80(0.40) = 32$ $\sigma^2 = n\pi(1 - \pi) = 80(0.40)(0.60) = 19.2$

   $\sigma = \sqrt{19.2} = 4.38$

   a. $z = (25.5 - 32)/4.38 = -1.48$ $0.4306 + 0.5000 = 0.9306$

   b. $z = (40.5 - 32)/4.38 = 1.94$ $0.5000 - 0.4738 = 0.0262$

   c. $z = (34.5 - 32)/4.38 = 0.57$ $0.5000 + 0.2157 = 0.7157$

   d. $z = (29.5 - 32)/4.38 = -0.57$

      $z = (36.5 - 32)/4.38 = 1.03$ $0.2157 + 0.3485 = 0.5642$

13. a. 0.0838   **b.** 0.7148

CHAPTER EIGHT

1. a **2.** d **3.** c **4.** b **5.** c **6.** d **7.** b **8.** c **9.** c **10.** c

11. a. Bowling Green, Cincinnati, Toledo

b. Miami, Ohio University, Youngstown

12. a. $\mu = \dfrac{6 + 4 + 7 + 3 + 2}{5} = \dfrac{22}{5} = 4.4$

$\sigma^2 = \dfrac{114 - \dfrac{(22)^2}{5}}{5} = \dfrac{17.2}{5} = 3.44$

$\sigma = \sqrt{3.44} = 1.8547$

b.

Sample	Items	Mean
1	6, 4, 7	5.6667
2	6, 4, 3	4.3333
3	6, 4, 2	4.0000
4	6, 7, 3	5.3333
5	6, 7, 2	5.0000
6	6, 3, 2	3.6667
7	4, 7, 3	4.6667
8	4, 7, 2	4.3333
9	4, 3, 2	3.0000
10	7, 3, 2	4.0000

c.

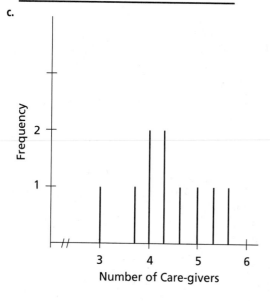

d. $\sigma^2_{\overline{X}} = \dfrac{199.3324 - \dfrac{(44)^2}{10}}{10} = 0.573$

$\sigma_{\overline{X}} = 0.757$ $\mu_{\overline{X}} = \dfrac{44}{10} = 4.40$

e. The mean of the sample means is the same as the population means. There is more dispersion in the population than in the sampling distribution.

CHAPTER NINE

1. a   **2.** a   **3.** a   **4.** a   **5.** a   **6.** c   **7.** c   **8.** b   **9.** b   **10.** b

11. a. $\sigma_{\overline{X}} = \dfrac{70}{\sqrt{70}} = \$8.37$

b. $510 \pm (1.96)\dfrac{70}{\sqrt{70}} = 510 \pm 16.40$; from \$493.60 up to \$526.40

12. a. $\dfrac{10}{\sqrt{50}} = 1.41$ years

b. $38 \pm 2.58(1.41) = 38 \pm 3.6$; from 34.4 up to 41.6 years

13. a. $\sigma_p = \sqrt{\dfrac{\pi(1-\pi)}{n}} = \sqrt{\dfrac{(0.7)(0.3)}{20}} = 0.10$

b. $0.65 \pm 2.58(0.10) = 0.65 \pm 0.26$; from 39 up to 91%

14. a. $\sqrt{\dfrac{(0.75)(0.25)}{500}} = 0.019$

b. $0.75 \pm 1.65(0.019) = 0.75 \pm 0.03$; from 0.72 up to 0.78

15. $n = \left(\dfrac{zs}{E}\right)^2 = \left[\dfrac{(1.96)(3000)}{200}\right]^2 = 865$

CHAPTER TEN

1. a   **2.** b   **3.** c   **4.** d   **5.** c   **6.** c   **7.** c   **8.** c   **9.** d   **10.** b

11. H_0: $\mu = 32.0$, H_1: $\mu \neq 32.0$. Reject H_0 if $z < -2.58$ or $z > 2.58$

$z = \dfrac{31.8 - 32.0}{0.4/\sqrt{36}} = -3.00$

Reject H_0. We conclude that the mean fill is not 32.0 oz.

12. a. H_0: $\mu_1 = \mu_2$ H_a: $\mu_1 \neq \mu_2$

b. If z is not between -1.81 and 1.81, reject H_0.

c. $z = \dfrac{5.21 - 4.65}{\sqrt{\dfrac{(2.3)^2}{40} + \dfrac{(1.9)^2}{30}}} = \dfrac{0.56}{0.50} = 1.11$

d. There is *no* difference in the mean length of stay in the two cities.

13. H_0: $\pi_1 = \pi_2$, H_1: $\pi_1 \neq \pi_2$, reject H_0 if $z < -1.65$ or $z > 1.65$.

$$\bar{p} = \frac{150 + 312}{200 + 400} = \frac{462}{600} = 0.77 \qquad p_1 = \frac{150}{200} = 0.75$$

$$p_2 = \frac{312}{400} = 0.78$$

$$z = \frac{0.75 - 0.78}{\sqrt{.77(1 - .77)\left(\dfrac{1}{200} + \dfrac{1}{400}\right)}} = \frac{-0.030}{0.036} = -0.83$$

Do not reject H_0. There is no difference in the proportion favoring the dance.

CHAPTER ELEVEN

1. d **2.** d **3.** b **4.** b **5.** d **6.** b **7.** d **8.** a **9.** b **10.** a **11.** c

12. H_0: $\mu \geq 98.6°$ F H_a: $\mu < 98.6°$ F $\alpha = 0.05$
df $= n - 1 = 25 - 1 = 24$
Reject H_0 if t is less than -1.711; otherwise, fail to reject.
$$t = \frac{\overline{X} - \mu}{s\sqrt{n}} = \frac{98.3 - 98.6}{0.64/\sqrt{25}} = -2.34$$
H_0 is rejected and H_a accepted. The evidence indicates unusually low body temperatures.

13. H_0: $\mu_d \leq 0$ H_a: $\mu_d > 0$ $\alpha = 0.01$ df $= n - 1 = 7 - 1 = 6$
Reject H_0 if t is greater than 3.143; otherwise, fail to reject H_0.

Before	After	d	d^2
180	160	20	400
156	164	−8	64
188	172	16	256
132	130	2	4
208	200	8	64
196	190	6	36
190	184	6	36
		50	860

$$\overline{d} = 50/7 = 7.14$$

$$s_d = \sqrt{\frac{860 - \dfrac{(50)^2}{7}}{6}} = 9.15$$

$$t = \frac{7.14}{9.15/\sqrt{7}} = 2.065$$

A significant reduction is not shown by these data.

14. H_0: $\mu_1 = \mu_2$, H_a: $\mu_1 \neq \mu_2$, $\alpha = 0.05$, df $= n + n - 2 = 5 + 8 - 2 = 11$
Reject H_0 if t is less than -2.201 or more than 2.201; otherwise, fail to reject H_0.

$$s_1 = \sqrt{\frac{7{,}734 - \frac{(178)^2}{5}}{5 - 1}} = 18.69 \qquad s_2 = \sqrt{\frac{7{,}102 - \frac{(208)^2}{8}}{8 - 1}} = 15.56$$

$$s_p = \sqrt{\frac{4(18.69)^2 + 7(15.56)^2}{11}} = 16.77 \qquad t = \frac{35.6 - 26}{16.77\sqrt{\frac{1}{5} + \frac{1}{8}}} = 1.00$$

Do not reject H_0. There is no evidence that one location is significantly better than the other.

CHAPTER TWELVE

1. a **2.** b **3.** c **4.** c **5.** c **6.** a **7.** b **8.** d **9.** d **10.** a

11. a. H_0: $\mu_1 = \mu_2 = \mu_3$ H_a: At least one is different.

b.

Source	Sum of Squares	df	Mean Square	F
Treatments	0.45	2	0.225	0.08
Within	25.80	9	2.867	
Total	26.25			

$$SST = \frac{(18)^2}{3} + \frac{(31)^2}{5} + \frac{(26)^2}{4} - \frac{(75)^2}{12}$$
$$= 0.45$$
$$SS\ total = 495 - \frac{(75)^2}{12} = 26.25$$
$$SSE = 26.25 - 0.45 = 25.80$$

c. The critical value is 4.26
d. Because the test statistic is only 0.08, we cannot reject H_0. The means all appear to be equal.
e. The p-value is greater than .05.

CHAPTER THIRTEEN

1. c **2.** d **3.** a **4.** c **5.** a **6.** b **7.** c **8.** a **9.** a **10.** c

11. a.

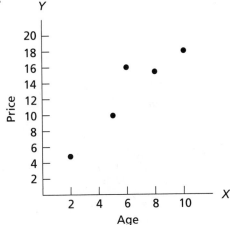

b.

X	Y	XY	X²	Y²
2	5	10	4	25
6	16	96	36	256
10	18	180	100	324
5	9	45	25	81
8	15	120	64	225
$\Sigma X = 31$	$\Sigma Y = 63$	$\Sigma XY = 451$	$\Sigma X^2 = 229$	$\Sigma Y^2 = 911$ $n = 5$

$$r = \frac{5(451) - (31)(63)}{\sqrt{[5(229) - (31)^2][5(911) - (63)^2]}} = 0.92$$

c. $r^2 = (0.92)^2 = 0.85$

d. $H_0: \rho = 0$ $H_a: \rho \neq 0$

With 0.05 level of significance and 3 df, the critical value is 3.182.

$$t = \frac{0.92\sqrt{5 - 2}}{\sqrt{1 - (.92)^2}} = 4.07 > 3.182. \text{ Reject } H_0.$$

e. The price of wine does change with age. The price and age are positively correlated with 85% of variation in one accounted for by variation of the other.

12. a.

Coaches	Sportswriters	d	d²
1	1	0	0
2	5	−3	9
3	4	−1	1
4	6	−2	4
5	2	3	9
6	3	3	9
7	10	−3	9
8	11	−3	9
9	7	2	4
10	12	−2	4
11	8	3	9
12	9	3	9
			$\Sigma d^2 = 76$

$$r = 1 - \frac{6(76)}{12(12^2 - 1)} = 0.73$$

b. A positive correlation does exist between ranking among coaches and ranking among sportswriters.

c. $t = 0.73\sqrt{\dfrac{(12 - 2)}{1 - (.73)^2}} = 3.38$

H_0: No correlation in population.

H_a: Positive rank correlation in population.

Reject H_0 if $t > 1.812$. Reject H_0. We can conclude that a significant correlation exists.

CHAPTER FOURTEEN

1. a   **2.** c   **3.** d   **4.** a   **5.** e   **6.** c   **7.** d   **8.** b   **9.** d

10. a.

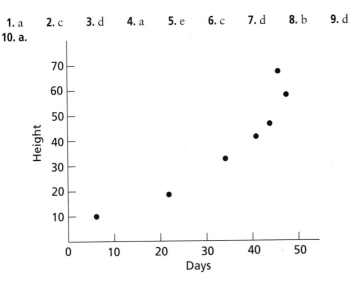

b.

Days X	Height Y	X^2	Y^2	XY
6	10	36	100	60
22	19	484	361	418
34	31	1,156	961	1,054
42	39	1,764	1,521	1,638
45	47	2,025	2,209	2,115
48	58	2,304	3,364	2,784
47	66	2,209	4,356	3,102
244	270	9,978	12,872	11,171

$$b = \frac{7(11,171) - 244(270)}{7(9,978) - (244)^2} = \frac{12,317}{10,310} = 1.195$$

$$a = \frac{270 - 1.195(244)}{7} = \frac{-21.58}{7} = -3.083 \qquad Y' = -3.083 + 1.195X$$

c. $s_{Y \cdot X} = \sqrt{\dfrac{12,872 - (-3.083)(270) - (1.195)(11,171)}{7 - 2}} = 8.43$

d. $Y' = -3.083 + 1.195(25) = 26.792$

$$26.792 \pm (2.015)(8.43)\sqrt{\frac{1}{7} + \frac{(25 - 244/7)^2}{9,978 - (244)^2/7}}$$

$$26.792 \pm 7.762 = 19.030 \text{ to } 34.554$$

CHAPTER FIFTEEN

1. b   **2.** c   **3.** a   **4.** a   **5.** a   **6.** c   **7.** b   **8.** d   **9.** b   **10.** b   **11.** a

12. $Y' = 5.00 + 2.00X_1 + 0.50X_2 + 0.30X_3 + 1.00X_4$

13. $F = \dfrac{50}{4} = 12.50$, which is significant even at the 0.01 level (4.18 critical value). So at least one of the net regression coefficients is not zero.

14. The test statistic corresponding to X_1 is only 2.0. That value may not be significant and could be dropped. For a two-sided test with a 0.05 level of significance, the critical values are ± 2.06. At this level each of the other three variables appears significant.

CHAPTER SIXTEEN

1. c   **2.** d   **3.** d   **4.** c   **5.** a   **6.** b   **7.** a   **8.** b   **9.** b   **10.** a

11. H_0: There is no relationship H_a: There is a relationship.
df $= (2 - 1)(3 - 1) = 2$
If $\chi^2 > 5.991$, reject H_0.

	Favor		Oppose		Undecided		
	f_o	f_e	f_o	f_e	f_o	f_e	Totals
Democrats	22	21	13	14	28	28	63
Republicans	8	9	7	6	12	12	27
	30		20		40		90

$$X^2 = \frac{(22 - 21)^2}{21} + \frac{(13 - 14)^2}{14} + \frac{(28 - 28)^2}{28} + \frac{(8 - 9)^2}{9} + \frac{7(7 - 6)^2}{6} + \frac{(12 - 12)^2}{12} = 0.40$$

We fail to reject H_0. There is no relationship.

12. H_0: $\pi_1 = \frac{1}{10}$, $\pi_2 = \frac{2}{10}$, $\pi_3 = \frac{3}{10}$, $\pi_4 = \frac{4}{10}$
H_a: At least one is different.
Reject H_0 if $\chi^2 > 11.345$.

	f_o	f_e	$f_o - f_e$	$(f_o - f_e)^2/f_o$
Full-sized	38	20	18	16.2
Medium-sized	62	40	22	12.1
Compact	41	60	-19	6.0
Subcompact	59	80	-21	5.5
	200			39.8

H_0 is rejected. The distribution of cars sold is not as suggested in H_0.

CHAPTER SEVENTEEN

1. b   **2.** b   **3.** a   **4.** a   **5.** d   **6.** c   **7.** d   **8.** b   **9.** a   **10.** c

11. Consult a binomial table with $n = 15$ and $\pi = 0.5$. The probability of 13 or more "+" signs is only 0.003, which is smaller than 0.01. So the researcher could reject the null hypothesis of no difference. The lung capacity of non-smokers is greater.

12.

Couple	d	Rank of Absolute Difference	R^+	R^-
A	12	10	10	
B	−4	4		4
C	2	2	2	
D	0			
E	8	9	9	
F	6	7	7	
G	3	3	3	
H	−1	1		1
I	−5	5.5		5.5
J	0			
K	7	8	8	
L	5	5.5	5.5	
M	0			
			44.5	10.5

The critical value is 3, so we fail to reject H_0. There is no significant difference between the sexes.

13.

Inner City		Suburban		Rural	
95	1	108	6.5	102	4
100	2.5	114	10	109	8
100	2.5	118	12	112	9
106	5	122	15	120	13.5
108	6.5	124	16	127	19
115	11	126	18	129	20
120	13.5	130	21.5	140	24
125	17	136	23		
130	21.5	142	25		
		150	26		
	80.5		173		97.5

$$H = \frac{12}{(26)(27)}\left[\frac{(80.5)^2}{9} + \frac{(173)^2}{10} + \frac{(97.5)^2}{7}\right] - 3(27)$$

$H = 5.68$

This is less than the critical value of 5.991, so we cannot conclude that a difference exists among the three groups.

CHAPTER EIGHTEEN

1. c   **2.** b   **3.** a   **4.** c   **5.** b   **6.** d   **7.** a   **8.** b   **9.** a   **10.** c.

11. $\bar{c} = \dfrac{14}{10} = 1.4$

UCL $= 1.4 + 3\sqrt{1.4} = 4.95$

LCL $= 1.4 - 3\sqrt{1.4} = 0$

The control limits are from 0 to 4.95 and all the points are within the limits.

12.

Time	$\bar{X}$	R
8	1.000	.02
9	1.008	.02
10	1.002	.03
11	1.004	.03
12	1.008	.03
1	0.992	.04

$\bar{\bar{X}} = \dfrac{6.014}{6} = 1.0023 \qquad \bar{R} = \dfrac{.17}{6} = .0283$

UCL $= 1.0023 + .577(0.0283) = 1.019$

LCL $= 1.0023 - .577(0.0283) = 0.986$

UCL $= (0.0283)(2.115) = .060$

All points are within limits for both the mean and the range.

13. $\bar{p} = \dfrac{40}{10(40)} = 0.10$

UCL $= .10 + 3\sqrt{\dfrac{.1(.9)}{40}} = 0.24$

LCL $= .10 - 3\sqrt{\dfrac{.1(.9)}{40}} = 0.100$

The control limits range from 0 to .24. All data points are within these limits. The day there were 7 students absent converts to .175, which is within the limits.

Answers to Even-Numbered Chapter Exercises

CHAPTER 1

2. a. Yes. The order of the categories could be changed.
   b. Yes. Each smoker is listed in one category but excluded from the other categories.
   c. Yes. Every smoker who responded appears in at least one category.
   d. The percentage of male smokers dropped from about 50 percent in 1965 to about 30 percent in 1995. Likewise the percentage of whites who smoked decreased from about 40 percent of the population in 1965 to less than 28 percent in 1995. And the percentage of blacks smoking declined from 43 percent to 33 percent in 1995.
4. a. Each degree is ranked higher than the previous one.
   b. Yes. Each graduate is placed in one category but excluded from any other category.
   c. Yes. Each graduate appears in at least one category.
6. A population is the entire group which you are studying. A sample is a subset taken from a population.
8. Categories are mutually exclusive if an object is included in only one class.
10. Answers will vary.
12. An overwhelming number of consumers tested (400/500, or 80%) believe this toothpaste is excellent.
14. False. The categories can be presented in any order.
16. False. Software sales increased from 1992 to 1996.

CHAPTER 2

2. a. 7 because $2^7 = 128$
   b. $10, found by $(68.73 - 2.15)/7$ and rounded up.
   c. $0
4.

Number of Rolls	Class Frequencies
0 up to 3	10
3 up to 6	23
6 up to 9	10
9 up to 12	3
12 up to 15	3
15 up to 18	1
Total	50

The number of visits to relatives ranges from 0 up to 18. The largest concentration is in the 3 up to 6 class.

6.

Number of Patrons	Class Frequencies
25 up to 35	3
35 up to 45	5
45 up to 55	6
55 up to 65	5
65 up to 75	4
75 up to 85	3
85 up to 95	4
Total	30

The number of patrons using the library ranged from 25 up to 95. The largest concentration is in the 45 up to 55 class.

8. a. 6

b. 3; 50, 51 and 51

c. 79

d. 63, median is 64

e. 30

10.

Stem	Leaf
0	678
1	2344889
2	01238
3	2468
4	0

12. a.

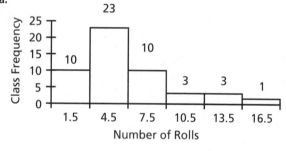

b.

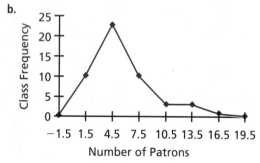

14.

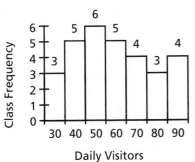

16. a. 80
  b. 50
  c. 7
  d. 50
  e. 275
  f. There are 30 observations between 200 and 250.
  g. 233

18. a.

Cumulative Seconds	Frequency
2 up to 3	200
3 up to 4	197
4 up to 5	190
5 up to 6	175
6 up to 7	146
7 up to 8	65
8 up to 9	15
9 up to 10	5

b. More-Than Cumulative Frequency Distribution:

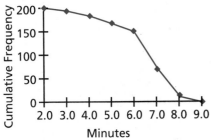

c. About 90% of the mice take more than 4.67 seconds to complete the maze.

20.

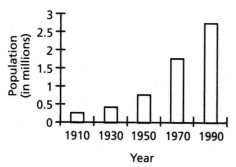

22.

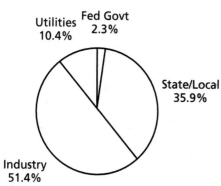

24. a. 5
   b. 10
   c. 50
   d.

Number of Changes	Frequency
50 up to 60	4
60 up to 70	5
70 up to 80	6
80 up to 90	2
90 up to 100	3
Total	20

   e. The range is 50 to 100, with 70 up to 80 changes on a typical day.

26.

Leading Digit	Trailing Digit
1	2479
2	112679
3	23456789
4	01235
5	37

28. a.

Percent of Births	Total
9.0 up to 14.0	5
14.0 up to 19.0	11
19.0 up to 24.0	18
24.0 up to 29.0	12
29.0 up to 34.0	3
34.0 up to 39.0	1

b.

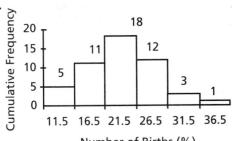

c.

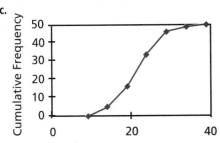

d. Approximately 18

30. a.

Age	Number	Percent of Total
15 up to 25	1	2.5
25 up to 35	2	5.0
35 up to 45	5	12.5
45 up to 55	10	25.0
55 up to 65	15	37.5
65 up to 75	4	10.0
75 up to 85	3	7.5
Total	40	100.0

The typical age is 57.

b. No

c. 25 out of 40 were in the 45 to 65 age bracket.

d. See part **a** above.

32. Below is a stem-and-leaf chart for the percent of flights on time for 12 airlines

3	76	149
3	77	
4	78	1
6	79	77
6	80	14
4	81	04
2	82	77

The lowest percent on time is 76.1 percent and the largest is 82.7 percent. The typical airline is on time about 79.7 percent of the time.

34. The following chart shows the growth in the number of GTE subscribers from 1990 to 1995.

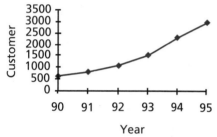

36. The following chart shows the average tuition cost for public and private colleges from the 1980–81 academic year to 1995–96.

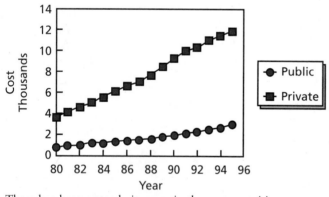

There has been a steady increase in the average tuition cost per year in both the public and the private schools.

38. In 1987 the price for a Toyota Camry and a Ford Taurus was about the same. The rate of increase for the Camry has been much faster since 1990.

40. a. The stem-and-leaf chart for the variable teacher salary is as follows

1	26	1
4	27	146
11	28	1449999
14	29	059
29	30	122344677889999
40	31	11222335889

(12)	32	112234677889
42	33	112446889
33	34	22446789
25	35	05567889
17	36	001
14	37	0179
10	38	0
9	39	13456
4	40	3
3	41	6
2	42	7
1	43	2

b. The stem-and-leaf chart for the percent of students passing the state proficiency examination is shown below.

1	2	8
3	3	34
4	3	7
5	4	0
9	4	7779
19	5	0000112344
27	5	55667999
39	6	001111223444
(18)	6	555666777888889999
37	7	11122233333444
23	7	5555778889
13	8	01123344
5	8	56
3	9	
3	9	58
1	10	0

CHAPTER 3

2. $21/5 = 4.2$

4. $96/10 = 9.6$

6. $-23/10 = -2.3$

8. $\overline{X} = \dfrac{24552}{5} = 4910.4$ (thousands) or 4,910,400

10. A sample mean is a statistic, but a population mean is a parameter.

12. a. $90/5 = 18$

   b. $(-13 - 18) + \cdots + (-28 - 18) = 0$

14. a. $\overline{X} = \dfrac{575}{7} = 82.1$

   b. $(76.0 - 82.1) + (78.4 - 82.1) \cdots + (87.4 - 82.1) = 0$

16. a. $\overline{X} = \dfrac{153.0}{8} = 19.125$

    b. $(2 - 19.125) + (2 - 19.125) + \cdots + (100 - 19.125) = 0$

18. $[15300(4.5) + 10400(3.0) + 150600(10.2)]/176300 = 9.28$

20. Weighted mean $= 28.2$ per 100,000, found by 225.3/8

22. $(2 + 10)/2 = 6$

24. 41 students

26. 3.38%, found by $(3.25 + 3.51)/2$.

28. $\overline{X} = \dfrac{4190}{60} = 69.833$

30. $\overline{X} = \dfrac{1310}{50} = 26.2$

32. a. 70 up to 80 class

    b. 70

    c. 10

    d. 28

    e. Median $= 70.87$

    f. Mode $= 75.0$

34. a. Median $= 35.77$ years

    b. Mode $= 35$

    c. You cannot find a midpoint for an open-ended class

36. a. interval

    b. interval

    c. ordinal

    d. nominal

38. a. $\overline{X} = \dfrac{60}{5} = 12.0$

    b. $\Sigma(X - \overline{X}) = (16 - 12) + \cdots + (15 - 12) = 0$

    c. Median age is 11.0

    d. There is no modal age

40. Fifty percent of the candidates take more than 8.5 years.

42. a. Mean $= 408/8 = 51.0$ percent

    b. Median $= 42.0$ percent

44. $\overline{X}_{\mu} = \dfrac{6(\$18) + 4(\$12) + 4(\$9)}{14} = \dfrac{\$192}{14} = \$13.71$

46. a. American: $\overline{X} = 270.08$ National: $\overline{X} = 264.18$

    b. American Median: $= 271.09$

    National Median: $= 266.83$

    c. American Mode $= 275$ National Mode $= 275$

    d. We cannot say that one league is "better" than the other with respect to batting. Perhaps the averages are slightly higher in the American League because pitchers do not bat.

48. Median $= 40 + \dfrac{101.9/2 - 44}{16.4}(10) = 44.24$ years

50. a. $\overline{X} = 31{,}553/28 = 1126.9$

    b. $\overline{X} = 31{,}600/28 = 1128.6$, using a lower limit of \$200 for the first class and an interval of \$400.

    c. Median $= 1000 + \dfrac{14 - 10}{11}(400) = 1145.45$

    The distribution of salaries is symmetrical, so the mean and the median are very close.

d. 4	0	4555
6	0	66
9	0	899
13	1	0011
(7)	1	2222333
8	1	45555
3	1	6
2	1	9
1	2	0

The median is 1100, according to the stem-and-leaf chart

52. Median $= -10.0 + \dfrac{109/2 - 34}{48}(10) = -5.73$

54. a. The mean selling price is $166,700 and the median selling price is $160,000, a difference of 6700. Either the mean or the median is a representative measure of central tendency.

b. The typical home has about 2155 square feet, 4 bedrooms, 2 baths, and is about 15 miles from the center of the city. Sixty-one percent of the homes have a pool and 68 percent have an attached garage.

c.
Attached Garage	Mean	Number
No	$138,100	24
Yes	$180,000	51

d. Home with attached garages sell for $42,000 more than those without attached garages.

CHAPTER 4

2. $\overline{X} = 3.8$ $MD = 2.16$
4. $\overline{X} = -2.3$ $MD = 8.24$
6. a. Range $= 91 - 58 = 33$
  b. $\overline{X} = 436/6 = 77.17$,
    $MD = 12.17$
8. a. The range is $82, found by $100 - $18
  b. $\overline{X} = 915/15 = 61.0$
  c. $MD = 314/15 = 20.93$
10. a. $\overline{X}_A = 142.0/10 = 14.2$ $\overline{X}_B = 159.0/10 = 15.9$
  b. For Pond A, the range is 4.5.
    For Pond B, the range is 8.5.
    The range is larger for Pond B than Pond A
  c. Pond A $MD = 10.4/10 = 1.04$
    Pond B $MD = 27.8/10 = 2.78$
    There is more variation in Pond B
12. a. $\overline{X} = 10$
    $s^2 = 8.571$
  b. $s = \sqrt{8.571} = 2.93$
14. a. $s^2 = 24.747$
  b. $s = \sqrt{24.747} = 4.97$
16. $s^2 = 3.88, s = 1.97$
18. $s^2 = 8.571$ $s = \sqrt{8.571} = 2.93$
20. $s^2 = 24.747, s = 4.97$

22. a. New Homes

$\overline{X} = 4800/6 = 660$ gallons $\qquad s^2 = 268/(8 - 1) = 38.29$

$s = \sqrt{38.29} = 6.19$ gallons

b. For older homes

$\overline{X} = 4000/5 = 800$ gallons $\qquad s^2 = 5200/(5 - 1) = 1300$

$s = \sqrt{1300} = 36.06$ gallons

c. Dispersion is greater for the older homes.

24. a. Monday:

$Q_1 = 14.33 \qquad Q_3 = 22.50$

$QD = (22.50 - 14.33)/2 = 8.17/2 = 4.085$

Friday:

$Q_1 = 12.62 \qquad Q_3 = 20.00$

$QD = (20.00 - 12.62)/2 = 3.69$

b. The dispersion on Monday is greater.

26. a. $Q_1 = 72.60 \qquad Q_3 = 90.60$

b. 9.00

28. Assume that the data is a population.

a. $\sigma^2 = 6.14 \qquad \sigma = 5.17$

$\sigma = (6.14)^2 = 37.6996 \qquad \sigma = (5.17)^2 = 26.73$

b. The dispersion on Friday is greater.

30. $s = 12.86$

32. $k = (575 - 500)/50 = 1.5;$ $\qquad$ at least $1 - \dfrac{1}{(1.5)^2} = 0.56$

34. a. $(825 - 450)/125 = 3.0$, about 89 percent

b. about 11%

36. The median is $253, the range is from $116 to $353. About 25 percent of the semi-private rooms are less than $214 and about 25% are above $304. The distribution is negatively skewed.

38. Salary: $CV = \dfrac{\$4000}{\$41,000}(100) = 9.76$

Length of employment: $CV = \dfrac{4}{15}(100) = 26.67$

There is relatively more dispersion in the length of employment.

40. Homeowner claims: $CV = \dfrac{\$425}{\$1260}(100) = 34\%$

Auto Policy: $CV = \dfrac{\$300}{\$875}(100) = 34\%$

The relative dispersion is the same for the two policies.

42. Slight negative skewness.

$sk = \dfrac{3(11.5 - 11.95)}{4.5} = -0.3$

44. $sk = \dfrac{3(\$376 - \$406)}{\$120} = -0.75$

Negative skewness

46. a. $390 - 134 = 256$

b. $MD = 50.4$

c. $\sigma^2 = 5874.82$

d. $\sigma = \sqrt{5874.82} = 76.6$

e. $sk = 0.407$

48. a. $\overline{X} = 11.346$

b. Median $= 11.316$

c. $s = 5.245$

d. $Q_1 = 7.33$ $Q_3 = 14.74$
  $QD = (14.74 - 7.33)/2 = 3.705$

50. a. Range $= 30 - 0 = 30$
  b. $Q_1 = 10.30$ $Q_3 = 18.90$
    Interquartile range $= 18.90 - 10.30 = 8.6$
    $QD = 8.6/2 = 4.3$
  c. $s = 5.8565$ $s^2 = 34.3$

52. a. $Q_1 = 18.00$ $Q_3 = 24.00$
  b. Interquartile range $= 24.00 - 18.00 = 6.00$
    $QD = 6/2 = 3$
  c. $s = 4.52$

54. $\mu = \dfrac{134,587}{24} = 5607.8$

  $\sigma = 13.010$

56. There are two outliers, one at about 40,000 and the other at about 55,000 deaths. The median number of deaths is about 1100, the first quartile is at 150 and the third is at 4000. About half the earthquakes resulted in between 150 and 4000 deaths.

58. The mean age is 55 years, the first quartile is 51 years and the third quartile is 58.5 years. There does appear to be one outlier, at 69 years of age, and the distribution of ages is somewhat positively skewed.

60. a. The mean teacher salary in the 94 districts is $33,181, the median is $32,708 and the standard deviation is $3549. The coefficient of variation and the coefficient of skewness are 10.7% and 0.40.
  b. The boxplot shows that the median salary is about $33,000, that the salaries range from about $26,100 up to $42,800. Fifty percent of the school districts have average salaries between $30,800 up to $35,000. There are two school districts that have average salaries above $40,000.
  c. The coefficient of variation and the coefficient of skewness for the percent of families on welfare in the district is 87.0% and 0.87.
  d. The boxplot shows that there are seven school districts that have 16 percent or more families on welfare. There are three extreme outliers, that have more than 25 percent of the families on welfare. The median district has about five percent on welfare and fifty percent of the school districts have between 3.7 and 8.6 percent of the families in the district on welfare.

CHAPTER 5

2. Two "heads", two "tails", or one of each
4. a. The election
  b. Democrat, Republican, Independent
  c. Male or female
6. a. Answers will vary
  b. 3/8
  c. Answers will vary
8. P(A or B) $= 0.4 + 0.1 = 0.5$
10. P(X or Y) $= 0.05 + 0.10 = 0.15$
12. a. P(Two primary) $= 0.614$
  b. P(Stomach or Pancreas) $= 0.162$
  c. P(Not lung) $= 0.674$
14. P(X or Y) $= 0.4 + 0.1 - 0.1 = 0.4$
16. P(X or Y) $= 0.55 + 0.35 - 0.20 = 0.70$
18. P(W or U) $= 0.23$
20. P(S or F) $= 0.80$

22. $P(X \text{ and } Y) = (0.8)(0.7) = 0.56$
24. a. $P(Y) = 4/10$
  b. $P(M|Y) = 1/4$
  c. $P(M \text{ and } Y) = 1/10$
26. a. $50/200 = 0.25$
  b. $50/200 = 0.25$
28. a. $(3)(2)(3) = 18$
  b. $1/18$
  c. Tree diagram not shown
30. $P(A|B) = P(A) = 0.6$
32. a. 0.25
  b. $1/3$
  c. $2/24$
  d. M and Y are independent because $P(M) = 8/24 = 1/3$ and $P(M|Y) = 2/6 = 1/3$ also.
  e. M and Y are not mutually exclusive because $P(M \text{ and } Y) = 2/24 > 0$.
34. a. $P(\text{Both finish}) = 0.36$
  b. Yes
  c. Special Rule of Multiplication
  d. Relative Frequency
  e. $P(\text{Neither}) = 0.16$
36. a. $(0.3)^3 = 0.027$
  b. $(0.7)^3 = 0.343$
38. $1 - 0.4 = 0.6$
40. $P(\sim Y) = 0.75$
42. $(0.80)(0.80) = 0.64$
44. $52 \times 51 = 2652$
46. $(5)(4)(6) = 120$
48. $10^4 = 10,000$
50. 1680
52. 24
54. 6720
56. 70
58. 125,970
60. 15
62. a. $P(\text{Business}) = .40$
  b. $P(3.0 \text{ and educ}) = .067$
  c. $P(2.0 \text{ or educ}) = .433$
64. $P(\text{none}) = .000125$
  $P(\text{at least one}) = 0.999875$
66. .10
68. A defective set with a probability of 0.25, or a good set with a probability of 0.75
70. a. Subjective
  b. Relative frequency
  c. Classical
72. a. Observing the sex of the child.
  b. Boy, boy; boy, girl; girl, boy; girl, girl
  c. Yes because a child is either "female" or "male", but never both.
  d. Yes seems reasonable if the gender of one child does not affect the gender of the other.
74. a. $P(\text{Black or Caucasian}) = 0.81$
  b. $P(\text{Not black or Caucasian}) = 1 - 0.81 = 0.19$
76. a. Let A = Afternoon and M = Morning
  $P(M \text{ or } A) = 0.50$
  b. No, because 20% subscribe to both

c. No, because $(0.30)(0.40) = 0.12$, not 0.20
78. a. .25   **b.** .30   **c.** .25
  d. No, only 38% of males, but 42% of females work full time, for example.
80. $(0.75)^4 = 0.316$
82. $P(H \text{ and } W) = P(H) \cdot P(W) = (0.50)(0.70) = 0.35$
84. a. $P(\text{Both fail}) = (0.20)(0.05) = 0.001$
  b. $P(\text{Primary fails}) \cdot P(\text{Backup works}) = (0.02)(0.95) = 0.019$
  c. The primary and backup sources are independent.
86. a. $P(\text{Neither}) = 0.808$
  b. $1 - P(\text{Neither}) = 0.192$
88. a. $450/670 = 0.672$
  b. $70/670 = .1040$
  c. $450/670 + 70/670 - 50/670 = 0.701$
  d. $50/670 = 0.075$
  e. $50/70 = .7140$
90. a. $P(\text{Male and Illiterate}) = P(M) \cdot P(I/M) = (0.49)(0.20) = 0.098$
    $P(\text{Female and Illiterate}) = P(F) \cdot P(I/F) = (0.51)(0.17) = 0.0867$
  b. 0.1847
  c. $P(M/I) = 0.098/0.1847 = 0.531$
92. a.

Welfare Group	Frequency	Mean Percent Passing
Low	43	72.0
Moderate	32	64.0
High	19	54.0
Total	94	66.0

  1. .66
  2. .72
  3. .54
  4. $P(\text{Moderate or Fail}) = 0.34 + .034 - 0.12 = 0.56$
  b.

Welfare	Small	Medium	Large	Size of District Total
Low	22	17	4	43
Moderate	9	18	5	32
High	5	7	7	19
Total	36	42	16	94

  1. $P(\text{small}) = .38$
  2. $P(\text{Low/small}) = .61$
  3. $P(\text{Small and low}) = .23$
  4. $P(\text{Small or low}) = .61$
  5. $P(3 \text{ medium}) = .0856$

CHAPTER 6

2. $\mu = 43.50$ $\sigma^2 = 102.75$
  $\sigma = 10.14$
4. $\mu = 1.14$ $\sigma^2 = 1.0804$ $\sigma = 1.0394$

6. $\mu = 1.7$
$\sigma = \sqrt{1.21} = 1.10$

8. $(.10)(\$150,000) = \$15,000$

10. $P(6) = 0.0100$

12. $P(3) = 0.23$

14. $P(0) = 0.2373$

16. a. $P(2) = 0.073$
b. $P(4) = 0.200$
c. $P(6) = 0.041$

18. a. $P(3) = 0.313$
b. $P(5) = 0.031$
c. $P(X > 1) = .969$

20. a. 0.201
b. 0.367
c. 0.055
d. $1 - 0.367 = 0.633$

22. $P(3) = 0.0284$

24. $P(0) = 0.0498$
$P(X > 0) = 0.9502$

26. a. $P(1) = 0.1637$
b. 0.0176

28. $\mu = 2.0 \qquad \sigma = 1.0$

30. $P(X = 3) + P(X = 4) = 0.0079$
$P(X \geq 1) = 0.3297$

32. $P(X = 0) = 0.000$
$P(X \geq 1) = 1.0 - P(X = 0) = 1.0$
$P(\text{more than half}) = 0.882$

34. a. $\mu = 0.70$
b. $\sigma^2 = 0.61$
$\sigma = \sqrt{0.61} = 0.78$
c. $P(0)P(0) = (0.50)(0.50) = 0.25$

36. $\mu = 1.3 \qquad \sigma = 1.1$

38. 1. Each outcome is either a "success"or a "failure".
2. It is a count of the number of successes.
3. Each trial is independent.
4. The probability of a success is the same from trial to trial.

40. $n = 15 \ \pi = 0.1, P(0) = 0.206$

42. a. 10.5, found by 15(0.70)
b. 0.206
c. $P(X \leq 10) = 0.485$

44. $n = 6, \pi = 0.20$
a. 0.262
b. 0.393
c. $P(X \geq 1) = 0.345$

46. $n = 15, \pi = 0.20$
$P(X \leq 4) = 0.836$

48. $\mu = (0.01)(100) = 1.00$
a. $P(X = 0) = 0.3679$
b. $P(X \geq 1) = 1 - 0.3679 = 0.6321$

50. $n = 14, \pi = 0.70$
$P(X > 10) = 0.355$

52. $n = 12, \pi = 0.70$
a. $P(8) = 0.231$

b. $P(X < 5) = 0.009$
c. $P(X \geq 10) = 0.253$
54. $n = 50$, $\pi = 0.80$, $\mu = 4$
   a. $P(2) = 0.1465$
   b. $P(0) = 0.0183$
      $P(X \geq 1) = 0.9817$
56. $\mu = 5.0$
   a. 0.0067
   b. $P(X < 5) = 0.4405$
   c. $[P(X < 5)][P(X < 5)] = 0.1940$
58. $n = 10$, $\pi = 0.30$
   $P(X \geq 6) = 0.047$
60. Using the binomial: $P(X = 1/n = 20, \pi = 0.05) = 0.377$
   Using the Poisson: $P(X = 1/\mu = 1.0) = 0.3679$
   Very little difference in the results.

CHAPTER 7

2. $z = (17 - 25)/8 = -1.0$
4. a. $90 \pm 2(10) = 90 \pm 20$; Time interval: 70 up to 110.
   b. $90 \pm 3(10) = 90 \pm 30$; Time interval: 60 up to 120.
6. a. $z = (X - \mu)/\sigma = (12 - 20)/5 = -1.6$
   b. $\mu \pm 3\sigma = 20 \pm 3(5) = 5$ up to 35
      Virtually all of the values fall between 5 and 35.

8. Plumbers: Carpenters:
   Joe earns \$14/hour.   Neil earns \$12/hour.
   $z = (\$14 - 15)/1.75 = -0.57$ $z = (\$12 - 10)/1.25 = 1.60$

While Joe earns more than Neil, Joe is 0.57 standard units below the mean for his trade whereas Neil is 1.6 standard units above average for his trade.
10. $\mu = 10$ $\sigma = 1.5$

Mrs. Stevens   **Mrs. Delwhiler**
$z = (12 - 10)/1.5 = 1.33$ $z = (9 - 10)/1.5 = -0.67$

Mrs. Stevens shopped 1.33 standard deviations more than the mean time, and Mrs. Delwhiler shopped 0.67 standard deviations less than the mean.
12. $z = (17 - 25)/8 = -1.0$
   $P(z \leq -1.0) = 0.5000 - 0.3413 = 0.1587$
14. a. $z = (13.0 - 10)/2 = 1.5$
      $P(z > 1.5) = 0.5000 - 0.4332 = 0.0668$
   b. $z = (9 - 10)/2 = -0.5$
      $z = (12.5 - 10)/2 = 1.25$
      $P(-0.5 \leq z \leq 1.25) = 0.1915 + 0.3944 = 0.5859$
16. a. $z = (26,000 - 28,000)/1500 = -1.33$
      $z = (27,000 - 28,000)/1500 = -0.67$
      $P(-1.33 \leq z \leq -0.67) = 0.4082 - 0.2486 = 0.1596$
   b. $z = (25,000 - 28,000)/1500 = -2.00$
      $P(z < -2.0) = 0.5000 - 0.4772 = 0.0228$
18. $(X - 25)/8 = \pm 0.52$ $X = 25 \pm 8(0.52) = 20.84$ and 29.16
20. $X = 40,000 - 1.65(3000) = 35,050$

22. $X = 1.0 - 1.28(0.05) = 0.936$

24. $\mu = (50)(0.3) = 15$ $\sigma^2 = (50)(0.3)(0.7) = 10.5$

26. $\mu = 150(0.08) = 12.00$
$\sigma^2 = 150(0.08)(0.92) = 11.04$
$\sigma = \sqrt{11.04} = 3.32$
$z = (4.5 - 12.00)/3.32 = -2.26$
$P(z > -2.26) = 0.5000 + 0.4881 = 0.9881$

28. $\mu = 500(0.05) = 25.0$
$\sigma^2 = 500(0.05)(0.95) = 23.75$
$\sigma = \sqrt{23.75} = 4.87$
$z = (19.5 - 25.0)/4.87 = -1.13$
$P(z < -1.13) = 0.5000 - 0.3708 = 0.1292$

30. **Using the Poisson Table**

a. $P[(X = 2)/(\mu = 4.0)] = 0.1465$

b. $P(X > 7) = 0.0298 + 0.0132 + \cdots$
$= 0.0511$

Using Normal Approximation
$\mu = 4$ $\sigma = \sqrt{4} = 2$

a. $z = (1.5 - 4.0)/2 = -1.25$
$z = (2.5 - 4.0)/2 = -0.75$
$P(-1.25 \le z \le -0.75)$
$= 0.3944 - 0.2734$
$= 0.1210$

b. $z = (7.5 - 4.0)/2 = 1.75$
$P(z > 1.75) = 0.5000 - 0.4599$
$= 0.0401$

32. a.

Test	μ	σ	Score	z-value
Sales	50	7	60	$(60 - 50)/7 = 1.43$
Personnel	120	25	150	$(150 - 120)/25 = 1.20$
Financial	85	5	90	$(90 - 85)/5 = 1.00$

b. Applicant did the best on the sales test; 1.43 standard deviations above the mean.

c. $P(z > 1.43) = 0.5000 - 0.4236 = 0.0764$

d. Sales, because that z score was the largest.

34. $z = (375 - 400)/60 = -0.42$
$z = (450 - 400)/60 = 0.83$

$P(z > 0.83) = 0.5000 - 0.2967 = 0.2033$
$P(-0.42 \le z \le 0.83) = 0.1628 + 0.2967 = 0.4595$

36. $\mu = n\pi = 50(0.80) = 40$
$z = (35.5 - 40)/2.8284 = -1.59$
$z = (45.5 - 40)/2.8284 = 1.94$

$\sigma = \sqrt{50(0.80)(0.20)} = 2.8284$
$P(z > -1.59) = 0.4441 + 0.5000 = 0.9441$
$P(z < 1.94) = 0.4738 + 0.5000 = 0.9738$

38. $z = (200 - 215)/20 = -0.75$
About 77% of the days.
$-1.28 = (X - 200)/20$
$X = 174.4$

$P(z > -0.75) = 0.2734 + 0.5000 = 0.7734$

40. $P(X > 5) = 1 - .8946 = .1054$

42. a. $z = (8000 - 7200)/600 = 1.33$
$P(z > 1.33) = 0.5000 - 0.4082 = 0.0918$
b. $z = (8000 - 8600)/560 = -1.07$
$P(z < -1.07) = 0.5000 - 0.3577 = 0.1423$

44. a. $z = (260 - 235.6)/36.3 = 0.67$
$P(z > 0.67) = 0.5000 - 0.2486 = 0.2514$
b. $z = (180 - 235.6)/36.3 = -1.53$
$P(z < -1.53) = 0.5000 - 0.4370 = 0.0630$
c. $z = (240 - 235.6)/36.3 = 0.12$
$z = (250 - 235.6)/36.3 = 0.40$

$P(0.12 \leq z \leq 0.40) = 0.1554 - 0.0478 = 0.1076$

d. $P(X < 240) = P(z < 0.12) = 0.5000 + 0.0478 = 0.5478$

46. a. $\sigma = 20/1.645 = 12.158$

b. $z = (100 - 120)/12.158 = -1.645$

$P(-1.645 \leq z \leq 0) = 0.4500$

c. $z = (150 - 120)/12.158 = 2.47$

$P(z > 2.47) = 0.5000 - 0.4932 = 0.0068$

d. 0 and 1.645

48. $\mu = 80(0.90) = 72 \qquad \sigma = \sqrt{80(0.90)(0.10)} = 2.6833$

$z = (69.5 - 72)/2.6833 = -0.93$

$P(z > -0.93) = 0.3238 + 0.5000 = 0.8238$

50. a. $z = (50 - 39.52)/6.29 = 1.67$

b. $z = (25 - 39.52)/6.29 = -2.31$

c. $P(z > 1.67) = 0.5000 - 0.4525 = 0.0475$

d. $P(z < -2.31) = 0.500 - 0.4896 = 0.0104$

e. $1.28 = \dfrac{X - 39.52}{6.29}$

$X = 39.52 + 8.0512 = 47.5712$ patients

52. $\mu = 12 \qquad \sigma = \sqrt{12} = 3.4641$

$z = (15.5 - 12.0)/3.4641 = 1.01$

$P(z > 1.01) = 0.5000 - 0.3438 = 0.1562$

54. $\mu = 15 \qquad \sigma = \sqrt{15} = 3.8730$

$z = (12.5 - 15.0)/3.8730 = -0.65$

$P(z > -0.65) = 0.2422 + 0.5000 = 0.7422$

56. $\mu = 80 \qquad \sigma = 12$

$z = (60 - 80)/12 = -1.67$

$P(z > -1.67) = 0.5000 + 0.4525 = 0.9525$

58. $\mu = 18 \qquad \sigma = 5.5$

a. $z = (30 - 18)/5.5 = 2.18$

$P(z < 2.18) = 0.5000 + 0.4854 = 0.9854$

b. 0.000

c. $z = (30.5 - 18)/5.5 = 2.27$

$z = (29.5 - 18)/5.5 = 2.09$

$P(2.09 \leq z \leq 2.27) = 0.4884 - 0.4817 = 0.0067$

60. Selling price,

$z = \dfrac{89,550 - 107,800}{20,000} = -0.91 \qquad P(z < -0.91) = 0.5000 - 0.3186 = 0.6844$

Time on the market,

$z = \dfrac{50 - 43}{14.5} = -0.48 \qquad P(z < -0.48) = 0.5000 - 0.1844 = 0.3156$

62. a. $-0.84 = \dfrac{X - 15}{1}$

$X = 15 - 0.84 = 14.16$

b. $1.28 = \dfrac{X - 15}{1}$

$X = 16.28$

64. a. $z = \dfrac{200.0 - 166.7}{35.68} = 0.93 \qquad P(z > 0.93) = 0.5000 - 0.3238 = 0.1762$

Using the normal distribution, we would expect about 18% of the homes, or about 13, to sell for more than $200,000. There were actually 14 homes that sold for more than $200,000, so the normal distribution was a good approximation.

b. $z = \dfrac{2100 - 2154.8}{207} = -0.26$ $P(z > -0.26) = 0.1026 + 0.5000 = 0.6026$

We expect about 60%, or about 45, of the homes to be larger than 2100 square feet. Forty-eight of the 75 homes, or about 64 percent, have more than 2100 square feet. The normal distribution does give a good approximation.

CHAPTER 8

2. The population is the six possible faces on the die. Perhaps you want to learn the mean or typical number of spots or the relative frequency of each number of spots. After throwing the die a few more times, you might find the sample mean.

4. a. Jackson, Zaborowski, and Rodgers.
  b. Answers will vary.

6. Huttner, Jackson, Parmar, Torok, Kimmel, Bertka, Holt, and Zaborowski.

8. Answers will vary.

10. a. $\mu = \dfrac{10 + 4 + 12 + 11 + 9 + 8}{6} = 9$

b.

Possible Sample	$\bar{X}$
10, 4, 12, 11	9.25
10, 4, 12, 9	8.75
10, 4, 12, 8	8.50
10, 4, 11, 9	8.50
10, 4, 11, 8	8.25
10, 4, 9, 8	7.75
10, 12, 11, 9	10.50
10, 12, 11, 8	10.25
10, 12, 9, 8	9.75
10, 11, 9, 8	9.50
4, 12, 11, 9	9.00
4, 12, 11, 8	8.75
4, 11, 9, 8	8.00
12, 11, 9, 8	10.00
4, 12, 9, 8	8.25

c.

Sample Mean	Probability
7.75	1/15
8.00	1/15
8.25	2/15
8.50	2/15
8.75	2/15
9.00	1/15
9.25	1/15
9.50	1/15
9.75	1/15
10.00	1/15
10.25	1/15
10.50	1/15

d.

Histogram of Sample Means

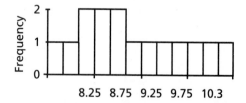

8.25 8.75 9.25 9.75 10.3

e.

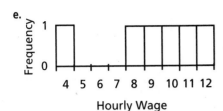

Hourly Wage

f. Both distributions have a mean of $9.00. However, the distribution of sample means has much less spread than the population distribution.

12. a. $\mu = \dfrac{18 + 18 + 12 + 18 + 14}{5} = 16$

b. $_5C_2 = \dfrac{5!}{2!3!} = 10$

c.

Sample Points	Mean
18, 18	18.0
18, 12	15.0
18, 18	18.0
18. 14	16.0
18, 12	15.0
18, 18	18.0
18, 14	16.0
12, 18	15.0
12, 14	13.0
18, 14	16.0

d. $\dfrac{18.0 + 15.0 + \cdots + 16.0}{10} = \dfrac{160}{10} = 16$

The mean of the sample means and the mean of the population are both 16.

e.

Sample Mean	Frequency
13	1
15	3
16	3
18	3
	10

The sample means range from 13 to 18, but the population values range from 12 to 18. There is more dispesion in the population values.

14. a. Select twenty *nine-digit* strings from a random number table. The first number might be 336-20-8848. Another could be 758-49-5206. These were read from the upper left corner of Appendix Table H.

b. Select twenty *eight-digit* strings from a random number table. Then attach zeros to the fronts of the first two, ones to the front of the next two, and so forth. Some possibilities are:
019-94-7728, 081-82-5014, 175-42-9426, 111-58-2108, 237-01-5843

16. Obtain a list of births from local hospitals or Bureau of Vital Statistics. Randomly select the desired sample from the list.

18. Population Characteristics

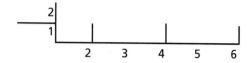

Population Mean: $\mu = \dfrac{2 + 4 + 6}{3} = 4.0$

$\sigma^2 = \dfrac{8}{3}$

Population standard deviation: $\sigma = 1.633$

Sample Characteristics	
Sample Points	$\bar{X}$
2,2	2
2,4	3
2,6	4
4,2	3
4,4	4
4,6	5
6,2	4
6,4	5
6,6	6

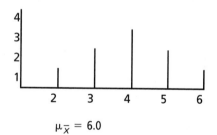

$\mu_{\bar{X}} = 6.0$

Frequency Distribution of Sample Means $(\bar{X})$			
$\bar{X}$	f	$f\bar{X}$	$f\bar{X}^2$
2	1	2	4
3	2	6	18
4	3	12	48
5	2	10	50
6	1	6	36
		36	156

$$\sigma_{\bar{x}}^2 = \frac{156 - \dfrac{(36)^2}{9}}{9} = 1.333$$

$$\sigma_{\bar{x}} = \sqrt{1.333} = 1.155$$

There is less variation in the distribution of sample means than in the population, 1.155 versus 1.633.

20. a. $p = \dfrac{1 + 1 + 0 + 0 + 1 + 1}{6} = \dfrac{4}{6} = 0.6667$

b. $_6C_2 = \dfrac{6!}{4!2!} = 15$

c.

Sample	Results	Proportion Good
1	1,1	1.00
2	1,0	0.50
3	1,0	0.50
4	1,1	1.00
5	1,1	1.00
6	1,0	0.50
7	1,0	0.50
8	1,1	1.00
9	1,1	1.00
10	0,0	0.00
11	0,1	0.50
12	0,1	0.50
13	0,1	0.50
14	0,1	0.50
15	1,1	1.00

$$\mu_{\bar{x}} = \frac{1.00 + 0.50 + \cdots + 1.00}{15} = 0.6667$$

The sample proportion and the population proportion are equal.

22. Answers to this question will vary. MINITAB was used to generate five samples of $n = 10$. The sample means were 3.5, 3.6, 3.8, 3.7, and 3.5. The mean of these sample means is 3.62, which is close to the population mean of 3.58.

CHAPTER 9

2. Regardless of the shape of the population, as the sample size is increased, the distribution of the sample means will approach the normal distribution.

4. The standard error of the sampling mean is the standard deviation of the distribution of sample means. It is equal to the standard deviation of the original population divided by the square root of the sample size.

6. $96.2 \pm (1.96)\dfrac{9.7}{\sqrt{45}} = 96.2 \pm 2.8$

8. $1.5 \pm (1.96)\dfrac{0.3}{\sqrt{36}} = 1.5 \pm 0.1$

10. $7.5 \pm (1.96)\dfrac{1.0}{\sqrt{400}} = 7.5 \pm 0.098$

12. $10.7 \pm (2.58)\dfrac{11.6}{\sqrt{100}} = 10.7 \pm 3.0$

14. a. $\sqrt{\dfrac{(0.11)(0.89)}{2200}} = 0.0067$

  b. $0.11 \pm (1.96)(0.0067) = 0.11 \pm 0.01$

16. $0.10 \pm 1.96\sqrt{\dfrac{(0.10)(0.90)}{250}} = 0.10 \pm 0.04$

18. a. $0.60 \pm (1.96)(0.0310) = 0.60 \pm 0.06$

  b. $0.60 \pm (2.58)(0.0310) = 0.60 \pm 0.08$

20. $0.30 \pm (1.96)\sqrt{\dfrac{(0.30)(0.70)}{840}} = 0.30 \pm 0.031$

$0.30 \pm (2.58)\sqrt{\dfrac{(0.30)(0.70)}{840}} = 0.30 \pm 0.041$

22. When the sample is more than five percent of the population.

24. $0.60 \pm (0.114)(0.896)$

0.60 ± 0.10

26. $8.4 \pm (0.7276)(0.9375)$

8.4 ± 0.68

28. The three factors that determine the size of the sample are: (1) population standard deviation, (2) level of confidence, and (3) the allowable error.

30. Choose any values for s and E, such as 100 and 10.

For 95%: $n = \left[\dfrac{1.96(100)}{10}\right]^2 = 384.16$ For 99%: $n = \left[\dfrac{2.58(100)}{10}\right]^2 = 665.64$

The sample is 1.73 times larger for a sample at the 99 percent level of confidence.

32. $n = \left[\dfrac{1.96(\$2.50)}{\$0.50}\right]^2 = 96.04$

34. $n = (0.3)(0.7)\left[\dfrac{1.96}{0.05}\right]^2 = 323$

36. $5.9 \pm (2.58)\dfrac{2}{\sqrt{100}} = 5.9 \pm 0.52$

38. $3.9 \pm (1.96)\dfrac{1.8}{\sqrt{80}} = 3.9 \pm 0.39$

40. $25.6 \pm (1.65)\dfrac{5.1}{\sqrt{200}} = 25.6 \pm 0.60$

42. a. $68.6 \pm (2.58)\dfrac{8.2}{\sqrt{50}} = 68.6 \pm 2.99$

b. The value suggested by the NCAA is included in the confidence interval, hence it is reasonable.

c. Changing the level of confidence to 95 would reduce the width of the interval. The value of 2.58 would change to 1.96.

44. a. $0.56 \pm (1.96)\sqrt{\dfrac{(0.56)(0.44)}{1000}} = 0.56 \pm 0.03$

b. The lower point of the interval is more than 0.50, so we conclude that a majority feel the President is doing a good job.

46. $69 \pm (1.96)\dfrac{49}{\sqrt{30}} = 69 \pm 17.53$

48. $n = (0.15)(0.85)\left[\dfrac{1.96}{0.02}\right]^2 = 1225$

50. $n = (0.5)(0.5)\left[\dfrac{2.58}{0.03}\right]^2 = 1849$

$n = (0.5)(0.5)\left[\dfrac{2.58}{0.05}\right]^2 = 666$ samples

$n = (0.5)(0.5)\left[\dfrac{2.58}{0.01}\right]^2 = 16,641$ samples

If the margin of error is 0%, then the sample size cannot be determined. A sample of 1850 seems like a reasonable sample size.

52. Deleting the four school districts in question, the mean percent on welfare is 6.197, with a standard deviation of 3.956, for a sample of 90. The 99 percent confidence level is:

$$6.197 \pm (2.58)\frac{3.956}{\sqrt{90}} = 6.197 \pm 1.076$$

So the endpoints of the confidence interval are 5.121 and 7.273. All four of the school districts are outside the interval. We conclude that these districts are different from the others.

CHAPTER 10

2. -1.44 and 1.44

4. a. Reject H_0 if z is less than -1.96 or z is greater than 1.96.

b. $z = \dfrac{12 - 10}{3/\sqrt{64}} = 5.33$

c. The null hypothesis is rejected.

d. p-value $= P(z < -5.33) + P(z > 5.33) = 2(0.5000 - 0.5000) = 2(0) = 0$

6. a. H_0: $\mu = \$1010$, H_a: $\mu \neq \$1010$

b. Reject H_0 if z is not between -2.58 and 2.58.

c. $z = \dfrac{\$1090 - \$1010}{\$300/\sqrt{50}} = 1.89$

So we reject H_0.

d. The p-value is $2(0.5 - 0.4706) = 0.0588$.

8. H_0: $\mu = 70$, H_a: $\mu \neq 70$, Reject H_0 if $z < -1.96$ or $z > 1.96$.

$$z = \frac{72 - 70}{2/\sqrt{121}} = 11.0$$

Reject H_0. The population mean is not 70.

10. H_0: $\mu = 125$, H_a: $\mu \neq 125$, Reject H_0 if $z < -1.96$ or $z > 1.96$.

$$z = \frac{126 - 125}{5/\sqrt{44}} = 1.33$$

H_0 is not rejected. There has not been a change in the mean daily revenue.

12. The critical value is -1.13.

14. a. Reject H_0 if $z < -1.75$

b. $z = \dfrac{215 - 220}{15/\sqrt{64}} = -2.67$

c. Reject H_0

d. p-value $= P(z < -2.67) = 0.5000 - 0.4962 = 0.0038$

16. H_0: $\mu \leq 30$ H_a: $\mu > 30$ Reject H_0 if $z > 1.65$

$$z = \frac{30.08 - 30.00}{0.50/\sqrt{36}} = 0.96$$

Do not reject H_0.

18. a. H_0: $\mu \leq 90$ H_a: $\mu > 90$

b. $z = \dfrac{\overline{X} - \mu}{\sigma/\sqrt{n}}$

c. If $z > 1.28$, reject H_0.

d. $z = \dfrac{92 - 90}{9/\sqrt{49}} = \dfrac{2}{1.29} = 1.56$

Reject H_0. There has been an increase in the warhead density.

e. The p-value is $0.5000 - 0.4406 = 0.0594$
20. a. One-tailed
b. Reject H_0 if $z > 1.75$
c. $z = \dfrac{5.67 - 5.35}{\sqrt{\dfrac{(0.75)^2}{65} + \dfrac{(0.63)^2}{45}}} = 2.42$
d. Reject H_0.
e. p-value $= P(z > 2.42) = 0.5000 - 0.4922 = 0.0078$
22. $H_0: \mu_e \leq \mu_s$ $\quad H_a: \mu_e > \mu_s$ $\quad$ Reject H_0 if $z > 1.65$.
$z = \dfrac{54{,}330 - 54{,}290}{\sqrt{\dfrac{(142)^2}{60} + \dfrac{(135)^2}{30}}} = 1.30$
Do not reject H_0.
24. $H_0: \mu_m = \mu_{al}$ $\quad H_a: \mu_m \neq \mu_{al}$
If z is not between -1.96 and 1.96, reject H_0.
$z = \dfrac{32.9 - 29.6}{\sqrt{\dfrac{(5.7)^2}{36} + \dfrac{(5.5)^2}{49}}} = 2.68$
H_0 is rejected. There is a difference in mean age.
p-value is $2(0.5 - 0.4963) = 0.0074$.
26. a. $H_0: \mu_1 = \mu_2$ $\quad H_a: \mu_1 \neq \mu_2$
b. If z is not between -2.58 and 2.58, reject H_0.
c. $z = \dfrac{20.44 - 19.45}{\sqrt{\dfrac{(2.96)^2}{30} + \dfrac{(2.13)^2}{30}}} = 1.49$
Do not reject the null hypothesis, no difference in the burn rates.
d. p-value is $2(.5000 - .4318) = .1362$.
28. a. Reject H_0 if $z < -1.65$ or $z > 1.65$
b. $z = \dfrac{0.29 - 0.40}{\sqrt{\dfrac{(0.40)(0.60)}{49}}} = -1.57$
c. H_0 is not rejected.
d. p-value $= 2(0.5000 - 0.4418) = 0.1164$
30. $H_0: \pi \geq 0.20$ $\quad H_a: \pi < 0.20$ $\quad$ Reject H_0 if z is less than -2.33
$z = \dfrac{\dfrac{60}{400} - 0.20}{\sqrt{\dfrac{(0.20)(1 - 0.20)}{400}}} = -2.5$
H_0 is rejected. Fewer than 20% of those receiving welfare payments are ineligible.
32. $H_0: \pi = 0.16$ $\quad H_a: \pi \neq 0.16$
Decision Rule: If z is not between -1.96 and 1.96, reject H_0.
$z = \dfrac{\dfrac{70}{600} - 0.16}{\sqrt{\dfrac{(0.16)(0.84)}{600}}} = -2.90$
Reject the null hypothesis. The difference between the two proportions is statistically significant. The p-value is $2(0.5 - 0.4981) = 0.0038$

34. a. H_0 is rejected if $z > 1.96$ or $z < -1.96$

b. $\bar{p} = \dfrac{110 + 170}{150 + 200} = \dfrac{280}{350} = 0.80$

c. $z = \dfrac{\dfrac{110}{150} - \dfrac{170}{200}}{\sqrt{0.80(1 - 0.80)\left(\dfrac{1}{150} + \dfrac{1}{200}\right)}} = -2.70$

d. H_0 is rejected.

e. p-value $= P(z < -2.70) + P(z > 2.70) = 2(0.5000 - 0.4965) = 0.0070$

36. $H_0: \pi_1 \le \pi_2 \qquad H_a: \pi_1 > \pi_2$

π_1 refers to young drivers. Reject H_0 if z is greater than or equal to 1.65

$\bar{p} = \dfrac{30 + 55}{100 + 200} = 0.283$

$z = \dfrac{0.30 - 0.275}{\sqrt{(0.283)(0.717)\left(\dfrac{1}{100} + \dfrac{1}{200}\right)}} = 0.45$

H_0 is not rejected. There is no difference in the proportion of drivers who take risks.

38. $H_0: \pi_1 = \pi_2 \qquad H_a: \pi_1 \ne \pi_2$

At a 0.05 level of significance, if z is not between -1.96 and 1.96, reject H_0.

$\bar{p} = \dfrac{10 + 25}{100 + 200} = \dfrac{35}{300} = 0.117$

$z = \dfrac{0.10 - 0.125}{\sqrt{(0.117)(0.883)\left(\dfrac{1}{100} + \dfrac{1}{200}\right)}} = \dfrac{-0.025}{0.039} = -0.64$

There is no difference in the proportions.

40. Population #1 is from the windowless schools.

$H_0: \mu_1 \le \mu_2 \qquad H_a: \mu_1 > \mu_2 \qquad$ Reject H_0 if $z > 2.33$

$z = \dfrac{94 - 90}{\sqrt{\dfrac{(8)^2}{100} + \dfrac{(10)^2}{80}}} = 2.91 \qquad$ Reject H_0.

There is significantly more anxiety in windowless schools.

42. $H_0: \mu \ge 30 \qquad H_a: \mu < 30 \qquad$ Reject H_0 if $z < -2.05$

$z = \dfrac{28.5 - 30}{5/\sqrt{40}} = -1.90$

Do not reject H_0.

44. a. $H_0: \mu_1 \ge \mu_2 \qquad H_a: \mu_1 < \mu_2$ where subscript one refers to Lake Anna.

b. $z = \dfrac{\overline{X}_1 - \overline{X}_2}{\sqrt{\dfrac{s_1^2}{n_1} + \dfrac{s_2^2}{n_2}}}$

c. If $z < -2.33$ reject H_0.

d. $z = \dfrac{20 - 40}{\sqrt{\dfrac{(4)^2}{50} + \dfrac{(8)^2}{40}}} = -14.4$

Reject H_0. The Lake Anna fish are smaller.

46. a. $H_0: \mu = 16.01 \qquad H_a: \mu \ne 16.01$

b. If z is not between -1.65 and 1.65, then reject H_0.

c. $z = \dfrac{15.97 - 16.01}{0.005/\sqrt{40}} = \dfrac{-0.04}{0.00079} = -50.6$

That is certainly significant!

48. a. $H_0: \mu_1 = \mu_2 \qquad H_a: \mu_1 \neq \mu_2$

b. If z is not between -1.65 and 1.65, then reject H_0.

c. $z = \dfrac{52.7 - 51.8}{\sqrt{\dfrac{(2.5)^2}{100} + \dfrac{(2.6)^2}{150}}} = \dfrac{0.9}{0.328} = 2.74$

Reject H_0. The means heights are significantly different.

50. a. Let the subscript "1" refer to men.

$H_0: \mu_1 \leq \mu_2 \qquad H_a: \mu_1 > \mu_2$

b. If z is greater than 2.05, reject H_0.

c. $z = \dfrac{353 - 315}{\sqrt{\dfrac{(18)^2}{80} + \dfrac{(21)^2}{80}}} = \dfrac{38}{3.092} = 12.29$

Reject H_0. Men are paid significantly more.

52. $H_0: \pi \leq 0.2 \qquad H_a: \pi > 0.2$

If $z > 1.65$ reject H_0.

$z = \dfrac{\dfrac{56}{200} - 0.20}{\sqrt{\dfrac{(0.20)(0.80)}{200}}} = 2.83$

Reject H_0. This development *does* have a greater proportion of movers.

54. a. $H_0: \pi = 0.65 \qquad H_a: \pi \neq 0.65$

b. If z is not between -1.65 and 1.65, then reject H_0.

c. $z = \dfrac{\dfrac{120}{200} - 0.65}{\sqrt{\dfrac{(0.65)(0.35)}{200}}} = -1.48$

d. Utah seems to follow the national pattern.

56. $H_0: \pi_I \leq \pi_F \qquad H_a: \pi_I > \pi_F$

Reject H_0 if $z > 2.33$

$\bar{p} = .3714$

$z = \dfrac{0.45 - 0.30}{\sqrt{(0.3714)(1 - 0.3714)\left(\dfrac{1}{722} + \dfrac{1}{794}\right)}} = 6.04$

H_0 is rejected. The divorce rate is higher among identical twins.

58. $H_0: \pi_1 = \pi_2 \qquad H_a: \pi_1 \neq \pi_2$

For a 5% significance level, if z is not between -1.96 and 1.96, reject H_0.

$z = \dfrac{\dfrac{8}{37} - \dfrac{7}{38}}{\sqrt{(0.2)(0.8)\left(\dfrac{1}{37} + \dfrac{1}{38}\right)}} = 0.35$

We fail to reject H_0.

60. $H_0: \pi_1 = \pi_2 \qquad H_a: \pi_1 \neq \pi_2$

Reject H_0 if z is less than -1.96 or greater than 1.96.

$$\bar{p} = \frac{200 + 110}{1000 + 500} = 0.207$$

$$z = \frac{0.20 - 0.22}{\sqrt{(0.207)(1 - 0.207)\left(\dfrac{1}{1000} + \dfrac{1}{500}\right)}} = -0.90$$

The difference is not significant in the proportions.

62. $H_0: \pi_1 \leq \pi_2 \qquad H_a: \pi_1 > \pi_2$

$\pi_1 =$ Second period. Reject H_0 if z is greater than 2.33.

$$\bar{p} = \frac{83 + 146}{420 + 423} = 0.272$$

$$z = \frac{0.345 - 0.198}{\sqrt{(0.272)(1 - 0.272)\left(\dfrac{1}{420} + \dfrac{1}{423}\right)}} = 4.80$$

There has been a significant increase in car pooling.

64. $H_0: \pi \geq 0.50 \qquad H_a: \pi < 0.50$

Reject H_0 if $z < -1.65$.

$$z = \frac{\dfrac{143}{300} - 0.50}{\sqrt{\dfrac{(0.50)(1 - 0.50)}{300}}} = -0.81$$

Do not reject H_0.

66. $H_0: \pi_n \leq \pi_o \qquad H_a: \pi_n > \pi_o$

Reject H_0 if $z > 2.33$

$$\bar{p} = \frac{46 + 28}{70 + 60} = 0.57$$

$$z = \frac{\dfrac{46}{70} - \dfrac{28}{60}}{\sqrt{(0.57)(1 - 0.57)\left(\dfrac{1}{70} + \dfrac{1}{60}\right)}} = 2.19$$

H_0 is not rejected.

68. $H_0: \pi \leq 0.10 \qquad H_a: \pi > 0.10$

Reject H_0 if $z > 1.65$

$$z = \frac{\dfrac{20}{150} - 0.10}{\sqrt{\dfrac{(0.10)(0.90)}{150}}} = \frac{0.0333}{0.0245} = 1.36$$

H_0 is not rejected.

70. a. $H_0: \mu \geq 35 \qquad H_a: \mu < 35$

Reject H_0 if $z < -1.65$

$$z = \frac{33.181 - 35.000}{3.549/\sqrt{94}} = \frac{-1.819}{0.366} = -4.97$$

H_0 is rejected. The mean salary is less than $25,000.

b. $H_0: \mu \leq 20.0 \qquad H_a: \mu > 20.0$

Reject if $z > 1.65$.

$$z = \frac{7.233 - 20.0}{6.522/\sqrt{94}} - 12.767$$

H_0 is not rejected. The mean percent on welfare is not greater than 20 percent.

c. H_0: $\mu \leq 0.90$ H_a: $\mu > 0.90$
H_0 is rejected if $z > 1.65$.

$$z = \frac{65.86 - 90.0}{13.609/\sqrt{94}} = -17.24$$

H_0 is not rejected. The mean percent of students passing per school district is not more than 90 percent.

CHAPTER 11

2. $t = -2.977$ and $t = 2.977$

4. H_0: $\mu \leq 4.3$ years H_a: $\mu > 4.3$ years
Reject H_0 if t exceeds 2.539.

$$t = \frac{4.6 - 4.3}{1.2/\sqrt{20}} = 1.12$$

H_0 is not rejected.

6. H_0: $\mu \geq 56$ H_a: $\mu < 56$
Reject H_0 if $t < -1.796$.

$$t = \frac{53.5 - 56.0}{2.00/\sqrt{12}} = -4.33$$

H_0 is rejected. The p-value is less than 0.005.

8. H_0: $\mu \leq 1.6$ H_a: $\mu > 1.6$
Reject H_0 if $t > 3.747$.
$\overline{X} = 2.2$ $s = 0.68$

$$t = \frac{2.2 - 1.6}{0.68/\sqrt{5}} = 1.97$$

The null hypothesis cannot be rejected.

10. H_0: $\mu \geq 40$ H_a: $\mu < 40$
Reject H_0 if $t < -1.895$.

$$t = \frac{38 - 40}{6.44/\sqrt{8}} = \frac{-2.00}{2.28} = -0.88$$

H_0 is not rejected. The mean age could be 40 years.

12. a. $df = 11 + 13 - 2 = 22$ If t is greater than 2.508, reject H_0.

b. $s_p = \sqrt{\dfrac{10(9)^2 + 12(15)^2}{22}} = 12.63$

c. $t = \dfrac{41 - 32}{12.63\sqrt{\dfrac{1}{11} + \dfrac{1}{13}}} = 1.74$

d. Do not reject H_0.

14. H_0: $\mu_1 \geq \mu_2$ H_a: $\mu_1 < \mu_2$
Reject H_0 if t is less than -2.552.

$$s_p = \sqrt{\frac{9(.8)^2 + 9(1.0)^2}{18}} = 0.906 \qquad t = \frac{2.5 - 3.1}{0.906\sqrt{\dfrac{1}{10} + \dfrac{1}{10}}} = -1.48$$

Do not reject the null hypothesis. The p-value is between 0.10 and 0.05.

16. H_0: $\mu_1 \leq \mu_2$ H_a: $\mu_1 > \mu_2$
Reject H_0 if t is greater than 1.895.

$$t = \frac{24.00 - 17.00}{4.07\sqrt{\frac{1}{4} + \frac{1}{5}}} = 2.56$$

H_0 is rejected. The mean expense is greater for sales executives (less for production).

18. a. Reject H_0 if t is greater than 2.624

b. $t = \dfrac{1.5}{0.9/\sqrt{15}} = \dfrac{1.5}{0.2324} = 6.45$

c. Yes, reject H_0.

d. Since $6.45 > 2.977$, the p-value < 0.005

20. H_0: $\mu_d \le 0$ H_a: $\mu_d > 0$ $\alpha = 0.01$ df $= n - 1 = 10 - 1 = 9$

Reject H_0 if t exceeds 2.821; otherwise, fail to reject H_0.

$$t = \frac{3.6}{5.46/\sqrt{10}} = 2.085$$

Do not reject the null hypothesis.

22. H_0: $\mu_d = 0$ H_a; $\mu_d \ne 0$

Reject H_0 if t is greater than 2.201 or less than -2.021.

$$t = \frac{-1.367}{4.336/\sqrt{12}} = -1.09$$

H_0 is not rejected. There is no difference in sales. The p-value is between 0.5 and 0.2.

24. H_0: $\mu \ge 20$ pounds overweight H_a: $\mu < 20$ pounds overweight

Reject H_0 if t is less than -1.761.

$$t = \frac{18 - 20}{5/\sqrt{15}} = -1.55$$

There is not sufficient evidence to doubt the claim in the newspaper.

26. H_0: $\mu \ge 26.0$ H_a: $\mu < 26.0$

Reject H_0 if t is less than -3.365.

$$t = \frac{25.03 - 26.0}{0.468/\sqrt{6}} = -5.08$$

H_0 is rejected. The car does not meet EPA specifications.

28. H_0: $\mu = 25$ H_a: $\mu \ne 25$ $\alpha = 0.10$ df $= 16 - 1 = 15$

H_0 is rejected if t is less than -1.753 or t is greater than 1.753.

$$t = \frac{25.938 - 25.000}{3.395/\sqrt{16}} = 1.11$$

H_0 is not rejected. The p-value is between 0.5 and 0.2.

30. H_0: $\mu_1 = \mu_2$ H_a: $\mu_1 \ne \mu_2$ $\alpha = 0.01$ df $= 7 + 6 - 2 = 11$

H_0 is rejected if t is less than -3.106 or greater than 3.106.

$$t = \frac{8.14 - 8.00}{1.3769\sqrt{\frac{1}{7} + \frac{1}{6}}} = 0.18$$

H_0 is not rejected. There is no difference in the mean weight gain of the two groups.

32. H_0: $\mu_d \le 0$ H_a: $\mu_d > 0$

Reject H_0 if t exceeds 2.718.

$$t = \frac{7.75}{10.83/\sqrt{12}} = 2.48$$

The evidence does not support the manufacturer's claim.

34. H_0: $\mu_d \le 0$ H_a: $\mu_d > 0$ df $= 8 - 1 = 7$

Reject H_0 if t exceeds 1.895.

$$t = \frac{0.61}{0.788/\sqrt{8}} = 2.19$$

H_0 is rejected. The sales training program is effective. The p-value is between 0.05 and 0.025.

36. H_0: $\mu \le 6.0$ H_a: $\mu > 6.0$ df = $6 - 1 = 5$ $\alpha = 0.01$

H_0 is rejected if $t > 3.365$.

$$t = \frac{7.2 - 6.0}{1.05/\sqrt{6}} = 2.80$$

H_0 is not rejected. The population mean still could be 6 homes started.

38. H_0: $\mu_d \le 0$ H_a: $\mu_d > 0$

H_0 is rejected if $t > 2.718$

$$t = \frac{3.375}{1.602/\sqrt{12}} = 7.30$$

H_0 is rejected. The asking price is more than the selling price.

40. H_0: $\mu \ge 1.00$ $\mu < 1.00$

H_0 is rejected if t is less than -1.833. Otherwise, fail to reject H_0.

$$t = \frac{0.988 - 1.00}{0.0358/\sqrt{10}} = -1.06$$

H_0 is not rejected.

42. H_0: $\mu_1 \le \mu_2$ H_a: $\mu_1 > \mu_2$ Reject H_0 if $t > 2.650$.

$$t = \frac{2.955 - 2.657}{0.3918\sqrt{\dfrac{1}{7} + \dfrac{1}{8}}} = 1.47$$

H_0 cannot be rejected. The evidence does not suggest there has been an increase in grade point average.

44. H_0: $\mu_d \le 0$ H_a: $\mu_d > 0$

H_0 is rejected if t is greater than 1.796.

$$t = \frac{3.58}{2.746/\sqrt{12}} = 4.52$$

H_0 is rejected. Drivers make more errors after drinking whiskey. The p-value is less than .005.

46. H_0: $\mu_d \le 0$ H_a: $\mu_d > 0$

Reject if $t > 1.895$

$$t = \frac{4.375}{5.58/\sqrt{8}} = 2.22$$

H_0 is rejected. There is a difference in the ratings.

48. H_0: $\mu_d \le 0$ H_a: $\mu_d > 0$

H_0 is rejected if t is greater than 1.796.

$$t = \frac{3.33}{161.45/\sqrt{12}} = 0.07$$

H_0 is not rejected. No reason to believe Model B is more expensive.

50. a. Selecting every fourth school district beginning with the first district the mean is $2833, the standard deviation is $1807, and $n = 24$. Assume a 0.05 significance level.

H_0: $\mu = 3000$ H_1: $\mu \ne 3000$ Reject H_0 if $t < -2.069$ or $t > 2.069$

$$t = \frac{2833 - 3000}{1807/\sqrt{24}} = -0.453$$

H_0 is not rejected. The mean amount spent on instruction is not different from $3000.

b. The information on the size of the schools and the amount spent on instruction is shown below:

District Size	Mean Amount	Standard Deviation	n
Small	2888.00	1723	36
Medium	2551.10	283.6	42
Large	2812.20	367.4	16

$H_0: \mu_s = \mu_1$　　　$H_a: \mu_s \neq \mu_1$
Reject H_0 if $t < -2.01$ or $t > 2.01$.

$$t = \frac{2888 - 2812.20}{1455.54\sqrt{\dfrac{1}{36} + \dfrac{1}{16}}} = 0.173$$

Do not reject H_0.

c. The information on the proportion of students on welfare and the amount spent on instruction is shown below.

Percent on Welfare	Mean	Standard Deviation	n
Low	2750.00	905	43
Medium	2515.70	25	32
High	3019.00	2001	19

$H_0: \mu_l = \mu_h$　　　$H_a: \mu_l \neq \mu_h$　　　$\alpha = 0.05$　　　df $= 43 + 19 - 2 = 60$
Reject H_0 if $t < -2.000$ or $t > 2.000$

$$t = \frac{2750 - 3019}{1332.11\sqrt{\dfrac{1}{43} + \dfrac{1}{19}}} = -0.733$$

Do not reject H_0. There is no difference in the mean amount spent on instruction between districts with low welfare and those with high welfare.

. . . .
CHAPTER 12

2. a. $F_{7,6,0.05} = 4.21$
　b. $F_{3,9,0.01} = 6.99$
4. $H_0: \sigma_b^2 = \sigma_a^2$　　　$H_a: \sigma_b^2 \neq \sigma_a^2$
　H_0 is rejected if F is greater than 2.33.
　$$F = \frac{225}{150} = 1.50$$
　H_0 is not rejected. The population variances are the same.
6. a. df $= k - 1 = 3 - 1 = 2$
　b. df $= n - k = 13 - 3 = 10$
　c. df $= n - 1 = 13 - 1 = 12$
　d. 7.56
8. $H_0: \mu_1 = \mu_2 = \mu_3$　　　H_0: Not all means are equal
　H_0 is rejected if F is greater than 5.14.
　$F = 13.43$
　H_0 is rejected. There is a difference in the mean waiting time of the banks. The p-value is less than 0.01.

10. H_0: $\mu_1 = \mu_2 = \mu_3 = \mu_4$ H_a: Not all means are equal
H_0 is rejected if $F > 5.29$.
$$F = \frac{136.72}{24.77} = 5.52$$
H_0 is rejected and H_a accepted. The ethical behavior of attorneys varies by region.

12. H_0: $\sigma_s^2 \le \sigma_a^2$ H_a: $\sigma_s^2 > \sigma_a^2$
H_0 is rejected if $F > 3.87$.
$$F = \frac{(4200)^2}{(2000)^2} = 4.41$$
H_0 is rejected. There is more variation in the sales group.

14. H_0: $\mu_1 = \mu_2 = \mu_3 = \mu_4$ H_a: Not all means are equal
Reject H_0 if F is greater than 3.24.
$$F = \frac{86.85}{37.20} = 2.33$$
H_0 cannot be rejected. The evidence does not suggest any differences in the reading assignments. The p-value is larger than 0.05.

16. H_0: $\mu_1 = \mu_2 = \mu_3 = \mu_4$ H_a: Not all means are equal
Reject H_0 if F is greater than 3.41.
$$F = \frac{39.01}{37.985} = 1.03$$
Do not reject H_0. There is no difference in time to solve the puzzle.

18. H_0: $\mu_1 = \mu_2 = \mu_3$ H_a: Not all means are equal
Reject H_0 if the computed F is greater than 6.93.
$$F = \frac{380.6}{25.3} = 15.03$$
H_0 is rejected and H_a is accepted. There is a difference among the three methods.

20. H_0: $\mu_1 = \mu_2 = \mu_3$ H_a: Not all means are equal
H_0 is rejected if F is greater than 4.10.
$F = 9.45$
H_0 is rejected. There is a difference in the mean score for the age groups. The p-value is less than 0.01.

22. H_0: $\mu_1 = \mu_2 = \mu_3 = \mu_4$ H_a: Not all means are equal
H_0 is rejected if $F > 2.97$ (estimated).
$F = 68.99$
H_0 is rejected. Some means are significantly different from the others.

24. H_0: $\mu_a = \mu_b$ H_a: $\mu_a \ne \mu_b$ Reject the null hypothesis if $F > 4.60$
$$F = \frac{86.3/1}{972.6/14} = 1.24$$
H_0 cannot be rejected

26. H_0: $\mu_1 = \mu_2 = \mu_3 = \mu_4 = \mu_5$ H_a: Not all means are equal
H_0 is rejected if F exceeds 4.31.
$$F = \frac{155.18}{135.31} = 1.15$$
H_0 cannot be rejected. No difference in the number of customers is shown.

28. H_0: $\sigma_1^2 \le \sigma_2^2$ H_a: $\sigma_1^2 > \sigma_2^2$
If $F > 3.68$ reject H_0.
$$F = \frac{(12.51)^2}{(4.56)^2} = 7.53$$
Reject H_0. There is more variability in the oil stocks.

30. a. H_0: $\mu_1 = \mu_2 = \mu_3$ H_a: Not all treatment means are equal
Using $\alpha = 0.05$, H_0 is rejected if F exceeds 3.10.
The computed value of F is 0.98, so H_0 is not rejected.

b. $H_0: \mu_1 = \mu_2 = \mu_3$ H_a: Not all treatment means are equal

Using $\alpha = 0.05$, H_0 is rejected if F exceeds 3.10.

The decision is to reject H_0, because the computed value of F is 4.47. There is a difference in the mean number of students depending on the level of welfare in the district.

CHAPTER 13

2. a.

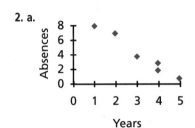

b. Absences decrease as years of employment increase.

4.

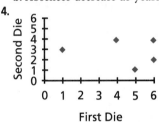

The two dice are not related.

6. $r = \dfrac{6(59) - (19)(25)}{\sqrt{[6(71) - (19)^2][6(143) - (25)^2]}} = -0.98$

A strong negative correlation.

8. a.

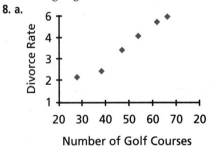

b. $r = \dfrac{6(1170.1) - (295)(22.1)}{\sqrt{[6(15{,}553) - (295)^2][6(88.19) - (22.1)^2]}} = 0.99$

c. The divorce rate increases as the number of golf courses increases.

10. 0.92, 92% explained.

12. .46, 46% explained

14. $H_0: \rho \le 0$ $H_a: \rho > 0$

Reject H_0 if $t > 2.552$.

$$t = \frac{0.40\sqrt{20 - 2}}{\sqrt{1 - (0.40)^2}} = 1.852$$

H_0 is not rejected.

16. H_0: $\rho \le 0$ H_a: $\rho > 0$

Reject H_0 if $t > 1.734$.

$$t = \frac{0.21\sqrt{20 - 2}}{\sqrt{1 - (0.21)^2}} = 0.911$$

H_0 is not rejected. The p-value is greater than 0.100.

18. a. $r_s = 1 - \frac{6(62.50)}{12[12^2 - 1]} = 0.781$

b. Assume a 0.05 significance level with 10 df. Critical value is 1.812.

H_0: No rank correlation. H_a: Positive rank correlation.

$$t = \frac{0.781\sqrt{12 - 2}}{\sqrt{1 - (0.781)^2}} = 3.95$$

Therefore, reject H_0.

20. a. $r = \frac{10(559) - (150)(35)}{\sqrt{[10(2406) - (150)^2][10(139) - (35)^2]}} = 0.670$

b. If you were to test the hypothesis of no correlation using a two-tailed 5% significance level test, the critical value is 2.306.

H_0: $\rho = 0$ H_a: $\rho \ne 0$

$$t = \frac{0.67\sqrt{10 - 2}}{\sqrt{1 - (0.67)^2}} = 2.55$$

There is correlation between food expenditure and family size.

22. a.

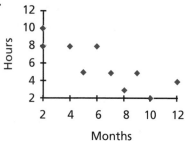

As the months owned increases the number of hours exercised decreases.

b. $r = \frac{10(313) - (65)(58)}{\sqrt{[10(523) - (65)^2][10(396) - (58)^2]}} = -0.827$

c. H_0: $\rho \ge 0$ H_a: $\rho < 0$ Reject H_0 if $t < -2.896$

$$t = \frac{-0.827\sqrt{10 - 2}}{\sqrt{1 - (0.827)^2}} = -4.16$$

Reject H_0. There is a negative association between months owned and hours exercised.

24. a. $r = 0.531$

b. H_0: $\rho \le 0$ H_a: $\rho > 0$ Reject H_0 if $t > 1.337$

$$t = \frac{0.531\sqrt{18 - 2}}{\sqrt{1 - (0.531)^2}} = 2.507$$

Reject H_0. There is a positive association between high temperature and precipitation.

26. a. $\Sigma d^2 = 44.5$

b.

c. $r_s = 1 - \dfrac{6(44.5)}{10(10^2 - 1)} = 0.73$

d. There is a strong direct relationship between order of completion and grade rank.

28. a. $r = \dfrac{-4821}{74(4.9)(36)} = -0.369$

b. $H_0: \rho \geq 0 \qquad H_a: \rho < 0$

$t = \dfrac{-0.369\sqrt{75 - 2}}{\sqrt{1 - (0.369)^2}} = -3.392$

H_0 is rejected. There is a negative association between distance and selling price.

30. $H_0: \rho \leq 0 \qquad H_a: \rho > 0 \qquad$ Reject H_0 if $t > 1.812$, using the .05 significance level

$t = \dfrac{0.225\sqrt{12 - 2}}{\sqrt{1 - (0.225)^2}} = 0.730$

H_0 is not rejected. The p-value is greater than 0.10.

32. a. The correlation between selling price and size of the home is 0.535.

$H_0: \rho \leq 0 \qquad H_a: \rho > 0 \qquad$ Reject H_0 if $t > 1.67$

$t = \dfrac{0.535\sqrt{75 - 2}}{\sqrt{1 - (0.535)^2}} = 5.410$

H_0 is rejected. There is a positive correlation between the two variables.

b. The correlation between selling price and distance is -0.373.

$H_0: \rho \geq 0 \qquad H_a: \rho < 0 \qquad$ Reject H_0 if $t < -1.67$

$t = \dfrac{-0.373\sqrt{75 - 2}}{\sqrt{1 - (-0.373)^2}} = -3.435$

Reject H_0. There is a negative correlation between distance and selling price. The further from the center of the city, the lower the selling price.

c. The correlation between the selling price and the number of bedrooms is 0.492.

$H_0: \rho \leq 0 \qquad H_a: \rho > 0 \qquad$ Reject H_0 if $t > 1.67$

$t = \dfrac{0.492\sqrt{75 - 2}}{\sqrt{1 - (0.492)^2}} = 4.828$

H_0 is rejected. There is a positive association between selling price and the number of bedrooms.

CHAPTER 14

2. a. $Y' = a + bX$

b. Approximately -100 employees.

c. Approximately 0.20 minorities per non-minority, estimated by using the points (1000, 100) and (2000, 300); hence $[(300 - 100)/(2000 - 1000) = 0.20]$.

d. $Y' = -100 + 0.20(3000) = 500$ minority employees.

e. The negative value for a indicates that smaller firms have no minority employees. The slope b means they generally hire 2 minorities for every 10 nonminorities.

4. a. $Y' = 39 - 0.002(7000) = 25$

b. The Y-intercept is 39.

c. The slope is -0.002.

d. For each $1000 increment in median annual income, the birthrate decreases by 2 per thousand.

6. a.

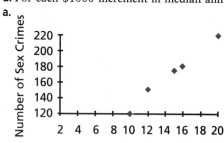

b. $b = \dfrac{5(12,945) - (73)(848)}{5(1125) - (73)^2} = 9.53$ $a = \dfrac{848 - (9.53)(73)}{5} = 30.46$

$Y' = 30.46 + 9.53X$

c. If there were no theaters there would be about 30 crimes. Each additional theater adds 9.53 crimes.

d. $Y' = 30.46 + 9.53(18) = 202$

8. a.

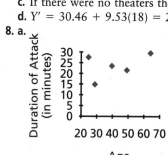

b. $b = \dfrac{5(5160) - (210)(119)}{5(9850) - (210)^2} = 0.16$ $a = \dfrac{119 - (0.16)(210)}{5} = 17.08$

$Y' = 17.08 + 0.16X$

c. Ten years of age appears to add 1.6 minutes to an attack's duration. The length of the attack is at least 17.08 minutes.

d. $Y' = 17.08 + 0.16(42) = 23.8$

10. $s_{(Y \cdot X)} = 3.582$

12. a. $b = \dfrac{6(51,030) - (31)(6780)}{6(239) - (31)^2} = 202.96$ $a = \dfrac{6780 - (202.96)(31)}{6} = 81.37$

$Y' = 81.37 + 202.96X$

b. $s_{(Y \cdot X)} = 20.39$

14. $Y' = -25.7373 + 1.044(80) = 57.7827$

a. 57.7827 ± 5.21

b. 57.7827 ± 11.23

16. a. $Y' = 81.37 + 202.96(6) = 1299.13$

1299.13 ± 18.21

b. $1299.13 \pm (43.47)(1.0842) = 1299.13 \pm 47.13$

18. a. $b = \dfrac{6(5153) - (440)(69)}{6(32,482) - (440)^2} = \dfrac{558}{1292} = 0.432$ $a = \dfrac{69 - (0.432)(440)}{6} = -20.180$

   b. $s_{(Y \cdot X)} = 0.575$

20. a.

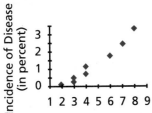

Age of Incarceration

Age at First Contact

   b. $b = \dfrac{7(1799) - (90)(141)}{7(1192) - (90)^2} = \dfrac{-97}{244} = -0.40$ $a = \dfrac{141 - (-0.40)(90)}{7} = 25.29$

   c. $s_{(Y \cdot X)} = 1.32$
   d. $Y' = 25.29 - 0.40(14) = 19.69$
      19.69 ± 3.69

22. a.

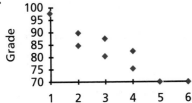

Incidence of Disease (in percent)

Length of exposure (in weeks)

   b. $b = \dfrac{8(66.1) - (37)(10.6)}{8(203) - (37)^2} = \dfrac{136.6}{255} = 0.536$ $a = \dfrac{10.6 - (0.536)(37)}{8} = -1.15$

   c. $s_{(Y \cdot X)} = 0.200$
   d. $Y' = -1.15 \pm 0.536(5) = 1.53$
   e. 1.53 ± 0.79

24. a.

Grade

Number of Absences

   $b = \dfrac{9(2372) - (30)(743)}{9(120) - (30)^2} = \dfrac{-942}{180} = -5.233$ $a = \dfrac{743 - (-5.233)(30)}{9} = 100$

   b. $s_{(Y \cdot X)} = 3.658$
   c. $Y' = 100 - 5.233(3) = 84.3$
      84.3 ± 2.37
   d. 84.3 ± 7.33

26. a.

b. $b = \dfrac{25(151,113) - (1653)(2219)}{25(117,181) - (1653)^2} = 0.5571$

$a = \dfrac{2219 - (0.5571)(1653)}{25} = 51.92$

$Y' = 51.92 + 0.5571(100) = 107.63$

c. $s_{(Y \cdot X)} = 9.47$

107.63 ± 21.33

d. $r = 0.737$

$t = \dfrac{0.737\sqrt{25 - 2}}{\sqrt{1 - (0.737)^2}} = 5.23$

$H_0: \rho = 0$ versus $H_a: \rho > 0$ with 23 degrees of freedom has a critical value of 1.714

Thus we can reject the null hypothesis. There is a positive association between the variables.

28. Chicago:

$Y' = 0.7002 + 0.0368X$ $\qquad s_{Y \cdot X} = 0.5019 \qquad r^2 = 0.685$

LA:

$Y' = 11.5740 - 0.1767X$ $\qquad s_{Y \cdot X} = 0.4545 \qquad r^2 = 0.821$

We conclude that the association between temperature and the amount of rainfall is fairly strong in both cities. In Chicago when the temperature increases, rainfall increases, but in L.A. when the temperature increases rainfall decreases.

30. a. The regression equation is $Y' = -699.2 + 391.71X$. Forty-three percent of the variation in the number of students in the system is explained by the variation percent of families on welfare. For an increase of one percent in the percent on welfare the number of students in the system can be expected to increase by about 392 students. The larger school systems have the higher percent of students on welfare. A system with 10 percent of the students on welfare can expect to have 3218 students. The 95 percent confidence interval is from 2559 up to 3876. The 95 percent prediction interval is from −2691 students up to 9127 students. Logically, the end point of the prediction interval cannot be less than zero.

b. The independent variable regarding the attendance rate for the comprehensive exam explains 44.6 percent of the variation in the percent of students passing the comprehensive exam. The regression equation is $Y' = -718.7 + 8.235$. An increase of 1 percent in the attendance rate will result in an increase of 8.235 in the percent passing the exam. A school system with 90 percent attendance rate can be expected to have passing rate of 22.45 percent. The 95 percent confidence interval is from 12.2 to 32.7, and the 95 percent prediction interval is from 0 to 45.16

CHAPTER 15

2. a. $Y' = 60.0$, found by $Y' = 55.0 - 3.2(20) + 2.3(30)$

b. It decreases by 3.20.

4. a. $Y' = 145.15$, found by $Y' = 30 + 1.05(3) + 6.5(8) + 0.005(12,000)$

b. Another year without a child adds $6.5(1) = 6.5$. An additional \$5000 in income adds $0.005(5000) = 25$, so the additional income contributes more to the satisfaction.

6. a. $n = 26$

  b. 5

  c. $R^2 = \dfrac{60}{200} = 0.30$

  d. $s_{Y \cdot 12345} = \sqrt{7.00} = 2.646$

8. a.

Source	DF	SS	MS	F
Regression	5	1500	300.00	9.00
Error	15	500	33.33	
Total	20	2000		

  b. $H_0: \beta_1 = \beta_2 = \beta_3 = \beta_4 = \beta_5 = 0$ H_a: Not all β are 0
    H_0 is rejected if $F > 2.90$. At least one of the β_i is not equal to 0.

  c. In each case, H_0 is rejected if $t < -2.131$ or $t > 2.131$. In this case for X_1, $t = 4.0$; for X_2, $t = -0.83$; for X_3, $t = 1.0$; for X_4, $t = -3.0$; and for X_5, $t = 4.0$. Variables X_2 and X_3, are not significantly different from zero.

10. a. 22.25

  b. 2.75 points.

12. a. 50.7 or \$50,700

  b. \$8890

  c. $0.32(10) = 3.2$, or \$3200

14. a. 0.447. It is likely that the correlation would be positive.

  b. Bonus has the stronger correlation with sales.

  c. $R^2 = \dfrac{10,740.8}{11,659.1} = 0.921$ $S_{Y \cdot 12} = \sqrt{76.5} = 8.746$

  d. $H_0: \beta_1 = \beta_2 = 0$ H_a: Not all regression coefficients are 0.
    H_0 is rejected if $F > 3.89$. H_0 is rejected.

  e. H_0 is rejected in both cases if $t < -2.179$ or $t > 2.179$. Both t-values (5.52 and 6.90) are greater than 2.179, so H_0 is rejected in both cases. Both variables should be included.

16. a.

	Sub	Popul	Adv
Popul	0.643		
Adv	0.709	0.357	
Income	0.710	0.254	0.441

  The correlation between Adv and Income (0.441) is relatively high for those two variables to be called independent.

  b. The regression equation is:
    $Y' = -5.733 + 0.007537$ Popul $+ 0.05088$ Adv $+ 1.0974$ Income

  c. $F = 35.38$ At least one of the net regression coefficients does not equal 0.

  d.

Predictor	Coef	StDev	t-ratio	p
Constant	−5.733	8.427	−0.68	0.504
Popul	0.007537	0.001813	4.16	0.000
Adv	0.05088	0.01414	3.60	0.002
Income	1.0974	0.2450	4.48	0.000

  None of the variables should be deleted as their p-values are all 0.002 or smaller.

  e. The regression equation with the intercept set equal to zero is:
    $Y' = 0.00753$ Popul $+ 0.0547$ Adv $+ 0.932$ Income

18. Answers will vary

20. a. Using the three independent variables the coefficient of determination is 0.475. We interpret this as less than half of the variation in the dependent variable passing is accounted for by the three independent variables. Using the three independent variables the regression equation is:

$Y' = -646.1 - 0.218X_1 + 0.002X_2 - 7.436X_3$

The null hypothesis is that all the net regression coefficients are zero and the alternate is that at least one is not zero. The null hypothesis is rejected if the computed F is greater than 2.70. The computed F is 27.16, so the null hypothesis is rejected. Not all the net regression coefficients are zero.

For individual tests, the critical values of t are -1.99 and 1.99. For these test we conclude that the independent variable welfare is not necessary. The computed value of t is -0.89, which is not in the rejection region.

CHAPTER 16

2. The characteristics of the Chi-Square distribution are: it cannot assume negative values, it is positively skewed, and there is a family of distributions.

4. a. less than 0.05

b. less than 0.01

c. greater than 0.10

d. less than 0.10

6. H_0: The die is fair. H_a: The die is not fair.
H_0 is rejected if χ^2 is greater than 11.070
$\chi^2 = 11.20$
H_0 is rejected. The die is not fair.

8. a. H_0: $\pi_1 = \pi_2 = \cdots = \pi_9$
  H_a: At least one proportion is different.

b. The expected frequencies are $\dfrac{200}{10} = 20$

c. The critical value is 21.666

d. $\chi^2 = 9.70$
  We cannot reject the null hypothesis.

10. H_0: There has been no change in the distribution.
H_a: There has been a change in the distribution.
H_0 is rejected if χ^2 is greater than 9.210.
$\chi^2 = 8.46$
H_0 cannot be rejected.

12. H_0: The distribution is unchanged
H_a: There was a change.
Reject H_0 if $\chi^2 > 5.991$
$\chi^2 = 4.67$
H_0 is *not* rejected. The distribution is unchanged.

14. H_0: $\pi_f = \pi_j = \frac{1}{4}$; $\pi_s = \frac{1}{2}$ H_a: At least one is different
Reject H_0 if $\chi^2 > 5.991$.
$\chi^2 = 2.00$
We fail to reject H_0.

16. df $= (2 - 1)(4 - 1) = 3$

18. H_0: Party affiliation and opinion are independent.
H_a: Affiliation and opinion are related.
H_0 is rejected if $\chi^2 > 5.991$.
$\chi^2 = 67.103$
Reject H_0. Party affiliation affects opinion.

20. H_0: Party affiliation distributions are the same for all districts
H_a: They are different by district.
H_0 is rejected if $\chi^2 > 13.277$.
$\chi^2 = 20.93$
We can reject the null hypothesis. The distributions are different among the districts.

22. H_0: $\pi_s = 0.50$, $\pi_r = \pi_e = 0.25$
H_a: Distribution is not as given above.
Reject H_0 if $\chi^2 > 4.605$.
$\chi^2 = 3.52$
H_0 is not rejected.

24. H_0: $\pi_s = \pi_c = \pi_e$
H_a: The proportions are not equal.
Reject H_0 if $\chi^2 > 5.991$.
$\chi^2 = 5.6679$
H_0 is not rejected. There is not a preference for the gifts.

26. H_0: There is no difference with respect to social class.
H_a: There is a difference with respect to social class.
If $\chi^2 > 9.488$ reject H_0.
$\chi^2 = 214.65$
H_0 is rejected.

28. H_0: This group follows the given distribution
H_a: This group is different
If $\chi^2 > 7.815$ reject H_0.
$\chi^2 = 8.73$
H_0 is rejected. This group follows a different distribution.

30. H_0: Years of education follows the expected distribution
H_a: Years of education does not follow the expected distribution.
If $\chi^2 > 7.815$ reject H_0.
$\chi^2 = 23.25$
H_0 is rejected. The observed frequencies differ from the expected frequencies.

32. H_0: Age and income are not related
H_a: Age and income are related.
H_0 is rejected if $\chi^2 > 13.277$
$\chi^2 = 6.85$
H_0 cannot be rejected. There is no relationship shown between age and income.

34. H_0: The two distributions are the same.
H_a: The distributions are different.
H_0 is rejected if $\chi^2 > 16.812$.

	1990		1996		Total
	f_0	f_e	f_0	f_e	
Liberal Democrat	250	274.7	300	275.3	550
Socialist	112	98.4	85	98.6	197
Clean Government	58	56.9	56	57.1	114
Dem. Socialist	38	32.0	26	32.0	64
Communist	26	26.0	26	26.0	52
New Liberal	8	7.0	6	7.0	14
Independents	19	16.0	13	16.0	32
Total	511		512		1032

$$\chi^2 = \frac{(250 - 274.7)^2}{274.7} + \cdots + \frac{(13 - 16)^2}{16} = 11.9$$

We fail to reject H_0. There has been no change in the distribution between 1990 and 1996.

36. H_0: The five brands are of equal quality.

H_a: The brands are different.   df = 4

Decision rule: H_0 is rejected if $\chi^2 > 9.488$.

$\chi^2 = 9.778$

Reject H_0. Some of the brands are significantly different.

38. a. H_0: Pool and township are not related.

H_a: Pool and township are related.

H_0 is rejected if $\chi^2 > 9.488$

$\chi^2 = 4.571$

The null hypothesis is not rejected. We have not shown there to be a relationship between pool and township.

b. H_0: Attached garage and township are not related

H_a: Attached garage and township are related.

H_0 is rejected if $\chi^3 > 9.488$

$\chi^2 = 1.768$

The null hypothesis is not rejected.

CHAPTER 17

2. H_0: $\pi = 0.50$ H_a: $\pi \neq 0.50$ $n = 6$

$P(X = 0) + P(X = 6) = 0.016 + 0.016 = 0.032$

H_0 is rejected if the signed differences are all $+$, or all $-$. Since there are five instances where there are more $+$ signs, H_0 cannot be rejected. The evidence does not suggest a salary difference.

4. H_0: $\pi = 0.50$ H_a: $\pi > 0.50$ Reject H_0 if there are 11 or more plus signs.

$P(X \geq 11) = 0.029$

Reject H_0. St. Vincent is busier.

6. H_0: $\pi = 0.50$ H_a: $\pi \neq 0.50$ $n = 17$, because there are 3 differences of 0.

H_0 is not rejected if z is in the interval between -1.96 and 1.96.

$$z = \frac{11.5 - 8.5}{\sqrt{17(0.5)(0.5)}} = 1.46$$

H_0 is not rejected. There is no difference in optimism.

8. H_0: $\pi = 0.50$ H_a: $\pi < 0.50$

If $z < -1.65$, reject H_0.

$$z = \frac{5.5 - 6.0}{\sqrt{12(0.5)(0.5)}} = -0.29$$

This is not enough evidence to say comprehension is increased.

10. H_0: $\pi = 0.50$ H_a: $\pi < 0.50$

Reject H_0 if $z < -1.65$.

$$z = \frac{(3 + 0.50) - 7}{1.87} = -1.87$$

Reject H_0. Significantly more "near misses" occurred on the West Coast.

12. H_0: Sum of the ranks is the same.

H_a: Sum of the ranks is different.

Reject H_0 if the smaller of the sums is less than or equal to 25. Since $R^+ = 46$, H_0 cannot be rejected.

14. H_0: The two distributions are the same.

H_a: The St. Vincent distribution is larger.

Reject H_0, if $R^+ \leq 25$. Do not reject H_0 because $R^+ = 28.5$

16. H_0: There is no difference in the distributions.
   H_a: There is a difference in the distributions.
   Do not reject H_0 if z is between -1.65 and 1.65.

$$z = \frac{88.5 - \dfrac{8(8 + 10 + 1)}{2}}{\sqrt{\dfrac{(8)(10)(8 + 10 + 1)}{12}}} = 1.11$$

   Do not reject H_0.

18. H_0: There is no difference in the two brands.
   H_a: There is a difference in the two brands
   Do not reject H_0 if z is between -1.96 and 1.96.

$$z = \frac{53 - \dfrac{8(17)}{2}}{\sqrt{\dfrac{(8)(8)(17)}{12}}} = -1.58$$

   We fail to reject H_0.

20. H_0: There is no difference in the distributions.
   H_a: There is a difference in the distributions.
   If z is not between -1.96 and 1.96, reject H_0.

$$z = \frac{97.5 - \dfrac{9(9 + 8 + 1)}{2}}{\sqrt{\dfrac{(9)(8)(9 + 8 + 1)}{12}}} = 1.59$$

   We cannot reject H_0.

22. H_0: Populations are the same. H_a: Populations are not the same.
   Reject H_0 if H exceeds 11.345.
   $H = 5.09$
   H_0 cannot be rejected.

24. H_0: The three distributions are the same. H_0: At least one is different.
   Reject H_0 if H exceeds 4.605.
   $H = 7.62$
   Reject H_0. At least one of the reaction time distributions is different.

26. H_0: $\pi \le 0.50$ H_a: $\pi > 0.50$
   There are eight workers, so $n = 8$. The $P(+ \ge 7) = 0.035$. Hence, the decision rule: Do not reject H_0 if the number of plus signs is 6 or fewer. Since there are 7 plus signs, H_0 is rejected. The production is larger with music.

28. H_0: $\pi = 0.50$ H_a: $\pi < 0.50$
   Reject H_0 if $z < -1.65$.

$$z = \frac{2.5 - 5}{1.58} = -1.58$$

   We fail to reject the null hypothesis.

30. H_0: $\pi = 0.50$ H_a: $\pi < 0.50$
   $P(X = 0) + P(X = 1) = 0.011$
   Decision Rule: If the number of plus signs is fewer than 2, reject H_0. There is only one plus sign (subject 8), so the null hypothesis is rejected.

32. H_0: There is no difference in the number of cigarettes smoked.
   H_a: There is a decrease in the number of cigarettes smoked.
   Reject H_0 if R^+ is less than or equal to 10.
   The null hypothesis is rejected because $R^+ = 2.5$. A significant decrease in smoking is shown.

34. H_0: $\pi \le 0.5$ H_a: $\pi > 0.5$

Because of the tie $n = 24$. The null hypothesis is rejected if z is greater than 2.33.

$$z = \frac{18.5 - 12}{\sqrt{6}} = 2.65$$

Reject the null hypothesis. We conclude that police officer A gives more tickets than B.

36. H_0: The two distributions are the same.

H_a: R^- is smaller.

If $R^- \leq 5$, reject H_0 at the 0.05 significance level.

$R^- = 6$.

We fail to reject H_0. These data show no significant change in productivity.

38. H_0: The two distributions are identical H_a: R^- is smaller.

If $R^- < 91$, reject H_0. $R^- = 27$

Reject H_0. Police officer A gives more traffic tickets.

40. H_0: Amounts awarded by the judges are the same.

H_a: Amounts awarded by the judges are not the same.

H_0 is not rejected if z is in the interval between -2.58 and 2.58.

$$z = \frac{55 - \dfrac{8(11 + 8 + 1)}{2}}{\sqrt{\dfrac{(11)(8)(11 + 8 + 1)}{12}}} = -2.06$$

The null hypothesis cannot be rejected.

42. H_0: Murder rates are the same for the three groups.

H_a: Murder rates are not the same.

H_0 is rejected if $H > 5.991$.

$H = 2.028$

H_0 cannot be rejected.

44. H_0: Gas mileage is the same ($\pi \leq 0.50$).

H_a: Super mileage is greater ($\pi > 0.50$).

Using the binomial table with $\pi = 0.50$ and $n = 14$, the $P(X \geq 12) = 0.006 + 0.001 = 0.007$, which is as close as possible to 0.01 without exceeding it. H_0 is rejected if the number of plus signs is 12 or more out of 14. There are 10 plus signs, so H_0 cannot be rejected.

46. H_0: Gas mileage is the same.

H_a: Super mileage is greater

The decision rule is to reject H_0 if R^- is equal to or less than 15. Since R^- is 36.5, H_0 cannot be rejected. No evidence to conclude that mileage is different.

48. H_0: The distributions are the same.

H_a: The distributions are not the same.

Reject the null hypothesis if the computed value is greater than 5.991.

$$H = \frac{12}{24(25)}\left[\frac{(43.5)^2}{8} + \frac{(106.5)^2}{8} + \frac{(150)^2}{8}\right] - 3(25) = 14.33$$

H_0 is rejected. The distributions are not the same.

50. a. H_0: The distributions are the same.

H_a: The distributions are not the same.

Reject H_0 if $H > 5.991$.

$H = 10.95$

H_0 is rejected. We conclude that there is a different amount spent per student.

b. H_0: The distributions are the same.

H_a: The distributions are not the same.

Reject H_0 if $\chi^2 > 5.991$.

The computed value of χ^2 is 39.99, so H_0 is rejected. We conclude that there is a difference in the distribution of the salaries.

CHAPTER 18

2.

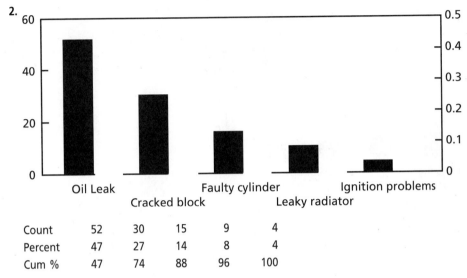

Count	52	30	15	9	4
Percent	47	27	14	8	4
Cum %	47	74	88	96	100

Almost 75 percent of the complaints relate to the oil leaks and cracked blocks.

4. Attribute charts are the result of classifying a product or service as acceptable or not acceptable. Variable charts are based on some type of measurement or reading.

6. $3\dfrac{s}{\sqrt{n}} = A_2\overline{R}$

$3\dfrac{s}{\sqrt{n}} = (0.577)(0.50)$

$3s = (\sqrt{5})(0.577)(0.50)$

$s = 0.215$

8. a. $UCL = D_4\overline{R} = 2.282(6.67) = 15.22$

$LCL = D_3\overline{R} = 0$

b. Oven is operating within bounds except for the first range (8 A.M.) is outside the control limits.

10. $\overline{c} = \dfrac{24}{15} = 1.6$

1.6 ± 3.79

The control limits are from 0 to 5.39 tears. Each sample roll is within these limits.

12. $\overline{c} = \dfrac{36}{10} = 3.6$

3.6 ± 5.69

There is no case where the number of voids is greater than 9, so the process is in control.

14.

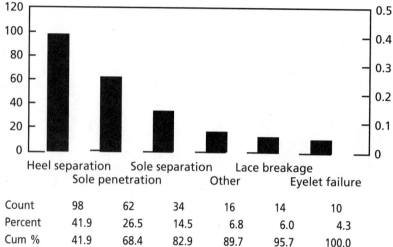

Pareto Chart
Defect in Jogging Shoe

	Heel separation	Sole penetration	Sole separation	Other	Lace breakage	Eyelet failure
Count	98	62	34	16	14	10
Percent	41.9	26.5	14.5	6.8	6.0	4.3
Cum %	41.9	68.4	82.9	89.7	95.7	100.0

16. a. $\overline{\overline{X}} = 87.36$, $UCL = 88.11$, $LCL = 86.61$
$\overline{R} = 1.3$, $UCL = 2.75$, $LCL = 0$

d. Mean outside diameter is 87.36. Mean range is 1.3. All sample means between 86.61 and 88.11; range between 0 and 2.75.

18. $\overline{X} = \dfrac{551.34}{10} = 55.14$ $\overline{R} = \dfrac{355}{10} = 35.5$

$UCL = 55.14 + (0.577)(35.5) = 75.62$ Upper control limit for mean
$LCL = 55.14 - (0.577)(35.5) = 34.66$ Lower control limit for mean
$UCL = 2.115(35.5) = 75.08$ Upper control limit for range

The last bank officer, Simonetti, is below the mean amount loaned for the group.

20. a. $\overline{p} = \dfrac{93}{15(200)} = 0.031$

b. $3\sqrt{\dfrac{0.031(1 - 0.031)}{200}} = 0.037$

$UCL = 0.031 + 0.037 = 0.068$
$LCL = 0.031 - 0.037 = 0$

All the packages are within the limits.

c. 10 packages mean $\overline{p} = 0.05$, which is well within the limits.

22. a. $\overline{p} = \dfrac{86}{12(50)} = 0.1433$

$3\sqrt{\dfrac{0.1433(1 - 0.1433)}{50}} = 0.15$

$UCL = 0.14 + 0.15 = 0.29$
$LCL = 0.14 - 0.15 = 0$

b.

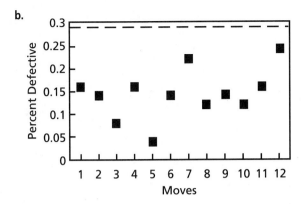

c. None of the moves are out of control in terms of the number of complaints.

24. $\bar{c} = \dfrac{60}{20} = 3.0$

$UCL = 3.0 + 3.0\sqrt{3} = 3.00 + 5.20 = 8.20$
$LCL = 3.0 - 3.0\sqrt{3} = 3.00 - 5.20 = 0$

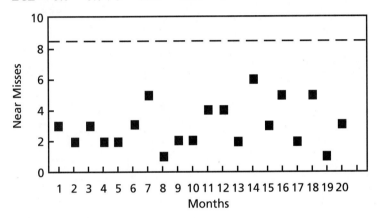

All the points are within the control limits.

Index

A

addition rules for probability 151–155

alternate hypothesis A claim about the population parameter that is accepted if the null hypothesis is rejected, 299–300, 322

analysis of variance (ANOVA) technique 368, 372–384, 395, 470–471

arithmetic mean The sum of all the values in the sample divided by the number of values, 70–71, 73–74, 81
  weighted 76

assignable variation 560

assumptions
  about ANOVA 372–373
  about multiple regression 464
  about regression 444–445

attribute control chart 566, 575–579

attribute variable 11

autocorrelation 464

average 8, 18–19, 70, 141

B

bar chart 49–50

biased sampling 246

bimodal data 81

binomial distribution 190–191, 226–229, 241

binomial formula 192

binomial probability distribution 189–193

box plot 122

C

c bar chart 577–579, 588

cause-and-effect diagram 562–564

census An observation of the entire population, 249

central limit theorem If all samples of a fixed size are selected from any population, the sampling distribution of the sample means is approximately a normal distribution. This approximation improves with larger samples, 268–275, 331

central tendency 70

chance variations 560

charts 37, 49, 53, 56, 560–570, 575–579

Chebyshev's theorem For any set of observations (sample or population), the minimum proportion of the values that lie within k standard deviations of the mean is at least $1 - \frac{1}{k^2}$ where k is any constant greater than 1, 120–121

chi-square distribution 496, 587

chi-square test statistic 498

class 28
  frequency 28
  limits 31

classical concept of probability The number of favorable outcomes divided by the number of possible outcomes, 148–149, 241

coefficient of correlation A measure of the strength of the association between two variables, 403–408, 411–414

coefficient of determination The proportion of the total variation in one variable that is explained by the other variable, 410, 491

coefficient of multiple determination The proportion (percentage) of the total variation in the dependent variable, Y, that is explained by the independent variables, 467–468

coefficient of rank correlation 415

coefficient of skewness 128–129

coefficient of variation The standard deviation divided by the mean, 126

combination One particular arrangement of a group of objects or persons selected from a larger group without regard to order, 170–172

complement rule 165

computer applications 20–21, 56–58

conditional probability The likelihood that an event will occur, given that another event has already occurred, 156–157

confidence interval A range of values constructed from sample data so that a parameter occurs within that range at a preselected probability. The preselected probability is termed the "level of confidence," 276–277, 279–284, 286, 331, 445–446, 455

contingency table Frequency data resulting from the simultaneous classification of *more than one* variable or trait of the observed item, 508–511, 587

continuous probability distribution 241

continuous random variable 184

continuous variable 11

control chart for the mean 566–569, 588

control chart for the range 569–570, 588